2nd International Conference on Precision Machinery and Manufacturing Technology (ICPMMT 2017)

AA003536

MATEC Web of Conferences Volume 123 (2017)

Kending, Taiwan
19 - 21 May 2017

Editors:

Gow-Yi Tzou

ISBN: 978-1-5108-4811-5

Printed from e-media with permission by:

Curran Associates, Inc.
57 Morehouse Lane
Red Hook, NY 12571

Some format issues inherent in the e-media version may also appear in this print version.

Printed by Curran Associates, Inc. (2017)

For additional information, please contact EDP Sciences – Web of Conferences at the address below.

EDP Sciences – Web of Conferences
17, Avenue du Hoggar
Parc d'Activité de Courtabœuf
BP 112
F-91944 Les Ulis Cedex A
France

Phone: +33 (0) 1 69 18 75 75
Fax: +33 (0) 1 69 28 84 91

contact-edps@webofconferences.org

Additional copies of this publication are available from:

Curran Associates, Inc.
57 Morehouse Lane
Red Hook, NY 12571 USA
Phone: 845-758-0400
Fax: 845-758-2633
Email: curran@proceedings.com
Web: www.proceedings.com

TABLE OF CONTENTS

WELDING OF TITANIUM ALLOYS 1
Pasang Tim, Tao Yuan, Azizi Mana, Kamiya Osamu, Mizutani Masami, Misiolek Wojciech

COMPARISONS OF COMPRESSION CHARACTERISTICS IN COMPRESSION FORMING OF POROUS CYLINDER BASED ON SLAB METHOD AND FINITE ELEMENT METHOD UNDER CONSTANT SHEAR FRICTION 9
Tzou Gow-Yi, Hwang Yeong-Maw, Pan Sai-Chih

MEASUREMENT OF WISDOM FORGING FORCE USING PIEZOELECTRIC MATERIALS 16
Hsia Shao-Yi, Chou Yu-Tuan, Chang Chen-Chung, Hsu Ling-Yu

CASE STUDY OF VIRTUAL REALITY IN CNC MACHINE TOOL EXHIBITION 21
Kao Yung-Chou, Lee Chung-Shuo, Liu Zhi-Ren, Lin Yu-Fu

RIDE RESPONSES OF MACPHERSON SUSPENSION SYSTEMS 27
Yu Cheng-Chi, Chiu Chih-Chi

DEVELOPMENT OF A XYθ_Z 3-DOF NANOPOSITIONING STAGE WITH LINEAR DISPLACEMENT AMPLIFICATION DEVICE 37
Chu Chih-Liang, Chen Hung-Chi, Sie Ming-Han

EFFECTS OF ELASTIC BALL AND COATING ON PURE SQUEEZE EHL MOTION FOR CONSTANT LOAD WITH COUPLE STRESSES USING FDM METHOD 43
Chu Li-Ming, Chang Yuh-Ping, Lin Jaw-Ren

SIMULATION ANALYSIS OF TURBINE BLADE IN 3D PRINTING AQUARIUM 49
Chen Dyi-Cheng, Lai Bo-Yan, Gao Fu-Yuan

A HISTOGRAM STATISTICAL METHOD FOR THE DETECTION OF LOCALIZED FAULTS IN DEEP GROOVE BALL BEARING 54
Lin Shang-Chih, Huang Yennun

APPLYING VIRTUAL REALITY TO REMOTE CONTROL OF MOBILE ROBOT 62
Chen Chin-Shan, Lui Ching-Wen

EFFECTS OF PARTICLE SHAPE AND TEMPERATURE ON COMPACTION OF COPPER POWDER AT MICRO SCALE 68
Chang Chao-Cheng, Wu Ming-Ru

APPLYING DYNAMIC MOLD TEMPERATURE CONTROL TO COSMETIC PACKAGE DESIGN 74
Hsiao Shih-Wen, Yen Chih-Huang, Lee Chu-Hsuan

CAD-BASED INTELLIGENT ROBOT SYSTEM INTEGRATED WITH 3D SCANNING FOR SHOE ROUGHING AND CEMENTING 81
Chiu Cheng-Chang, Wang Wen-Teng, Yin Wan-Shan

WARPAGE BEHAVIOR ANALYSIS IN PACKAGE PROCESSES OF EMBEDDED COPPER SUBSTRATES 88
Hwang Yeong-Maw, Ou Tsang-Hung, Su Wei-Hung

RESEARCH ON ENERGY RECYCLE PERCENTAGE OF SERVO MOTOR 94
Tseng Yen-Ming, Chen Li-Shan, Pan Jeng-Shyang, Huang Jing

A COMPACT LOW INSERTION LOSS BANDPASS FILTER BASED ON MEANDERED SELF-COUPLED RING RESONATOR 101
Cheng Ching-Chien, Cheng Kong-Xin, Kung Huang-Kuang, Wang Chin-Yu, Jeng Yng-Huey

A STUDY OF SMALL SIZED QUARTZ TUNING FORK USING FINITE ELEMENT ANALYSIS 108
Tsai Gwo-Chung

PIERCING AND SURFACE-CRACK DEFECTS IN COMBINATION EXTRUSION OF FASTENERS 113
Lin Heng-Sheng, Lee Chien-Yu, Li Wen-Shun, Wu Cheng-Tsung

EXPERIMENTAL AND FEM ANALYSIS OF THREE CARBON STEEL CHARACTERIZATION UNDER QUASI-STATIC STRAIN RATE FOR BUMPER BEAM APPLICATION 119
Alemayehu Dawit Bogale, Huang Song-Jeng, Koricho Ermias Geberekidan

AN ARTIFICIAL FLEXIBLE ROBOT ARM BASED ON PNEUMATIC MUSCLE ACTUATORS 125
Renn Jyh-Chyang, Li Bo-Han

AN AUTOMATIC ENERGY-SAVING AND THERMAL MONITORING/CONTROLLING SYSTEM FOR A POND 130
Cheng Ching-Chien, Chiu Min-Chie, Chiu Che-Min, Kung Huang-Kuang, Wang Chin-Yu, Gao Zi-Jie

HOW IT OUTSOURCING IMPACTS ON VALUE CREATION OF MANUFACTURER? 135
Chen Li-Shan, Tseng Yen-Ming

OPTIMUM DESIGN AND RESEARCH ON NOVEL VEHICLE HYBRID EXCITATION SYNCHRONOUS GENERATOR 141
Liu Zhong-Shu
HIGH CURRENT DENSITY ON ELECTROPLATING SMOOTH ALKALINE ZINC COATING 147
Bai Allen, Yang Kai-Lin, Chen Hao-Long, Hong Yu-hong, Chang Shyue-Bin
A ROBUST AND FAST CONTROL TECHNOLOGY OF AC POWER CONDITIONING FOR HIGH-SPEED MICROMACHINING 154
Chang En-Chih
THE MEASURING TECHNIQUE DEVELOPED TO EVALUATE THE THERMAL DIFFUSIVITY OF THE MULTI-LAYERED THIN FILM SPECIMENS 160
Li Tse-Chang, Li Yang-Ru, Lin Jen-Fin
PREPARATION AND PROPERTIES OF THE (SR,BA)NB_2O_6 THIN FILMS BY USING THE SPUTTERING METHOD 170
Diao Chien-Chen, Yang Chen-Fu, Wu Chia-Ching
DESIGN OF BOOKS INVENTORY WITH RFID ANTENNA IN LIBRARY MANAGEMENT SYSTEM 176
Cheng Ching-Chien, Shih Sung-Tsun, Cheng Kong-Xin, Kung Huang-Kuang, Wang Chin-Yu
QUALITY PREDICTION MODELING FOR MULTISTAGE MANUFACTURING BASED ON CLASSIFICATION AND ASSOCIATION RULE MINING 184
Kao Hung-An, Hsieh Yan-Shou, Chen Cheng-Hui, Lee Jay
INFLUENCE OF THIN FILM THICKNESS OF WORKING ELECTRODES ON PHOTOVOLTAIC CHARACTERISTICS OF DYE-SENSITIZED SOLAR CELLS 190
Lai Yeong-Lin, Hsu Hung-Ru, Lai Yeong-Kang, Zheng Chun-Yi, Chou Yung-Hua, Hsu Nai-Kun, Lung Guan-Yun
FABRICATION OF WORKING AND COUNTER ELECTRODES ON PLASTIC SUBSTRATES FOR FLEXIBLE DYE-SENSITIZED SOLAR CELLS 195
Lai Yeong-Lin, Lai Yeong-Kang, Zheng Chun-Yi, Xu Guo-Hui, Wang Yi-Ming, Chen Shan-Ching
MANAGING SELF CONGRUITY TO INFLUENCE BEHAVIORAL INTENTION IN ORGANIC FOOD CONTEXTS IN FUJIAN PROVINCE, CHINA 200
Yen Tsai-Fa
FEM SIMULATION ANALYSIS OF WIRE ROD DRAWING PROCESS USING THE ROTATING DIE UNDER COULOMB FRICTION 205
Tzou Gow-Yi, Chai Un-Chin, Hsu Chao-Ming, Hsu Hung-Ying
THE IMPLEMENT OF A REMOTE CONTROL SYSTEM FOR AUTOMATED WAREHOUSE BY ZIGBEE NETWORKING 209
Cheng Ho-Chih, Chiu Min-Chie
A DESIGN OF TOXIC GAS DETECTING SECURITY ROBOT CAR BASED ON WIRELESS PATH-PATROL 216
Cheng Ho-Chih, Chiu Min-Chie, Zeng Kun-Fu, Chiu Che-Min
Author Index

MATEC Web of Conferences **123**, 00001 (2017)
DOI:10.1051/matecconf/201712300001
ICPMMT 2017

Welding of titanium alloys

Tim Pasang[1,*], *Yuan* Tao[1], *Mana* Azizi[1], *Osamu* Kamiya[2], *Masami* Mizutani[3], and *Wojciech* Misiolek[4]

[1]Department of Mechanical Engineering, Auckland University of Technology, Auckland, New Zealand
[2]Department of System Design Engineering, Faculty of Engineering Science, Akita University, Japan
[3]Joining & Welding Research Institute, Osaka University, Osaka, Japan
[4]Institute for Metal Forming, Department of Materials Science and Engineering, Lehigh University, Bethlehem, PA18015, USA

Abstract. This article briefly summarises previous reports as well as some recent results on welding of various titanium alloys. Titanium is available in a wide range, hence, their properties and characteristics (including welding characteristics) may also varied significantly. The alloys studied represented three major areas; they are (1): unalloyed or commercially pure titanium (CP Ti), (2): near α and α+β alloys, and (3): β alloys. From our preliminary results, it can be reported that the structure of fusion zone (FZ) and heat affected zone (HAZ) of alloys from area (1) may change but their hardness remained, arguably, the same as the as-received material. Alloys from area (2) showed an increase in hardness in the FZ and HAZ areas due to the formation of α-prime or martensite, while alloys from area (3) experienced a reduction in hardness associated with dissolution of α phase leaving only retained β in the FZ. Further investigations are required to confirm these findings. Such information are very useful so that appropriate post welding heat treatment (PWHT) can be applied to improve their properties and performance.

1 Introduction

Titanium and its alloys are used in many different areas such as aerospace, automotive, sporting equipment, chemical, marine and medical due to their excellent high strength-to-weight ratio, good creep resistance, excellent corrosion resistance, good biocompatibility, etc [1, 2]. In general, titanium and its alloys can be divided into (i) unalloyed (also known as commercially pure – CP Ti), (ii) α and near α alloys, (iii) α+β alloys, and (iv) β alloys. The characteristics including welding of these alloys are shown in Figure 1. Welding characteristics are crucial because a lot of titanium are available in sheets and plates forms. Depending on alloy classification, titanium and its alloys have moderate to excellent weldability [1]. Figure 1 indicates a better weldability of unalloyed or CP Ti, α-alloys and near-α alloys compared with α/β alloys or β alloys.

Titanium can be welded with arc welding i.e. high heat/energy input and low power density such as gas tungsten arc welding (GTAW), gas metal arc welding (GMAW), plasma arc welding (PAW), or with high-energy beam technique such as laser beam welding (LBW) and electron beam welding (EBW) [1, 2]. Depending on the method, the weld could result in a very narrow or fairly wide fusion zone (FZ) and heat affected zone (HAZ) due to the heat input experienced by the work piece.

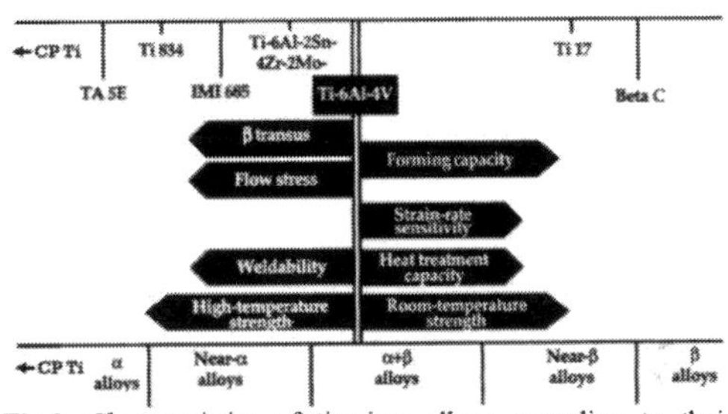

Fig.1. Characteristics of titanium alloys according to their group [1].

1.1 Unalloyed titanium - Commercially Pure Titanium (CP Ti)

The strengths of CP Ti are relatively lower than most other titanium alloys, but it has the best weldability. Typical yield strengths ranged from 170MPa to 480Ma [1]. Its strengths come from the strengthening elements i.e. oxygen and nitrogen in the form of α phase accompanied by some minor β phase due to the presence of Fe [2]. Welding characteristics of CP Ti have been reported by various authors. Karpagaraj et al. [3] performed GTAW on CP Ti sheets with a thickness of 1.6 and 2 mm containing equiaxed alpha phase grains with the presence of fine and disperse precipitation of beta phase. They pointed out that the interface between

* Corresponding author: tpasang@aut.ac.nz

DOI:10.1051/matecconf/201712300001

the BM and the HAZ where the thermal cycles occurred during the welding process. Due to this complex thermal cycle, the equiaxed grains had grown in larger size in the HAZ. Compared to the microstructure of the HAZ, there is an inhomogeneous growth of microstructure at the FZ. FZ of the weld bead obtained in GTAW is fairly wide and it is clearly differentiable. Based on the microstructural studies, the degree of grain coarsening decreases as one moves farther from the FZ to the BM. Gorinin et al. [4] reported a huge difference in microstructures between fast-cooled (500°/sec) and slow-cooled (20°/sec) of welded CP Ti samples. The fast-cooled sample had needle (martensite-like structure) and lamellar-shaped α-phase, while the slow-cooled sample showed a coarse acicular α. The HAZ consists of equiaxed transformed β grains which increase in size as the FZ is approached. Within these grains, colonies of α-phases are present. Cui Li at al. [5] also observed a significant grain coarsening occurred in the HAZ immediately adjacent to the base metal and the degree of grain coarsening increases as the grains move towards the centre of the fusion zone. They also pointed out that higher welding speeds (5 m/min vs. 4m/min) resulted in fine-grained acicular alpha in the weld centreline because the higher welding speed leads to higher cooling rate. Lathabai et al. [6] reported columnar grains in the fusion zone (FZ) which were significantly coarser than those observed in the HAZ and base metal. The columnar grains in the fusion zone were a result of epitaxial growth from the HAZ. The HAZ consisted of equiaxed transformed β grains which increase in size as the FZ is approached. Colonies of α-phases were present within these grains. They also reported that hardness across the BM-HAZ-FZ was comparable. Note that all the tensile samples fractured at the weld area. Mitchell et al. [7] conducted electron beam welding on a 1.6 mm thick CP Ti and found the hardness to be comparable between the weld zones (HAZ and FZ) and the BM (Figure 2). The base metal had a yield strength (YS) of 630 MPa, ultimate tensile strength (UTS) of 695 MPa and elongation of around 27%. Following welding the YS dropped to 530 MPa, UTS reduced to 660 MPa and elongation to 22%. The tensile tested samples fractured at the BM suggesting reliable weld joints. In brief, typically CP Ti welds have coarse columnar grains in the fusion zones (FZ) compared with the heat affected zones (HAZ) and the base metal (BM) [2-7]. In the HAZ, the degree of grain coarsening increases as one moves towards the centre of the FZ.

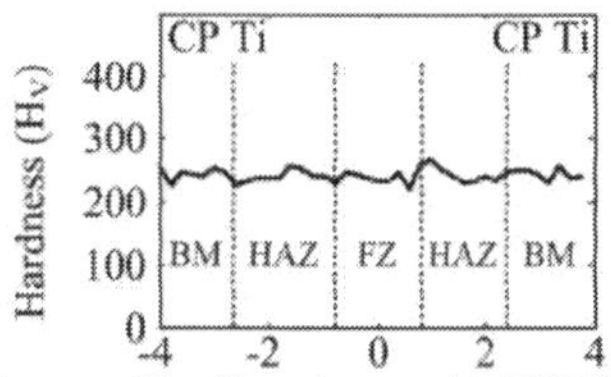

Fig. 2. Hardness profile of laser beam welded CP Ti [7].

1.2 Near α-alloy

Chamanfar et al. [8] investigated properties and microstructure of a laser beam welded (LBW) near α titanium alloy sheet known as Ti–6Al–2Sn–4Zr–2Mo (Ti6242). The base metal (BM) had a bimodal microstructure with primary α and transformed β. The HAZ microstructure had a higher primary α volume fraction in the HAZ region near base metal while the FZ contained acicular α-prime (martensite) (Figure 3). The BM had a yield strength (YS) of 880 MPa, ultimate tensile strength (UTS) of 1120 MPa and elongation of 19.4%. In the as-welded condition, the average YS was around 775 MPa, UTS was 977 MPa and elongation of 13.4%. So, there was a slight reduction in strength and ductility from the BM by about 12%, 13% and 31%, respectively. Electron probe microanalysis (EPMA) demonstrated a uniform distribution of the main alloying elements across the weld. The hardness increased from the BM toward the HAZ and FZ (Figure 4) mainly due to a higher α volume fraction in HAZ and α-prime formation in the FZ.

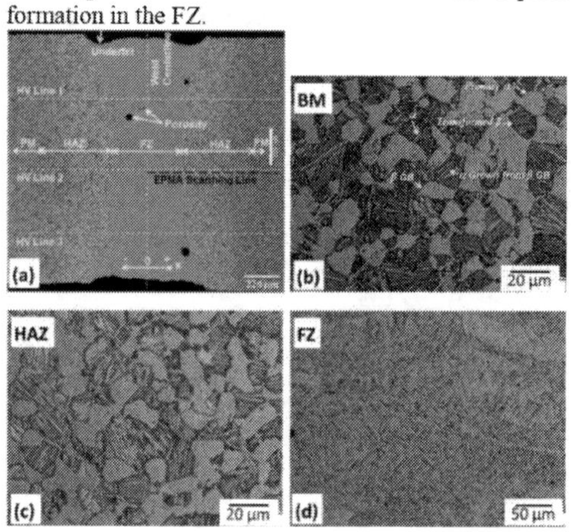

Fig. 3. (a) Weld profile of LBW Ti6242, (b) microstructure of BM showing bimodal structure of α and transformed β, (c) HAZ area showing larger grain size bimodal structure, and (d) α-prime or martensite structure in the FZ area [8].

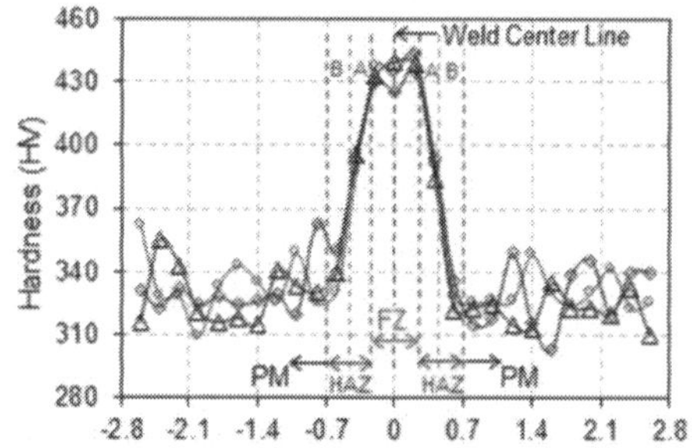

Fig. 4. Hardness profile of EBW Ti6242 showing an increased hardness in the HAZ and FZ due to the formation of α prime – martensite [8].

1.3 α + β alloy (Ti6Al4V)

Ti6Al4V (Ti64) is, arguably, the most widely used titanium among all titanium alloys. It is commonly used as a reference against other titanium alloys. Ti64 has a great combination of properties such as strength and toughness as well as excellent corrosion resistance [1]. Upon fast cooling, this alloy is capable of producing martensite, hence, its strength provider. Weldability of Ti6Al4V (Ti64) has been investigated by various researchers. Akman et al. [9] investigated a laser welding of Ti64 and suggested a high cooling rates caused the formation of α-prime in the FZ and HAZ region. While the FZ was predominantly occupied by α-prime (martensite), the HAZ consisted of α-prime (martensite) and acicular α. The micro hardness profile across the weldment indicates that hardness in the FZ is higher than both the HAZ and BM. Mazumder and Steen [10] also conducted laser butt welding on Ti64 with 2kW power. The BM had an annealed structure with dark β in a bright α matrix. The HAZ had a mixture of α-prime (martensite) and primary α. The grains were larger in the near HAZ (near FZ) than in the far HAZ. The FZ consisted mainly of α-prime within prior β-grain boundary. Sundaresan et al. [11] reported that rapid cooling and subsequent α-prime transformation are effective strengthening methods for many titanium alloys. The observed large increase in hardness in laser welded Ti64 is due to high cooling rate associated with laser beam welding. The FZ of α/β alloys contains coarse, columnar prior beta grains. These grains may have originated from the near HAZ adjacent to the fusion boundary during solidification [11]. The grain structures of Ti64 welds showed the presence of a small amount of acicular α, a larger amount of α-prime in the HAZ and α-prime covering the entire FZ [9,11]. Huiquang et al. [12] reported columnar grains in the FZ. They pointed out that the amount of lamellar structure increases (decreasing quantity of equiaxed grains) with heat input. Hardness profiles showed an increase in hardness due to the presence of α-prime in the FZ compared with the BM. They suggested that the increase in hardness, hence less ductile, coupled with the presence of micropores contributed to the fracture in the weld region. The tensile tested samples fractured transgranularly with fine dimples on the fracture surface. Most recently, LBW of Ti64 were performed under conduction regime [18] and under keyhole regime [19], where both modes leading to butt welds with similar tensile properties than the unwelded base metal. In fact, these welded samples broke at the BM, confirming that the HAZ and FZ are stronger than the BM in the welded samples [18]. A study by Mitchell et al. [7] on welding of Ti64 by electron beam welding (EBW) suggested a slight increase in strength in the weld zone (HAZ and FZ). This is reflected in the increase of the tensile strengths where the unwelded base metal YS of 941 MPa, UTS of 983 MPa and elongation of around 13% compared to as-welded with YS around 960 MPa, UTS 1010 MPa but a slight reduction in elongation to 8.5%. In brief, the microstructure was principally α-prime in the FZ and acicular α with α-prime in the HAZ (1, 9-12].

1.4 Metastable β and β alloys

An alloy is considered to be a β alloy if it contains sufficient β stabilizer alloying element to retain the β phase without transformation to martensite on quenching to room temperature [1,2,13]. A metastable β alloys have to have a minimum Mo_{eq} value of 10 *(see formula below)* to retain the β-phase at room temperature [13]. The higher the value of the Mo_{eq} of an alloy the higher the volume retention of the β phase for the same cooling conditions. Metastable β titanium alloys that can retain an apparently fully β-phase microstructure, are candidates for use in many demanding structural applications where light weight, impact toughness and high-strength are required [1,2].

$$Mo_{eq}\ (wt.\%) = 1.0Mo + 0.67V + 0.44W + 0.28Nb + 0.22Ta + 2.9Fe + 1.6Cr - 1.0Al \quad (1)$$

Although relatively limited, investigations on weldability of β-titanium alloys have been conducted by various researchers. Becker and Baeslack [14] studied the properties and microstructure of welded Ti-15V-3Cr-3Al-3Sn (Ti15-333), Ti8V-7Cr-3Al-4Sn-1Zr (Ti87341) and Ti-8V-4Cr-2Mo-2Fe-3Al (Ti84223) and revealed lower strength in the weld zones (HAZ and FZ) compared with the BM. They also suggested that the alloys were weldable. Fracture modes of the weldments indicated transgranular microvoid coalescence. Liu et al. [15] studied the weldability of Beta-21S sheet using laser welding technique. They reported that the FZ and HAZ were narrow with fine retained β grain structures. Epitaxial grain growth was observed in the narrow HAZ through the fusion boundary into the FZ. The FZ had transitioned from a solidification mode of a cellular-type along the fusion boundary to a completely cellular-dendritic (or columnar dendritic) solidification mode at the weld centreline. Baeslack et al. [16] conducted welding investigations on Beta-C™ implementing gas tungsten arc welding (GTAW or TIG) and observed epitaxial growth from the near-HAZ into the FZ, which solidified with a cellular mode and progressively formed a complete columnar-dendritic grain structure at the weld centreline. The rapid cooling coupled with high-beta stability-containing elements ensure the retained β-phase in the weld region. The presence of retained β-phase is supported by the phase diagram *(see Figure 9)*, wherein the β alloys (Ti5553, Ti–10–23, and Ti–13–11–3) sit just beyond the martensite start where its formation is suppressed. A portion of the HAZ was heated above the β transus during welding causing α precipitates to go back into solution. β-phase was fully retained upon fast cooling from above the β transus, leaving an α-free portion of the HAZ. The remainder of the HAZ was heated high enough in the α+β phase field to cause a reduction in volume fraction of the precipitated α-phase at the transition from the BM to the HAZ. Consequently, there are two distinct regions in the HAZ, one which has exceeded the β transus during welding (resulting in large grains) and another which has not.

Recently, a metastable β titanium alloy known as Ti-5Al-5V-5Mo-3Cr (Ti5553) has gained significant interest. Apart from its high strength, excellent hardenability and fracture toughness, this alloy also offers high fatigue resistance making it an ideal material for aircraft landing gear application [1,2]. A few other niche applications in the aerospace area have, since, been specified. For these applications, there may be a need to join them, e.g. by welding, and therefore its weldability needs to be investigated. Mitchell et al. [7] was the first to initiate welding investigation of Ti5553 by using electron beam welding (EBW). The hardness profile is shown in Figure 5 where lower hardness occurred in the HAZ and FZ. This is the opposite of hardness profile reported for $\alpha + \beta$ titanium such as Ti64. From tensile tests, it was revealed that all fractures took place in the FZ. This work was continued by Sabol et al [17], and is reported in the following section.

The tensile properties, strain distribution, fracture mechanisms, and microstructure of electron beam welded (EBW) Ti-5Al-5V-5Mo-3Cr (Ti5553) in the as-welded condition were investigated by Sabol et al. [17] in order to evaluate the weldability of this alloy [17]. Rolled sheets of Ti5553 were electron beam welded (EBW) perpendicular to the rolling direction and sub-size tensile specimens were machined from the sheets. Microstructural examination showed the FZ and HAZ consisted of large grains of retained β phase whereas the BM consisted of α surrounding small β grains containing dispersed α precipitates (Figure 6).

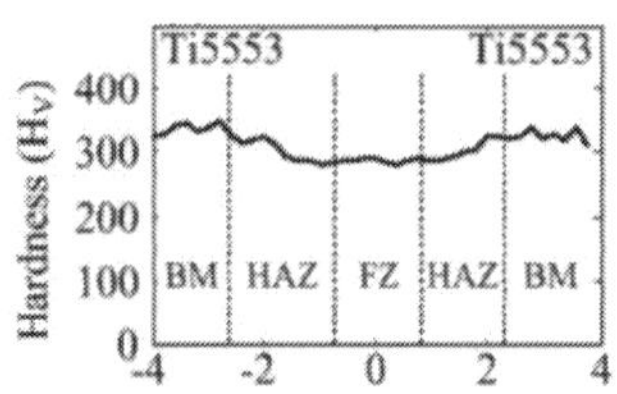

Fig. 5. Hardness profile of as-welded Ti5553 using laser beam welding [7].

Tensile tests revealed that elongation at fracture in the as-welded condition was comparable to un-welded Ti5553 while also revealing that the tensile strength was lower. Hardness profiles across the welds showed a decrease in average hardness from the base metals to the fusion zone (Figure 7). The decrease in hardness can be attributed to the large grain size, retained β phase, and loss of Al during welding. Fracture occurred in the weld zone and was primarily due to microvoid coalescence (Figure 8). These results indicate that Ti5553 is readily weldable and displays reasonable properties in the as-welded condition.

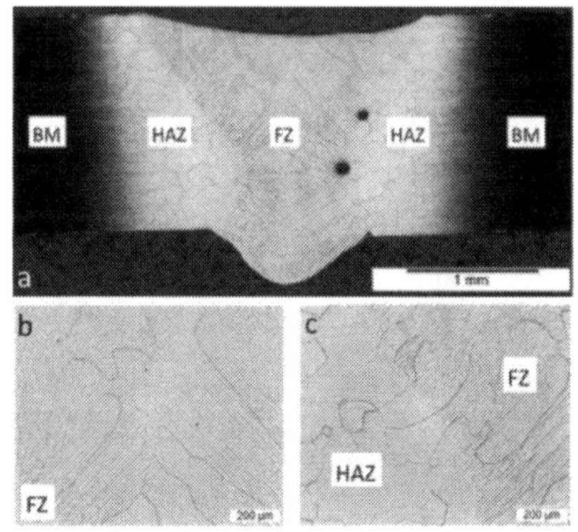

Fig. 6. (a) Weld profile of Ti5553 after EBW, and (b-c) micrographs showing grain structures in the HAZ and FZ [17].

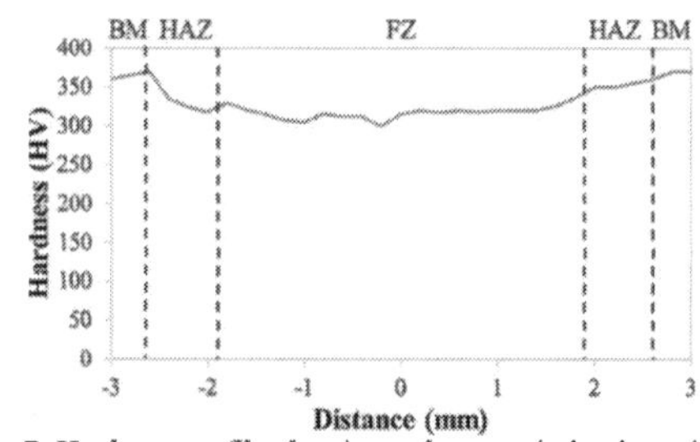

Fig. 7. Hardness profile showing a decrease in hardness in the HAZ and FZ area [17].

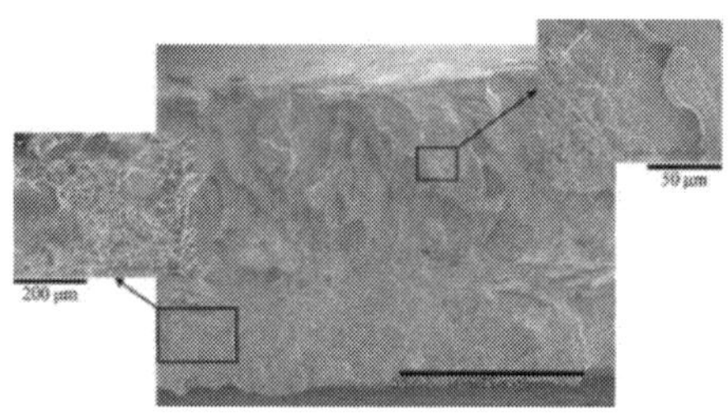

Fig. 8. SEM images showing dimples features on the fracture surface [17].

2 Current investigations

Apart from a brief summary on previous works, this paper also present the more recent results from an on-going research on welding of titanium alloys at Auckland University of Technology. Figure 9 illustrates a simplified phase diagram showing locations of some common titanium alloys with respect to the titanium classification described in the introduction including those used in this investigation. In this article we reported the on-going welding results using gas tungsten arc welding (GTAW) or commonly known as tungsten inert gas (TIG) welding on a few different titanium across the phase diagram. The materials were CP Ti, Ti6al4V (Ti64) and Ti-5Al-5V-5Mo-3Cr (Ti5553) which represent the unalloyed (CP Ti), $\alpha + \beta$ alloys and β alloy, respectively. Their chemical composition are presented in Table 1, and the microstructure prior to welding are given in Figure 10. The CP Ti had nearly equiaxed α grains with an average size of 20 μm in the longitudinal direction. It also contained disperse β phase (dark). The presence of β phase is associated with the addition of small amounts of Fe in CP Ti. The Ti64 alloy had elongated primary α grains in the α/β matrix, while Ti5553 alloy showed typical α/β microstructure with globular α particles distributed within the β matrix. These α particles have an average size of less than 5 μm. Figure 11 illustrates the sampling from a welded material.

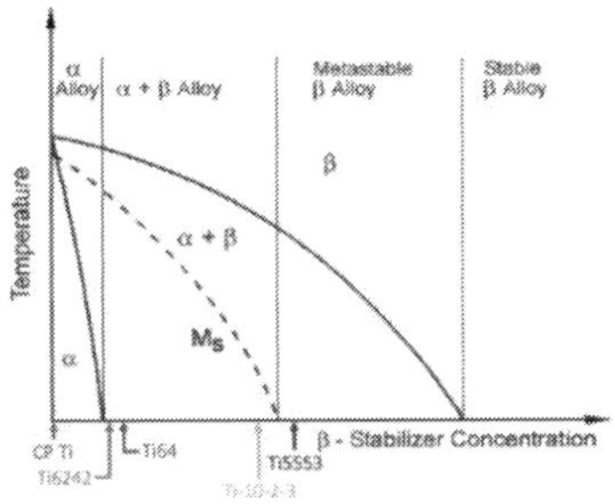

Fig. 9. A simplified phase diagram showing locations of various titanium alloys (modified from [1]).

Table 1: Chemical composition (wt.%) of the three alloys used in this investigation

	Ti	Al	V	Mo	Cr	Fe	O	N
CP Ti	Bal	0.16	0.01	0.01	0.01	0.22	0.28	0.01
Ti64	Bal	6.08	3.85	0.01	0.02	0.17	0.05	0.01
Ti5553	Bal	5.03	5.10	5.06	2.64	0.38	0.14	0.01

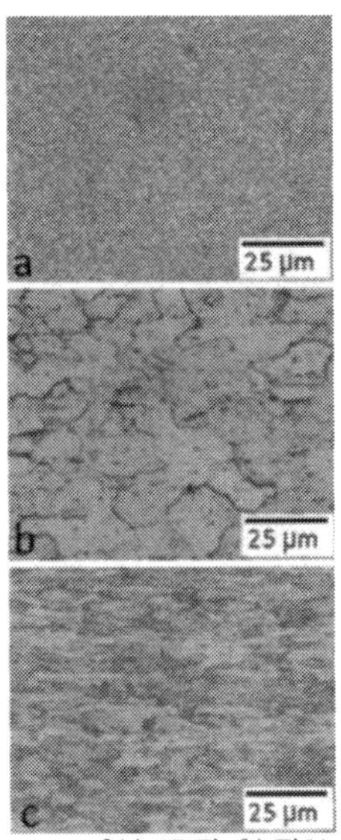

Fig. 10. Microstructure of (a) CP Ti, (b) Ti64 and (c) Ti5553.

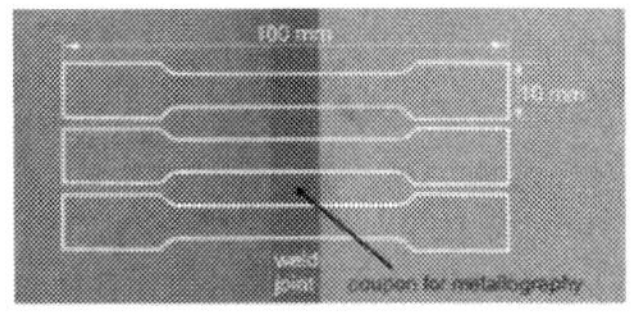

Fig. 11. Schematic diagram showing location for of the tensile and metallography samples, i.e. weld profiles and hardness

2.1 Unalloyed titanium - Commercially Pure Titanium (CP Ti)

The weld profile of the CP Ti-CP Ti butt joint using GTAW is given in Figure 12. The width of the FZ is around 8 mm. This is much wider than those of EBW or LBW, i.e. around 2 mm. The higher magnification micrographs showed BM material with equiaxed α grains – a typical structure in the annealed condition. The HAZ also had equiaxed grains with acicular α phase starting as smaller grains near BM and gets larger as it moves towards FZ area (Figures 12 and 13). The grains in the FZ area can be as big as 200μm consisting predominantly of acicular α phase (Figure 13). The hardness values were was consistent throughout the BM-HAZ-FZ-HAZ-BM, i.e. between 250 to 270 HV (Figure 14a). From tensile testing, it is clear that fracture took place at the BM (Figure 14b). The UTS was around 690 MPa and 560 MPa for the BM and as-welded respectively. The fracture surface showed the presence of dimples indicating ductile fracture mechanism (Figure 15).

Fig. 12. Weld profile of CP Ti-CP Ti from GTAW from BM-HAZ-FZ-HAZ-BM.

Fig. 13. Micrographs showing equiaxed grains in the BM and acicular α in both HAZ and FZ, but larger grains in the FZ.

DOI:10.1051/matecconf/201712300001

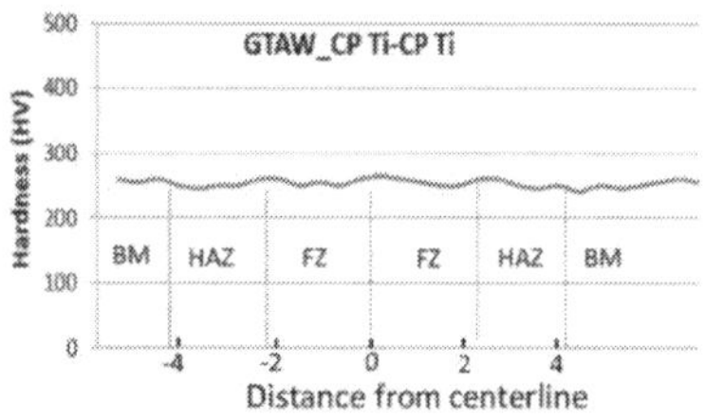

a

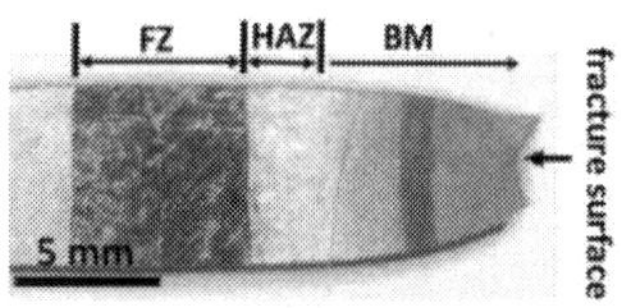

b

Fig. 14. (a) Hardness profile showing comparable hardness from BM-HAZ-FZ-HAZ-BM, and (b) fracture location of the tensile tested CP Ti-CP Ti showing fracture location at the BM.

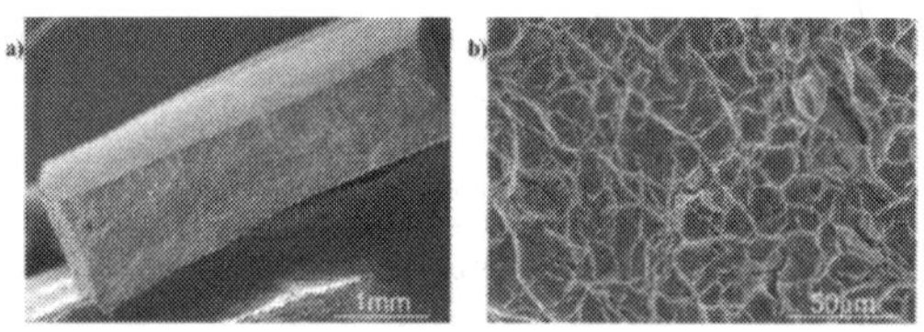

Fig. 15. SEM images showing (a) fracture location of CP Ti welds in the BM, and (b) dimples features on the fracture surface.

2.2 α + β alloy (Ti6Al4V or Ti64)

The weld profile of the Ti64-Ti64 butt joint using GTAW is given in Figure 16. The width of the FZ is around 8 mm. The higher magnification micrographs showed BM material with fine α grains aligned in the longitudinal direction indicating the material was, most likely, in the as-rolled condition. The HAZ had equiaxed grains with acicular α phase and some evidence of the presence of α-prime (martensite). The grains were smaller near BM and gets larger as it moves towards FZ area (Figures 16 and 17). The grains in the FZ area were large, and predominantly populated by α-prime and acicular α to a minor extent (Figure 17). The hardness profile showed an increase in the HAZ and FZ, i.e. from around 330 HV in the BM to around 380 HV in the FZ (Figure 18a). From tensile testing, it is clear that fracture took place at the fusion line, i.e. border between HAZ and FZ (Figure 18b). The UTS was around 980 MPa and 890 MPa for the BM and as-welded respectively. Dimples were also observed on the fracture surface (Figure 19) but much smaller and shallower compared to those observed in the CP Ti fracture surface.

Fig. 16. Weld profile of Ti64-Ti64 from GTAW.

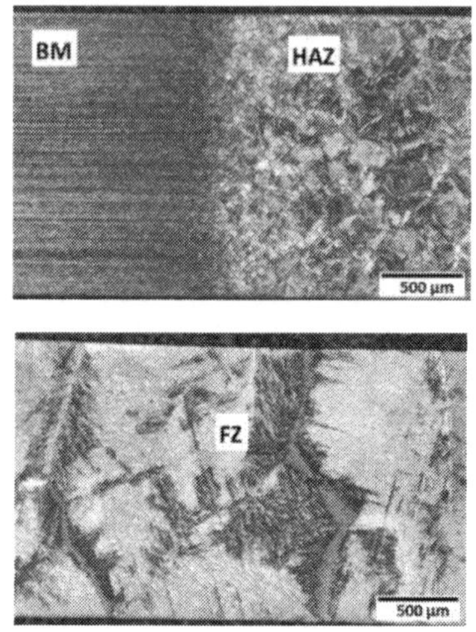

Fig. 17. Micrographs showing very fine grains in the BM, acicular α in the HAZ and mainly α-prime in the FZ. Grains in the FZ were also larger.

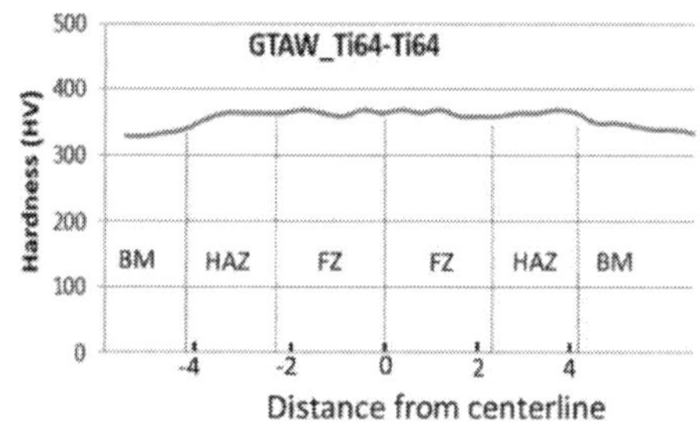

a

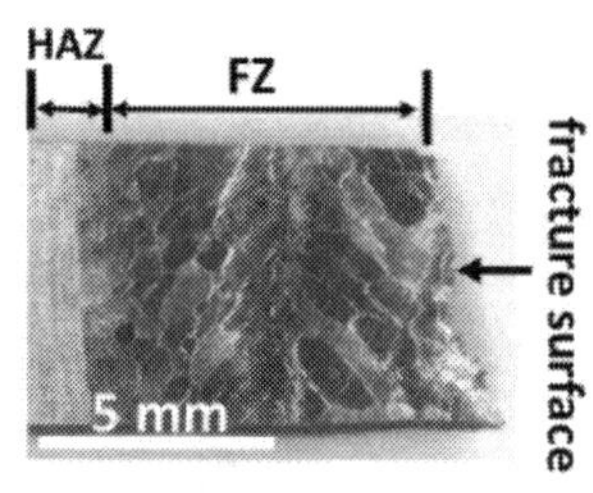

b

Fig.18. (a) Hardness profile of Ti64-Ti64 GTAW, and (b) fracture location of the tensile tested Ti64-Ti64 GTAW.

DOI:10.1051/matecconf/201712300001

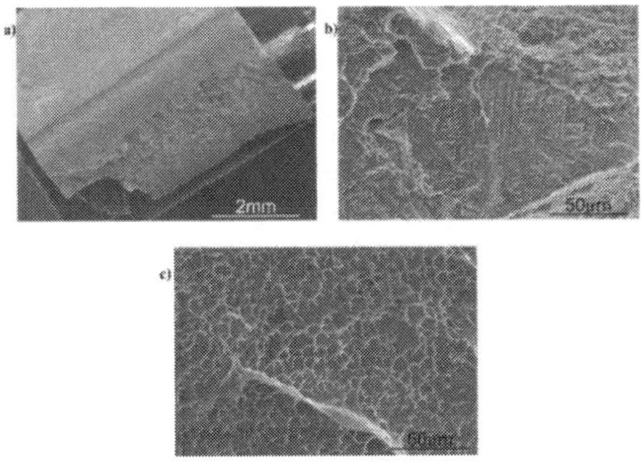

Fig. 19. SEM images showing fracture surface of a tensile tested Ti64-Ti64 butt weld.

2.3 β alloy (Ti-5Al-5V-5Mo-3Cr or Ti5553)

The weld profile of the Ti5553-Ti5553 butt joint using GTAW is given in Figure 20. The width of the FZ is around 7 mm on the top surface and 4 mm at the bottom surface. This V shape is due to the relatively thin sheet used in this experiment, i.e. around 1.6 mm compared with around 2 mm and 2.4 mm thick for CP Ti and Ti64, respectively. Although not very clear in the image, but the BM had a very fine grains resulting from forging. The HAZ is decorated with retained, equiaxed β grains with larger grains at the fusion boundary and smaller grains towards BM. Epitaxial grain growth from the HAZs into the FZ was clearly observed (Figure 21). The FZ consisted of a columnar dendritic-typed grain morphology indicating a high concentration of β-stabilizing elements. The hardness profile showed a decrease in the HAZ and FZ, i.e. from around 380 HV in the BM to around 300 HV in the FZ (Figure 22a). From tensile testing, it appears that fracture occurred in the FZ (Figure 22b). The UTS was around 1050 MPa and 720 MPa for the BM and as-welded respectively. This is consistent with the result from hardness which indicating low hardness in the FZ. Low magnification SEM image showed a brittle-type fracture, but at higher magnification, it is obvious that dimples were present on the fracture surface (Figure 23).

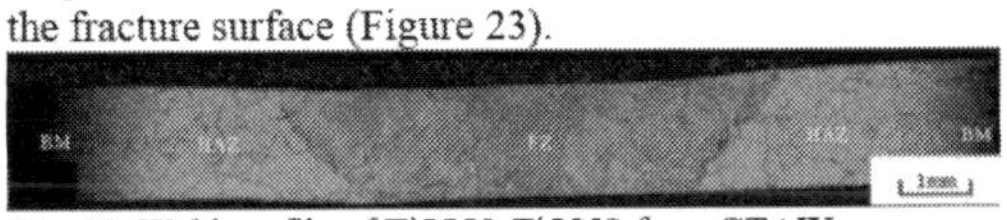

Fig. 20. Weld profile of Ti5553-Ti5553 from GTAW.

Fig. 21. Micrographs showing very fine grains in the BM (most likely as-forged), equiaxed grains with retained β α in the HAZ and columnar dendritic grains in the FZ.

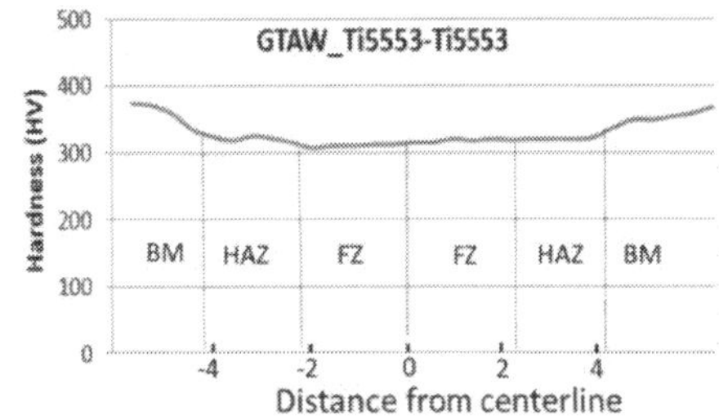

a

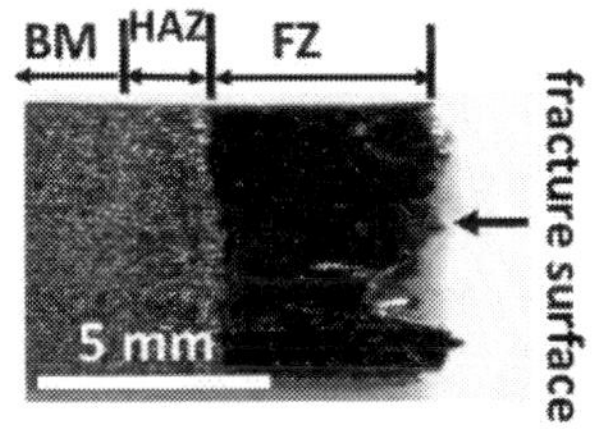

b

Fig. 22. (a) Hardness profile of Ti5553-Ti5553 GTAW, and (b) fracture location of the tensile tested Ti5553-Ti5553 GTAW.

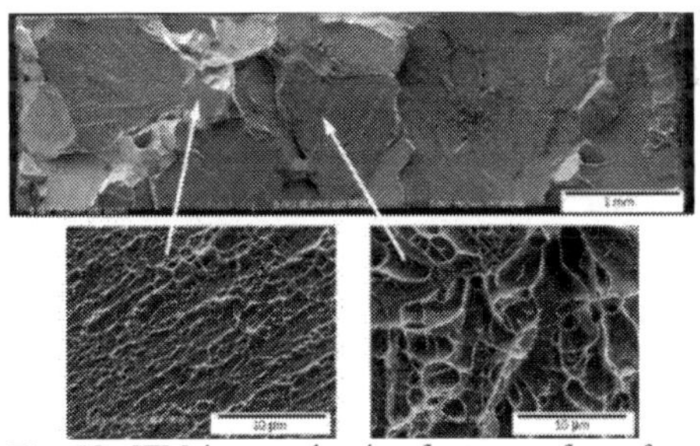

Fig. 23. SEM images showing fracture surface of a tensile tested Ti5553-Ti5553 butt weld.

3 Summary

A schematic diagram showing hardness profiles of as-welded titanium alloys is given in Figure 24 below. Although it is not comprehensive, it indicates the strength and weakness of certain areas across the welds. For unalloyed titanium, hardness remained constant throughout. Near α alloys and α + β alloys showed an increase of hardness in the HAZ and FZ, while the β alloys showed a decrease of hardness, hence, low strength in the HAZ and FZ area. This implies a careful consideration when welding titanium alloys for an intended applications. It also suggested a proper post welding heat treatment (PWHT) is required to minimise the effect of strengthening or softening in the HAZ and FZ to avoid premature failures.

DOI:10.1051/matecconf/201712300001

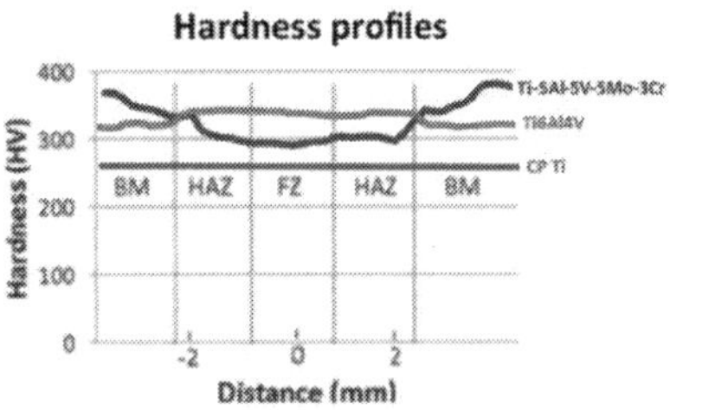

Fig. 24. General hardness profiles of various titanium alloys.

The author (TP) would like to thank the Loewy Foundation for providing support for the Institute for Metal Forming, Lehigh University where some of these research have been carried out. TP would also like to thank the Japan Society for the Promotion of Science (JSPS), and the Joining and Welding Research Institute (JWRI), Osaka University for the funding through JSPS Fellow to conduct laser welding at JWRI.

References

1. M.J. Donachie Titanium: A Technical Guide second Edition, *ASM International* (2004)
2. G. Lütjering, J.C. Williams, Titanium. Second edition, Berlin Springer (2003)
3. A. Karpagaraj, N. Sivashanmugam, *Materials Science & Engineering*, **A640**, 180-189 (2015)
4. I.V. Gorinin, J.B. Florinsky, B.B. Chechukin, The Science, Technology and Application of Titanium, Pergamon Press, 843-848 (1968)
5. L. Cui, M. Kutusna, T. Simizu, K. Horio, F, *Materials and Design*, **30**, 109-114 (2009)
6. S. Lathabai, B.L. Jarvis, K.J. Barton, *Materials Science and Eng.*, A299, 81-93 (2001)
7. R. Mitchell, A. Short, T. Pasang, G. Littlefair, Structural Integrity and Failure - SIF, University of Auckland (2010)
8. A. Chamanfar, T. Pasang, A. Ventura, W.Z. Misiolek, *Materials Science & Engineering*, **A663**, 213-224 (2016)
9. E. Akman, A. Demir, T. Canel, T. Sınmazcelik, *Journal of Material Processing Tech.*, **209**, 3705-3713 (2009)
10. J. Mazumder and W. M. Steen, *Metallurgical Transactions A*, **13a**, 865 (1982)
11. S. Sundaresan, G.D. Janaki Ram, G. Madhusudhan, *Mater Sci Eng A*, **262**:88–100 (1999)
12. W. Huiqiang, F. Jicai, H. Jingshan, *Bulletin of Materials Science*, **27**, 387-392 (2001)
13. P.J. Bania (1993), Beta titanium alloys in the 1990's, TMS, Warrendale, PA, 3-14 (1993)
14. D. Becker, W.A. Baeslack III, *Welding Journal*, **59**, 85-93 (1980)
15. P. Liu, K.H. Hou, W.A. Baeslack III, J. Hurley, Titanium '92, 1477 (1993)
16. W.A. Baeslack III, P.S. Liu, D.P. Barbis, J.R. Schley J.R. Wood, *Titanium '92*, 1469-1476 (1993)
17. J.C. Sabol, T. Pasang, W.Z. Misiolek, J.C. Williams, *Journal of Materials Processing Technology*, **212**, 2380-2385 (2012)
18. J.M. Sánchez-Amaya, M. R. Amaya-Vázquez, F.J. Botana, Laser welding of light metal alloys: aluminium and titanium alloys (Ed. S. Katayama), AMAZON (UK), 215-254, 2013
19. X-L. Gao, L-J. Zhang, J. Liu, J-X Zhang. A comparative study of pulsed Nd:YAG laser welding and TIG welding of thin Ti6Al4V titanium alloy plate. *Materials Science and Engineering: A*, **559**, 14–21, 2013

Comparisons of compression characteristics in compression forming of porous cylinder based on slab method and finite element method under constant shear friction

Gow-Yi Tzou[1,*], *Yeong-Maw* Hwang[2], and *Sai-Chih* Pan[2]

[1]Department of Mechanical and Automation Engineering, Chung Chou University of Science and Technology, Taiwan, R.O.C.
[2]Department of Mechanical and Electro-Mechanical Engineering, National Sun Yat-Sen University, Taiwan, R.O.C.

Abstract. The study assumes constant shear friction to perform mechanics analysis and comparisons of compression characteristics in compression forming of porous cylinder based on slab method and finite element method. The vertical stress (p), radial stress (q), compression force (P), average radius, and density after forming can be systematically explored. Because the volume in forming of porous cylinder cannot keep constant, just only keep the mass constancy. Therefore, the derivation is more complicated. The mechanics analytical model established by the slab method compares with the FEM using DEFORM-2D analysis software each other to realize the variations of both models; the maximum average error for the compressive load (P) under various reduction ratios is 14.8% between the slab method and the finite element method. The results with a good agreement are shown in the study. These models can be offered to the reference of porous cylinder compression in industries.

1 Introduction

The porous materials have a wide range of applications in the diverse industries. They are important materials for lubricating bearings, metal filters, metal electrodes, and light-weight parts for the automotive industry. There have been many relevant studies discussed in several aspects.

Yang et al. [1] explored the porous materials on the relationship of bulging deformation, forging force and density change during open die forging. By FEM analysis and other experiment results, Yang et al. [1] investigated the applicability of FEM simulations as well. Lee et al. [2] emphasized the importance of the empty holes effect and the density distribution in the study of porous materials. In addition to their criteria on plastic yielding, they also compared with other criteria on yielding in various studies. Based on their yielding condition, Gurson, Shirizly et al. [3] used experiments in macro and micro scales to prove the accuracy of FEM in predicting the density of porous materials during compression processing. Doege et al. [4] predicted the density variations of the porous metal ring in the compacting process, and proposed an analytic method of ordinary differential equations (ODE), which is easier than Hill's partial differential equations (PDE). As for the yielding criteria and the plastic behaviour [5-9], Huang et al. [5] used the sintered powder compacts to perform uniaxial tension and compression tests under various frictional conditions, and the results were used to compare to FEM each other. They described the plastic behavior of porous materials in Gurson-Tvergaard's criterion on yielding. Green [6] used rigorous mathematics to establish the plasticity theory for porous solids, and proposed the differential equations and the relationship of yield criterion on the stress-strain and the void ratio. This was the basis for the derivation of the slab method which is proposed in this paper. Using the stress-strain curve data in the sintered copper experiment, Shima et al. [7] proposed the plasticity theory for porous metals and applied it to a friction-free closed-dies compression experiment to verify its practicability. Doraivelu et al. [8] reviewed many yield criteria proposed by various scholars. They aimed at the compressible powder metallurgy materials to propose a new yielding function, and used aluminum alloy (A7091) uniaxial compression experiment to test the accuracy and practicability. Dias et al. [9] used the plasticity equation to analyze the parameters of the sintered porous materials after the forging process. They confirmed the experiment results agreeable to the theoretical analysis of the proposed model, which can be applied to the cold forging of sintered performs. Jha et al. [10] detailed the experiments on forming characteristics and failure mechanisms from the start of spraying iron metal powder to the completion of sintering processing. In the forming properties, they discussed the size of powder particles in details, and also classified the types of crack to the failure mechanisms. Nguyen et al. [11] investigated the mechanical behaviour of porous functionally graded (FG) beams; they indicated that it is necessary to take the effect of porosities into consideration in the vibration analysis of FG beams traversed by moving forces.

* Corresponding author: johnsontzou@gmail.com

2 Simulation analysis and modelling

2.1 Finite element method

In this study, DEFORM-2D is used to perform the simulation analysis on the compression forming of porous cylinder. Based on the finite element method, the DEFORM-2D analyzes two-dimensional flow in the process of metal forming. It is often used to analyze of forging, extrusion, drawing, upset forging etc. in forming process used by industrial and academic groups. The software settings in DEFORM-2D include: porous materials; the function of flow stress, σ=Kε^n (K=500MPa, n=0.3); Young's modulus, E=210GPa; Cylindrical height to diameter ratio (Hi/Di) 15/12=1.25, 12/12=1, 9/12=0.75; initial relative density before compression, ρ_i=0.7; numbers of element=3000; constant shear friction factor between workpieces and dies (m) are taken as 0.1, 0.3, 0.5; compression speed of upper die =1 mm/s.

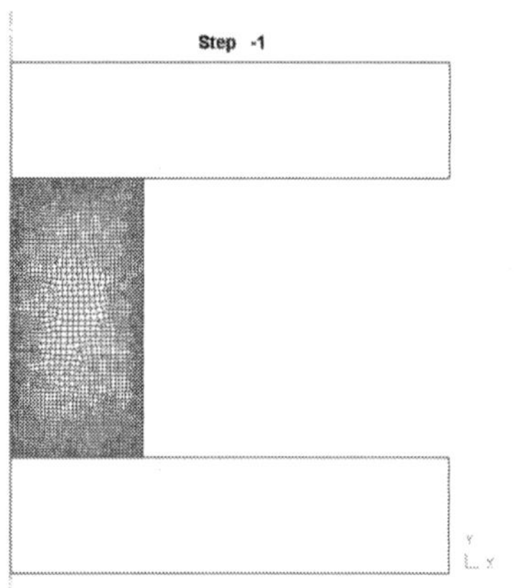

Fig. 1. The diagram of pre-process in compression forming of porous cylinder in DEFORM-2D.

2.2 Slab method

In general, before the processing effects are unknown, the slab method can analyze the influential parameters on the processing effects in advance, and probe into the conceptual discussion. By the slab method, the stress distribution and compressive force of the porous cylindrical material in compression forming are explored and then compare to the simulation results of DEFORM-2D.

To simplify the complexity of derivations, the study establishes a mechanics analytical model for the porous cylinder material between upper and lower die, and assumes that the friction between the interface of the dies and cylinder is constant shear friction. With the cylinder material is porous, the compressive pressure distribution and compressive force of the porous cylinder material during the compression processing are deduced by the theory of slab method.

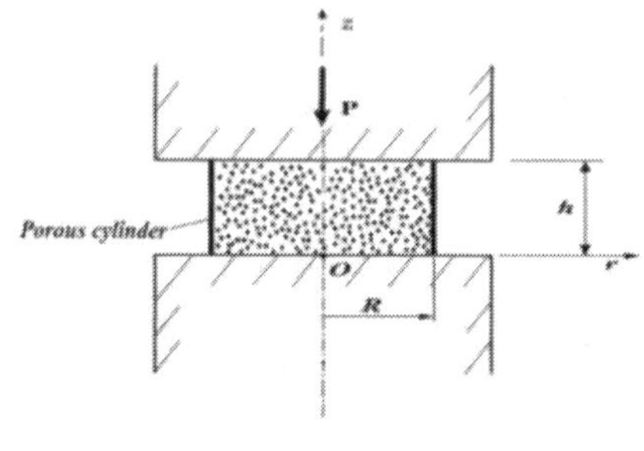

Fig. 2. Schematic diagram of porous cylinder in compression.

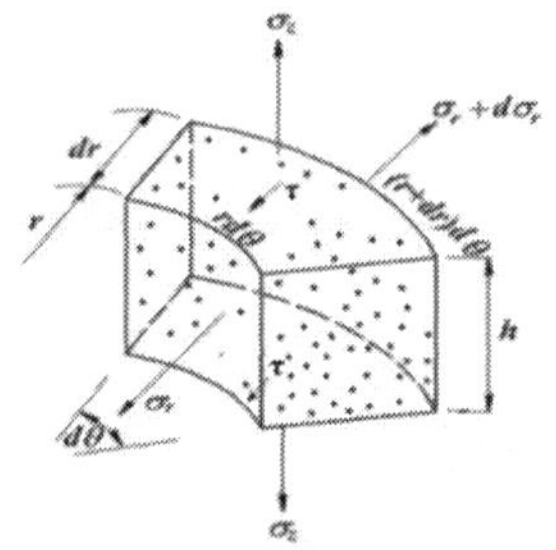

Fig. 3. Stress element diagram of porous cylinder in compression.

In order to derive the compression forming analytical model, the following basic assumptions are required:

(1)Assuming that the cylinder is a porous material, and the pores are evenly distributed in the cylindrical material.

(2)The mechanics analysis is based on axisymmetry.

(3)The stresses on the elements are distributed uniformly; radial (σ_r) and vertical (σ_z) stresses are considered as principal stresses.

(4)The interfacial friction between the die and the porous cylinder is assumed to be constant shear friction; τ is utilized in the radial direction of the cylinder material.

(5)The bulging effect on the free surface is ignored.

(6)Since the porous material has holes, the incremental deformation makes the material denser and the void space less. The conservation of mass can be applied to the porous material; the conservation of the volume is unable to be used.

Therefore, the yielding criterion of the porous material should be considered first before setting up the slab method for an analytical model of compression forming of porous material.

2.2.1 Green's yielding criterion

The Green's yielding criterion [4] is expressed as follows:

$$G \cdot J_2' + BI_1^2 = DY_i^2 = Y_R^2 \quad (1)$$

where $G=3$, J_2' is the second stress deviation tensor, I_1 is the first stress constant tensor, $D=\rho_r^5$, ρ_r is the relative density after the compression(ρ_r=1-f , f is porosity rate), and Y_i is the initial yield strength of porous material, $B=\left(\frac{1}{9}\right)\left(2.49(1-\rho_r)^{0.514}\right)^2$

From Eq.(1), when $\rho_r=1$, it indicates that the materials has no void, i.e. it is a sound material (f=0); then $3J_2'=Y_i^2$ is the yielding criterion for sound material. As $\rho_r \neq 1$, it indicates that the material is a porous material:

$$J_2' + \frac{B}{3}I_1^2 = \frac{DY_i^2}{3} = D\left(\frac{Y_i}{\sqrt{3}}\right)^2 = Dk_i^2 = (\sqrt{D}k_i)^2 = k_e^2 = \frac{Y_R^2}{3} \quad (2)$$

Where Eqs. (1) and (2) are the yielding criteria for porous material proposed by Doege et al. [4].

2.2.2 Force equilibrium

Using cylindrical coordinate system, the axisymmetric is satisfied. The radial force equilibrium equation is $\sum F_r = 0$.

$$\frac{d\sigma_r}{dr} - \frac{2\tau}{h} = 0 \quad (3)$$

where $\tau = mk_D = m(\sqrt{D}k_i) = m\sqrt{\rho_r^5}k_i$

Let $\sigma_r = q,\ \sigma_z = p,\ \sigma_r = \sigma_\theta$,

$$\frac{dq}{dr} - \frac{2\ m\ k_D}{h} = 0 \quad (4)$$

The radial stress (q) can be obtained.

$$q = \frac{2\ m\ k_D}{h} r + c \quad (5)$$

where c is an integral constant, it can be obtained by the boundary condition.

2.2.3 Boundary condition

The radial stress is zero on free surface of cylinder, that is $q=0,\ at\ \ r=R$, Substitute this boundary condition into Eq. (5); then the integral constant c is determined.

$$c = -\frac{2m\ k_D}{h} R \quad (6)$$

2.2.4 Radial stress

The radial stress can be expressed as

$$q = \frac{2\ m\ k_D}{h}(r-R) \quad (7)$$

2.2.5 Compression stress (Vertical stress)

Due to the axisymmetric problem, the radial stress equals to the hoop stress, i.e. $\sigma_r = \sigma_\theta$, and $\sigma_{r\theta} = \sigma_{\theta z} = \sigma_{zr} = 0$,

so

$$\begin{aligned} J_2' &\equiv \frac{1}{6}\left[(\sigma_r-\sigma_\theta)^2 + (\sigma_\theta-\sigma_z)^2 + (\sigma_z-\sigma_r)^2\right] + \sigma_{r\theta}^2 + \sigma_{\theta z}^2 + \sigma_{zr}^2 \\ &= \frac{1}{6}\left[(q+p)^2 + (p+q)^2\right] + 0^2 + 0^2 + 0^2 \\ &= \frac{1}{3}(p+q)^2 \end{aligned} \quad (8)$$

Let $\sigma_z = -p$

$$I_1 = \sigma_r + \sigma_\theta + \sigma_z = 2\sigma_r + \sigma_z = 2q - p \quad (9)$$

Substitute J_2' and I_1 into Eq. (2)

$$J_2' + \frac{B}{3}I_1^2 = k_D^2$$

Obtain

$$\frac{1}{3}(p+q)^2 + \frac{B}{3}(2q-p)^2 = k_D^2$$

Then

$$p^2 + \frac{2(1-2B)q}{(1+B)}p + \frac{1}{(1+B)}\left[(1+4B)q^2 - 3k_D^2\right] = 0$$

It can be rearranged as

$$p = a_1 q + \sqrt{a_4 - b_1 q^2} \quad (10)$$

where

$$a_1 = \frac{2B-1}{1+B},\ a_2 = \left(\frac{2B-1}{1+B}\right)^2,\ a_3 = \frac{1+4B}{1+B},\ a_4 = \frac{3k_D^{\ 2}}{1+B}$$

$$b_1 = a_3 - a_2\ ,\ b_2 = \frac{2mk_D}{h}\ ,\ b_3 = \sqrt{\frac{a_4}{b_1\ b_2^{\ 2}}}$$

2.2.6 Compression force

The compression force can be found by integrating the vertical stress over the contact surface

$$P = \int_0^R p dA = 2\pi \int_0^R p \cdot r dr$$

After the cumbersome and complicated integration, the compression force can be expressed as

$$P = \frac{-\pi a_1 b_2 R^3}{3} + \frac{2\pi\sqrt{b_1}\ b_2}{3}\left(\left(b_3^2 - R^2\right)^{3/2} - b_3^3\right)$$

$$+\pi R\sqrt{b_1}\, b_2\left(R\sqrt{b_3^2-R^2}+b_3^2 \sin^{-1}\frac{R}{b_3}\right) \quad (11)$$

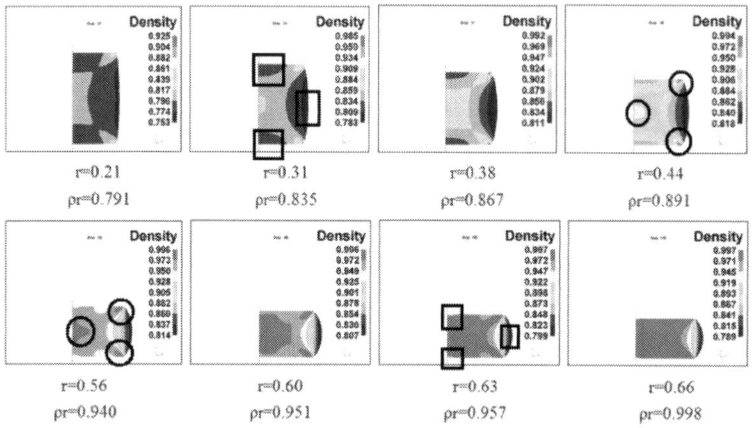

Fig. 4. The variations of relative density (ρ_r) under various reduction ratios.

2.2.7 The mass conservation law

The cylinder with porosity is compacted with the increase of deformation to reduce the void. Therefore, the volume constancy is not suitable for the porous material; however the mass conservation law can be satisfied. With the mass conservation law, the mass of materials before deformation is equal to one after deformation, i.e. $m_i = m_r$. As the strain in height, the initial density, and relative density of deformation are given, the deformed cylinder radius R can be derived from the mass conservation law as follows:

$$R=\sqrt{\frac{\rho_i \cdot h_i}{\rho_r \cdot h}}R_i \quad (12)$$

3 Results and discussions

Figure 4 shows the variations of relative density (ρ_r) under various reduction ratios. The reduction ratios from 21% to 66% are selected for discussions, and the relative density distributions in porous cylinder are not distributed uniformly. The areas of higher density (circled in the graphs) are located at the upper right corner, the lower right corner, and the left-middle side (i.e. peripheral on top and bottom of cylinder, and center of cylindrical axis). The areas of lower density (framed in the graphs) are located at the upper left corner, the lower left corner, and right-middle side (i.e. two ends of cylindrical axis, and the middle part of cylinder outside peripheral). The peripheral on top and bottom of the cylinder and the center of the cylinder are compactly compressed at first (i.e. ρ_r approached to the value of 1.0 rapidly); contrarily, the center on top and bottom of the cylinder and the bulging area of the cylindrical periphery are densely compressed at last (i.e. ρ_r approached to the value of 1.0 slowly).

Figure 5 shows the velocity field distributions under different reductions, it shows the flow direction of the material around the cylindrical axis is axially downward, and then flow to the radial direction. The flow rate is the highest on the cylinder top and the lowest near the cylinder bottom. The flow direction on the cylindrical periphery inclines to the right, and the radial direction will be more rightward from top to bottom.

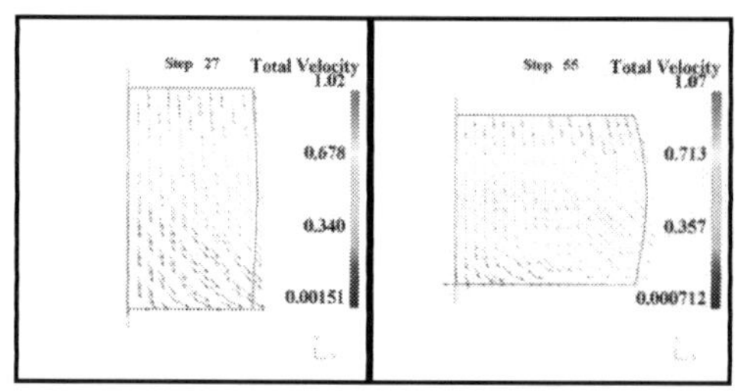

Fig. 5. The velocity field distribution under different reductions.

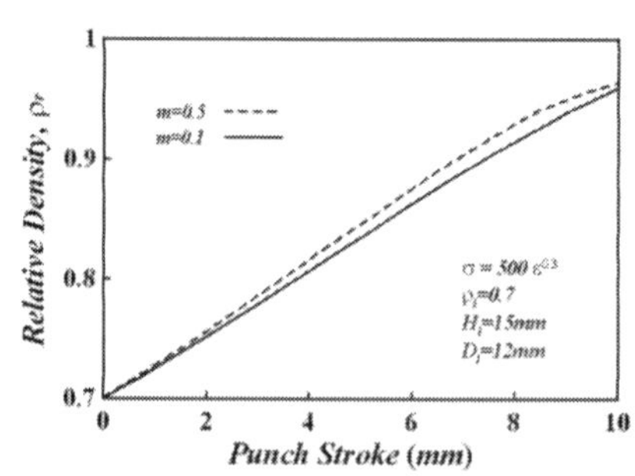

Fig. 6. Variations of the relative density with punch strokefor different frictional factors.

Figure 6 shows variations of the relative density with punch stroke for different frictional factors. Under the same punch stroke, the greater the interface frictional factor (m) between workpiece and die is, the higher the relative density (ρ_r) is; however, the difference is not evident. Figure 7 shows variations of the relative density with punch stroke for various aspect ratios. In the same punch stroke, the small aspect ratio corresponds to the large relative density. Namely, the compression force for a small aspect ratio will result in high relative density due to high compression force. As shown in Figure 7, the effect on the variation of aspect ratio to ρ_r is greater than on the variation of frictional factor to ρ_r. In Figure 6, as the frictional factor increases, ρ_r only changes slightly.

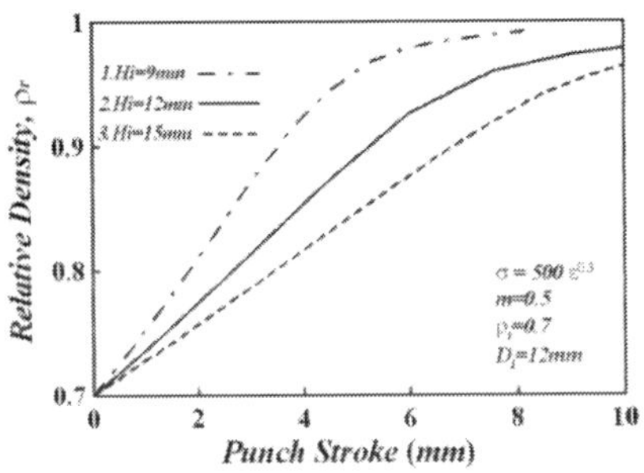

Fig. 7. Variations of the relative density with punch stroke for various aspect ratios.

From Figure 8 - 10, the distributions of vertical stress (*p*) and radial stress (*q*) under reduction ratio of 40% based on the slab method and the finite element method under various frictional factors (m=0.1, 0.3, 0.5). The figures show that the absolute value of vertical and radial stresses will increase when the frictional factor increases. In the figures, the curve interval is fluctuated larger when the frictional factor increased; the inaccuracy would be more within the results of two methods when the frictional factor is increased. The vertical stress (*p*) of the finite element method is higher than that of the slab method, and the radial stress (*q*) of the finite element method is lower than that of the slab method. As a result of increasing frictional factor, the vertical stress (*p*) near free surface end by the finite element is lower than by the slab method.

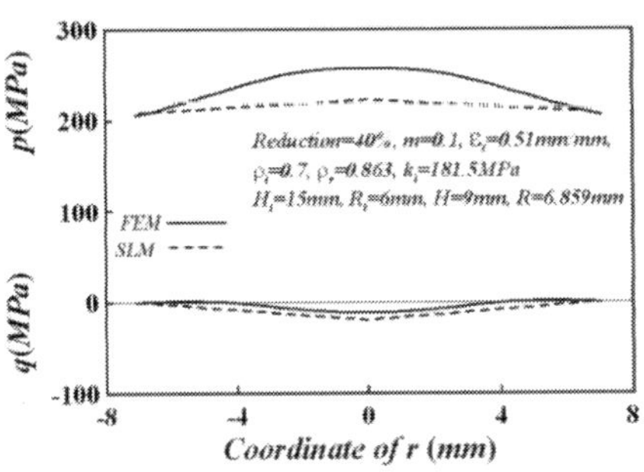

Fig. 8. Vertical stress and radial stress for m=0.1.

Fig. 11 and 12 show the average radius under various forming conditions. Figure 11 shows that the average radius decreases when the frictional factor increases because of the higher frictional factor causing much restraint on the top and bottom of cylinder (i.e. the contact area with die); the average radius in slab method is higher than in finite element method. When the frictional factor m=0.1, the curves of slab method and finite element method are close (max. error is 3.24%), and when m=0.5, the curve of FEM fluctuates downward largely (max. error is 6.26%). It can be known that the slab method is not sensitive to the change of frictional factor on predicting the average radius after cylinder forming. In Figure 11, the average radius after cylinder forming decreases when the frictional factor increases.

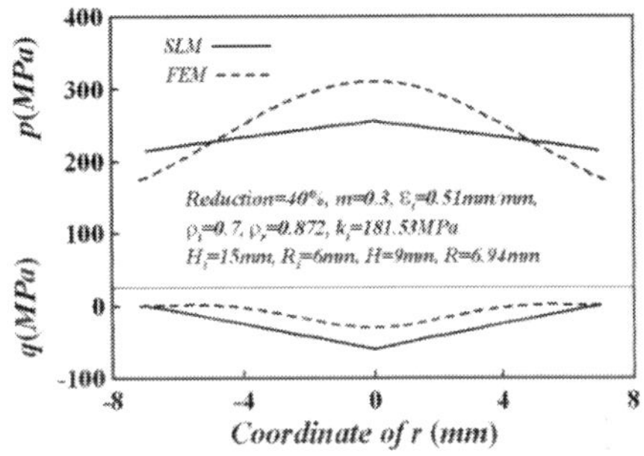

Fig. 9. Vertical stress and radial stress for m=0.3.

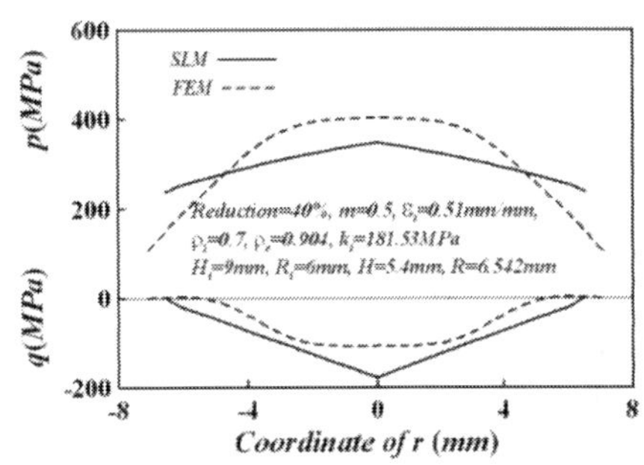

Fig. 10. Caption of the Figure 1. Below the figure.

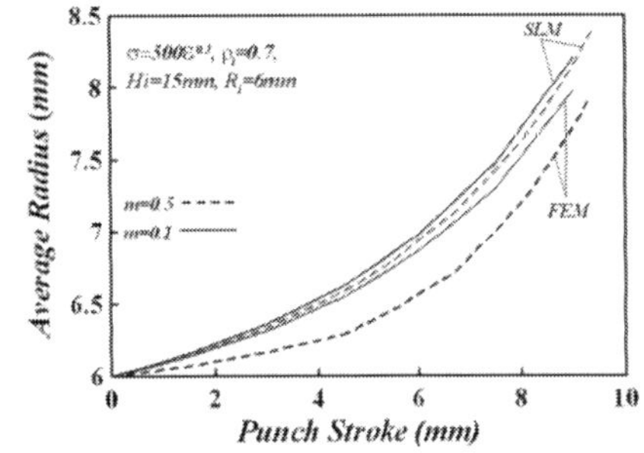

Fig. 11. Variations of average radius with punch stroke for different frictional factors.

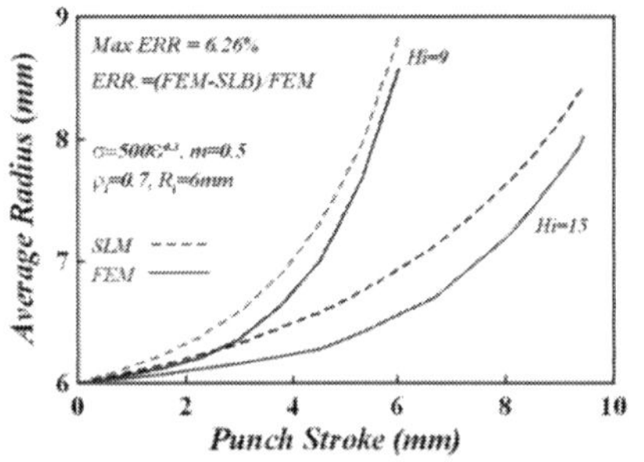

Fig. 12. Variations of average radius with punch stroke for different aspect ratios.

Because of the conservation of mass, Figure 12 shows the workpiece with lower aspect ratio in the same punch stroke has a larger average radius. It is also shown that the slab method is equally well as the finite element method on predicting the average radius after cylinder forming, and the maximum error between both methods is 6.26%only.The bulging ratio indicates the severity of cylinder bulge, and the bulging ratio is obtained by formula (13) and Fig. 13 with the radius value on upper, middle, and lower locations. The bulging ratio can be defined as below:

$$\textbf{Bulging Ratio} = [R_{max} - (R_t + R_b)/2] \,/\, R_i \quad (13)$$

In the finite element method, the bulging ratio can be obtained in various aspect ratios. Three different aspect ratios are compared together in Figure 14. It shows that a larger aspect ratio (Hi=15mm) corresponds to a higher bulging ratio; conversely, a smaller aspect ratio (Hi=9mm) conforms to a lower bulging ratio. That is to say a more slender workpiece after forming will have a more bulging ratio.

Figure 15 shows variations of bulging ratio with reduction for different frictional factors. The top and bottom radii of workpiece are constrained from increasing the frictional factor upon which the bulge will be intensified. The larger the frictional factor is, the greater the bulging ratios. Thus, the sufficient lubrication may decrease the bulge. Although the larger frictional factor causes more bulging, the increment of the bulge in the middle of workpiece is not more noticeable than the decrease of the radius on the contact area of die. As for the average radius, it can decrease as the frictional factor increases.

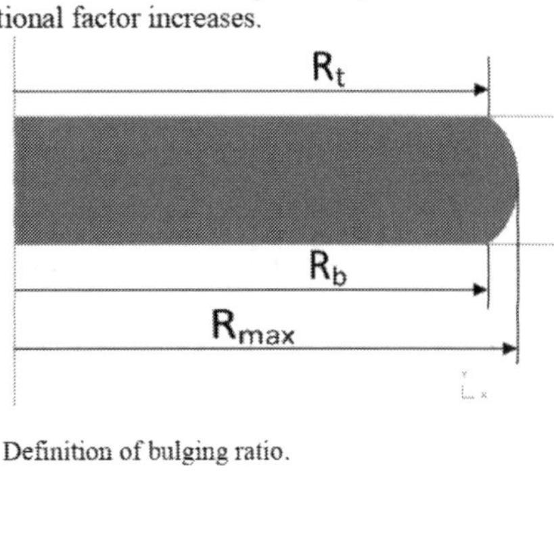

Fig. 13. Definition of bulging ratio.

Figure 16 shows variations of the compression force with punch stroke between FEM and SLM for different aspect ratios, under the same punch stroke, the maximum error between SLM and FEM is around 14.8%. It is noted that the one with larger aspect ratio (Hi=15mm, slender) requires less compression force (*P*), power, and energy.

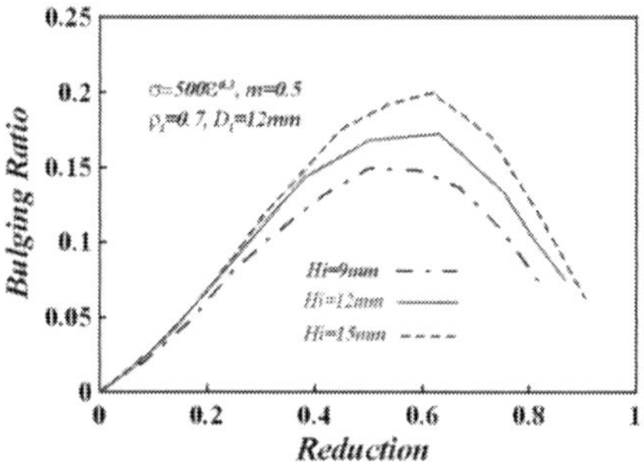

Fig. 14. Variations of bulging ratio with reduction for different aspect ratios.

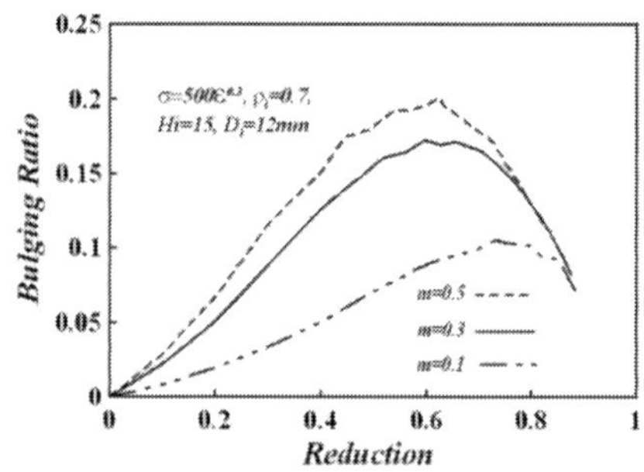

Fig. 15. Variations of bulging ratio with reduction for different frictional factors.

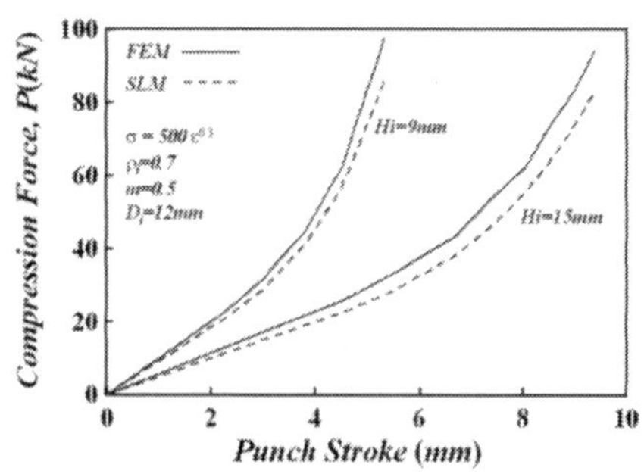

Fig. 16. Variations of the compression force with punch stroke between FEM and SLM for different aspect ratios.

Figure 17 shows variation of the compression force with punch stroke between FEM and SLM for different frictional factors. As for SLM and FEM, the larger interface friction (m) causes more compression force(P) under same punch stroke. In addition, FEM has a larger value compared to SLM on predicting the compression force ($P_{FEM} > P_{SLM}$). Conversely, FEM has a smaller value than SLM on prediction the average radius (AVG R_{FEM} < AVG R_{SLM}).

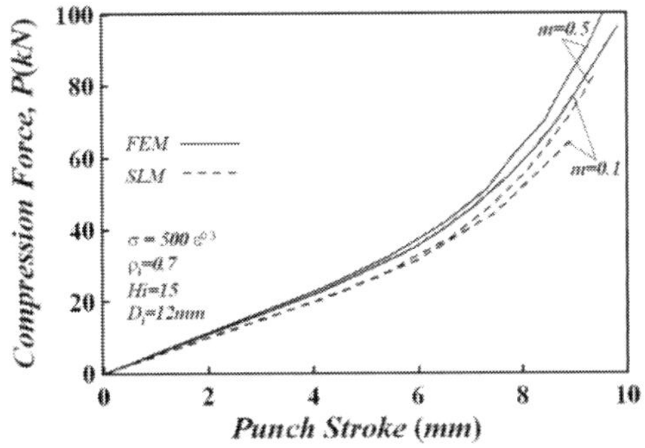

Fig. 17. Variation of the compression force with punch stroke between FEM and SLMfor different frictional factors.

4 Conclusions

Through a series of analyses based on the slab method and FEM simulation in this study, the results for both models are also compared to realize the variations, and the major conclusions are as:

(1)The maximum error for the compression force (P) under various reduction ratios is around 14.8% between the slab method and the finite element method.

(2)The variation in relative density is not uniform during the whole compression process. The peripheral on top and bottom of the cylinder and the center of the cylinder are compressed compactly first, and the value of ρ_r will reach 1 earlier. The center on top and bottom of the cylinder and the bulging area of the cylindrical periphery are compressed slowly, and the value of ρ_r is slow to reach 1.

(3)Since the increment of frictional factor (m) will increase the bulging rate, the sufficient lubrication will decrease the bulge. However, the increment of frictional factor increases the compression force slightly; the great difference only can be seen in a larger punch stroke.

(4)In geometric view, the larger aspect ratio causes the considerable bulge and the low compression force (ρ_r slowly approach to 1); the compression force will be smaller under same punch stroke.

(5)By the distributions of velocity field, vertical stress, and radial stress, the major change in the compression process is the vertical direction; therefore, the radial stress (q) is far less than the vertical stress (p), radial velocity field is also less than axial velocity field.

(6)The relative density (ρ_r) of porous material forming will increase intensely and slow down before reaching the value of 1.

(7)When the reduction rate is 0.66, the relative density of cylinder is 0.998, which is very close to the density of sound material ($\rho_r = 1$). At the larger reduction ratio or punch stroke, the process is similar to the compression forming of sound material and produces more compression force, which is more than at the initial stage of porous material forming.

(8)When the reduction ratio increases, the material deformation and the work hardening increase. Thus, the compression force and the hardness will increase while the reduction ratio increases.

Acknowledgment

The authors want to extend their thanks to the Ministry of Science and Technology. The advice and financial support of MOST are gratefully acknowledged.

References

1. T. S. Yang, Y. C. Hsu, *Journal of Materials Processing Technology*, **177**, 154 (2006)
2. D. N. Lee, H. S. Kim, Powder Metallurgy **35**, 275 (1992)
3. A. Shirizly, J. Tirosh, L. Rubinsky, *Journal of Materials Science and Engineering*, **249**, 55 (1998)
4. E. Doege, A Bagaviev, *International Journal of Mechanical Sciences*, **39** (10), 1151 (1997)
5. C. C. Huang, J. H. Cheng, *Journal of Materials Processing Technology*, **44**, 489 (2002)
6. R. J. Green, *International Journal of Mechanical Sciences*, **14**, 215 (1972)
7. S. Shima, M. Oyane, *International Journal of Mechanical Sciences*, **18**, 285 (1976)
8. S. M .Doraivelu, H. L. Gegel, J. S. Gunasekera, J. C. Malas, J. T. Morgan, *International Journal of Mechanical Sciences*, **26**, 527 (1984)
9. A. D. M. Dias, L. Schaeer, *International Journal of Machining Tools and Manufacture*, **33**, 577 (1993)
10. A. K. Jha, S. Kumar, *International Journal of Machine Tool Design & Research*, **26**, 369 (1986)
11. D.K.Nguyen, T. T. Tran,and S. G. Buntara,and V. T. Bui, *International Journal of Engineering and Technology Innovation*, **6** (3), 173 (2016)

Measurement of wisdom forging force using piezoelectric materials

Shao-Yi Hsia[1,*], *Yu-Tuan* Chou[2], *Chen-Chung* Chang[1], and *Ling-Yu* Hsu[1]

[1]Department of Mechanical & Automation Engineering, Kao-Yuan University, Kaohsiung City, 821, Taiwan
[2]Department of Applied Geoinformatics, Chia Nan University of Pharmacy & Science, Tainan City, 717, Taiwan

Abstract. Aiming at forging machines for bolts and nuts with up to 61% market share, adding piezoelectric sensing units to the mold forging is discussed in this study. In the research process, it attempts to develop a sensing element with low cost, practicable limited space, acceptable performance stability, and real-time measurement of forging load by a five-stage forming system for special nuts matched with the theory of plastic forming, the CAD/CAE design and numerical analysis, and the installation of a piezoelectric unit. It aims to assist fastener businesses in the intelligentization, networking, and systemization of machines and further integrate into the Internet service manufacturing system to enhance the overall competitiveness of fastener industry.

1 Introduction

Intelligent manufacturing is the inevitable product of mechanized, automated, and informatized application to the mature stage. The competitive elements of manufacturers therefore change from previous materials and energy to the resources of materials, energy, and information as the core competitiveness. Intelligent manufacturing refers to the establishment of digitized factories and intelligentized factories/manufacturers, through various intelligent manufacturing systems (or equipment) and automated or intelligentized production lines, to present multiple sensing devices for avoiding possible risks and reaching zero worry for enterprises [1,2]. Apparently, "intelligent manufacturing" or "Industry 4.0" stresses more on the application of automation intelligentized equipment, Internet of machines, Cyber-Physical System (CPS), cloud, and big data analysis. Nevertheless, fastener industry in Taiwan still exists in following technology gaps [1], when developing intelligent equipment.

(1) Sensing technology for molds, principal production equipment (e.g. forming machines, rolling machines), and sieve sorters has not been developed to real-time monitor the quality of fasteners in the process.
(2) M2M connection technology and communication protocol as well as the standards for equipment like forming machines have not been established.
(3) Technology for real-time mass data processing, analyzing, and predicting has not been established.
(4) There has not been a solution for production schedule CPS integration systems that the virtual and the reality could not be automatically integrated.
(5) There has not been a large manufacturer preceding the integration to provide a complete multidisciplinary service environment.

Finite Element Method could simultaneously analyze mechanical problems derived from all interface of blanks and molds under plastic deformation. Theoretically, there is almost no restriction to geometric complexity. The boom of technology in past years further accelerates the calculation with computers; besides, the development of simulation software has computer-assisted simulation technology become mature and stable. Finite Element Method therefore presents more superior use efficiency and accuracy. The application of Finite Element Method and the relevant technology to forging forming is reviewed as below.

O'Connell *et al.* [3] developed Slab method and Finite Element Method DEFORM-2D in 1996 to estimate the forming load of hexagon bolts trimless closed-die forging for selecting suitable forming machines, avoiding energy waste, and reducing the production speed. Hyunkee *et al.* [4], in 1996, designed the forming process sequence with a component by using FORMEX and Finite Element software DEFORM, to establish the program database for the process and effectively reduce the time for forming process sequence and mold design in the product production. MacCormack *et al.* [5] used Finite Element Analysis software DFFORM for the multiple-stage cold forging forming analysis in 2002 and observed the strain distribution, with three-stage damage and flow, after the product was formed in order to understand the failure value and the effective strain value. In the same year, Behrens *et al.* [6] utilized Finite Element Analysis software MSC.SuperForm 2002 for the failure analysis of various production processes and discussed the material breaks in cold forging and warm forging. The results revealed that the surface cracks could be observed under two situations and the material damage could be reduced by slightly modifying the geometric shape of the key workpiece.

* Corresponding author: syhsia@cc.kyu.edu.tw

Su [7] indicated in 2014 that the idea of sensors proposed by Rosenbla in 1957 became a primary part in the research on neural network. Moreover, Professor John Holland in University of Michigan, in order to develop artificial intelligence systems, developed Genetic Algorithm in 1960 by referring to the solution for the mathematical optimization.

Hsia *et al.* [8], in 2016, attempted to experiment multiple-stage fastener cold forging forming machines with accelerometers and vibration testing software by capturing the background vibration at the feeding system and starting the feeding system but not feeding materials as well as the vibration signals of time for the feeding system starting to feed materials. MTALAB software was further used for the FFT analysis to view the distribution of frequency spectra.

Although there have been several researchers proposing various studies on fastener forging forming to solve the practical problems on sites, the introduction of sensing technology to the promotion of intelligent manufacturing, attempting to proceed real-time monitoring and diagnosis technology on fastener molds and principal production equipment (e.g. forming machines, rolling machines) matched with sensing units, is still at the beginning stage. In this study, the theory of plastic forming is matched with the CAD/CAE design and numerical analysis to design and develop a piezoelectric measurement unit, expecting to real-time measure the forging load to assist fastener businesses in promoting intelligentized manufacturing and to further enhance the overall competitiveness of the industry.

2 Basic principle

2.1 Cold forging forming

Metal forming is an extremely complicated deformation process, on which material characteristics, temperature conditions, friction conditions, lubrication, blank size, and mold shape present certain degree of influence. During theoretical analyses, people could merely change difficult mechanical problems in mathematics into engineering problems through some hypotheses and experimental or empirical data because of material nonlinearity (nonlinear relationship between stress and strain) and geometric nonlinearity (nonlinear relationship between strain and displacement) as well as the complexity of boundary conditions and the difficulty in mathematics. Various analyses or mathematical models with different approximation and application are therefore generated, including principal stress method, slip line method, upper bound method, lower bound method, and Finite Element Method. With Finite Element Method, materials are divided into several elements (triangular, quadrilateral) to calculate the strain and stress of elements with multiple simultaneous equations so as to be analyzed with computers and applied to differnt field in past years. Finite Element forming software DEFORM 3D is utilized in this study for the simulation and analysis.

2.2 Piezomaterial unit

In addition to the dielectric properties and elastic behavior of general materials, piezoelectric ceramics also present piezoelectric properties. Numerous performance parameters are the primary basis of piezoelectric ceramics being broadly applied.

(1) Piezoelectric constant

Piezomaterial presents both mechanical and electrical characteristics that the constitutive law is denoted as below.

$$\sigma_{ij}= C^{E}_{ijkl}\varepsilon_{kl}-e_{mij}E_{m} \quad (1)$$

$$D_{n,n}=e_{mkl}\varepsilon_{kl}+\mu^{\sigma}_{mn}E_{m} \quad (2)$$

where σ_{ij}: stress tensor

ε_{kl}: strain tensor

E_m: electric field vector

C^{E}_{ijkl}: stiffness tensor of constant electric field

e_{mkl}: piezoelectric tensor

μ^{σ}_{mn}: permittance tensor of constant elastic stress

The mechanical characteristics should conform to the equation of motion of an elastomer.

$$\sigma_{ij,j}+fi=\rho\ddot{u}_i \quad (3)$$

Meanwhile, when the linear elastic behavior of small deformation is taken into account, the relationship between displacement and strain is denoted as below.

$$\varepsilon_{ij}=\frac{1}{2}(u_{i,j}+u_{j,i}) \quad (4)$$

Regarding the electrical characteristics, it is assumed that the magnetic effect caused by electric fields is extremely small that it could be ignored. Meanwhile, the electric charge in piezomaterial is little that it could be ignored. The Maxwell electromagnetic equation therefore could be simplified as

$$D_{n,n}=0 \quad (5)$$

In real situations, the measured and applied physical quantity is voltage that the electric field is transformed into voltage.

$$Ei=-\phi_{,i} \quad (6)$$

ϕ: electric potential

(2) Electromechanical coupling coefficient

The electromechanical coupling coefficient k is a parameter comprehensively reflecting piezomaterial properties, presenting the coupling effect between the mechanical energy and the electrical energy of piezomaterial. The electromechanical coupling coefficient could be defined as

$$k^2=\frac{\text{Electrical energy converted into mechanical energy}}{\text{Input electrical energy}} \quad (7)$$

or

$$k^2=\frac{\text{Mechanical energy converted into electrical energy}}{\text{Input mechanical energy}} \quad (8)$$

3 Research method and architecture

3.1 Research method

The forging force required for fasteners generally relies on the experience of manufacturers. For instance, it requires about five stages to gradually complete a special nut diagonal of about 11.33mm. A five-mold forging machine, LS-C-NF-14B-5F, of a company could forge the nut diagonal below 14mm, and the machine could bear the load about 850kN. However, no-one knows whether the amount is just enough or would have allowance. Matching the simulation of forming software could acquire the forging force at each stage; nonetheless, measuring with a commercial load cell would be restricted to the exceeding forging force at a single stage. In this case, a larger size load cell is necessary. Nevertheless, merely a certain size load cell could be placed between molds at each stage because of the limited space in the forging machine.

Accordingly, the plastic forming analysis is combined with a piezoelectric sensing unit in this study to develop a sensor which is suitable for the fastener forming site. The developed sensing unit presents the following advantages.

(1) The expense for the developed piezoelectric sensor is far lower than imported or domestically developed load cells or signal sensors.
(2) The modified total forging force analysis model could assist in the selection of forging machine specifications for innovative fasteners and avoid over investment or overload operation. For an equipment manufacturer, it could accurately quote the price and enhance the product price-performance ratio. For a fastener manufacturer, on the other hand, it could correctly acquire the forging force to accurately control the production process and make failure diagnoses and exclusion in advance.
(3) The forging force waveform could be correctly monitored to judge the degradation of molds. When ISO2372 standards are matched, a mold could be replaced at proper time.

AISI-1010 materials are selected for the experiment in this study. Five-stage special nuts, Figure 1, which requires five stages for the completion step by step, are the research target. SolidWorks is matched to construct the geometric design, and Finite Element forming software is introduced for the simulation to acquire the forging force at each stage. The research process is shown in Figure 2. SolidWorks is matched in the beginning to establish the geometric figure, and then the forming software is used for the simulation and analysis, which is further verified with the measuring results.

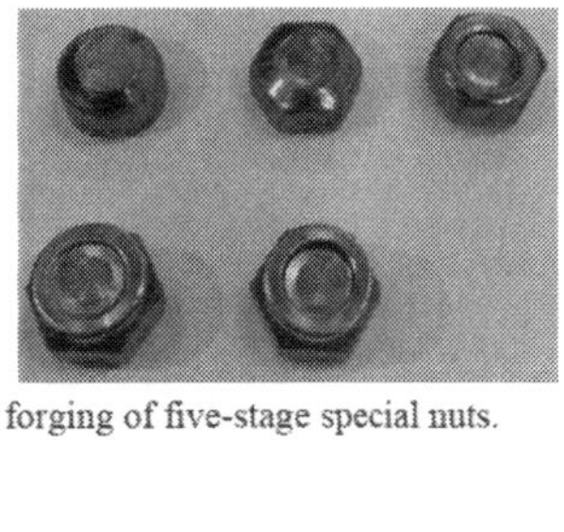

Fig. 1. Cold forging of five-stage special nuts.

3.2 Steps

(1) Introduction of machine and fastener

Special nuts with simpler geometric shape (Figure 1) are analyzed in this study. The product is preceded from wire cutting, material arranging, and preforming to the final finished product through five-mold cold forging (Figure 3, five-stage nut forging machine), and the forging force at each stage is acquired with simulation.

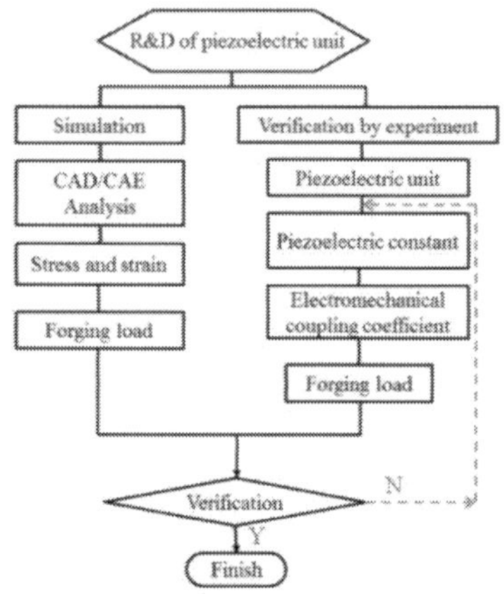

Fig. 2. Flow chart of this study.

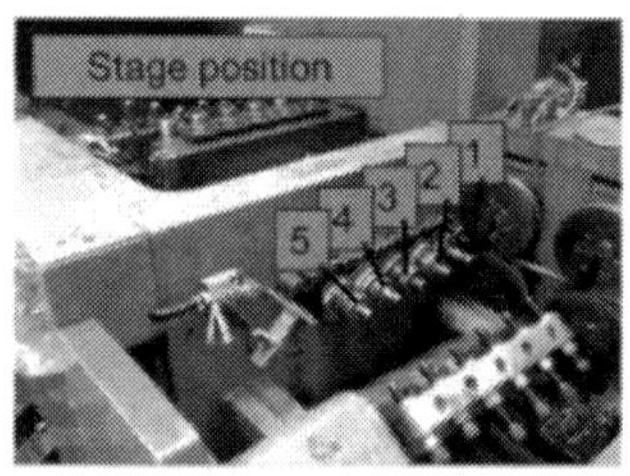

Fig. 3. Five-stage nut forging machine.

(2) Manufacturing and use of piezoelectric unit

Current piezoelectric sensors embedded in most forging machines could merely be used for the comparison with continuous signals, but do not show absolute relationship with forging force. General piezoelectric units, because of low prices, could not provide the sensitivity (mv/g) of common acceleration probes or the conversion coefficient (mv/N) of load cells, and the piezoelectric unit properties can hardly maintain the approximate stability. For this reason, it is important to acquire conversion coefficients and maintain certain performance curve. The piezoelectric unit utilized in this study, Figure 4, is embedded at the second to the fifth stages, Figure 5.

(3) Application of CAD/CAE forming design and analysis software

Before the cold forging forming simulation, the mechanical properties of materials could be analyzed to conform to the actual production conditions, and then the cylindrical test is preceded and analyzed the results, e.g. flow stress. SolidWorks, the computer-assisted design software, is further utilized for drawing the production molds on site, including punch pin, workpiece, die, and punch, and transforming into STL files for the simulation with Finite Element forming analysis software DEFORM. Convergence Analysis is used in the research process for acquiring proper number of grids to remain the forming stability as well as shorten the time for computer simulation and calculation.

DOI:10.1051/matecconf/201712300003

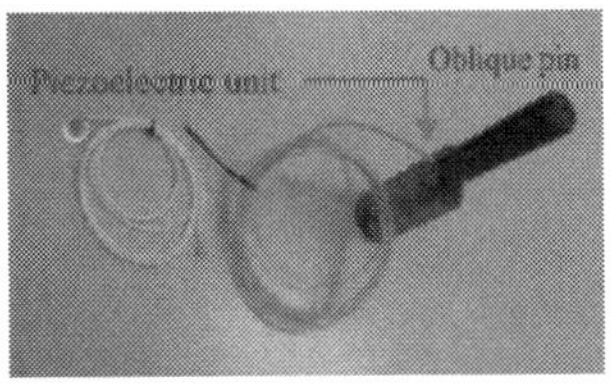

Fig. 4. Piezoelectric unit embedded in the oblique pin.

Fig. 5. Four sets piezoelectric unit embedded at the second to the fifth stages.

4 Results and discussion

Figure 6 shows the effective stress distribution from the first to the fifth stages, where the first and the second stages are the preforming that the born stress is smaller; the fourth and the fifth stages are the principal forming stage. Figure 7 displays the effective strain distribution from the first to the fifth stages, where the strain mostly appears on the fourth and the fifth stages and concentrates on the central deformation area.

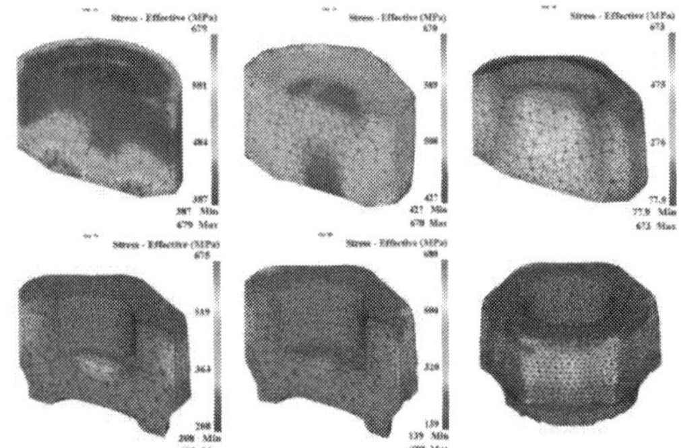

Fig. 6. Effective stress distribution from the first to the fifth stages and full scale.

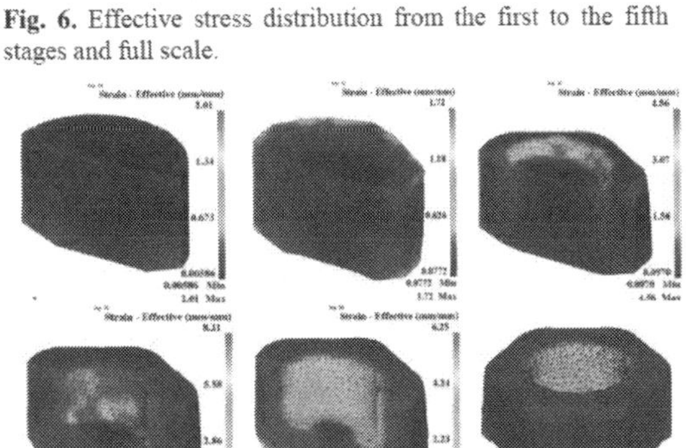

Fig. 7. Effective strain distribution from the first to the fifth stages and full scale.

This section analyzes the forging load from the first stage to the fifth stage, as displayed in Fig. 8. In the preforming of the workpiece in the first stage, the forging load is 121kN around 14.37% of the total punch loads in the five stages. In forming in the second stage, the forging load is 185kN approximately 21.97%. In the third stage, the forging load, 188kN approximately 22.32%, is larger than in any other stage. In the fourth stage, the forging load, 185kN around 21.97%, has the same value as the second stage in this study. Overall, the total forging load for forming special nut from the first stage to the fourth stage was 842kN.

Figure 9 and Figure 10 present the forging load at the third and fifth stages measured by the embedded piezoelectric unit, where (a) is the simulation results and (b) is the magnification of a single signal acquired from the measurement. By comparing (a) and (b), the simulation results and the measured signals exist in certain relationship. When the piezoelectric constant of the piezoelectric unit and the electromechanical conversion coefficient could be studied and discussed in the future, the relationship between electrical signals and forging force would be discovered to assist fastener businesses in the machine intelligentization, networking, and systemization as well as to enhance the overall competitiveness of the fastener industry.

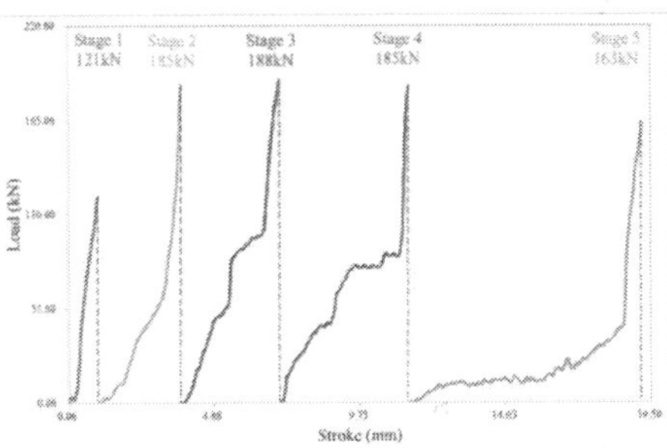

Fig. 8. Forging load from the first to the fifth stages with simulation method.

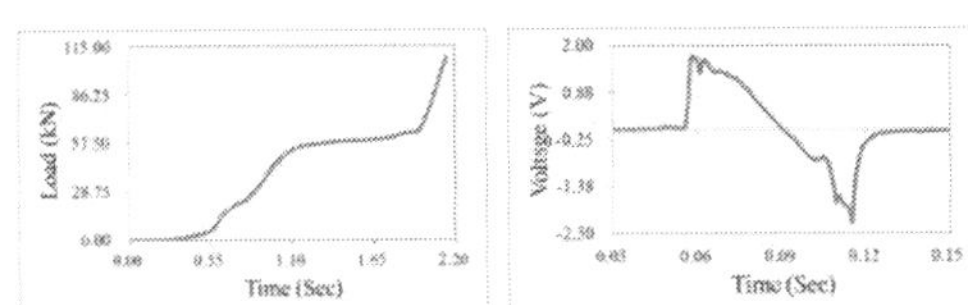

(a) (b)

Fig. 9. Forging load at the third stage measured by the embedded piezoelectric unit: (a) simulation results, (b) magnification of a single signal.

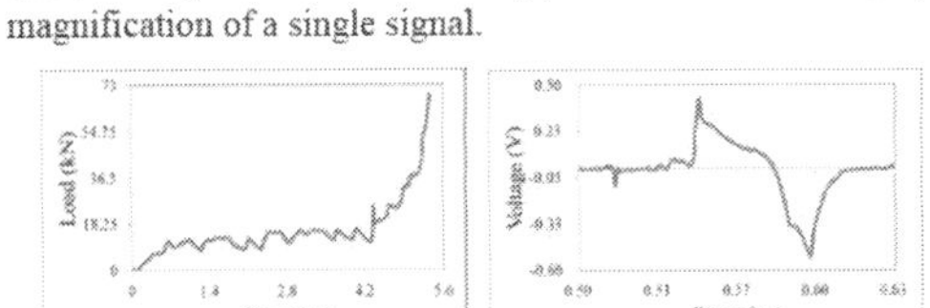

(a) (b)

Fig. 10. Forging load at the fifth stage measured by the embedded piezoelectric unit: (a) simulation results, (b) magnification of a single signal.

5 Conclusions

This study could assist businesses in verifying the accuracy of the software analysis by the acquisition of forging force. In the future, the development of any new-style molds could be analyzed and modified with the forming software before contracting for manufacturing so as to achieve the best economic benefits. The relevant conclusions are summarized as following.

(1) Finite Element Analysis software DEFORM-3D matched with experimental data for the simulation and analysis could shorten the time for mold testing to reduce human expenses and production costs as well as save materials and reduce equipment loss.

(2) Using the piezoelectric sensing unit for measuring forging force could assist fastener businesses in the machine intelligentization, when acquiring the research data like piezoelectric conversion coefficients, so that "the intelligent manufacturing system with introspective prediction could avoid risks in advance and present zero worry for enterprises."

References

1. C.Y. Chen, C.C. Yeh, Application of intelligent manufacturing to fastener industry, ITIS research topic, Metal Industries Research & Development Centre, (2016)
2. H.-Y. Chi, Outlook and review of 2015 fastener industry in Taiwan. Industry Report, 82-84 (2016)
3. M. O'Connell, B. Painter, G. Maul, T. Altan, Flashless closed-die upset forging-load estimation for optimal cold header selection, *Journal of Materials Processing Technology*, **59**(1-2), 81-94 (1996)
4. H. Kim, T. Altan, Cold forging of steel-practical examples of computerized part and process design, *Journal of Materials Processing Technology*, **59**, 122-131 (1996)
5. C. Maccormack and J. Monaghan, 2D and 3D finite element analysis of a three stage forging sequence, *Journal of Materials Processing Technology*, **127**, 48-56 (2002)
6. A. Behrens and H. Just, Extension of the forming limits in cold and warm forging by the FE based fracture analysis with the integrated damage model of effective stresses, *Journal of Materials Processing Technology*, **125-126**, 235-241 (2002).
7. S.Su, Shape Optimization design of cold forging punch, (2014)
8. S.Y. Hsia, Y.C. Li, H.-H. Hsu, Y.T. Chou, and C.C. Yang, Vibration analysis of multiple-stage fastener cold forging forming machine feeding system, 2016 TSTP Taiwan Society for Technology of Plasticity, No. B2-6(29) (2016)
9. Ultrasonic flaw detection editors, Ultrasonic flaw detection, Industry Report, Hydroelectric power press, 75-90 (1980)
10. H.S. Huang, S.Y. Hsia, and S.K. Yang, 2016, Optimal design of ultrasonic transducer system applied on biolipid extraction from Chlorella, *Sensors and Materials*, **28**(5), 511–522 (2016)
11. K. Nakamura, Ultrasonic transducers-Materials and design for sensors, actuators and medical applications, Woodlead publishing (2012)

Case study of virtual reality in CNC machine tool exhibition

Yung-Chou Kao[1,*], *Chung-Shuo* Lee[2], *Zhi-Ren* Liu[3], and *Yu-Fu* Lin[4]

[1]Advanced Institute of Manufacturing with High-tech Innovations, National Chung Cheng University, No.168, Sec. 1, University Rd., Min-Hsiung Township, Chia-Yi County 621, Taiwan (R.O.C.)
[2] Advanced Institute of Manufacturing with High-tech Innovations, National Chung Cheng University, No.168, Sec. 1, University Rd., Min-Hsiung Township, Chia-Yi County 621, Taiwan (R.O.C.)
[3]Department of Mechanical Engineering, National Chung Cheng University, No.168, Sec. 1, University Rd., Min-Hsiung Township, Chia-Yi County 621, Taiwan (R.O.C.)
[4]Micro/Meso Mechanical Manufacturing R&D Department, Metal Industry Research and Development Center, 1001 Kaonan Highway, Kaohsiung, Taiwan 811 (R.O.C.)

Abstract. Exhibition and demonstration are generally used in the promotion and sale-assistance of manufactured products. However, the transportation cost of the real goods from the vender factory to the exposition venue is generally expensive for huge and heavy commodity. With the advancement of computing, graphics, mobile apps, and mobile hardware the 3D visibility technology is getting more and more popular to be adopted in visual-assisted communication such as amusement games. Virtual reality (VR) technology has therefore being paid great attention in emulating expensive small and/or huge and heavy equipment. Virtual reality can be characterized as 3D extension with Immersion, Interaction and Imagination. This paper was then be focused on the study of virtual reality in the assistance of CNC machine tool demonstration and exhibition. A commercial CNC machine tool was used in this study to illustrate the effectiveness and usability of using virtual reality for an exhibition. The adopted CNC machine tool is a large and heavy mill-turn machine with the width up to eleven meters and weighted about 35 tons. A head-mounted display (HMD) was attached to the developed VR CNC machine tool for the immersion viewing. A user can see around the 3D scene of the large mill-turn machine and the operation of the virtual CNC machine can be actuated by bare hand. Coolant was added to demonstrate more realistic operation while collision detection function was also added to remind the operator. The developed VR demonstration system has been presented in the 2017 Taipei International Machine Tool Show (TIMTOS 2017). This case study has shown that young engineers and/or students are very impressed by the VR-based demonstration while elder persons could not adapt themselves easily to the VR-based scene because of eyesight issues. However, virtual reality has successfully being adopted and integrated with the CNC machine tool in an international show. Another machine tool on laser-assisted milling machine motion simulation has also been successfully conducted to show the expandability of the VR based technology. One can conclude that VR will be adopted and paid more and more attention in the future in helping CNC machine tool promotion. Further study could be extended to education and training system, and also for the maintenance system, too.

1 Introduction

In response to the needs of small-volume/large-variation of products nowadays, the processing applications have a lot of customization requirements. The mechanical equipment manufacturers and the processing application engineers usually need to spend a lot of times in the communications and repeated changes to determine which equipment is the most suitable for customers. There are many factors need to be considered such as processing methods, product size, and other special needs, to name only a few. For the equipment manufacturers, how to demonstrate the characteristics of the device and to satisfy the needs of the customers is an important factor in determining the customer's motivation to buy.

So far, a variety of international machine tool exhibition, like EMO in Hannover, Germany, JIMTOF in Tokyo, Japan, IMTS in Chicago, United States of America, CIMT in Beijing, China, TIMTOS in Taipei, Taiwan, etc. have been held regularly. When a region held an international machine tool exhibition, the nearby machinery and equipment manufacturers will generally gather together and show in the exhibition their equipment characteristics and advantages, but are limited by the space of the site and the booth. Therefore, the machine tools that can be moved to the exhibition hall are limited. So the equipment manufacturers will be subjected to this limitation when negotiating with their potential customers. They can only provide with the specifications of a series of products or a similar

* Corresponding author: imeyckao@ccu.edu.tw

machine tool in the exhibition hall. Another limitation is the cost of the transportation, the equipment manufacturers of huge machinery such as the heavy lathe will spend more cost on transportation by freighter, also on assembling, disassembling and labour cost. It will cost more than other lighter machine tools.

In order to attract more customers, the manufacturers will exhibit the best product and technology of the year. There are often three ways of demonstrations .The first is showing the technology by operating the machine tool in the exhibition, but it is expensive. The second is using videos or other digital technology to present the new functions of the machine, but it has the disadvantage that the customers cannot experience the product clearly. The third is the most conventional way to present the machined product by the machine tool, the customers can only understand the machine tool by catalogue, and the customers cannot view the real motion of the machine tool.

With the advancement of computer graphics, 3D visualization technology is getting more and more inexpensive and popular [1,2]. Amer and Peralez used low cost material and smart phone to make the VR Head-mounted Display (HMD) [1]. Zielinski et al. [2] developed the 3D visualization by using the Oculus Rift SDK and using MATLAB to transform 2D desktop display to 3D display in VR robotics collaboration. Besides the application of the virtual reality in games, applications have also been gradually introduced to research and development of other industries [3], e.g., the equipment manufacturing exhibition of 3D interactive media [4], motion of machine tool [5], machine tool-aided design [6], home design [7], smart manufacturing [8], robotic arm dynamic and simulation [9], creating new museum experiences [10], reducing work accidents [11] and education [12,13,14,15]. In the application to education, the user experience has been studied [12]. With the development of VR technology, it has been used for industry skills training[13], and for virtual vehicle and virtual production in automotive engineering education [14]. The 3D virtual reality experiments have been developed for circuit experiments in the electrical engineering education [15].

Therefore, this paper aims for enhancing the communication between customers and machine tool makers through the development of a demonstration-assisted virtual reality software with case study to ease realization. Exhibitors and Salesman can make a strong first impression with customers, and convert the 2D catalogue to interactive 3D experience. In order to make the virtual environment to real world, the customers can touch virtual panel to control the virtual machine tool by sensor of hand motions. Besides the motion simulation, the function of input NC programs and detection of collision are also developed. Using virtual reality technology on the demonstration could avoid the transportation of heavy machine tool to the exhibition for reducing the cost, it also could let customers view the machine tool from different angles.

2 System structure

The 3D model of a huge mill-turn VR machine tool is shown in Figure 1 with its width 11 meters. Figure 2 is the kinematic chain of the machine tool. In this kinematic chain, O is the starting origin, O* indicates that there is no relative motion with O. C represents the rotation axis in C-axis, W means workpiece, Z represents the linear axis in the Z-axis direction, X represents the linear axis in the X-axis direction, B represents the rotation axis in B-axis, Y represents the linear axis in the Y-axis direction, S represents the rotation axis in sub-spindle, T represents the rotation axis in tool, T* represents the tip of tool and there is no relative motion relationship with T.

Fig 1. 3D model of the VR mill-turn machine tool.

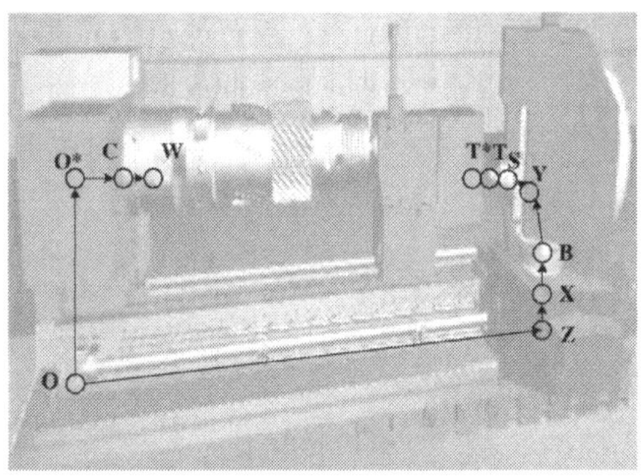

Fig 2. Kinematic chain of the developed VR mill-turn machine.

In VR machine tool, in order to simulate the machine movement during the operation, both the Virtual and/or Real CNC Controllers can be employed to read the NC code. The controller can interpret the NC code into motion information of each axis and then sends the information to the VR Client program. The VR Client is connected to the VR Server of the VR Machine Tool by sending the information to the Virtual Controller. After that, the Virtual Controller issues commands to move the Machine Tool Model. The system structure of the proposed VR Machine Tool is shown in Figure 3.

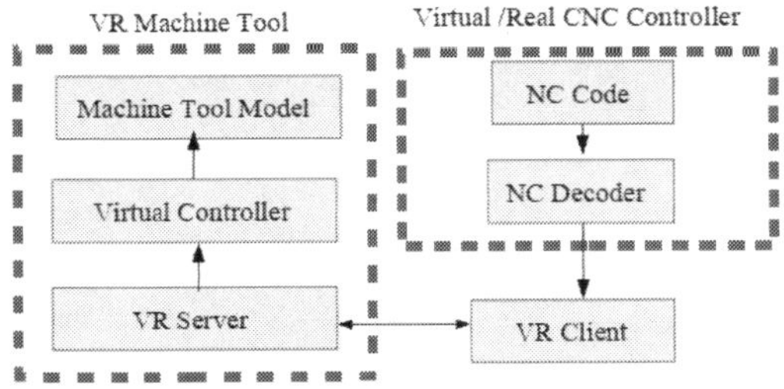

Fig 3. System Structure of the proposed VR machine tool.

3 System implementation

To build a VR machine tool, the 3D model of the machine tool was first constructed with a CAD system, according to the kinematic model of the machine tool. The machine components were split into two groups: movable compoments and fixed compoments. Then, the compoments were classified by functions, for the mill-turn machine tool. There are the sheet metal parts, the base, the various moving linear axes, the rotation axes, the tailstock, the turret, and so on. The 3D model files of each component were exported and converted into the formats that were supported by the Unity 3D™ VR development environment.

After imported the 3D model file to VR development environment, according to the heirachical structure of the VR machine tool motion (as shown in Figure 4.) to establish the linkage of the various components for the machine motion, the motion functions were programmed in C# scripts. Then, a Virtual Controller was developed with a program allowing for the user to operate the VR Machine Tool.

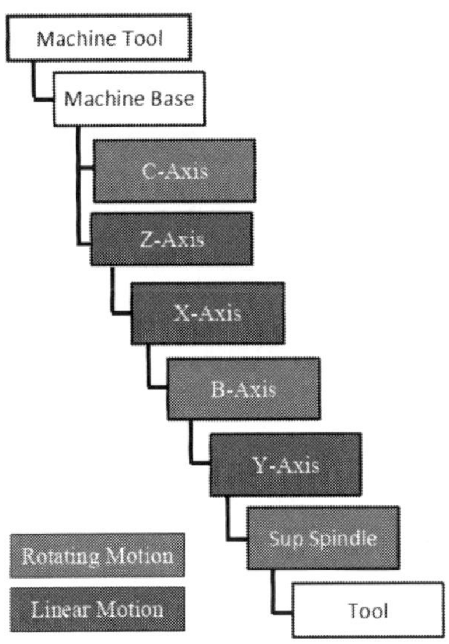

Fig 4. Heirachical structure of the VR machine tool motion.

To make the VR machine tool motions more authentic, a collision detection fuction was added. This is accomplished by adding a collision-check object outside the 3D model, the object can be represented by rectangle, sphere, cylindr or in accordance with the mesh of 3D model. It would be most accurate by using the mesh geometry of 3D model to determine the collision, but the computational loading is also much heavier and costly. This loading could be lessened by using simple parts as bounding boxes, i.e. rectangle, sphere or cylinder, can greatly improve the efficiency and improve performance in collision check. In the VR machine tool, the color of the collided compoment will be changed to red and a warning audio will be played to remind the user when the collision is detected.The collision detection examples are shown in Figure 5.

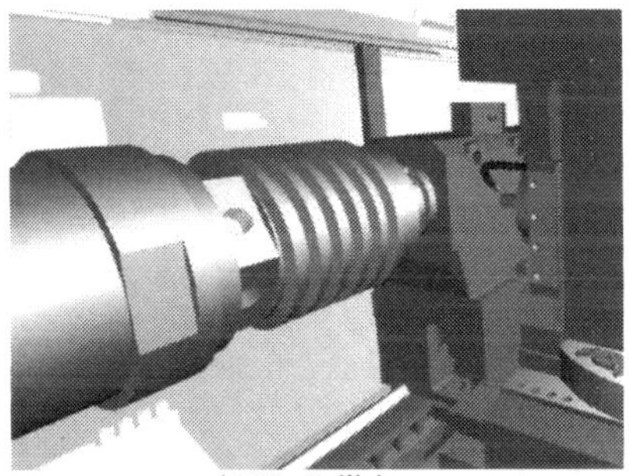

(a) No Collision.

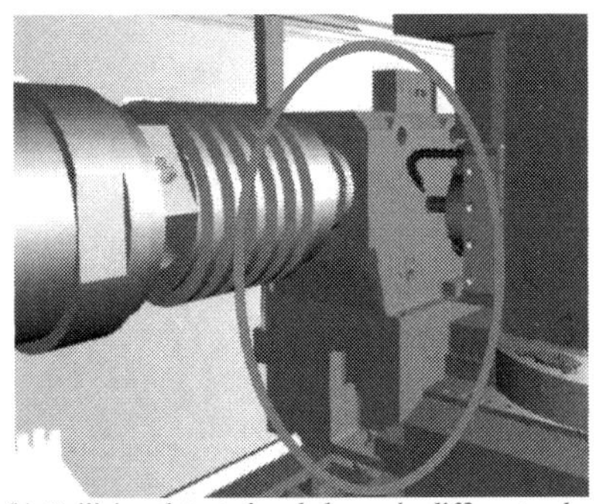

(b) Collision detected and shown in different color.

Fig 5. Examples for collision detection..

In order to allow for the users to have more realistic experences when operating the machine tool, a gesture sensor was applied outside the VR HMD, thus the user's gestures were grabbed for controlling the control panel, as shown in Figure 6.

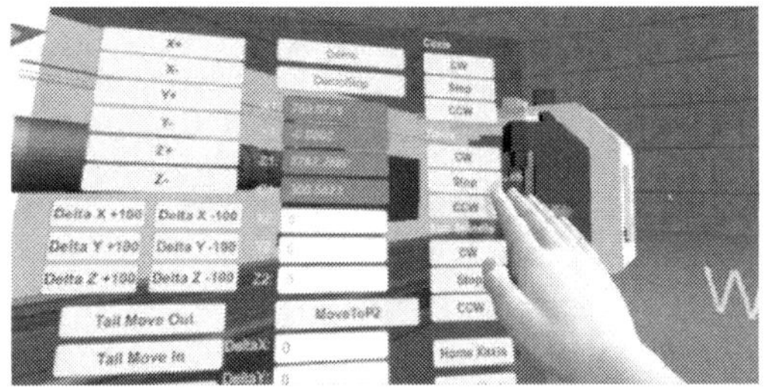

Fig 6. Using hand to touch the control panel.

DOI:10.1051/matecconf/201712300004

In the control panel, it can be divided into 3 parts accroding to NC code functions:

(1) Related to G code: Mechine coordinates display, Movement by incremental coordinates, Movement by absolute coordinates, Origin return of axes, Rotation angle of rotary axes.
(2) Related to M code: Clockwise and couter clockwise of spindles, Automatic tool change, Coolant spay.
(3) Others: Tailstock movement, Network connection for real and virtual controllers.

The user's viewing angle can be changed by using the develoepd VR controller, as shown in Figure 7.

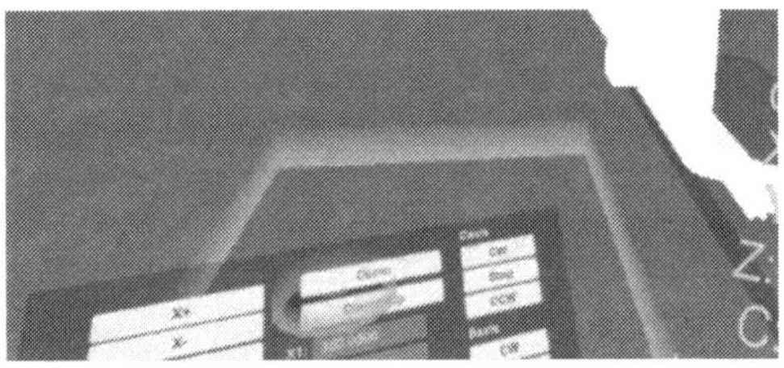

Fig 7. Move in scene to change the viewing angle.

Furthermore, in order to be more realistic, the coolant function was added to the simulation. Figure 8 shows the scene with spraying coolant.

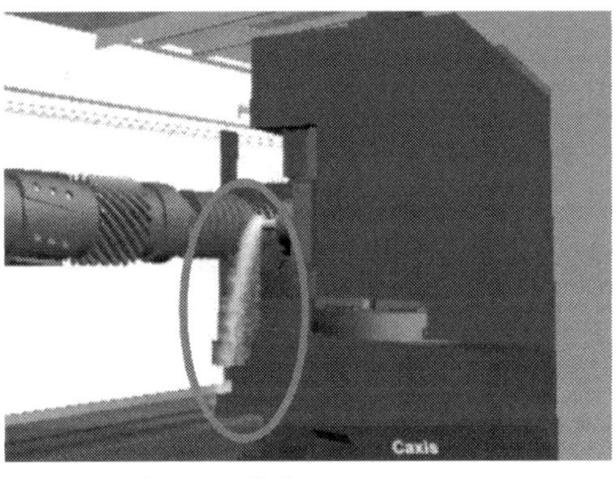

Fig 8. The coolant spray display.

4 Results and discussions

The developed VR demonstration system has been presented in the 2017 Taipei International Machine Tool Show (TIMTOS 2017), as shown in Figure 9. Figure 10 shows the screen seen by the user when the VR machine tool simulates the machine tool.

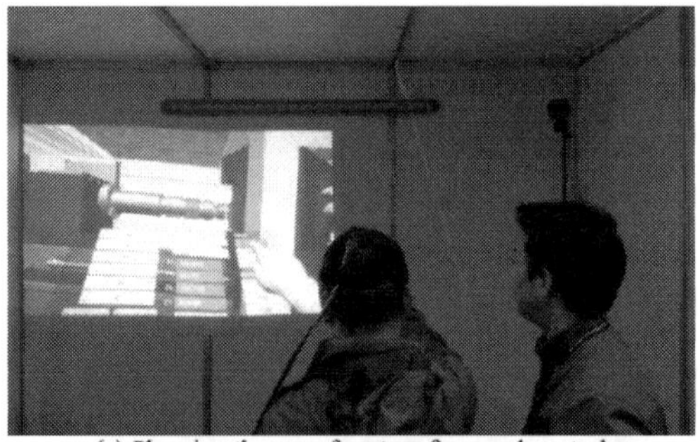

(a) Showing the use of gesture for panel control

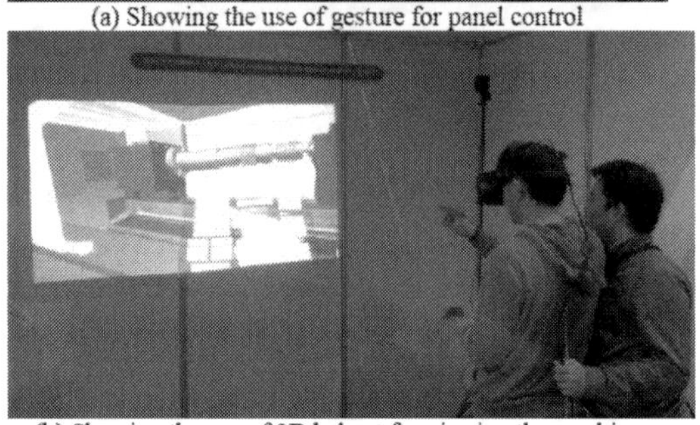

(b) Showing the use of 3D helmet for viewing the machine

Fig 9. System demonstration at TIMTOS 2017.

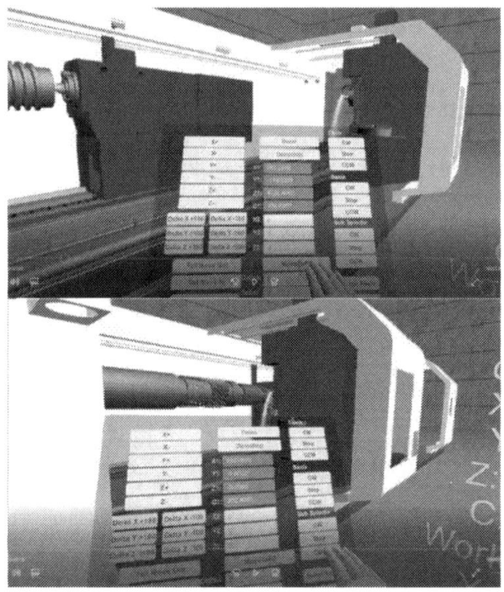

Fig 10. The screen seen by the user when the VR machine tool simulates the machine tool.

This demonstrated VR machine tool has the following functions: machine tool axis motion, tool change, reference position return, coolant spray, collision detection and NC decoding for controlling the machine. At the time of the exhibition, young engineers or students were much impressed by the interoperable VR machine tool. They learned how to use VR-based the system via hands-on experiences. However, it might be difficult for the elderly to adapt to the VR scene because of eyesight issues such as Presbyopia and Astigmatism. Overall, the integration of virtual reality and machine tools has been successfully accepted and experienced at an international fairs.

In addition to the establishment of the huge mill-turn machine tool VR-based demonstration system, a virtual laser-assisted milling machine motion simulation system is also under development. Figure 11 is a screen snapshot showing the machine tool definition. Figure 12 shows the presentation of the machine tool in the VR demonstration system. The axes of this laser-assisted milling machine are completely different from the huge mill-turn machine. It has been proven that different configurations of the machine tool can also be successfully constructed by using the VR technology.

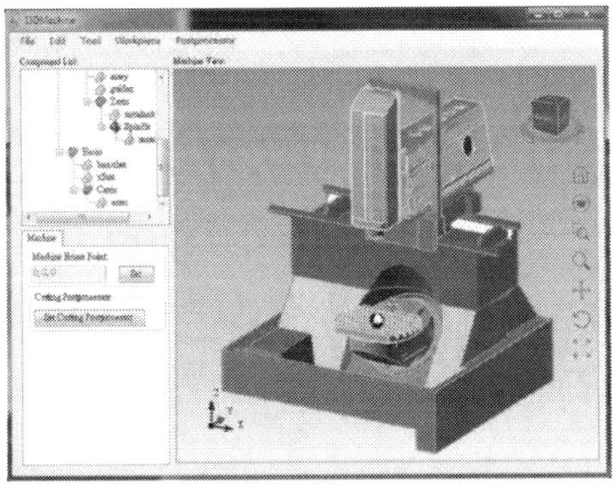

Fig 11. Virtual laser-assisted milling machine.

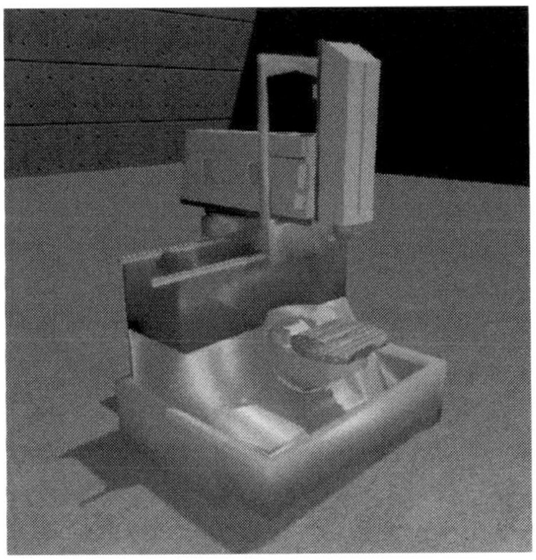

Fig 12. VR-based laser-assisted milling machine.

5 Conclusion and future work

One of the developed VR CNC machine tool is a large and heavy mill-turn machine. A head-mounted display (HMD) has been successfully attached to the developed VR CNC machine tool for the immersion viewing. From the results, it demonstrated that a user can see around the 3D scene of the huge mill-turn machine and the operation of the virtual CNC machine can be actuated by bare hand. The coolant display and collision detection functions had demonstrated the realistic operations for reminding the operator. The development of VR machine tool with laser-assisted milling has also been successfully under development. It showed the expandability of the proposed VR based technology for complicated processes and machine movement. In addition to the machine tool motion simulation, VR technology can also be used in the visualization of machining process in the future. It has also been successfully conducted to show the expandability of the VR-based technology for different applications.

One can conclude that VR will be adopted and paid more and more attention in the future in helping CNC machine tool promotion. Further study could be extended to the develpment of CNC education and training system, and also to the remote maintenance system for a complicated machine tools, too.

Acknowledgment

The authors appreciate the grant supported by the Metal Industry Research and Development Centre.

References

1. Ahmed Amer, Phillip Peralez, "Affordable altered perspectives: Making augmented and virtual reality technology accessible", IEEE Global Humanitarian Technology Conference (GHTC), 603-608 (2014)
2. David J. Zielinski, Ryan P. McMahan, Solaiman Shokur, Edgard Morya, Regis Kopper, "Enabling closed-source applications for virtual reality via OpenGL intercept-based techniques", IEEE 7th Workshop on Software Engineering and Architectures for Realtime Interactive Systems (SEARIS), 59-64 (2014)
3. Jiri Kovar, Katerina Mouralova, Filip Ksica, Jiri Kroupa, Ondrej Andrs , Zdenek Hadas, "Virtual Reality in Context of Industry 4.0 Proposed Projects at Brno University of Technology", 17th International Conference on Mechatronics – Mechatronika (ME), 1-7 (2016)
4. Feng Tian, Pan Wang, ChunYan Dong, Jie Gao, Haifeng Tang, Li Qian, Haojun Xu, "3D Design Tools for Equipment Manufacturing and Exhibition Based on Internet", International Conference on

Audio, Language and Image Processing (ICALIP), (2016) http://ieeexplore.ieee.org/document/7846539/
5. Luigi Pelliccia, Philipp Klimant, Marco Schumann, Franziska Pürzel, Volker Wittstock, Matthias Putz, "Energy visualization techniques for machine tools in virtual reality", Procedia CIRP, **41**, 329-333 (2016)
6. R. Neugebauer, P. Klimant, M. Witt, "Realistic Machine Simulation with Virtual Reality", Procedia CIRP, **3** , 103 – 108 (2012)
7. Wu Man, Zhang Qun, "The Deconstruction and Reshaping of Space: the Application of Virtual Reality in Living Space", 9th International Conference on Measuring Technology and Mechatronics Automation (ICMTMA), 410-413 (2017)
8. Windo Hutabarat, John Oyekan, Christopher Turner, Ashutosh Tiwari, Neha Prajapat, Xiao-Peng Gan, Anthony Waller, "Combining virtual reality enabled simulation with 3D scanning technologies towards smart manufacturing", Winter Simulation Conference (WSC),2774-2785 (2016)
9. Ahmed R. J. Almusawi, L. Canan Dülger, Sadettin Kapucu, "Robotic arm dynamic and simulation with Virtual Reality Model (VRM)", International Conference on Control, Decision and Information Technologies (CoDIT), 335-340 (2016)
10. Wolfgang Hürst, Bibi de Boer, Wouter Florijn, Xhi Jia Tan, "Creating new museum experiences for virtual reality", IEEE International Conference on Multimedia & Expo Workshops (ICMEW), 1-6 (2016)
11. Nedel L, de Souza VC, Menin A, Sebben L, Oliveira J, Faria F, Maciel A, "Using Immersive Virtual Reality to Reduce Work Accidents in Developing Countries.", IEEE Computer Graphics and Applications, **36,** issus:2, 36-46 (2016)
12.Catherine Ball, Kyle Johnsen, "An Accessible Platform for Everyday Educational Virtual Reality", IEEE 2nd Workshop on Everyday Virtual Reality(WEVR), 26-31 (2016)
13.Koreen Pagano, Angela Haddad, Tai Crosby "Virtual Reality—Making Good on the Promise of Immersive Learning: The effectiveness of in-person training, with the logistical and cost-effective benefits of computer-based systems", IEEE Consumer Electronics Magazine, **6**, issue:1, 45-47 (2017)
14.Irina Makarova, Rifat Khabibullin, Eduard Belyaev, Angelina Bogateeva, "The application of virtual reality technologies in engineering education for the automotive industry", International Conference on Interactive Collaborative Learning (ICL), 536-544 (2015)
15. M. Travassos Valdez, C. Machado Ferreira, Maria João M. Martins, F. P. Maciel Barbosa, "3D virtual reality experiments to promote electrical engineering education", International Conference on Information Technology Based Higher Education and Training (ITHET), 1-4 (2015)

Ride responses of macpherson suspension systems

Cheng-Chi Yu[1] and *Chih-Chi* Chiu[2,*]

[1]Department of Vehicle Engineering, National Taipei University of Technology, Taiwan
[2]Department of Vehicle Engineering, National Taipei University of Technology, Taiwan

Abstract. The main purpose of this study is to obtain more correct vehicle ride responses by using a nonlinear ride model considering the effect of Macpherson suspension geometry. Traditional ride model applied to analysis and controller design uses a two degree of freedom linear model, which includes sprung mass and unsprung mass and a spring and a damper vertically connect them. In fact, suspension components do not vertically position above the tire. The motions of body and tire are not going straight up and down. Therefore, the analysis results obtained by the simple model are often different from the experimental values of the actual vehicle. Because of the difference between simple model and actual vehicle, the control strategy almost cannot apply to actual vehicle. In order to understand the effect of suspension geometry on the vehicle ride responses and design a more practical control strategy, a nonlinear model including the geometric parameters of the suspension is constructed in this study. To estimate the initial equilibrium position of the suspension assembly under load, the static equilibrium analysis and mechanism motion analysis are synchronous implemented at the same time. The nonlinear model describes not only the relative position and velocity but also the force transmission between body and tire. Furthermore, by linearize this nonlinear model the development of control strategy for subsequent (semi) active suspension system could be expected.

1 Introduction

Currently, traditional suspension mathematical models are prevalent in analyzing and controlling the ride quality of vehicles [1-4], as shown in Fig. 1 [5]. However, suspension components such as springs, shock absorbers, and hydraulic cylinders are not installed directly above the vehicle tires or positioned perpendicular to the horizontal plane of the vehicle, nor do the vehicle body and tires exhibit a straight up-and-down motion (Fig. 2). Therefore, a simple straight up-and-down suspension model cannot produce analysis results concurring with those of experiments with real vehicles. Consequently, control strategies formulated using such a model cannot attain favorable effects.

Several studies have considered adopting the wheel center rate, which is derived from the geometric effect of the suspension, to replace the spring rate in a simplified model [6-7], thereby reducing the difference between the simplified model and real situations. However, few studies have mentioned the equivalent damping coefficients or equivalent driving force of shock absorbers, which are related to the control strategies of active and semi-active suspension systems. To accurately determine the effect of suspension components (e.g., springs, shock absorbers, and hydraulic cylinders) on vehicle dynamic behavior and design practicable control strategies, this study constructed a nonlinear model incorporating suspension-related geometric parameters. Analyses on the static balance and mechanism kinematics were also performed to estimate the initial balanced positions of suspension components subjected to various loads. This enabled comprehensively describing how the relative positions (spring rate), velocity (damping force), and force transmission (active suspension) of suspension components affect the motion of the vehicle body (ride quality) and tires (grip). Finally, the nonlinear model was linearized to design the control scheme of semi-active suspension systems.

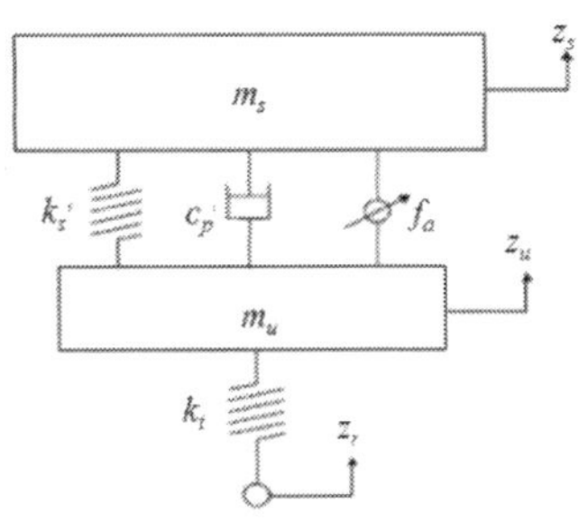

Fig. 1. Traditional suspension mathematical model [5].

* Corresponding author: z175657@gmail.com

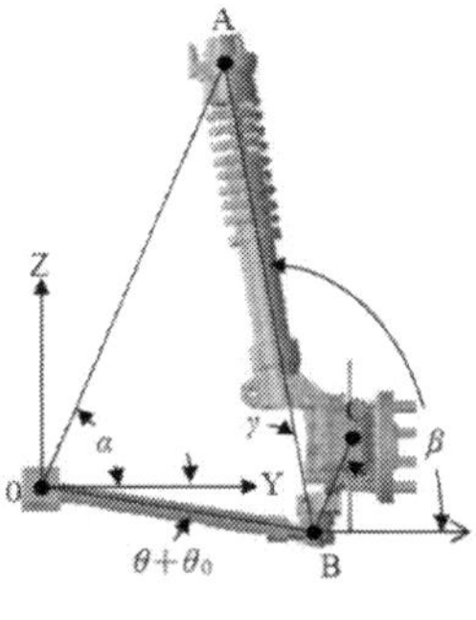

Fig. 2. Relative positions of suspension components in practice.

2 Ride quality models

2.1 The 1/4 vehicle suspension models

Four 1/4 vehicle suspension models were constructed, namely, straight up-and-down (hereafter referred to as traditional), nonlinear, nonlinear linearized, and ADAMS 1/4 vehicle suspension models. In addition, changes in road surface conditions were used as input to compare the differences between the models in terms of various ride quality indicators, namely, vehicle body vertical acceleration, spring deformation, and tire deformation.

2.1.1 Traditional suspension model

Fig. 1 presents the schematic of a traditional suspension model, whose dynamic equation is expressed as Eq. (1). m_s and m_u denote the mass of the vehicle body (sprung mass) and wheel (unsprung mass), respectively. k_s and k_t represent the stiffness of the spring and tire, respectively. c_p is the damping coefficient of the shock absorber. z_r signifies the road surface deflection. z_s and z_u denote the vertical displacements of the vehicle body and tire, respectively.

$$\begin{aligned} m_s\ddot{z}_s &= -k_s(z_s - z_u) - c_p(\dot{z}_s - \dot{z}_u) \\ m_u\ddot{z}_u &= k_s(z_s - z_u) + c_p(\dot{z}_s - \dot{z}_u) - k_t(z_u - z_r) \end{aligned} \tag{1}$$

Rearranging Eq. (1) yields Eq. (2), where $x=[x_1\ x_2\ x_3\ x_4]^T$ represents the state vector, $x_1 = z_s - z_u$ represents the suspension deformation, $x_2 = \dot{z}_s$ represents the vehicle body vertical velocity, $x_3 = z_u - z_r$ represents the tire deformation, and $z_4 = \dot{z}_u$ represents the vertical velocity of the unsprung mass.

$$\dot{x} = \mathbf{A}_q x + \mathbf{B}_q \dot{z}_r \tag{2}$$

$$\mathbf{A}_q = \begin{bmatrix} 0 & 1 & 0 & -1 \\ -\dfrac{k_s}{m_s} & -\dfrac{c_p}{m_s} & 0 & \dfrac{c_p}{m_s} \\ 0 & 0 & 0 & 1 \\ \dfrac{k_s}{m_u} & \dfrac{c_p}{m_u} & -\dfrac{k_t}{m_u} & -\dfrac{c_p}{m_u} \end{bmatrix}$$

$$\mathbf{B}_q = \begin{bmatrix} 0 & 0 & -1 & 0 \end{bmatrix}^T$$

2.1.2 Nonlinear model

Fig. 2 illustrates the suspension components in a balanced state. θ_0 and β_0 respectively denote the initial angle between the lower control arm and horizontal line, and that between the shock absorber and horizontal line. l_0 represents the initial length of the shock absorber.

When the lower control turns counterclockwise by $\theta°$, the displacement of the vehicle body (Point A), lower control arm (Point B), and tire center (Point C) relative to the initial balanced positions can be determined. Specifically, the displacements of Points A, B, and C are defined using (y_A, z_A), (y_B, z_B), and (y_C, z_C), respectively, as shown in Eq. (3), where l_A, l_B, and l_C respectively indicate the lengths of $\overline{OA}$, $\overline{OB}$, and $\overline{BC}$, α represents the angle between l_A and the horizontal line, and γ is the angle between l_C and l. When l_A, l_B, and the angles between the lower control arm and horizontal line (θ_0 and θ) are known, the vector loop method can be used to determine the angles between the shock absorber and horizontal line (β_0 and β).

$$\begin{aligned} y_A &= 0 \\ z_A &= z_s \\ y_B &= l_B(\cos(\theta-\theta_0) - \cos(-\theta_0)) \\ z_B &= z_s + l_B(\sin(\theta-\theta_0) - \sin(-\theta_0)) \\ y_C &= l_C(\cos(\beta-\gamma) - \cos(\beta_0-\gamma)) + y_B \\ z_C &= l_C(\sin(\beta-\gamma) - \sin(\beta_0-\gamma)) + z_B \\ \beta_0 &= \tan^{-1}\left(\frac{l_A\sin\alpha + l_B\sin\theta_0}{l_A\cos\alpha - l_B\cos\theta_0}\right) \\ \beta &= \tan^{-1}\left[\frac{l_A\sin\alpha + l_B\sin(\theta_0\text{-}\theta)}{l_A\cos\alpha - l_B\cos(\theta_0\text{-}\theta)}\right] \end{aligned} \tag{3}$$

The law of cosines is used to calculate the initial length (l_0) of the shock absorber and the length (l) after the lower control arm turns θ, as shown in Eq. (4). Subsequently, the spring deformation (Δl) and the compression rate of the shock absorber spring ($\Delta\dot{l}$) are derived, where $\alpha' = \alpha + \theta_0$.

$$\begin{aligned} l_0 &= \left(l_A^2 + l_B^2 - 2l_Al_B\cos\alpha'\right)^{1/2} \\ l &= \left[l_A^2 + l_B^2 - 2l_Al_B\cos(\alpha'-\theta)\right]^{1/2} \\ (\Delta l)^2 &= (l_0 - l)^2 \\ \Delta\dot{l} &= \dot{l}_0 - \dot{l} \end{aligned} \tag{4}$$

Equation (5) indicates using the Lagrange's Equation to construct the vehicle dynamic model, where q represents the system parameter, T represents the system kinetic energy, V represents the system potential energy, D represents the system dissipating function (e.g., the damping force), and Q represents the generalized force. Substituting the displacements and velocities of the mass points, which are respectively determined in Eq. (3) and

Eq. (4), into Eq. (5) yields Eq. (6), or the vehicle dynamic function with the vehicle body vertical displacement (z_s) and lower control arm turning angle (θ) as the system parameters.

$$\frac{d}{dt}\frac{\partial T}{\partial \dot{q}}+\frac{\partial V}{\partial q}+\frac{\partial D}{\partial \dot{q}}=Q \tag{5}$$

where,

$$\begin{aligned}T&=\frac{1}{2}m_s\dot{z}_s^2+\frac{1}{2}m_u\left(\dot{y}_C^2+\dot{z}_C^2\right)\\&=\frac{1}{2}(m_s+m_u)\dot{z}_s^2+\frac{1}{2}m_u[l_C^2\beta_\theta^2\dot{\theta}^2+l_B^2\dot{\theta}^2\\&+2l_Cl_B\beta_\theta\dot{\theta}^2\sin(\beta-\gamma)\sin(\theta-\theta_0)\\&+2\dot{z}_sl_C\beta_\theta\dot{\theta}\cos(\beta-\gamma)+2\dot{z}_sl_B\dot{\theta}\cos(\theta-\theta_0)\\&+2l_Cl_B\beta_\theta\dot{\theta}^2\cos(\beta-\gamma)\cos(\theta-\theta_0)]\\V&=\frac{1}{2}k_s(\Delta l)^2+\frac{1}{2}k_t(z_C-z_r)^2\\&=\frac{1}{2}k_s\{2a_1-b_1[\cos\alpha'+\cos(\alpha'-\theta)]\\&-2[a_1^2-a_1b_1(\cos\alpha'+\cos(\alpha'-\theta)]\\&+b_1^2[\cos\alpha'\cos(\alpha'-\theta)]^{1/2}\}+\frac{1}{2}k_t(z_C-z_r)^2\\D&=\frac{1}{2}c_p(\Delta\dot{l})^2=\frac{c_pb_1^2\sin^2(\alpha'-\theta)\dot{\theta}^2}{8[a_1-b_1\cos(\alpha'-\theta)]}\end{aligned}$$

where,

$$a_1=l_A^2+l_B^2\ ;\ b_1=2l_Al_B$$

$$\beta_\theta=\frac{\partial\beta}{\partial\theta}=\frac{l_A^2\text{-}l_B^2}{2\left[l_A^2+2\cos(\alpha\text{-}\theta_0+\theta)l_Al_B+l_B^2\right]}-\frac{1}{2}$$

When $q=z_s$,

$$\begin{aligned}&(m_s+m_u)\ddot{z}_s+[m_ul_C\beta_\theta\cos(\beta-\gamma)+m_ul_B\cos(\theta-\theta_0)]\ddot{\theta}\\&+m_ul_C\beta_{\theta2}\dot{\theta}^2\cos(\beta-\gamma)-m_ul_C\beta_\theta^2\dot{\theta}^2\sin(\beta-\gamma)\\&-m_ul_B\dot{\theta}^2\sin(\theta-\theta_0)+k_t(z_C-z_r)=0\end{aligned}$$

When $q=\theta$,

$$\begin{aligned}&[m_ul_C^2\beta_\theta^2+m_ul_B^2+2m_ul_Cl_B\beta_\theta\cos(\beta-\gamma-\theta+\theta_0)]\ddot{\theta}\\&+[m_ul_C\beta_\theta\cos(\beta-\gamma)+m_ul_B\cos(\theta-\theta_0)]\ddot{z}_s\\&+2m_ul_C^2\beta_\theta\beta_{\theta2}\dot{\theta}^2+2m_ul_Cl_B\beta_{\theta2}\dot{\theta}^2\cos(\beta-\gamma-\theta+\theta_0)\\&-2m_ul_Cl_B\beta_\theta\dot{\theta}^2\sin(\beta-\gamma-\theta+\theta_0)(\beta_\theta-1)\\&+m_u\dot{z}_sl_C\beta_{\theta2}\dot{\theta}\cos(\beta-\gamma)-m_u\dot{z}_sl_C\beta_\theta^2\dot{\theta}\sin(\beta-\gamma)\\&-m_u\dot{z}_sl_B\dot{\theta}\sin(\theta-\theta_0)+\frac{c_pb_1^2\sin^2(\alpha'-\theta)\dot{\theta}}{4[a_1-b_1\cos(\alpha'-\theta)]}\\&+k_t(z_C-z_r)[l_C\cos(\beta-\gamma)\beta_\theta+l_B\cos(\theta-\theta_0)]\\&-\frac{1}{2}k_s\sin(\alpha'-\theta)[b_1+\frac{d_1}{[c_1-d_1\cos(\alpha'-\theta)]^{1/2}}]=0\end{aligned} \tag{6}$$

Where,

$$c_1=a_1^2-a_1b_1\cos\alpha'\ ;\ d_1=a_1b_1-b_1^2\cos\alpha'$$

$$\beta_{\theta\theta}=\frac{d\beta_\theta}{d\theta}=\frac{l_Al_B\sin(\alpha-\theta_0+\theta)(l_A^2-l_B^2)}{\left[l_A^2+2l_Al_B\cos(\alpha-\theta_0+\theta)+l_B^2\right]^2}$$

The state vector of the dynamic function of the nonlinear suspension model is $x=[\ x_1\ x_2\ x_3\ x_4\ ]=\left[z_s\ \ \dot{z}_s\ \ \theta\ \ \dot{\theta}\right]^T$, and the state functions are expressed as Eq. (7).

$$\begin{aligned}\dot{x}_1&=x_2\\\dot{x}_2&=f_1(x_1,x_2,x_3,x_4,z_r)\\\dot{x}_3&=x_4\\\dot{x}_4&=f_2(x_1,x_2,x_3,x_4,z_r)\end{aligned} \tag{7}$$

where,

$$\begin{aligned}f_1&=\frac{-1}{g_1g_3-g_2g_4}[g_3g_5+k_t(z_C-z_r)(g_3-g_2g_6)-g_2g_7]\\f_2&=\frac{-1}{g_2g_4-g_1g_3}[g_4g_5+k_t(z_C-z_r)(g_4-g_1g_6)-g_1g_7]\\g_1&=(m_s+m_u)\\g_2&=(m_ul_C\beta_\theta\cos(\beta-\gamma)+m_ul_B\cos(\theta-\theta_0))\\g_3&=m_ul_C^2\beta_\theta^2+m_ul_B^2+2m_ul_Cl_B\beta_\theta\cos(\beta-\gamma-\theta+\theta_0)\\g_4&=m_ul_C\beta_\theta\cos(\beta-\gamma)+m_ul_B\cos(\theta-\theta_0)\\g_5&=m_ul_C\beta_{\theta2}\dot{\theta}^2\cos(\beta-\gamma)-m_ul_C\beta_\theta^2\dot{\theta}^2\sin(\beta-\gamma)\\&-m_ul_B\dot{\theta}^2\sin(\theta-\theta_0)\\g_6&=l_C\cos(\beta-\gamma)\beta_\theta+l_B\cos(\theta-\theta_0)\\g_7&=2m_ul_C^2\beta_\theta\beta_{\theta2}\dot{\theta}^2+2m_ul_Cl_B\beta_{\theta2}\dot{\theta}^2\cos(\beta-\gamma-\theta+\theta_0)\\&-2m_ul_Cl_B\beta_\theta\dot{\theta}^2\sin(\beta-\gamma-\theta+\theta_0)(\beta_\theta-1)\\&+m_u\dot{z}_sl_C\beta_{\theta2}\dot{\theta}\cos(\beta-\gamma)-m_u\dot{z}_sl_C\beta_\theta^2\dot{\theta}\sin(\beta-\gamma)\\&-m_u\dot{z}_sl_B\dot{\theta}\sin(\theta-\theta_0)+\frac{c_pb_1^2\sin^2(\alpha'-\theta)\dot{\theta}}{4[a_1-b_1\cos(\alpha'-\theta)]}\\&-\frac{1}{2}k_s\sin(\alpha'-\theta)[b_1-\frac{d_1}{[c_1-d_1\cos(\alpha'-\theta)]^{1/2}}]\end{aligned}$$

2.1.3 Linearization of the nonlinear model

A nonlinear model in a balanced state can be linearized, where the balanced state is expressed as $x_e=(x_{1e},x_{2e},x_{3e},x_{4e})=(0,0,0,0)$. The linearized state function is displayed as Eq. (8). Because the linearized function is overly long, the matrix elements expressed as partial differential equations are not displayed herein to conform to the space limitation of this manuscript.

$$\dot{x}=\mathbf{A}_{ql}x+\mathbf{B}_{ql}\dot{z}_r \tag{8}$$

$$\mathbf{A}_{ql} = \begin{bmatrix} 0 & 1 & 0 & 0 \\ \frac{\partial f_1}{\partial x_1} & \frac{\partial f_1}{\partial x_2} & \frac{\partial f_1}{\partial x_3} & \frac{\partial f_1}{\partial x_4} \\ 0 & 0 & 0 & 1 \\ \frac{\partial f_2}{\partial x_1} & \frac{\partial f_2}{\partial x_2} & \frac{\partial f_2}{\partial x_3} & \frac{\partial f_2}{\partial x_4} \end{bmatrix}_{x=x_e}$$

$$\mathbf{B}_{ql} = \begin{bmatrix} 0 & \frac{\partial f_1}{\partial z_r} & 0 & \frac{\partial f_2}{\partial z_r} \end{bmatrix}^T_{z_r=0}$$

2.1.4 Adams 1/4 vehicle suspension model

Fig. 3 depicts a 1/4 vehicle suspension model constructed using the ADAMS-View module. This model serves as the reference to be compared with other models.

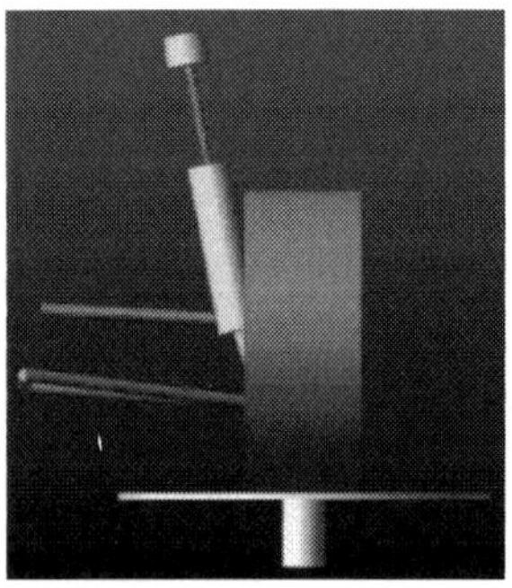

Fig. 3. ADAMS 1/4 vehicle suspension model.

2.2 The 1/2 vehicle suspension models

1/2 vehicle suspension models are more informative than their 1/4 counterparts. As shown in Fig. 4, data such as the vertical displacement (z_s) at the vehicle center of mass and the pitch angle (ϕ) are also available in a 1/2 model.

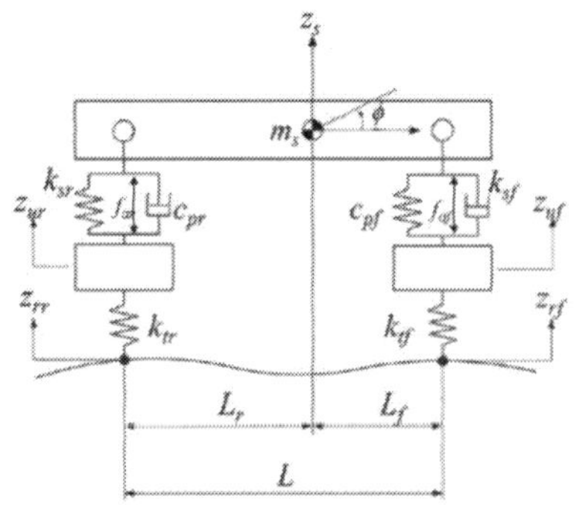

Fig. 4. 1/2 vehicle suspension model.

2.2.1 Traditional suspension model

Dynamic equations for determining z_s and ϕ are expressed as Eq. (9) and Eq. (10), respectively, where F_f and F_r respectively denote the reaction forces that the front and rear axle suspensions exert on the vehicle body; L_f and L_r respectively indicate the horizontal distances of the vehicle center of mass to the front and rear axles; L is the distance between the front and rear axles; I_P represents the inertia of mass moment in the pitch direction of the vehicle body; and z_{sf} and z_{sr} respectively denote the vertical displacements of the front and rear axle sprung mass when the rear tires bounce because of an uneven road surface. Equations (9) and (10) yield the dynamic vertical accelerations of the front and rear axle sprung mass ($\ddot{z}_{sf}$ and $\ddot{z}_{sr}$), as shown in Eq. (11). The vertical accelerations of the front and rear axle unsprung mass $\ddot{z}_{uf}$ and $\ddot{z}_{ur}$ are calculated using the same expressions for the 1/4 vehicle suspension model, as shown in Eq. (12). Equations (11) and (12) can be rearranged into a matrix form, as shown in Eq. (13), where $x = \begin{bmatrix} x_1 & x_2 & x_3 & x_4 & x_5 & x_6 & x_7 & x_8 \end{bmatrix}^T$ is the 1/2 vehicle state vector; $x_1 = z_{sf} - z_{uf}$ is the deformation of the front axle suspension; $x_2 = \dot{z}_{sf}$ is the vertical velocity of the front axle sprung mass; $x_3 = z_{uf} - z_{rf}$ is the deformation of the front axle tire; $x_4 = \dot{z}_{uf}$ is the vertical velocity of the front axle unsprung mass; $x_5 = z_{sr} - z_{ur}$ is the deformation of the rear axle suspension; $x_6 = \dot{z}_{sr}$ is the vertical velocity of the rear axle sprung mass; $x_7 = z_{ur} - z_{rr}$ is the deformation of the rear axle tire; and $z_8 = \dot{z}_{ur}$ is the vertical velocity of the rear axle unsprung mass. Finally, $\dot{z}_{hr} = \begin{bmatrix} \dot{z}_{rf} & \dot{z}_{rr} \end{bmatrix}$ represents the road surface input, and $f_{ha} = \begin{bmatrix} f_{af} & f_{ar} \end{bmatrix}$ denotes the driving force input.

$$\begin{aligned} m_s\ddot{z}_s &= \left[-k_{sf}(z_{sf} - z_{uf}) - c_{pf}(\dot{z}_{sf} - \dot{z}_{uf}) + f_{af}\right] \\ &\quad + \left[-k_{sr}(z_{sr} - z_{ur}) - c_{pr}(\dot{z}_{sr} - \dot{z}_{ur}) + f_{ar}\right] \\ &= F_f + F_r \\ I_p\ddot{\phi} &= \left[-k_{sr}(z_{sr} - z_{ur}) - c_{pr}(\dot{z}_{sr} - \dot{z}_{ur}) + f_{ar}\right]L_r \\ &\quad - \left[-k_{sf}(z_{sf} - z_{uf}) - c_{pf}(\dot{z}_{sf} - \dot{z}_{uf}) + f_{af}\right]L_f \\ &= F_r L_r - F_f L_f \end{aligned} \tag{9}$$

$$\begin{aligned} z_s &= \frac{L_r z_{sf} + L_f z_{sr}}{L} \\ \phi &= \frac{-z_{sf} + z_{sr}}{L} \end{aligned} \tag{10}$$

$$\begin{aligned} \ddot{z}_{sf} &= \left(\frac{1}{m_s} - \frac{L_f L_r}{I_p}\right)F_r + \left(\frac{1}{m_s} + \frac{L_f^2}{I_p}\right)F_f \\ \ddot{z}_{sr} &= \left(\frac{1}{m_s} + \frac{L_r^2}{I_p}\right)F_r + \left(\frac{1}{m_s} - \frac{L_f L_r}{I_p}\right)F_f \end{aligned} \tag{11}$$

$$\begin{aligned} m_{uf}\ddot{z}_{uf} &= k_{sf}(z_{sf} - z_{uf}) + c_{pf}(\dot{z}_{sf} - \dot{z}_{uf}) \\ &\quad - k_{tf}(z_{uf} - z_{rf}) - f_{af} \\ m_{ur}\ddot{z}_{ur} &= k_{sr}(z_{sr} - z_{ur}) + c_{pr}(\dot{z}_{sr} - \dot{z}_{ur}) \\ &\quad - k_{tr}(z_{ur} - z_{rr}) - f_{ar} \end{aligned} \tag{12}$$

$$\dot{x} = \mathbf{A}_h x + \mathbf{B}_{hz}\dot{z}_{hr} + \mathbf{B}_{hf} f_{ha} \tag{13}$$

$$\mathbf{A}_h = \begin{bmatrix} \mathbf{A}_{ha} & \mathbf{A}_{hb} \\ \mathbf{A}_{hc} & \mathbf{A}_{hd} \end{bmatrix}$$

$$\mathbf{B}_{hz} = \begin{bmatrix} 0 & 0 & -1 & 0 & 0 & 0 & 0 & 0 \\ 0 & 0 & 0 & 0 & 0 & 0 & -1 & 0 \end{bmatrix}^T$$

$$\mathbf{B}_{hf} = \begin{bmatrix} \mathbf{B}_{hfa} & \mathbf{B}_{hfb} \end{bmatrix}^T$$

Where,

$$\mathbf{A}_{ha} = \begin{bmatrix} 0 & 1 & 0 & -1 \\ -k_{sf}\left(\frac{1}{m_s}+\frac{L_f^2}{I_p}\right) & -c_{pf}\left(\frac{1}{m_s}+\frac{L_f^2}{I_p}\right) & 0 & c_{pf}\left(\frac{1}{m_s}+\frac{L_f^2}{I_p}\right) \\ 0 & 0 & 0 & 1 \\ \frac{k_{sf}}{m_{uf}} & \frac{c_{pf}}{m_{uf}} & -\frac{k_{tf}}{m_{uf}} & -\frac{c_{pf}}{m_{uf}} \end{bmatrix}$$

$$\mathbf{A}_{hb} = \begin{bmatrix} 0 & 0 & 0 & 0 \\ -k_{sr}\left(\frac{1}{m_s}-\frac{L_f L_r}{I_p}\right) & -c_{pr}\left(\frac{1}{m_s}-\frac{L_f L_r}{I_p}\right) & 0 & c_{pr}\left(\frac{1}{m_s}-\frac{L_f L_r}{I_p}\right) \\ 0 & 0 & 0 & 0 \\ 0 & 0 & 0 & 0 \end{bmatrix}$$

$$\mathbf{A}_{hc} = \begin{bmatrix} 0 & 0 & 0 & 0 \\ -k_{sf}\left(\frac{1}{m_s}-\frac{L_f L_r}{I_p}\right) & -c_{pf}\left(\frac{1}{m_s}-\frac{L_f L_r}{I_p}\right) & 0 & c_{pf}\left(\frac{1}{m_s}-\frac{L_f L_r}{I_p}\right) \\ 0 & 0 & 0 & 0 \\ 0 & 0 & 0 & 0 \end{bmatrix}$$

$$\mathbf{A}_{hd} = \begin{bmatrix} 0 & 1 & 0 & -1 \\ -k_{sr}\left(\frac{1}{m_s}+\frac{L_r^2}{I_p}\right) & -c_{pr}\left(\frac{1}{m_s}+\frac{L_r^2}{I_p}\right) & 0 & c_{pr}\left(\frac{1}{m_s}+\frac{L_r^2}{I_p}\right) \\ 0 & 0 & 0 & 1 \\ \frac{k_{sr}}{m_{ur}} & \frac{c_{pr}}{m_{ur}} & \frac{-k_{tr}}{m_{ur}} & \frac{-c_{pr}}{m_{ur}} \end{bmatrix}$$

$$\mathbf{B}_{hfa} = \begin{bmatrix} 0 & \left(\frac{1}{m_s}+\frac{L_f^2}{I_p}\right) & \frac{-1}{m_{uf}} & 0 \\ 0 & \left(\frac{1}{m_s}-\frac{L_f L_r}{I_p}\right) & 0 & 0 \end{bmatrix}$$

$$\mathbf{B}_{hfb} = \begin{bmatrix} 0 & \left(\frac{1}{m_s}-\frac{L_f L_r}{I_p}\right) & 0 & 0 \\ 0 & \left(\frac{1}{m_s}+\frac{L_r^2}{I_p}\right) & 0 & \frac{-1}{m_{ur}} \end{bmatrix}$$

2.2.2 Nonlinear model

Similar to the 1/4 nonlinear model, the 1/2 nonlinear model also adopts Eq. (3) and Eq. (4) to determine the displacements and velocities of the sprung and unsprung mass in the front and rear axles as well as the spring deformation and compression rate of the shock absorber. $\dot{z}_s$ and $\dot{\phi}$ are determined using Eq. (10) and then substituted into the kinetic energy, potential energy, and dissipating functions, as expressed in Eq. (14) .

$$\begin{aligned} T &= \frac{1}{2}m_s\dot{z}_s^2 + \frac{1}{2}I_p\dot{\phi}^2 + \frac{1}{2}m_{ur}\left(\dot{y}_{Cr}^2 + \dot{z}_{Cr}^2\right) \\ &\quad + \frac{1}{2}m_{uf}\left(\dot{y}_{Cf}^2 + \dot{z}_{Cf}^2\right) \\ V &= \frac{1}{2}k_{sr}(\Delta l_r)^2 + \frac{1}{2}k_{tr}\left(z_{Cr} - z_{rr}\right)^2 + \frac{1}{2}k_{sf}\left(\Delta l_f\right)^2 \\ &\quad + \frac{1}{2}k_{tf}\left(z_{Cf} - z_{rf}\right)^2 \\ D &= \frac{1}{2}c_{pr}(\Delta\dot{l}_r)^2 + \frac{1}{2}c_{pf}(\Delta\dot{l}_f)^2 \end{aligned} \tag{14}$$

The Lagrange's Equation, Eq. (5), is used to construct the vehicle dynamic motion with the vertical displacements of the front and rear axle sprung mass (z_{sf} and z_{sr}) and lower control arm rotation (θ_f and θ_r) as the system parameters, as shown in Eq. (15). Solving Eq. (15) determines the functional relationship of $\ddot{z}_{sf}$, $\ddot{z}_{sr}$, $\ddot{\theta}_f$, and $\ddot{\theta}_r$ with the system states. The state function of the nonlinear model is expressed as Eq. (16), where the state vector $x = [\ x_1\ x_2\ x_3\ x_4\ x_5\ x_6\ x_7\ x_8\ ]^T =$ $[\ z_{sf}\quad \dot{z}_{sf}\quad \theta_f\quad \dot{\theta}_f\quad z_{sr}\quad \dot{z}_{sr}\quad \theta_r\quad \dot{\theta}_r\ ]^T$, $f_3 = \ddot{z}_{sf}$, $f_4 = \ddot{\theta}_f$, $f_5 = \ddot{z}_{sr}$, and $f_6 = \ddot{\theta}_r$.

When $q = z_{sf}$,

$$\begin{aligned} &\frac{1}{L^2}\left(m_sL_r^2 + I_p + m_{uf}L^2\right)\ddot{z}_{sf} + \frac{1}{L^2}\left(m_sL_fL_r\right. \\ &\left.- I_p\right)\ddot{z}_{sr} + \left[m_{uf}l_{Cf}\beta_{\theta f}\cos\left(\beta_f - \gamma_f\right)\right. \\ &\left.+ m_{uf}l_{Bf}\cos\left(\theta_f - \theta_{0f}\right)\right]\ddot{\theta}_f \\ &+ m_{uf}l_{Cf}\beta_{\theta 2f}\dot{\theta}_f^2\cos\left(\beta_f - \gamma_f\right) \\ &- m_{uf}l_{Cf}\beta_{\theta f}^2\dot{\theta}_f^2\sin\left(\beta_f - \gamma_f\right) \\ &- m_{uf}l_{Bf}\dot{\theta}_f^2\sin\left(\theta_f - \theta_{0f}\right) + k_{tf}\left(z_{Cf} - z_{rf}\right) = 0 \end{aligned} \tag{15a}$$

When $q = \theta_f$

$$\begin{aligned} &\left[m_{uf}l_{Cf}^2\beta_{\theta f}^2 + m_{uf}l_{Bf}^2 + 2m_{uf}l_{Cf}l_{Bf}\beta_{\theta f}\cos(\beta_f\right. \\ &\left.- \gamma_f - \theta_f + \theta_{0f})\right]\ddot{\theta}_f + \left[m_{uf}l_{Cf}\beta_{\theta f}\cos(\beta_f - \gamma_f)\right. \\ &\left.+ m_{uf}l_{Bf}\cos(\theta_f - \theta_{0f})\right]\ddot{z}_{sf} + 2m_{uf}l_{Cf}^2\beta_{\theta f}\beta_{\theta 2f}\dot{\theta}_f^2 \\ &+ 2m_{uf}l_{Cf}l_{Bf}\beta_{\theta 2f}\dot{\theta}_f^2\cos(\beta_f - \gamma_f - \theta_f + \theta_{0f}) \\ &- 2m_{uf}l_{Cf}l_{Bf}\beta_{\theta f}\dot{\theta}_f^2\sin(\beta_f - \gamma_f - \theta_f + \theta_{0f})(\beta_{\theta f} \\ &- 1) + m_{uf}\dot{z}_{sf}l_{Cf}\beta_{\theta 2f}\dot{\theta}_f\cos(\beta_f - \gamma_f) \\ &- m_{uf}\dot{z}_{sf}l_{Cf}\beta_{\theta f}^2\dot{\theta}_f\sin(\beta_f - \gamma_f) \\ &- m_{uf}\dot{z}_{sf}l_{Bf}\dot{\theta}_f\sin(\theta_f - \theta_{0f}) \\ &+ \frac{c_{pf}b_{1f}^2\sin^2(\alpha'_f - \theta_f)\dot{\theta}_f}{4\left[a_{1f} - b_{1f}\cos(\alpha'_f - \theta_f)\right]} \\ &+ k_{tf}\left(z_{Cf} - z_{rf}\right)\left[l_{Cf}\cos(\beta_f - \gamma_f)\beta_{\theta f} + l_{Bf}\cos(\theta_f\right. \\ &\left.- \theta_{0f})\right] - \frac{1}{2}k_{sf}\sin(\alpha'_f - \theta_f)[b_{1f} \\ &- \frac{d_{1f}}{\left[c_{1f} - d_{1f}\cos(\alpha'_f - \theta_f)\right]^{1/2}}] + l_{Bf}f_{af} = 0 \end{aligned} \tag{15b}$$

When $q = z_{sr}$,

$$\begin{aligned}
&\frac{1}{L^2}\left(m_s L_f^2 + I_p + m_{ur} L^2\right)\ddot{z}_{sr} + \frac{1}{L^2}\left(m_s L_f L_r\right. \\
&\left. - I_p\right)\ddot{z}_{sf} + \left[m_{ur} l_{Cr} \beta_{\theta r} \cos(\beta_r - \gamma_r)\right. \\
&\left. + m_{ur} l_{Br} \cos(\theta_r - \theta_{0r})\right]\ddot{\theta}_r \\
&+ m_{ur} l_{Cr} \beta_{\theta 2r} \dot{\theta}_r^2 \cos(\beta_r - \gamma_r) \\
&- m_{ur} l_{Cr} \beta_{\theta r}^2 \dot{\theta}_r^2 \sin(\beta_r - \gamma_r) \\
&- m_{ur} l_{Br} \dot{\theta}_r^2 \sin(\theta_r - \theta_{0r}) + k_{tr}\left(z_{Cr} - z_{rr}\right) = 0
\end{aligned} \tag{15c}$$

When $q = \theta_r$,

$$\begin{aligned}
&\left[m_{ur} l_{Cr}^2 \beta_{\theta r}^2 + m_{ur} l_{Br}^2 + 2 m_{ur} l_{Cr} l_{Br} \beta_{\theta r} \cos(\beta_r\right. \\
&\left. - \gamma_r - \theta_r + \theta_{0r})\right]\ddot{\theta}_r + \left[m_{ur} l_{Cr} \beta_{\theta r} \cos(\beta_r - \gamma_r)\right. \\
&\left. + m_{ur} l_{Br} \cos(\theta_r - \theta_{0r})\right]\ddot{z}_{sr} + 2 m_{ur} l_{Cr}^2 \beta_{\theta r} \beta_{\theta 2r} \dot{\theta}_r^2 \\
&+ 2 m_{ur} l_{Cr} l_{Br} \beta_{\theta 2r} \dot{\theta}_r^2 \cos(\beta_r - \gamma_r - \theta_r + \theta_{0r}) \\
&- 2 m_{ur} l_{Cr} l_{Br} \beta_{\theta r} \dot{\theta}_r^2 \sin(\beta_r - \gamma_r - \theta_r + \theta_{0r})(\beta_{\theta r} - 1) \\
&+ m_{ur} \dot{z}_{sr} l_{Cr} \beta_{\theta 2r} \dot{\theta}_r \cos(\beta_r - \gamma_r) \\
&- m_{ur} \dot{z}_{sr} l_{Cr} \beta_{\theta r}^2 \dot{\theta}_r \sin(\beta_r - \gamma_r) - m_{ur} \dot{z}_{sr} l_{Br} \dot{\theta}_r \sin(\theta_r \\
&- \theta_{0r}) + \frac{c_{pr} b_{1r}^2 \sin^2(\alpha_r' - \theta_r)\dot{\theta}_r}{4\left[a_{1r} - b_{1r} \cos(\alpha_r' - \theta_r)\right]} \\
&+ k_{tr}\left(z_{Cr} - z_{rr}\right)\left[l_{Cr} \cos(\beta_r - \gamma_r)\beta_{\theta r} + l_{Br} \cos(\theta_r\right. \\
&\left. - \theta_{0r})\right] - \frac{1}{2} k_{sr} \sin(\alpha_r' - \theta_r)[b_{1r} \\
&- \frac{d_{1r}}{\left[c_{1r} - d_{1r} \cos(\alpha_r' - \theta_r)\right]^{1/2}}] + l_{Br} f_{ar} = 0
\end{aligned} \tag{15d}$$

$$\begin{aligned}
\dot{x}_1 &= x_2 \\
\dot{x}_2 &= f_3(x_1, x_2, x_3, x_4, x_5, x_6, x_7, x_8, z_{rf}, z_{rr}, f_{af}, f_{ar}) \\
\dot{x}_3 &= x_4 \\
\dot{x}_4 &= f_4(x_1, x_2, x_3, x_4, x_5, x_6, x_7, x_8, z_{rf}, z_{rr}, f_{af}, f_{ar}) \\
\dot{x}_5 &= x_6 \\
\dot{x}_6 &= f_5(x_1, x_2, x_3, x_4, x_5, x_6, x_7, x_8, z_{rf}, z_{rr}, f_{af}, f_{ar}) \\
\dot{x}_7 &= x_8 \\
\dot{x}_8 &= f_6(x_1, x_2, x_3, x_4, x_5, x_6, x_7, x_8, z_{rf}, z_{rr}, f_{af}, f_{ar})
\end{aligned} \tag{16}$$

2.2.3 Linearization of the nonlinear model

The nonlinear model in the balanced state is linearized, with the balanced state expressed as $x_e = (x_{1e}, x_{2e}, x_{3e}, x_{4e}, x_{5e}, x_{6e}, x_{7e}, x_{8e})$ =(0,0,0,0,0,0,0,0). Equation (17) yields the state function of the linearized model.

$$\dot{x} = \mathbf{A}_{hl} x + \mathbf{B}_{hlf} z_{rf} + \mathbf{B}_{hlr} z_{rr} + \mathbf{B}_{hlff} f_{af} + \mathbf{B}_{hlfr} f_{ar} \tag{17}$$

$$\mathbf{A}_{hl} = \begin{bmatrix}
0 & 1 & 0 & 0 & 0 & 0 & 0 & 0 \\
\frac{\partial f_3}{\partial x_1} & \frac{\partial f_3}{\partial x_2} & \frac{\partial f_3}{\partial x_3} & \frac{\partial f_3}{\partial x_4} & \frac{\partial f_3}{\partial x_5} & \frac{\partial f_3}{\partial x_6} & \frac{\partial f_3}{\partial x_7} & \frac{\partial f_3}{\partial x_8} \\
0 & 0 & 0 & 1 & 0 & 0 & 0 & 0 \\
\frac{\partial f_4}{\partial x_1} & \frac{\partial f_4}{\partial x_2} & \frac{\partial f_4}{\partial x_3} & \frac{\partial f_4}{\partial x_4} & \frac{\partial f_4}{\partial x_5} & \frac{\partial f_4}{\partial x_6} & \frac{\partial f_4}{\partial x_7} & \frac{\partial f_4}{\partial x_8} \\
0 & 0 & 0 & 0 & 0 & 1 & 0 & 0 \\
\frac{\partial f_5}{\partial x_1} & \frac{\partial f_5}{\partial x_2} & \frac{\partial f_5}{\partial x_3} & \frac{\partial f_5}{\partial x_4} & \frac{\partial f_5}{\partial x_5} & \frac{\partial f_5}{\partial x_6} & \frac{\partial f_5}{\partial x_7} & \frac{\partial f_5}{\partial x_8} \\
0 & 0 & 0 & 0 & 0 & 0 & 0 & 1 \\
\frac{\partial f_6}{\partial x_1} & \frac{\partial f_6}{\partial x_2} & \frac{\partial f_6}{\partial x_3} & \frac{\partial f_6}{\partial x_4} & \frac{\partial f_6}{\partial x_5} & \frac{\partial f_6}{\partial x_6} & \frac{\partial f_6}{\partial x_7} & \frac{\partial f_6}{\partial x_8}
\end{bmatrix}_{x = x_e}$$

$$\mathbf{B}_{hlf} = \begin{bmatrix} 0 & \frac{\partial f_3}{\partial z_{rf}} & 0 & \frac{\partial f_4}{\partial z_{rf}} & 0 & \frac{\partial f_5}{\partial z_{rf}} & 0 & \frac{\partial f_6}{\partial z_{rf}} \end{bmatrix}_{z_{rf} = 0}$$

$$\mathbf{B}_{hlr} = \begin{bmatrix} 0 & \frac{\partial f_3}{\partial z_{rr}} & 0 & \frac{\partial f_4}{\partial z_{rr}} & 0 & \frac{\partial f_5}{\partial z_{rr}} & 0 & \frac{\partial f_6}{\partial z_{rr}} \end{bmatrix}_{z_{rr} = 0}$$

$$\mathbf{B}_{hlff} = \begin{bmatrix} 0 & \frac{\partial f_3}{\partial f_{ar}} & 0 & \frac{\partial f_4}{\partial f_{ar}} & 0 & \frac{\partial f_5}{\partial f_{ar}} & 0 & \frac{\partial f_6}{\partial f_{ar}} \end{bmatrix}_{f_{af} = 0}$$

$$\mathbf{B}_{hlfr} = \begin{bmatrix} 0 & \frac{\partial f_3}{\partial f_{ar}} & 0 & \frac{\partial f_4}{\partial f_{ar}} & 0 & \frac{\partial f_5}{\partial f_{ar}} & 0 & \frac{\partial f_6}{\partial f_{ar}} \end{bmatrix}_{f_{ar} = 0}$$

2.2.4 Adams full vehicle suspension model

Fig. 5 illustrates the full vehicle suspension model built with the ADAMS-Car module. This model serves as the reference to be compared with the other models.

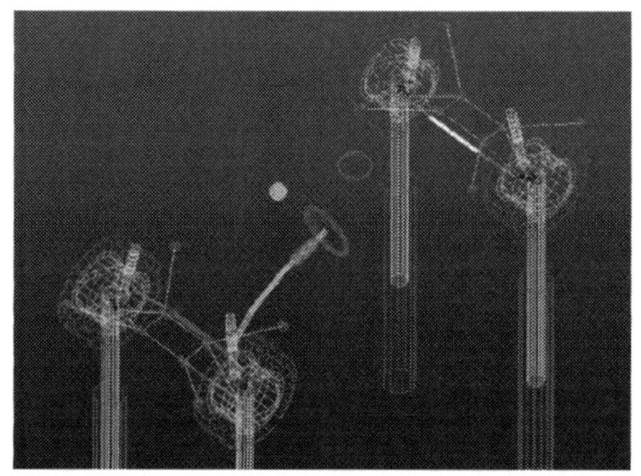

Fig. 5. ADAMS-Car full vehicle suspension model.

3 Initial balanced positions of suspension components

The nonlinear model of suspension geometry induces distinct system responses (nonlinear behavior) under dissimilar swing positions of the lower control arm. Therefore, the initial positions of a loaded suspension system must be identified prior to determining the system responses accurately.

In this study, a static balance analysis was conducted (Fig. 6), in conjunction with with a kinematics analysis performed using the vector loop method (Fig. 5)[8], to

DOI:10.1051/matecconf/201712300005

determine the initial angle of the lower arm (θ_0) under various loads. F_W represents the force that the road surface exerts on the suspension (tires), F_B represents the reaction force that the lower control arm exerts on the suspension along the $\overline{AB}$ direction, and F_C represents the force that the shock absorber spring exerts on the suspension. The functional relationships of the static balance functions and kinematics functions are expressed as (18) and (19).

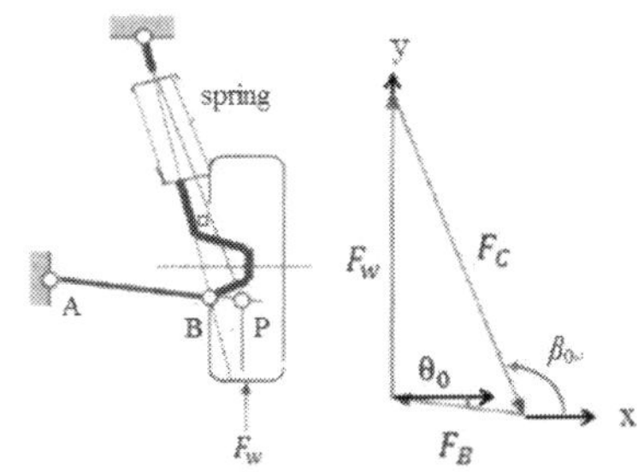

Fig. 6. Static balance analysis of the suspension system.

$$F_W + F_C + F_B = 0$$
$$F_x = -F_B \cos(-\theta_0) - F_C \cos\beta_0 = 0 \qquad (18)$$
$$F_y = -F_B \sin(-\theta_0) - F_C \sin\beta_0 = F_W$$

$$\theta_0 = \cos^{-1}\left(\frac{l_A^2 + l_B^2 - r_3^2}{2 l_A l_B}\right) - \alpha$$
$$\beta_0 = \tan^{-1}\left(\frac{l_A \sin\alpha + l_B \sin\theta_0}{l_A \cos\alpha - l_B \cos\theta_0}\right) \qquad (19)$$

4 Simulation and discussions

4.1 Basic suspension parameters

The parameter settings are described as follows: front and rear sprung mass (m_{sf}=143 kg, m_{sr} = 90 kg), front and rear tire mass (m_{uf}= m_{ur}= 10 kg), stiffness (k_{tf}= k_{tr}= 200 kN/m); interaxle distance (L = 1.8 m), and inertia of mass moment in the vehicle body pitch direction (I_P = 1100 kg · m^2). Fig. 7 depicts the spring stiffness; the springs were compressed by 10 mm when installed. Fig. 8 illustrates the damping characteristics of the shock absorbers. Suspension geometry was configured according to Fig. 2 and the parameters defined in Section 2.1.2: l_A = 0.545 m, l_B = 0.336 m, l_C = 0.107 m, α = 65.965°, and γ = 36.15°; before the system was loaded, the parameters were set as l_0 = 0.5554 m, β_0 = 101.52°, and θ_0 = −7.964°.

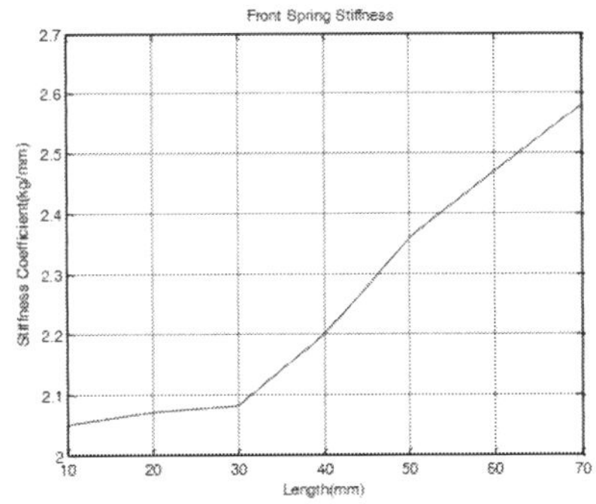

(a) Front axle

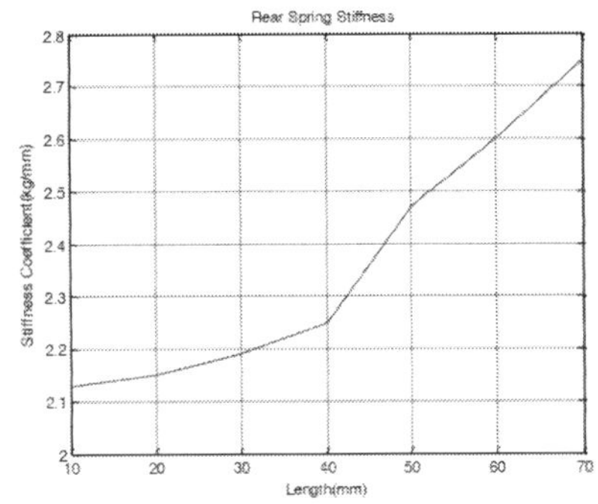

(b) Rear axle

Fig. 7. Spring stiffness.

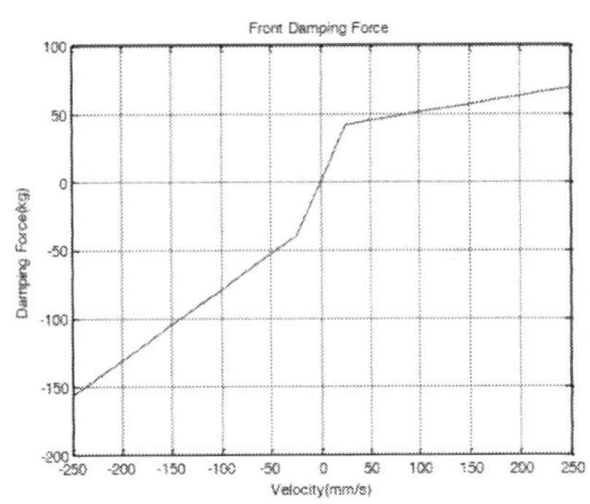

(a) Front axle

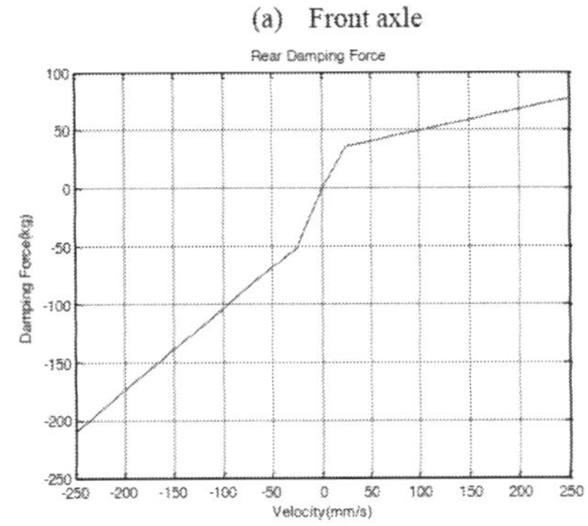

(b) Rear axle

Fig. 8. Damping characteristics of the shock absorbers.

 DOI:10.1051/matecconf/201712300005

4.2 Verification of the initial balanced positions of the suspension system

To verify the initial balanced positions of the suspension system through the proposed mathematical model, the front axle suspension parameters (defined in Section 4.1) were substituted into (18) and (19) ; this enabled determining the initial lower control arm angle and spring deformation of the loaded suspension system in the balanced state. The identical parameter settings were applied to the ADAMS software tool to conduct simulation and obtain the simulated initial lower control arm angle and spring deformation.

In the mathematical model, the lower control arm angle shifted from −7.964° (the unloaded state) to 2.467°, turning counterclockwise by 10.431°. The shock absorber length decreased by 0.059 m from 0.5554 m to 0.4964 m. Conversely, the ADAMS model determined that the lower control arm turned counterclockwise by 10.47°, and that the shock absorber length decreased by 0.05973 m. Therefore, the two models attained similar results (Fig. 1), verifying the feasibility of the mathematical model proposed in this study.

Table 1. Comparison of the initial balanced positions determined with the proposed and ADAMS models.

	Change in the lower control arm angle	Change in the shock absorber length
ADAMS model	10.470°	59.73mm
Mathematical model	10.431°	59.0mm

4.3 Verification of the 1/4 vehicle suspension mathematical models

A chirp signal, which is a type of time-variant string signal available in MATLAB Simulink, with constant amplitude was adopted as the road input. In this study, the signal frequency was set as 0–20 Hz to conduct 40 s of simulation, as shown in Fig. 9. The vehicle body vertical acceleration, spring deformation, and tire deformation determined using the traditional, nonlinear, and linearized nonlinear models under specific road inputs were compared with those simulated using the ADAMS model.

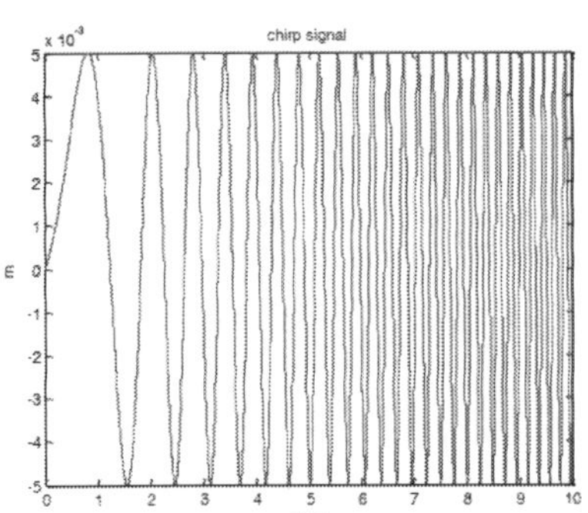

Fig. 9. Road input as a chirp signal.

Table 2. Mean values of the spring stiffness and shock absorber damping coefficient.

	k (kg/mm)	c (kg · s/mm)
Front axle	2.6088	0.628
Rear axle	2.8566	0.822

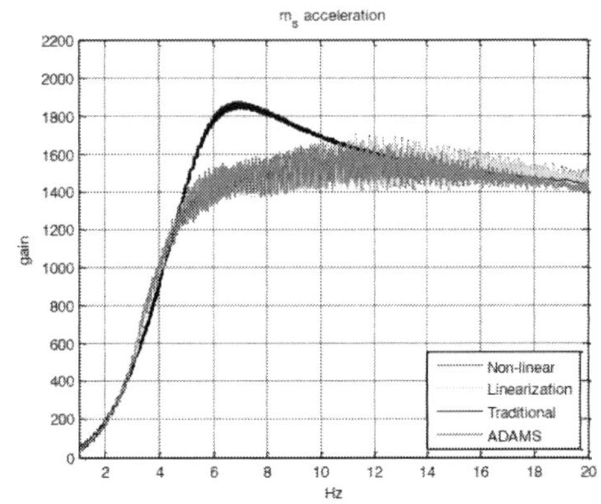

(a) Vehicle body vertical acceleration

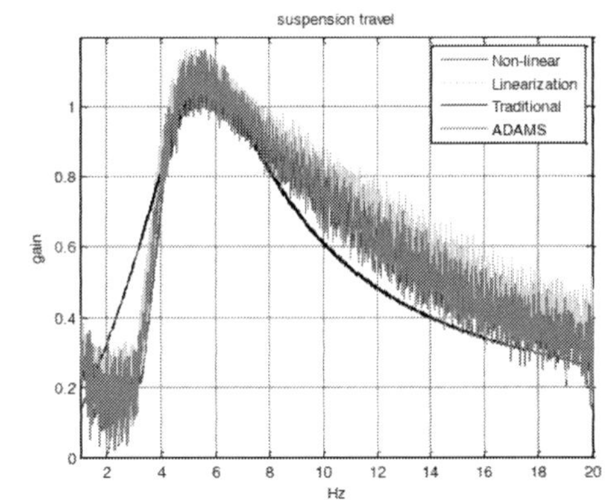

(b) Spring deformation

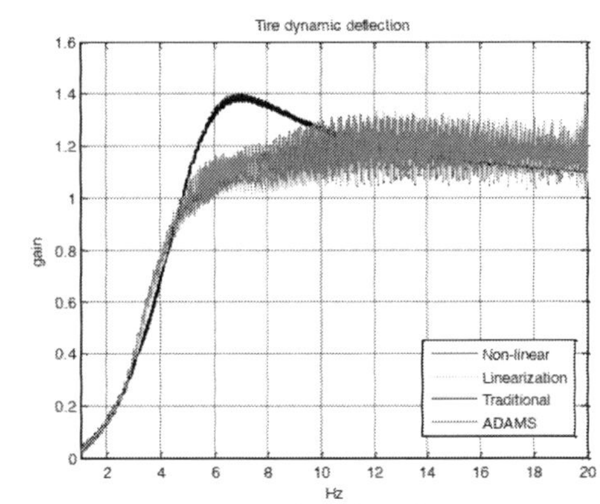

(c) Tire deformation

Fig. 10. Frequency responses resulting from the chirp signal road input.

The front axle parameters were adopted. Specifically, the nonlinear, linearized nonlinear, and ADAMS model used the spring stiffness and shock absorber damping coefficients respectively displayed in Figs. 7(a) and 8(a). By contrast, the spring stiffness value of the traditional model yielded the mean tire rate obtained using the parameter simulation with the ADAMS suspension platform (Section 4.1), and the shock absorber damping coefficient was the mean value acquired from Fig. 8; Table 2 lists these parameter values. The nonlinear and linearized nonlinear models required determining the initial balanced positions before conducting simulation, whereas the time domain simulation involved the fast Fourier transform to obtain the frequency response (Fig. 10).

The simulation results (Fig. 10) showed that compared with the traditional model, the nonlinear and linearized nonlinear models demonstrated dynamic responses more similar to those of the ADAMS model, verifying that the proposed nonlinear mathematic model is more reliable than the traditional model. The vibration displayed in the simulation results resulted from the jump resonance of the nonlinear model, in which slight variations in the input frequency caused substantial fluctuation in the output amplitude. Although the nonlinear frequency responses are not applicable to comparing the response values under different frequencies, they still correspond with the results of the time domain analysis.

4.4 Verification of the 1/2 vehicle suspension mathematical models

Fig. 11 presents the random road surface input built with the ADAMS software tool. Model verification was conducted using vehicle speeds of 10 and 15 m/s.

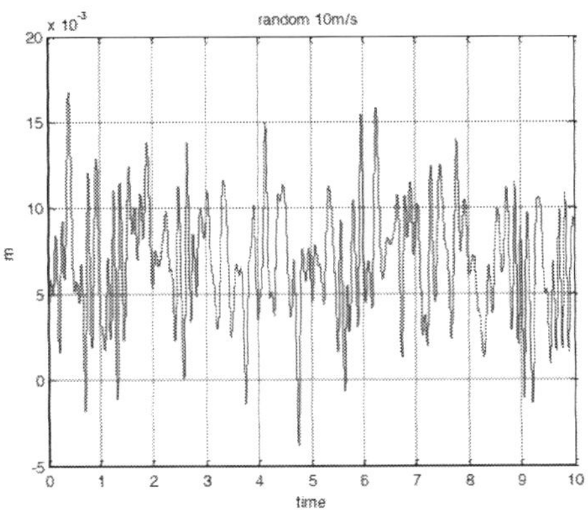

(a) Vehicle speed of 10 m/s

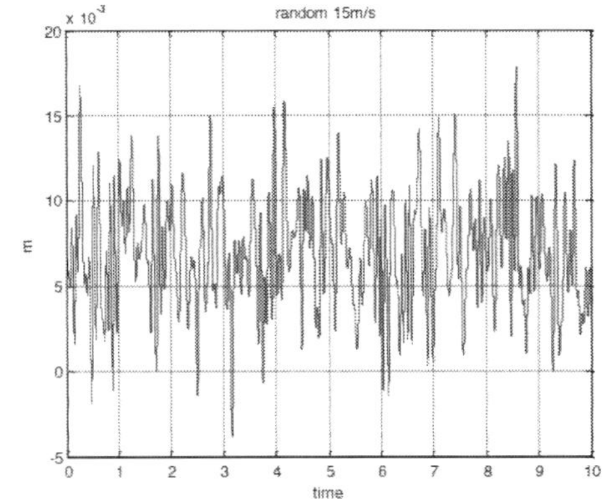

(b) Vehicle speed of 15 m/s

Fig. 11. Random road surface input.

The front and rear parameters of the nonlinear models were based on those defined in Section 4.1. The traditional model used the spring stiffness and shock absorber damping coefficient values displayed in Table 2. Figs. 12 and 13 illustrate the simulation results.

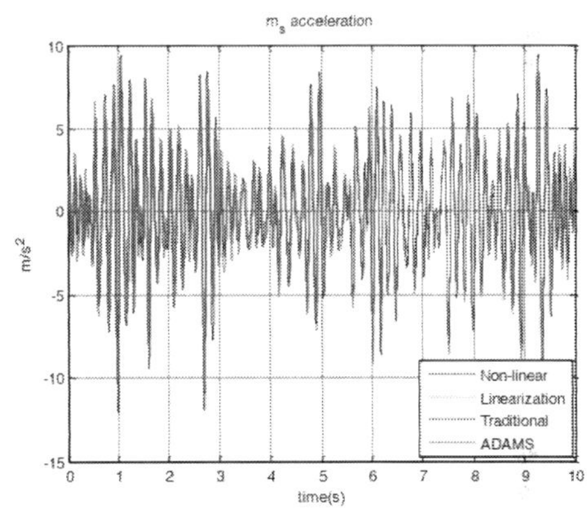

(a) Vertical acceleration at the vehicle center of mass

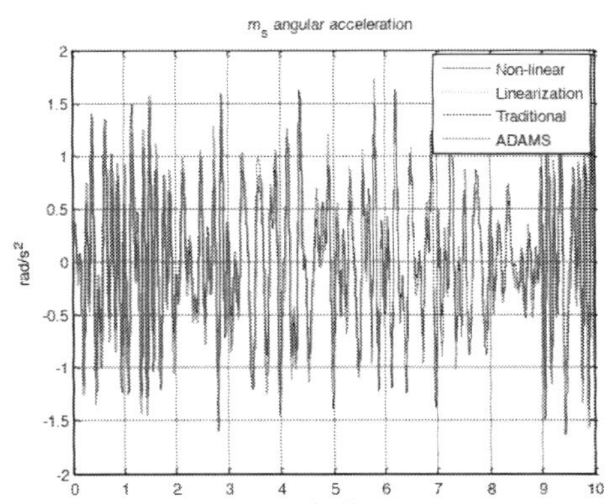

(b) Angular acceleration in the vehicle pitch direction

Fig. 12. Temporal responses at the vehicle speed of 10 m/s.

DOI:10.1051/matecconf/201712300005

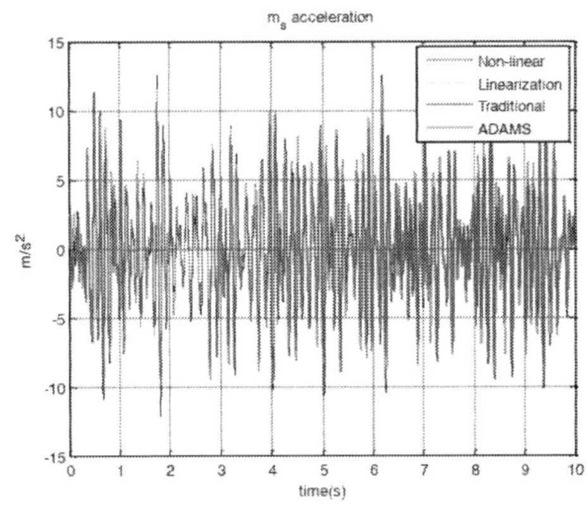

(a) Vertical acceleration at the vehicle center of mass

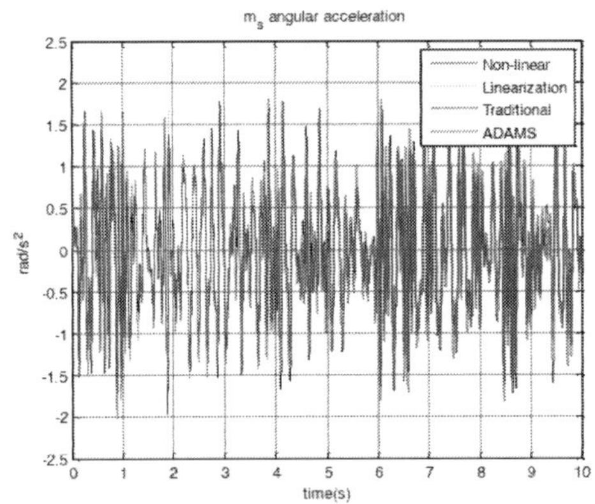

(b) Angular acceleration in the vehicle pitch direction

Fig. 13. Temporal responses at the vehicle speed of 15 m/s.

To evaluate the accuracy of the three models, the mean absolute differences (MADs) of the temporal responses obtained using the models were calculated and compared with those of the ADMAS model (Table 3). The MAD formula is expressed as (20).

$$\text{MAD} = \frac{1}{n}\sum_{i=1}^{n}\left(\left|\text{ADAMS}_i - \text{Math Model}_i\right|\right) \quad (20)$$

Table 3. Mean Absolute Difference.

		Non-Linear	Linearization	Traditional
Vertical acceleration at the vehicle center of mass (m/s^2)	10 m/s	0.8836	0.8822	1.0365
	15 m/s	0.9914	0.9864	1.2196
Angular acceleration in the vehicle pitch direction (rad/s^2)	10 m/s	0.1284	0.1297	0.1572
	15 m/s	0.1598	0.1592	0.2191

As shown in Table 3, the nonlinear and linearized nonlinear model attained results similar to those of the ADAMS model. By contrast, the results of the traditional model differed substantially from those of the ADAMS model, and the difference increased with vehicle speed. Comparing the 1/2 models verified that the proposed nonlinear mathematical model is more accurate than the traditional model in reflecting the actual dynamic behavior of vehicles.

5 Conclusions

This study constructed a nonlinear model composed of suspension-related geometric parameters, and adopted static balance analysis and kinematics analysis to estimate the initial balanced positions of the suspension components subjected to various loads. This enabled comprehensively describing how the relative positions, velocity, and force transmission of the suspension components affect the motion of the vehicle body and tires. The nonlinear model was also linearized to develop control strategies for semi-active suspension systems. The analysis results showed that compared with the traditional model, the nonlinear and linearized nonlinear models attained simulation results more similar to those of the ADAMS model, verifying the accuracy of the proposed model. Furthermore, when the nonlinearity of the suspension system increased, the nonlinear and linearized nonlinear models effectively predicted the variations in the dynamic responses. Therefore, the proposed model can facilitate understanding the effect of suspension components on vehicle dynamic behavior to propose practicable control strategies.

References

1. R. Ramajani, *Vehicle Dynamics and Control*, Springer Science, New York, pp. 267-354, (2006)
2. J. Y. Wong, Theory of Ground Vehicles, John Wiley, pp. 461-520, (2008)
3. M. M. S. Kaldas, K. Çalışkan, R. Henze and F. Küçükay, "Development of a Semi-Active Suspension Controller Using Adaptive-Fuzzy with Kalman Filter" SAE Paper No. 2011-01-0431, (2011)
4. S. Rasal, J. Jaganmohan, S. Agashe, K. P. Wani "Implementation of Fuzzy Logic Control in Semi-active Suspension for a Vehicle Using MATLAB SIMULINK" SAE Paper No. 2016-28-0035, (2016)
5. K. S. Hong, D. S. Jeon, W. S. Yoo, H. Sunwoo, S. Y. Shin, C. M. Kim, and B. S. Park, "A New Model and an Optimal Pole-Placement Control of the Macpherson Suspension System," SAE Technical Paper Series, 1999-01-1331, (1999)
6. W. F. Milliken and D. L. Milliken, Race Car Vehicle Dynamics, Society of Automotive Engineers, Inc., pp. 579-605, (1995)
7. M. Blundell and D. Harty, Multibody Systems Approach to Vehicle Dynamics, Elsevier Butterworth-Heinemann, pp. 185-334, (2015)
8. K. J. Waldron and G. L. Kinzel, *Kinematics, Dynamics, and Design of Machinery*, John Wiley & Sons, pp. 319-354, (2016).

Development of a XY θ_Z 3-DOF nanopositioning stage with linear displacement amplification device

Chih-Liang Chu[1,*], *Hung-Chi* Chen[1], and *Ming-Han* Sie[1]

[1]Department of Mechanical Engineering, Southern Taiwan University of Science and Technology, Tainan, Taiwan, R.O.C.

Abstract. The main purpose of this study is to develop a $XY\theta_Z$ 3-DOF nanopositioning stage with linear displacement amplification device. This stage is used by three groups of linear displacement amplification device to composition, and thus achieves precise $XY\theta_Z$ 3-DOF movement. The linear displacement amplification device makes use of a symmetrical layer mechanism, toggle amplification mechanism and parallel guiding spring to composition, and with amplification capability. Then it uses an equilateral triangle way to set the three groups of linear displacement amplification device. Overall design uses the finite element software ANSYS 12.0 to analysis the displacement, stress and dynamic response of nanopositioning stage. Experiment demonstrated takes the laser interferometer as the standard of displacement measurement. The result of experiment measurement shows that the maximum XY axis displacement and θ_Z rotational angle travel of the stage is 36.5μm, 32μm and 265 arc sec.

1 Introduction

In recent years, nanopositioning has become a key technology within many fields, including electron microscopy and x-ray lithography. These fields have an increasing demand for precision positioning systems capable of providing a wide operating range with high-accuracy and nanometer scale resolution, and minimum lateral and tilting motion. Therefore, developing systems that achieve precise positioning with a nanometer scale resolution has become a crucial requirement if highly precise micro-applications are to be developed.

When developing high-precision positioning mechanisms, it is common to employ piezoelectric actuators, which are capable of providing highly accurate nanoscale displacements. The stack-piezoelectric actuator effect produces an internal material twisting piezoelectric effect, which enables incremental displacements with a very high level of accuracy, delivers a high-resolution displacement output, provides a huge motive force, yields an efficient response, and enables a high mechanical switching efficiency. However, the displacement output of the piezoelectric actuator is limited, and consequently it is frequently necessary to adopt some form of amplification arrangement to extend its range. The displacement resolution of the actuator is further enhanced by means of a flexure hinge, which eliminates friction during mechanical motion of the actuator, and hence prevents undesirable heating effects.

Positioning systems play a fundamental role within nanotechnology applications, and therefore they are attracting an ever-increasing amount of interest within academic circles and in a variety of research units [1-9]. Physic Instrumente [10] provides products of nanopositioning stages to meet different applications. Xu and King [11] studied aspects of the design of flexure hinges for piezoactuator displacement amplifiers. Chang and Du [12] adopted the traditional Scott-Russell's straight-line mechanism method in developing a nanopositioning stage. In their design, a series connection method was employed to combine two Scott-Russell mechanisms in order to amplify the displacement output. In their device, the displacement output caused movement of a parallel guide spring, and it was shown that the straight line and angular deviation were minimized successfully. Their results indicated that the device was capable of a dynamic displacement in excess of 100 μm with a displacement resolution of 0.04 μm and an angular deviation of 31.1 μrad.

Ryu, Gweon, and Moon [13] successfully developed a $XY\theta_z$ micro-movement positioning stage. The stage uses three piezoelectric actuators and a monolithic flexure hinge mechanism, which is designed to provide large θ motion. Their study also provided an excellent source of reference for each of the elements of the stage. The micro-movement positioning stage provided displacements of 41.5 μm and 47.8 μm in the x- and y-directions, respectively, and demonstrated a yaw motion of approximately 322.8 arcsec.

When using a layer mechanism as the amplification device for a positioning stage, the greater the magnitude of the lateral offset error, the greater the angular deviation caused by the layer effect, and the greater the difference of the lateral deviation. Therefore, the present study develops a positioning stage that employs a stack-piezoelectric actuator, a flexure hinge, a layer mechanism, a toggle amplification mechanism, and a

* Corresponding author: cliang@stust.edu.tw

guide device. The aims of this study are to design and fabricate a XYθ_z 3-DOF nanopositioning stage with linear displacement amplification device.

2 Design principle

As shown in Fig. 1, the adoption of a layer mechanism in the conventional positioning stage will result in a lateral offset. A piezoelectric actuator input force of F results in a lateral offset displacement of Δ_X at the output end, which induces an error, and hence indirectly affects the accuracy of the device.

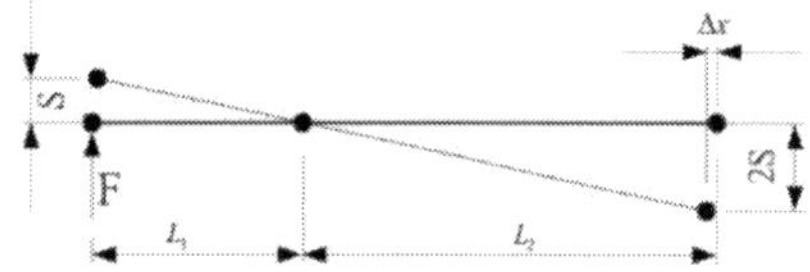

Fig. 1. A schematic drawing of a layer mechanism.

Fig. 2 presents a typical toggle amplification mechanism, in which A is the input and C is the output. When the toggle amplification mechanism is displaced, a lateral offset error is caused at the output end. For an input displacement of A_X, it is straightforward to determine the offset of the output. It is noted that the greater the layer angle, the greater the lateral offset of the output, and hence the greater the error. In the traditional design, two piezoelectric actuators are employed to simultaneously push the toggle amplification mechanism from either end in order to reduce the lateral offset error. However, this approach is not only more expensive, but also actually increases the lateral offset error because of the two different actuator forces.

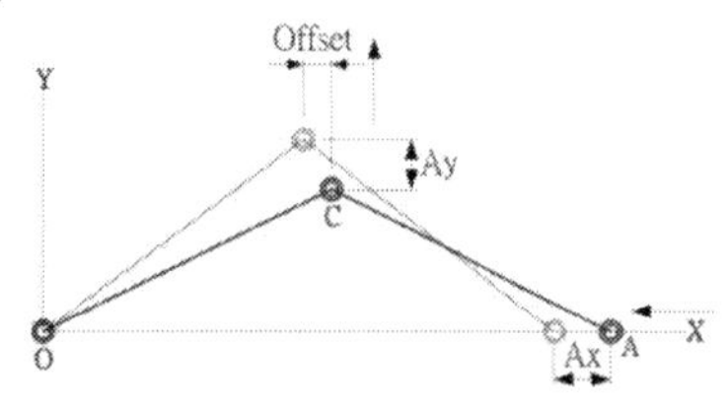

Fig. 2. A schematic demonstration of single direction input toggle amplification mechanism.

To overcome these limitations, the present study develops the positioning stage presented in Fig. 3. The proposed design combines the layer mechanism and a toggle amplification mechanism, and requires only one stack-piezoelectric actuator to supply the necessary driving input. When an electrical voltage is applied to the actuator, the corresponding length expansion, d_p, undergoes the first stage of amplification via a layer mechanism. Subsequently, the lateral offset of the layer mechanism is supplied as the input to the toggle amplification mechanism, where the second displacement amplification process is then performed. This two-step approach enables the lateral offset error caused by the layer mechanism amplification stage to be eliminated, and provides a two-stage displacement amplification function, hence increasing the output displacement of the positioning stage. Finally, the proposed design employs a 120° flexible pivot in order to ensure a precise straight-line displacement output.

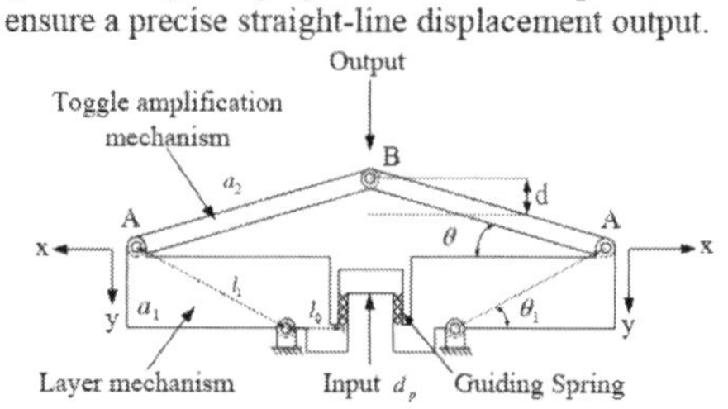

Fig. 3. A schematic demonstration of linear displacement amplification device [14].

From a rigid-body motion analysis of the proposed design, it can be shown that the relationship between the final output displacement and the amplification is given by:

$$d = a_1(\cos\theta_1 + a_2 \sin\theta_1)d_p \tag{1}$$

where $a_1=l_1/l_0$ is the amplification of the layer mechanism, $a_2=1/tan\ \theta$ is the amplification of the toggle, d_p is the actuator input displacement, and d is the final output displacement.

An illustration of the develop linear displacement amplification device is presented in Fig. 4. The motion of the components in the linear displacement amplification device is described as follows. The length of the piezoelectric actuator, component 1, changes due to the driving voltage between the electrodes of the piezoelectric actuator. This length change pushes or pulls the input terminal of the layer mechanism, component 3, and causes displacements of the output terminal of the mechanism. The toggle amplification mechanism, component 4, is driven by the output displacements of component 3. Finally, the platform, component 7, moves due to the output motion of the flexible pivot, component 5. The parallel guide spring, component 6, joined to the platform can offer reaction force to prevent the platform from rotating and moving laterally.

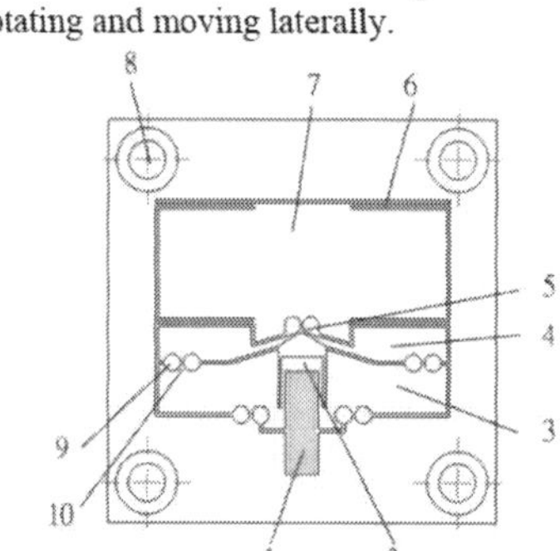

1. piezoelectric actuator, 2. layer guiding device, 3. layer mechanism, 4. toggle amplification mechanism, 5. flexible pivot, 6. parallel guide spring, 7. platform, 8. fixed screw holes, 9. cut-out slot to form flexure hinges, 10. flexure hinges.

Fig. 4. Schematic representation of the linear displacement amplification device.

Then use an equilateral triangle way to set the three groups of linear displacement amplification device presented in Fig. 5. This $XY\theta_z$ 3-DOF nanopositioning stage by three groups of linear displacement amplification device to composition, and thus achieve X, Y, θ_z movement presented in Fig. 6.

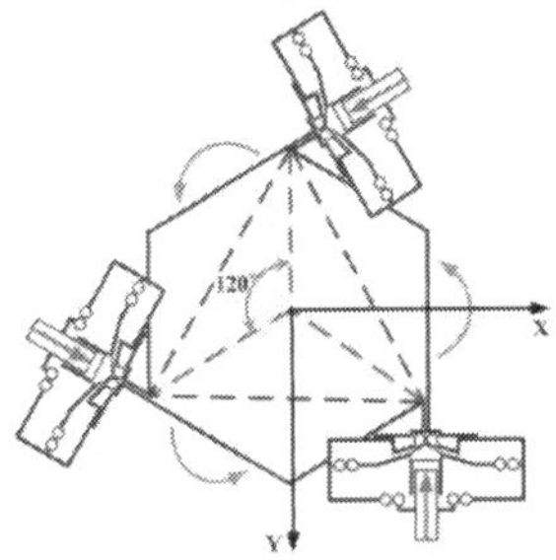

Fig. 5. Design schematic of the $XY\theta_z$ 3-DOF nanopositioning stage.

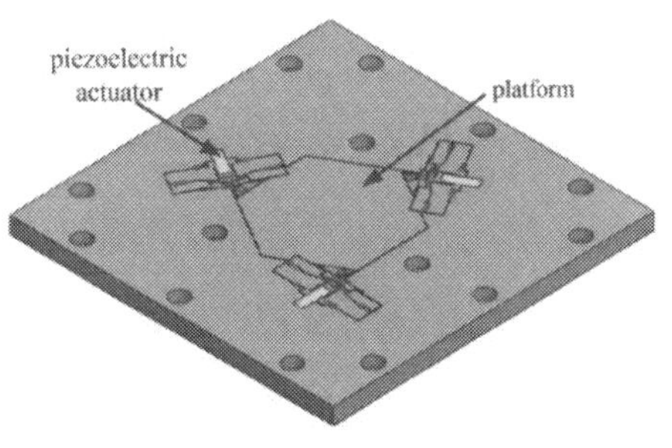

Fig. 6. Overall schematic of the $XY\theta_z$ 3-DOF nanopositioning stage.

3 Analysis and simulation

This study used the finite element method to develop the mathematical model of the stage, and then ANSYS software to simulate its operation. The design presented above was first modelled with limited elements. The simulation model adopted 20-nodes solid tetrahedron elements, SKD11 model steel as the structure material, and a structure thickness of 10mm. The Young's modulus was specified as 207×10^9N/m^2, and the Poisson's ratio as 0.35. The simulation results are presented in Fig. 7.

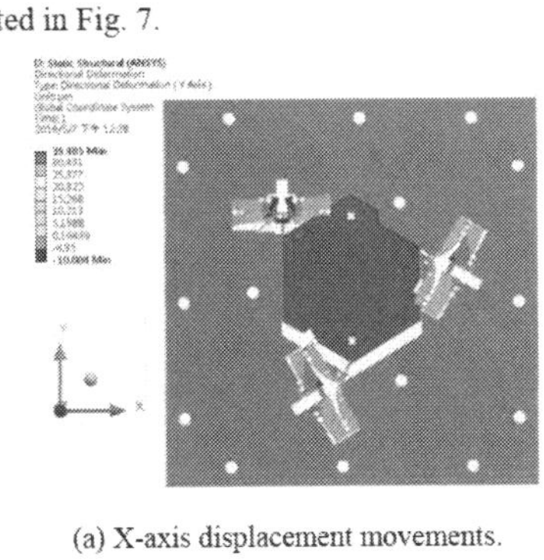

(a) X-axis displacement movements.

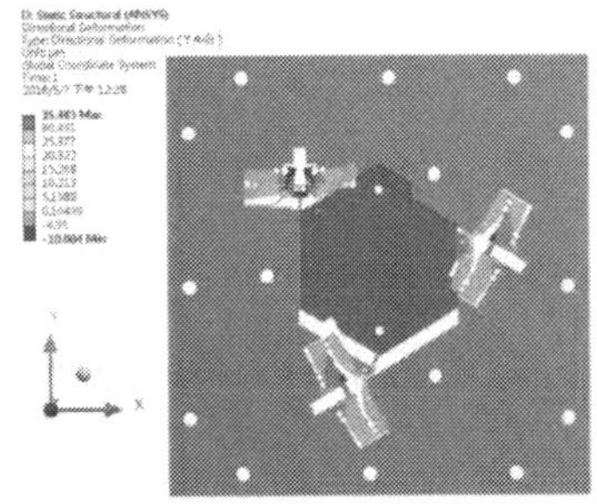

(b) Y-axis displacement movements

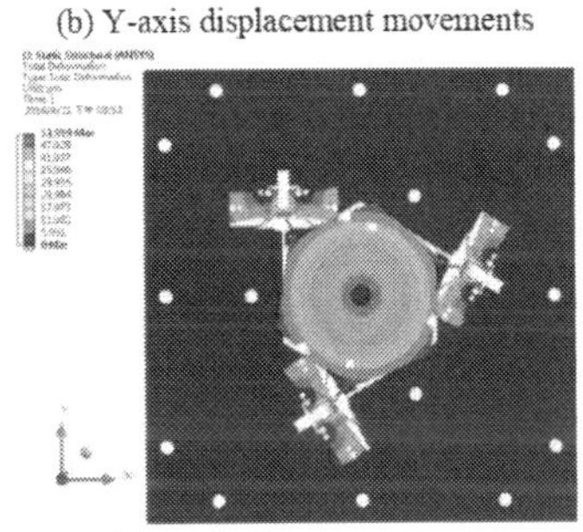

(c) θ_z-axis displacement movements

Fig. 7. ANSYS simulation result.

In the simulation process, a piezoelectric actuator input of 10μm was applied. This resulted in an output displacement of $XY\theta_z$ is 40.7μm, 35.4μm and 294.5 arcsec.

From the dynamic analysis presented in Fig. 8, it can be determined that the first natural frequency is 348.63 Hz, the second natural frequency is 349.34 Hz, and third natural frequency is 565.54 Hz. Analysis reveals that the stage is achieve $XY\theta_z$ 3-DOF movement.

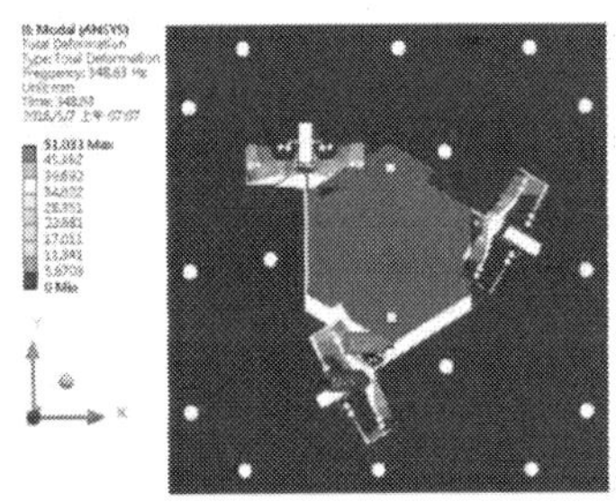

(a) The first natural frequency.

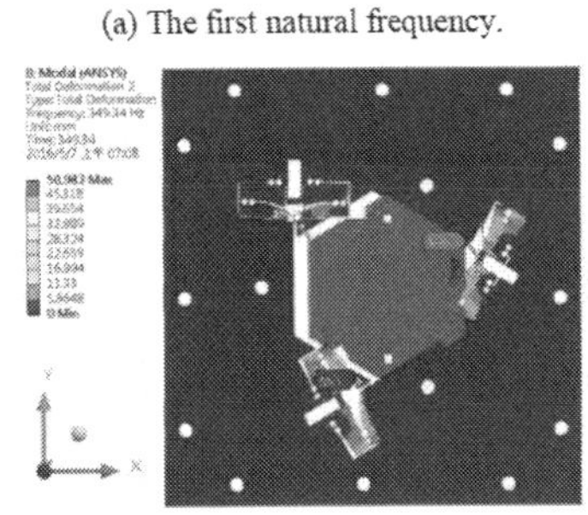

(b) The second natural frequency.

DOI:10.1051/matecconf/201712300006

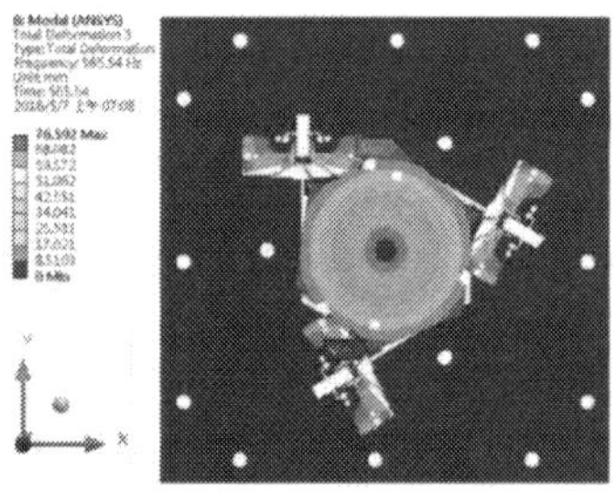

(c) The third natural frequency.

Fig. 8. ANSYS simulation results.

4 Experimental measurement

A precise Electro-Discharge Machining (EDM) process was used to fabricate the steel prototype (200mm×200mm×10mm) shown in Fig. 9. The corresponding engineering drawings were developed using solid modelling CAD software and then converted to DXF files as an input to the EDM machining. During the EDM process, all of the components were machined to a tolerance of ±5 μm.

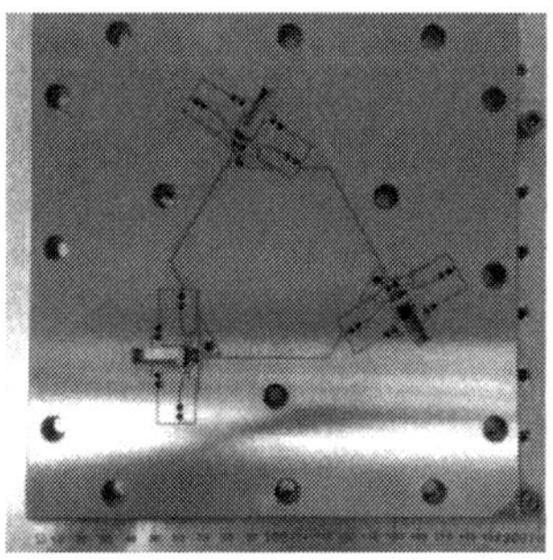

Fig. 9. Photograph of the nanopositioning stage.

Following assembly of the positioning stage, the piezoelectric actuator was installed at the pre-specified position and a series of experimental measurement tests were conducted in order to verify the functions of the stage and to determine its movement characteristics. The piezoelectric actuator, having a size of 5mm×5mm×18mm, is capable of travelling 20um and generating a force of 1800 N at 150 V [15].

As shown in Fig. 10, a non-contact laser interferometer (optodyne, model number MCV-500) with a resolution of 10 nm was used to investigate the performance of the stage. A mirror was attached to the movable stage such that the interferometer could detect the displacement of the stage as it was driven by the actuator. The function generator and a personal computer with a programmable AD/DA card were used to generate the desired waveform signal for testing the stage. The signals generated by the function generator or personal computer were amplified by a power amplifier to drive the piezoelectric actuator installed in the stage. The oscilloscope and personal computer were used to display, store, and analyse the experimental results in both time and frequency domains. The tested stage was placed on an air-bearing supported optical table and isolated from external vibrations.

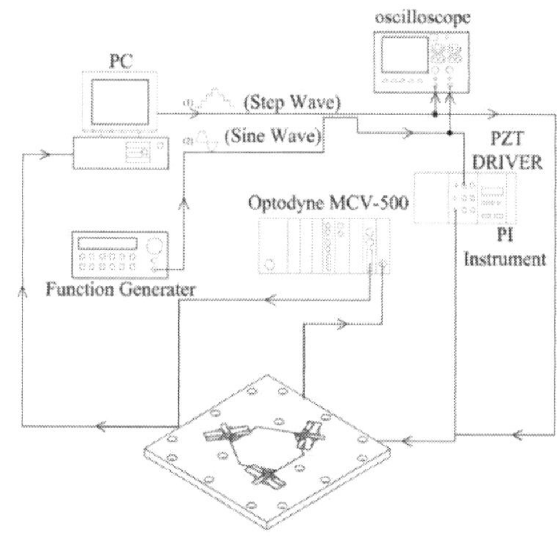

Fig. 10. Experiment set up flow chart.

4.1 Measurement of output displacement

To measure the XY-axis liner displacement, the mirror was initially located at the middle of the stage, as shown in Fig. 11. A voltage waveform was then applied, and the corresponding displacements measured by the laser interferometer. The mirror was positioning to a distance more than δ_x from the middle of the stage. Therefore, the maximum XY-axis liner displacement measurement was obtained for travel in both directions, and the straightness was then determined. Fig. 12, 13, 14 and 15 reveal that the X-axis liner displacement is 36.5 μm, the Y-axis liner displacement is 32 μm, and the XY-axis straightness is 0.3μm and 0.2μm.

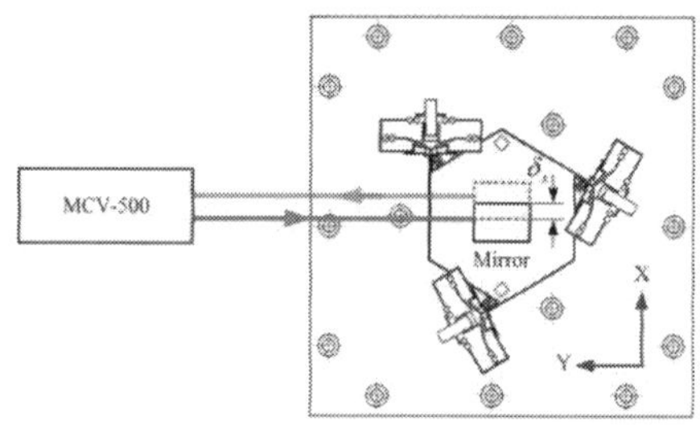

Fig. 11. Experimental arrangement for XY-axis liner displacement.

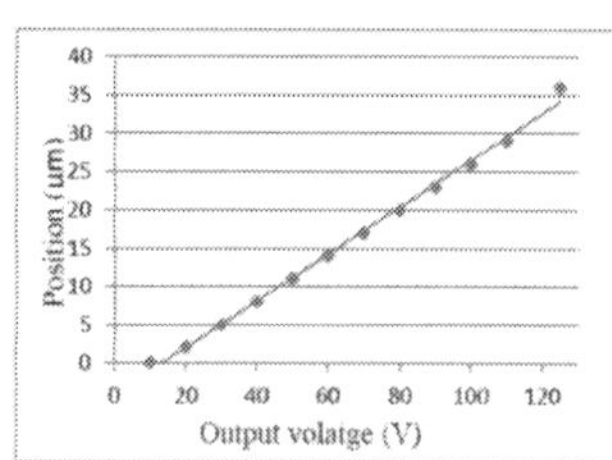

Fig. 12. X-axis liner displacement.

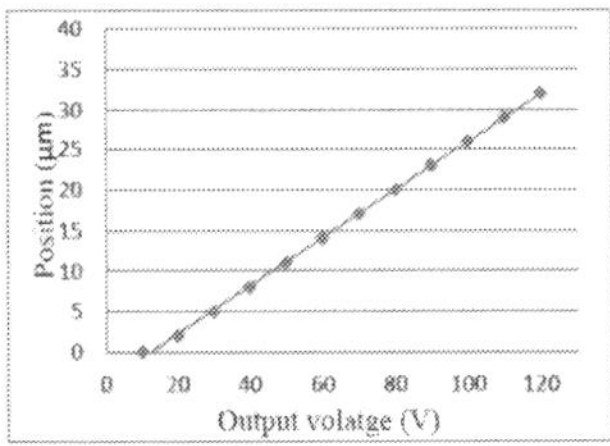

Fig. 13. Y-axis liner displacement.

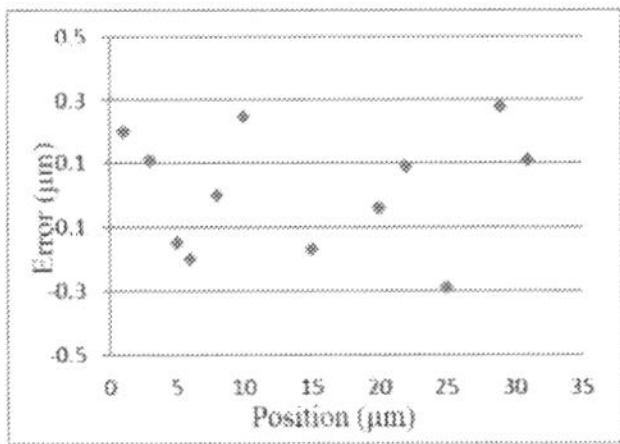

Fig. 14. X-axis straightness.

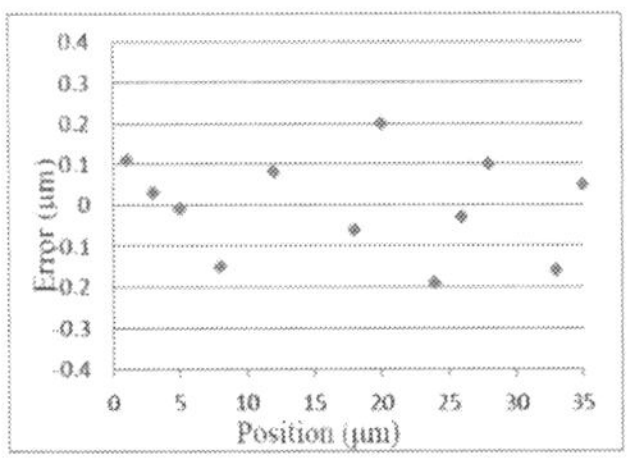

Fig. 15. Y-axis straightness.

To measure the θ_Z rotary displacement, the mirror was initially located at a distance more than 20 mm from the middle of the stage to the right side, as shown in Fig. 16. A voltage waveform was then applied, and the corresponding displacements measured by the laser interferometer. The mirror was then relocated to the symmetric position of the left side of the stage, and the experimental procedure was repeated. Hence, the maximum θ_Z rotary displacement measurement was obtained for travel in θ_Z directions. Fig. 17 and 18 reveal that the θ_Z rotary displacement is 265 arcsec, and that the θ_Z rotary angular error is ±10 arcsec.

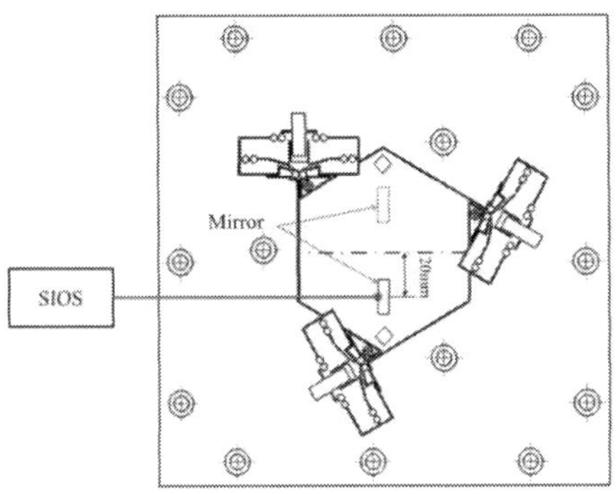

Fig. 16. Experimental arrangement for θ_Z rotary displacement.

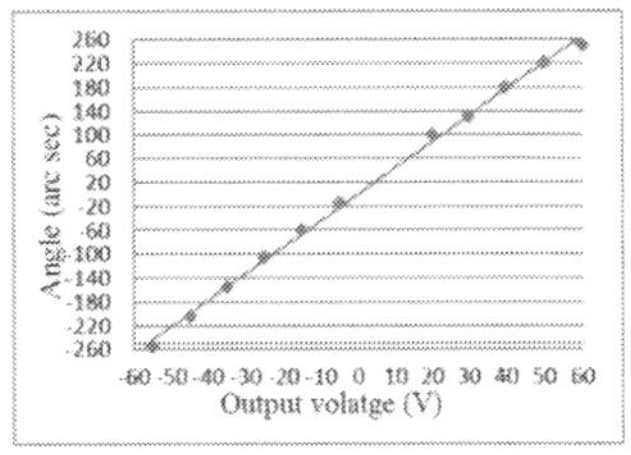

Fig. 17. θ_Z rotary displacement.

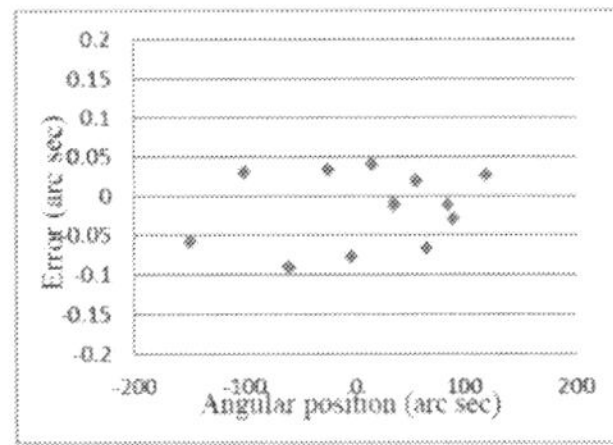

Fig. 18. θ_Z rotary angular error.

4.2 Frequency analysis

For a precision positioning device, a wide dynamic operating range is needed. Fig. 19 presents a schematic illustration of the experimental arrangement employed to analyse the frequency response of the proposed stage. An FTT Analyser (B&K, model 3560C) provided a sweep sinusoidal signal, which was amplified by the PI Instrument controller and then used to drive the piezoelectric actuator. An accelerometer was attached to the stage. Hence, providing the FFT Analyser with both the input signal of the piezoelectric actuator and the output signal of the accelerometer enabled the first natural frequency to be determined. As shown in Fig. 20, the first natural frequency is of the order of 349.8Hz, which is similar to the result obtained from the ANSYS analysis.

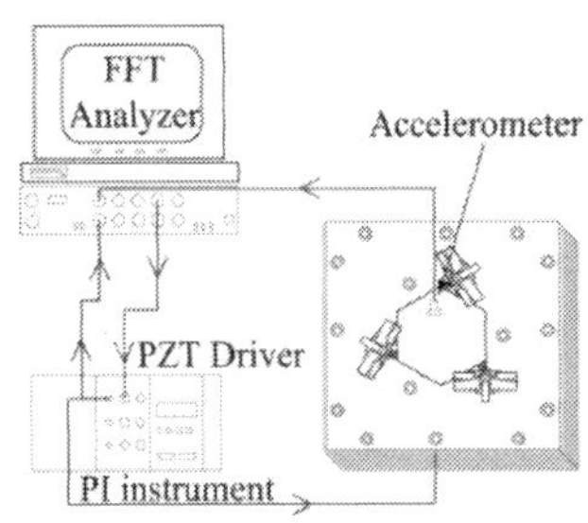

Fig. 19. Experimental arrangement for frequency analysis.

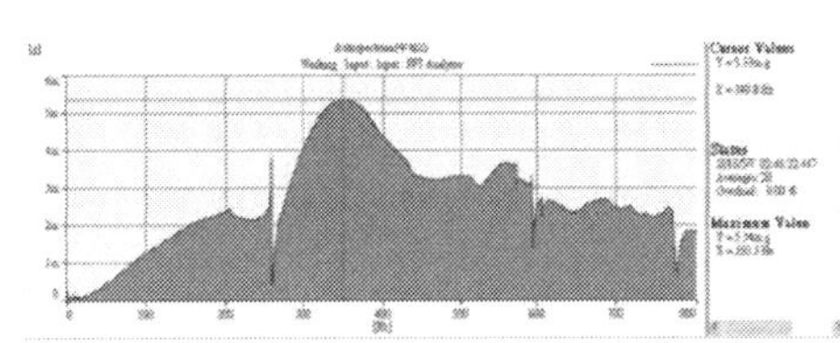

Fig. 20. The measured frequency response function of the nanopositioning stage.

5 Conclusion

This paper has presented the design and performance evaluation of a $XY\theta_z$ 3-DOF nanopositioning stage. The nanopositioning stage uses the minimum number of flexure hinges possible in order to reduce the angular elastic deformation. A layer mechanism is employed to provide the first stage of amplification, while the second amplification stage is provided by a toggle amplification mechanism, which eliminates the lateral offset effect of the preceding layer amplification.

The present experimental results have indicated that the θ_Z rotary displacement is 265 arcsec, that the θ_Z rotary angular error is less than ±10 arcsec, that the XY-axis liner displacement is approximately 36 μm, and that the XY-axis straightness is less than 0.3μm. Therefore, it has been demonstrated that the proposed $XY\theta_z$ 3-DOF nanopositioning stage.

Acknowledgement

The authors gratefully acknowledge the support to this project provided by the Ministry of Science and Technology, Taiwan, ROC under contract No. MOST 105-2622-E-218-009 -CC3 is gratefully acknowledged.

References

1. R. Yang, M. Jouaneh, R. Schweizer, Design and characterization of a low-profile micropositioning stage, *Precision Engineering*, **18**, 20-29 (1996)
2. F. E. Scire, E. C. Teague, Piezodrive 50-μm range stage with subnanometer resolution, *Rev. Sci. Instrum*, **42**, 1735-1740 (1978)
3. P. Gao, S. M. Swei, Z. Yuan, A new piezodriven precision micropositioning stage utilizing flexure hinge, *Nanotechnology*, **10**, 394-398 (1999)
4. S. H. Chang, C.K. Tseng, H. C. Chien, An Ultra-precision $XY\theta_Z$ Piezo-micropositioner-Part I: Design and Analysis, *IEEE Trans. Ultrasonics, Ferroelectrics, and Frequency Control*, **46**, 897-905 (1999)
5. S. H. Chang, C. K. Tseng, H. C. Chien, An Ultra-precision $XY\theta_Z$ Piezo-micropositioner-Part II: Experiment and Performance, *IEEE Trans. Ultrasonics, Ferroelectrics, and Frequency Control*, **46**, 897-905 (1999)
6. S. Yuki, P. Yuxin, K. Junji, A. Toyohiro, I. So, G. Wei, F.L. Tien, Design and construction of the motion mechanism of an XY micro-stage for precision positioning, *Sensors and Actuators A: Physical*, **201**, 395-406 (2013)
7. H. Tang, Y. Li, Design, analysis, and test of a novel 2-DOF nanopositioning system driven by dual mode, *IEEE Trans Robot*, **29**, 650-662 (2013)
8. L. Jianping, Z. Hongwei, Q. Xingtian, X. Z. Han Qu, F. Z. M. Zunqiang, F. Haishuang, Development of a compact 2-DOF precision piezoelectric positioning platform based on inchworm principle, *Sensors and Actuators A: Physical*, **222**, 87-95 (2015)
9. C. Leon, S. Bijan, B. Umesh, S. Julian, Z. Yongmin, Development and control of a two DOF linear–angular precision positioning stage, *Mechatronics*, **32**, 34-43 (2015)
10. Products for nanopositioning, Product Catalog of Physik Instrumente.
11. W. Xu, T. King, Flexure hinges for piezoactuator displacement amplifiers: flexibility, accuracy, and stress considerations, *Precision Engineering*, **19**, 4-10 (1996)
12. S. H. Chang, B. C. Du, A precision piezodriven micropositioner mechanism with large travel range, *Rev. Sci. Instrum*, **69**, 1785-1791 (1998)
13. J. W. Ryu, D. G. Gweon, K. S. Moon, Optimal design of a flexure hinge based $XY\theta_z$ wafer stage, *Precision Engineering*, **21**, 18-28 (1997)
14. C. L. Chu, S. H. Fan, A novel long-travel piezoelectric-driven linear nanopositioning stage, *Precision Engineering: The Journal of the International Societies for Precision Engineering and Nanotechnology*, **30**, 85-95 (2006)
15. Piezoelectrical and electrostrictive stack actuators and ring actuators, Product Catalog of Piezomechanik

Effects of elastic ball and coating on pure squeeze EHL motion for constant load with couple stresses using FDM method

Li-Ming Chu[1], *Yuh-Ping* Chang[2,*], and *Jaw-Ren* Lin[3]

[1]Department of Applied Science/Interdisciplinary Program of Green and Information Technology, National Taitung University, Taitung City 95092, Taiwan, R.O.C.
[2]Department of Mechanical Engineering, Kun Shan University, Tainan City 71070, Taiwan, R.O.C.
[3]Department of Mechanical Engineering, Nanya Institute of Technology, Taoyuan City 32091, Taiwan, R.O.C.

Abstract. In this paper, an elastic sphere approaching a lubricated elastic coating surface with couple stress lubricant is explored under constant load condition. The finite difference method (FDM), Gaussian elimination, and Gauss-Seidel iteration method are used to solve the transient modified Reynolds equation with its boundary condition, the elasticity deformation equations of ball and coating layer, load balance equation, and lubricant rheology equations simultaneously. The transient pressure profiles, film shapes, and elastic deformation during the pure squeeze process under various operating conditions in the elastohydrodynamic lubrication (EHL) regime are discussed.

1 Introduction

Many mechanical elements with contact pairs such as gear teeth, cams /followers, piston ring/cylinder, rolling element bearings, and the stretching process of metal sheets occurred elastic dimple at the center of the contact region due to the squeeze effects, reveal the elastohydrodynamic lubrication (EHL) problem. Surface coatings have long been used in industrial technologies, as they can reduce friction and wear, and thus improve the service life of mechanical components. Therefore, the characteristics of an isotropic elastic coating/elastic substrate with non-Newtonian lubricants in the EHL region need to be further investigated.

To solve the formation of the dimple in pure squeeze motion problems, many numerical solutions have been proposed, including those by Yang and Wen [1], and Chang [2]. In these studies, the ball dropping case is often used since it includes all the effects of pure squeeze motion which are of interest. Dowson and Wang [3] and Larsson and Högund [4] analyzed the bouncing of an elastic sphere on an oily plate. These analyses were restricted to normal motion in the first instance in order to develop the numerical technique and to relate the overall findings to the results presented by Safa and Gohar [5].

To solve the coupled hydrodynamic equation and elasticity equation of an EHL problem, the elasticity modulus and thickness of the coating are important parameters. Many studies have treated the coating and contact surfaces as linear elastic isotropic materials. Elsharkawy and Hamrock [6] used the deformation model developed by Johnson [7] to explore the Newtonian EHL of an elastic coated surface on a rigid cylinder and rigid substrate. Jaffar [8] derived a new set of explicit expressions for the contact pressure, total load, and penetration depth for the frictionless indentation problem of a spherical punch and a bounded thick elastic layer. Liu et al. [9] developed a coating EHL model for point contacts by combining the DC-FFT algorithm for the elastic deformation of a coated surface with the unified mixed EHL model. Habchi [10] developed a full system approach to analyze a thermal EHL problem in coated circular contacts with an equivalent cube under steady state condition. Chu et al. [11] analysed a rigid sphere approaching a lubricated flat surface with a layer of elastic coating on the elastic substrate is explored under constant load conditions. But, study of EHL in pure squeeze motion on coated surface with couple stress lubricant is lacking.

In this paper, pure squeeze EHL motion of circular contacts with coating is explored under constant load condition. The finite difference method and the Gauss-Seidel iteration method are used to solve the transient modified Reynolds equation, the elasticity deformation equation, load balance equation, and lubricant rheology equations simultaneously. The transient pressure profiles and film shapes during the pure squeeze process under various operating conditions in the EHL regime are discussed.

2 Theoretical analysis

2.1 Reynolds equation

* Corresponding author: ypc0318@mail.ksu.edu.tw

According to the Stokes micro continuum theory [12], the field equations of a coupled stress fluid in the absence of body forces and body couples are

$$\nabla \cdot \mathbf{V} = 0 \tag{1}$$

$$\rho D\mathbf{V} / Dt = -\nabla p + (\mu - \eta \nabla^2)\nabla^2 \mathbf{V} \tag{2}$$

where $\mathbf{V}$ is the velocity vector, ρ is the density, p is the pressure, μ is the classical viscosity coefficient, and η is a new material constant with the dimension of momentum responsible for the couple stress fluid property. Since the ratio η / μ has the dimensions of length squared, the dimension of $l = (\eta / \mu)^{1/2}$ characterizes the material length of couple stress fluids, and l is assumed to be a material constant in the present analysis.

For EHL problems, two spheres approach each other can be expressed as an equivalent sphere approaching a plane. In Figure 1, an elastic sphere of radius R is approaching an infinite rigid plate with a velocity under a constant load. The lubricant in the system is taken to be a compressible fluid. Under the usual assumption of EHL applicable to a thin film, the reduced momentum equations and the continuity equation governing the motion of the lubricant in polar coordinates can be obtained. Integrating the reduced momentum equations with the no-slip-boundary conditions, the velocity components are then obtained. Substituting velocity components into continuity equation and integrating across the film thickness with the boundary conditions of $v(r,z)$ *i.e.* $v(r,h)=\partial h/\partial t$, we can then derive the transient Reynolds equation in polar coordinates as:

$$\frac{\partial}{\partial r}\left(\frac{\rho r h^3}{\mu} f(l,h)\frac{\partial p}{\partial r}\right) = 12r\frac{\partial}{\partial t}(\rho h) \tag{3}$$

where

$$f(l,h) = 1 - 12(\frac{l}{h})^2 + 24(\frac{l}{h})^3 \tanh(\frac{h}{2l}) \tag{4}$$

or in dimensionless form as:

$$\frac{\partial}{\partial X}\left(\varepsilon X F \frac{\partial P}{\partial X}\right) = KX\frac{\partial}{\partial T}(\bar{\rho} H) \tag{5}$$

where $\varepsilon = \bar{\rho} H^3 / \bar{\mu}$, $K = 8\pi / W$

$$F(L,H) = 1 - 12(\frac{L}{H})^2 + 24(\frac{L}{H})^3 \tanh(\frac{H}{2L}) \tag{6}$$

The radial coordinate, X, is with its origin in the center of the contact. The boundary conditions for Eq. (5) are:

$$P(X \to \infty, T) = 0 \tag{7a}$$

$$\partial P(0,T) / \partial X = 0 \tag{7b}$$

$$P(R,T) \geq 0 \tag{7c}$$

2.2 Rheology equations

As the pressure increases with time, the elastic deformation and the effect of pressure on the viscosity and density cannot be neglected. This stage is denoted as the high-pressure stage. The pressure dependence of viscosity and density is an important issue of the present problems.

The viscosity of the lubricant is assumed to be the function of pressure only. The relationship between viscosity and pressure used by Roelands [13] can be expressed as:

$$\bar{\mu} = \exp\{(9.67 + \ln\eta_0)[-1 + (1 + 5.1\times 10^{-9} p)^{z'}]\} \tag{8}$$

where η_0 is the viscosity under ambient pressure, and z' is the pressure-viscosity index. According to Dowson and Higginson [14], the relationship between density and pressure is given as:

$$\bar{\rho} = \frac{\rho}{\rho_0} = 1 + \frac{0.6\times 10^{-9} p}{1 + 1.7\times 10^{-9} p} \tag{9}$$

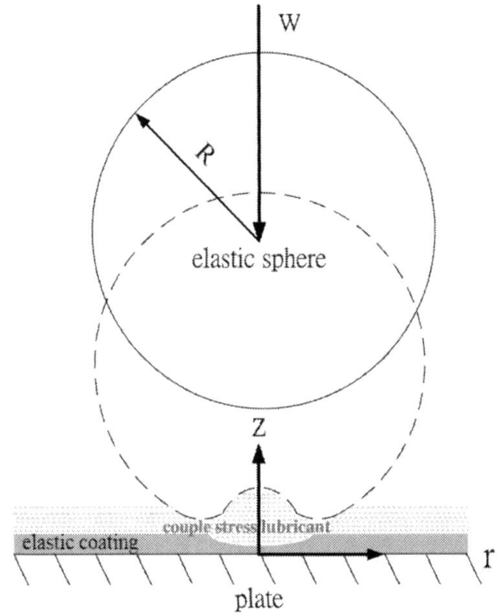

Fig. 1. Geometry of EHL under pure squeeze motion.

2.3 Elasticity equation

The film thickness within the EH conjunction can be written as:

$$h(r,t) = h_0(t) + r^2 / 2R + \delta(r,t) + \zeta(r,t) \tag{10}$$

To calculate the static deformation of elastic sphere caused by pressure distribution, influence coefficients D_{ij} are introduced. The deformation can thus be computed at discrete points i as a sum of the deformation contributions from all pressure points j :

$$\delta_i = \sum_{j=1}^{n} D_{ij} P_j \tag{11}$$

Within the frame of linear elasticity, the normal deformation of coating is given by [8]

$$\zeta(r,t)=\frac{2\alpha}{d}\int_0^a\int_0^\infty[L(\omega)J_0(\frac{r\omega}{d})J_0(\frac{s\omega}{d})d\omega]p(s)sds \quad (12)$$

Where

$L(\omega)=(2\kappa\sinh 2\omega-4\omega)/(2\kappa\cosh 2\omega+4\omega^2+k^2+1)$, $\alpha=(1-\nu^2)/E$, and $\kappa=3-4\nu$. J_0 is the Bessel function of the first kind of order zero and ν is Poisson's ratio.

The dimensionless film thickness between two elastic bodies with coating in circular contacts can be expressed as:

$$H_i=H_0+\frac{X_i^2}{2}+\sum_{j=1}^{n}D_{ij}P_j+\bar{\zeta}_i \quad (13)$$

where the influence coefficients, D_{ij} , are computed according to Yang and Wen [1] and Larsson and Höglund [4].

2.4 Load balance equation

The instantaneous load balance equation for a constant load condition is:

$$\int_0^\infty PXdX=1/3 \quad (14)$$

3 Numerical solution

The finite difference method (FDM), Gaussian elimination, and the Gauss-Seidel iteration method are used to solve the transient modified Reynolds equation with its boundary condition, the elasticity deformation equations of ball and coating layer, load balance equation, and lubricant rheology equations simultaneously. Equation (5) has to be discretized and solved at a finite number of locations along the radial coordinate axis X. The discretized form of the Equation (5) can be derived as:

$$(A_i+B_i)P_{i+1}-2B_iP_i+(B_i-A_i)P_{i-1}=$$
$$KX_i(\bar{\rho}\frac{H-H^{k-1}}{\Delta T}+H\frac{\bar{\rho}-\bar{\rho}^{k-1}}{\Delta T})_i \quad (15)$$

where

$A_i=(X\,\partial\varepsilon F/\partial X+\varepsilon F)_i/(2\Delta X)$, $B_i=(X\varepsilon F/\Delta X^2)_i$, ΔX is the mesh size, i. e., $\Delta X=X_i-X_{i-1}$, the superscript k, denoting discretization in time, ΔT is the time increment, i. e., $\Delta T=T^k-T^{k-1}$. The boundary condition (7b) at X=0 can be derived by fitting a second degree polynomial as:

$$3P_1-4P_2+P_3=0.0 \quad (16)$$

Apply boundary condition (7c) to the first three grid points at X=0. For the constant load case, the rigid separation is an unknown variable in each time step. It can be determined by solving the transient Reynolds equation with the load balance equation.

4 Results and Discussions

To discuss the effects of flow rheology and elastic deformation (δ) of spheres and coating (ζ) on the squeezing motion, the point contact EHL problems are discussed under the conditions of non-isoviscous, compressible lubricant, and constant load. Numerical solutions of film profiles (h) and pressure distributions (p) in pure squeeze motion are calculated using the parameters listed in Table 1. The initial falling height of the sphere is $20\mu m$. A typical problem with coating thickness, $\zeta=2a$, $W=5.24\times10^{-9}$, $G=3500$, E_b=220 GPa, $\nu_b=0.3$, and $\nu_c=0.3$ is solved.

Table 1. Computational data.

Inlet viscosity of lubricant, Pa-s	0.04
Inlet density of lubricant, kg/m^3	846
Pressure viscosity coefficient (α), 1/GPa	15.91
Pressure-viscosity index (Roelands)	0.4836
Equivalent radius of elastic ball, m	0.02
Elastic modulus of coating, GPa	110~330
Elastic modulus of balls, GPa	220
Poisson's ratio of ball	0.3
Poisson's ratio of coating	0.3
Density of balls, kg/m^3	7850

The upper limit of the computational region in the beginning is chosen as X_{max}=16.0. The Reynolds equation is discretized by the central difference technique in the space domain and the explicit technique in the time domain. When more than half of the region was cavitated, the maximum analyzed region X_{max} reduces to half of its initial region, and so on, until X_{max}=2.0. The grid was made up of 401 nodes, which are evenly distributed, in every calculating domain. The Gauss-Seidel iteration method is employed to calculate the film thickness and pressure distribution at each time step.

In the case of constant load conditions, Figure 2 shows the relative change in the p and h for a flexible sphere approaching a lubricated coating surface with couple stress lubricant. It is observed from Figure 2 that the p is quite flat at relatively large h, but it becomes steeper with decreasing h. It was found that the p_c gradually increases with decreasing h_c from 2.6ms to 15.7ms when the h_c decreases to 33nm. As time goes on, the h will decrease, and the contact region will reach the Hertz contact condition.

The p of couple stress lubricant is greater than that of Newtonian lubricant at central region. As the loading is constant, the integration of the pressure distribution over the loading area is a constant. Therefore, the p found reverse at outside central region. As seen in Figure 2, the h decreases gradually as the squeeze proceeds. The h

of couple stress lubricant is greater than that of Newtonian.

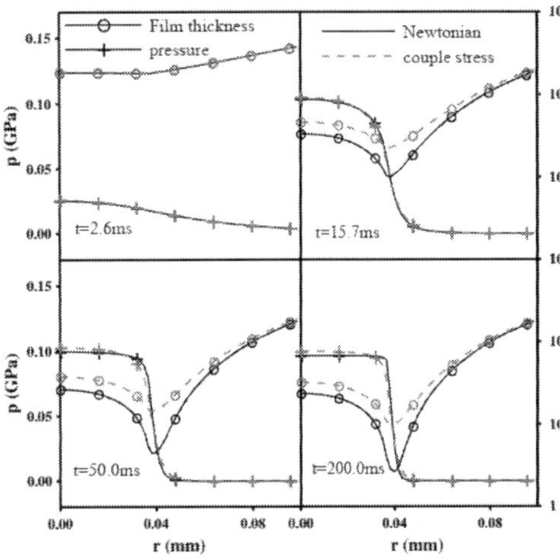

Fig. 2. p and h distribution versus time using different models

Figure 3 shows the central pressure and film thickness versus time under constant load condition. The p_c increases rapidly with time at the initial stage. Meanwhile, the p_c increases quickly to a maximum. Then the p_c decreases slowly to near the amplitude of the well-known Hertzian pressure with time at the final stage. This stage can be considered as the quasi-static condition. This stage keeps longer time. The h_c decreases rapidly with time at the initial stage. Then, they decrease slowly with time. The p_c with coating is smaller than that without coating. The h_c with coating is greater than that without coating. The h_{min} with coating is greater than that without coating at the initial stage. The h_{min} with coating is smaller than that without coating from 14ms to 60ms. The h_{min} with coating is greater than that without coating at the final stage.

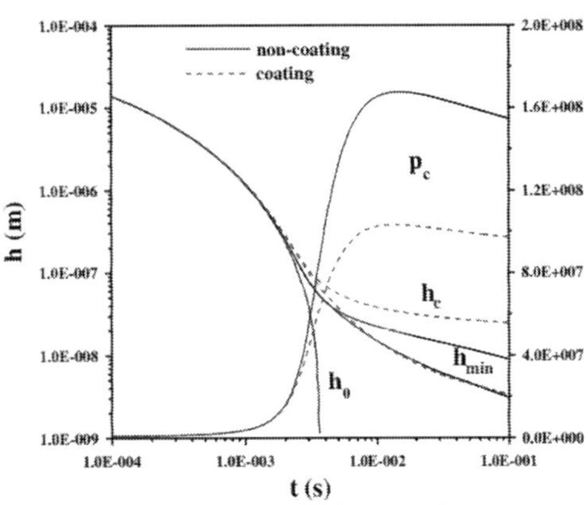

Fig. 3. p_c and h versus time with/without coating.

Figure 4 shows the central pressure and film thickness versus time under constant load condition for different *E* and ζ. It is observed from Figure 4 that the greater the *E*, the greater the p_c, and the smaller the h_c. The greater the ζ, the smaller the p_c, and the greater the h_c.

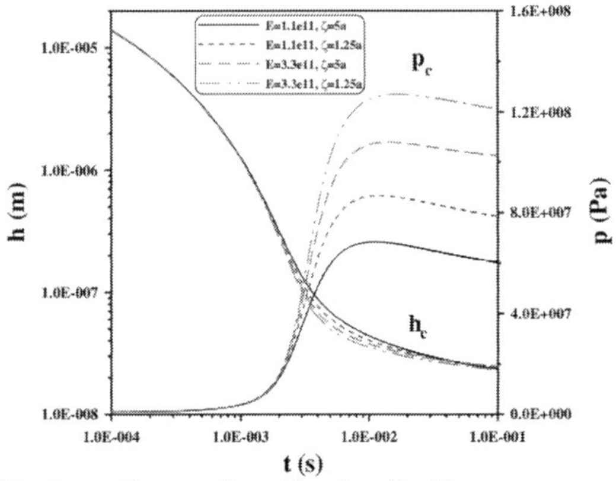

Fig. 4. p_c and h_c versus time with various *E* and ζ.

Figure 5 shows the *p* and *h* for different *E* in *r*-direction under constant load condition at t=0.1s. At central region, the greater the *E*, the greater the *p*, and the smaller the *h*. The film thicknesses and pressures are is found reverse at outside central region. The smaller the *E*, the greater the positions of minimum film thickness.

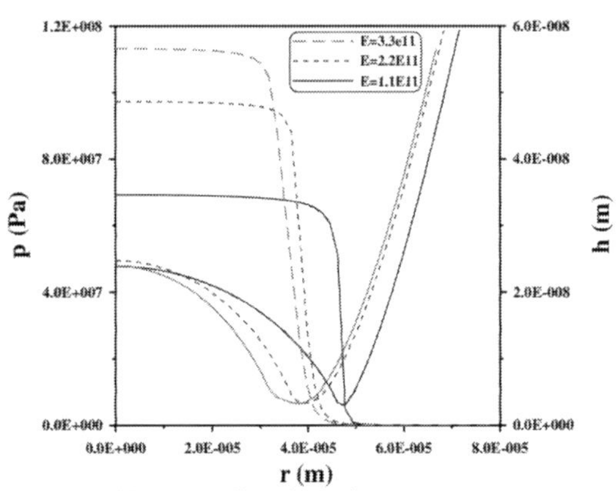

Fig. 5. p and h versus radius with various E.

Figure 6 shows the *p* and *h* for different ζ in *r*-direction under constant load condition at t=0.1s. At central region, the greater the ζ, the smaller the *p*, and the greater the *h*. The film thicknesses and pressures are found reverse at outside central region. The greater the ζ, the greater the positions of minimum film thickness.

Figure 7 shows the relationship of the central normal squeeze velocity (v_c) and the h_c using present models under constant load condition. The v_c decreased rapidly with decreasing h_c at the initial stage. The greater the *E*, the greater the v_c. The greater the ζ, the smaller the v_c.

Figure 8 shows the central pressure and film thickness versus time under constant load condition for different *l*. It is observed from Figure 8 that the greater the *l*, the greater the p_c, and the greater the h_c and h_{min}.

The time needed to achieve maximum p_c increases with increasing l. Similarly, the time needed to achieve the Hertzian pressure also increases with increasing l.

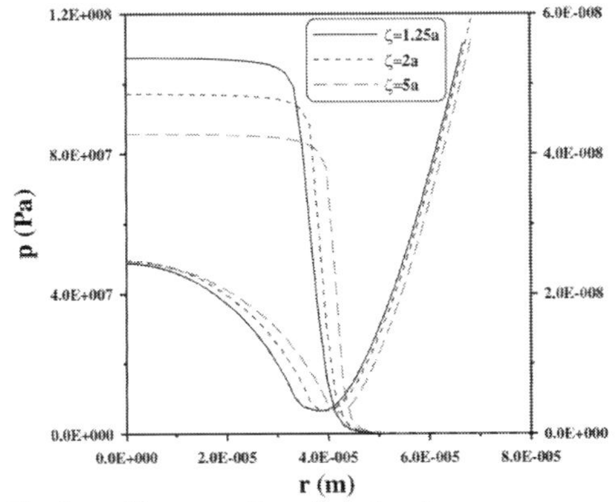

Fig. 6. p and h versus radius with various ζ.

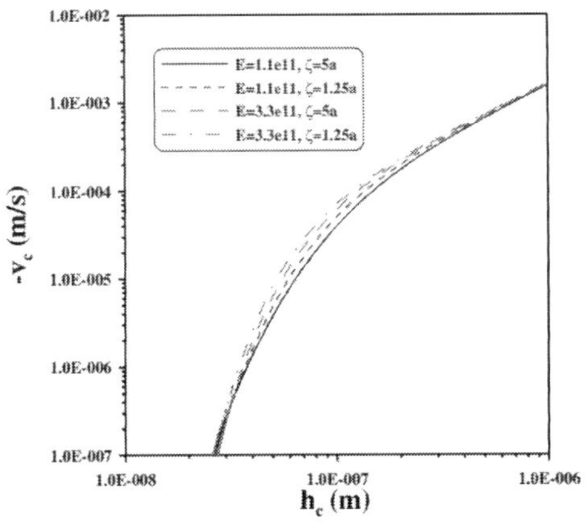

Fig. 7. Variation of v_c versus h_c various E and ζ.

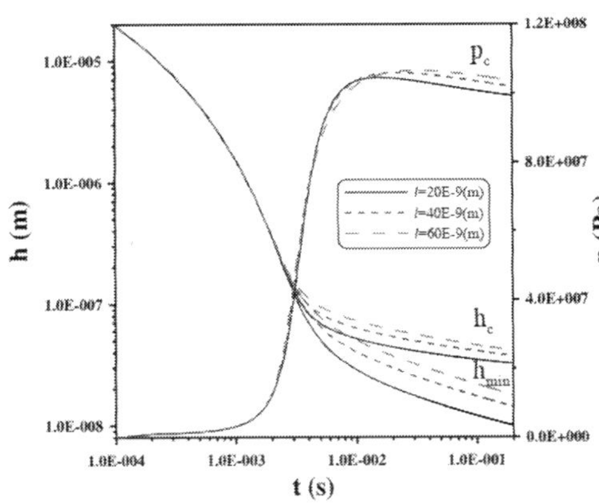

Fig. 8. p_c and h versus time with various l.

Figure 9 shows the p and h for different l in r-direction under constant load condition at t=0.1s. At central region, the greater the l, the greater the p, and the greater the h. The pressures are found reverse at outside central region.

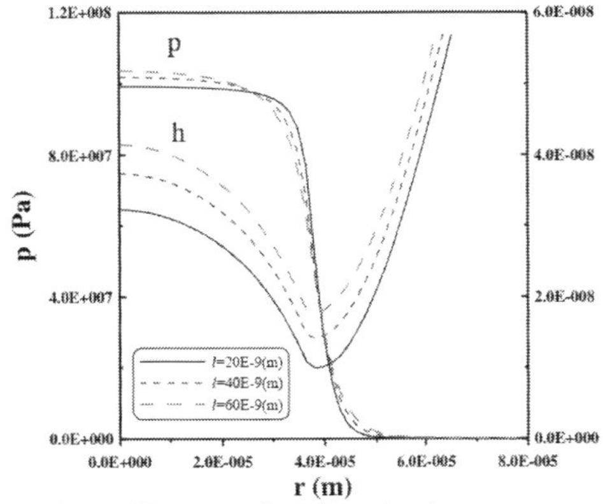

Fig. 9. p and h versus radius with various l.

5 Conclusions

In this study, a numerical method for general applications was developed to investigate the effects of the isotropic coating at pure squeeze EHL motion of circular contacts under constant load conditions using FDM. The conclusions from the main results can be summarized as follows:

1. At central region, the greater the E, the greater the p, and the smaller the h. The h and p is found reverse at outside central region.
2. At central region, the greater the ζ, the smaller the p, and the greater the h. The h and p is found reverse at outside central region.
3. The greater the l, the greater the h. At central region, the greater the l, the greater the p, the p is found reverse at outside central region.
4. The positions of minimum film thickness for smaller elastic modulus are greater than that for larger elastic modulus. The positions of minimum film thickness for greater coating thickness are greater than that for smaller coating thickness.
5. The greater the E, the greater the v_c. The greater the ζ, the smaller the v_c.

Acknowledgments

The authors would like to express their appreciation to the Ministry of Science and Technology (MOST 105-2221-E-143 -005) in Taiwan for financial support.

References

1. P. R. Yang, and S. Z. Wen, *Wear*, **142**(1), 17 (1991)
2. L. Chang, *ASME J. Applied Mech.*, **63**(2),347 (1996)
3. D. Dowson and D. Wang, *Wear*, **179**(1-2),29 (1994)
4. R. Larsson and E. Högund, *ASME J. Tribol.*, **117**(1), 94 (1995)

DOI:10.1051/matecconf/201712300007

5. M. M. A. Safa and R. Gohar, *ASME J. Tribol.*, **108**(1), 372 (1986)
6. A. Elsharkawy and B. Hamrock, *Proc. Inst. Mech. Eng. Part J. J. Eng. Tribol.*, **209**(2), 119 (1995)
7. K. L. Johnson, Contact Mechanics, Cambridge university press, Cambridge, UK, (1987)
8. M. J. Jaffar, Proc. *Inst. Mech. Eng. Part J. J. Eng. Tribol.*, **222** (1), 61 (2008)
9. Y. Liu, W. W. Chen, D. Zhu, S. Liu, and Q. J. Wang, *J. Tribol.-Trans. ASME*, **129**(3), 509 (2007)
10. W. Habchi, *Tribology International*, **73**, 57 (2014)
11. L. M. Chu, C. C. Yu, Q. D. Chen, W. L. Li, *Journal of Tribology*, ASME, **137** (1), 011503-1 (2015)
12. V. K. Stokes, Couple stresses in fluids, Phys. Fluids, **9**, 1709 (1966)
13. D. Dowson and G. R. Higginson, Elastohydrodynamic Lubrication, Pergamon, Oxford, UK, 88 (1966)
14. C. J. A. Roelands, J. C. Vlugter, and H. I. Watermann, *ASME J. Basic Eng.*, **85**(4), 601 (1963)

Simulation analysis of turbine blade in 3D printing aquarium

Dyi-Cheng Chen[1,*], *Bo-Yan* Lai[2], and *Fu-Yuan* Gao[3]

[1] Department of Industrial Education and Technology, National Changhua University of Education, Changhua 500, Taiwan
[2] Department of Industrial Education and Technology, National Changhua University of Education, Changhua 500, Taiwan
[3] Department of Industrial Education and Technology, National Changhua University of Education, Changhua 500, Taiwan

Abstract. 3D printing of the flexibility is the most admirable place, no matter when or where, as long as the machine can make the abstract design of finished products or difficult to process the finished product printed out as a sample. And in the product design, through the 3D print out the entity, to more specific observation of the advantages and disadvantages of finished products, which shorten the time of many creative research and development, but also relatively reduce the defective factors. As in recent years, 3D printing technology is progressing, material adhesion, precision and parts of the degree of cooperation has increased, coupled with many parts taking into account the cost, production and other issues, and then let a lot of light load small parts or special parts choose to use 3D to print the finished product to replace. This study focuses on the plastic turbine blades that drive water in the aquarium, but the 3D printing is done by stacking. However, the general stress analysis software can set the material to analyze the deformation results of the force, nor the 3D to analyze the software. Therefore, this study first analyzes the deformation of turbine blade by software, and then verifies the situation of 3D printing turbine blade, and then compares the actual results of software analysis and 3D printing. The results can provide the future of 3D product consider the strength factor. The study found that the spiral blade design allows the force points to be dispersed to avoid hard focus.

1 Introduction

In recent years, in the literature related research, Yuen [1] proposed a modelling method based on tectonics for the characteristics of blade design iteration, such as leaf design parameters, FDM production procedures, and to explore its use of PLA plastic material strength trends. Lee et al [2] fabricated helical vanes with laminated manufacturing techniques to explore their design parameters, such as leaf length, number of leaves, leaf turns, pitch height and other conditions, for the mixing efficiency of viscous liquids, to understand the feasibility of the laminated manufacturing technology spiral blades. Because of its simulation analysis software can set the material, set the pressure to simulate the material subjected to stress and strain results, because 3D printing is stacked, and when the stress is concentrated and parallel with the material stacking direction, even if the material strength. It is easy to break from the material stack. Therefore, this study will explore the use of the water tank in the 3D printing turbine blade its thickness to be able to withstand and avoid fracture.

1.1 Laminated manufacturing technology (3D printing)

Since the industry 4.0 reform, 3D printing technology has become a technology star in recent years, it is also the trend of cutting-edge manufacturing industry, due to 3D printing technology production process to escape the traditional framework. The idea can be achieved in short period of time. 3D printing technology will be 3D modeling parts into 2D layered slices, according to the layered slice of the material stacked to the molding position, repeated stacking action caused by the workpiece molding, different from the traditional manufacturing model to cut the workpiece material subtraction way, often cause unnecessary cost consumption. 3D printing technology using layers of stacking method of production, it can avoid unnecessary waste of material [1].

In the past, the product is in the development stage, subject to the design, open mold, trial production, testing and other processes, the time spent and human non-ordinary small and medium enterprises can afford, and the use of 3D printing technology for sample design and mold manufacturing. High complexity and difficult parts can be produced in a short time, in order to facilitate the designer and the user to modify and communicate, is conducive to product development and custom production [3].

* Corresponding author: dcchen@cc.ncue.edu.tw

3D print parts can be designed through complex control of the complex geometry of the geometry, the production of integrated spiral contact, translation joints and other confidential institutions, and confirm the parts gap, but in mechanical terms can still not achieve the required strength [4]. But the surface roughness of complex geometric parts is not easy to control and is not conducive to subsequent processing, so 3D printing technology in the manufacturing industry is generally used for the design of prototype, sample proofing. With 3D scanning technology developed, through the 3D scan for measurement and inspection, and numerical analysis of parts, can predict the type change and improve the error, and then control the deformation data and provide design parameters [5]. The 3D printing material is from the traditional engineering plastics (ABS, PLA), gradually evolved into composite metal powder material (stainless steel, titanium alloy), so that 3D printing parts gradually toward high strength, high density and both heat-resistant wear-resistant sintered alloy steel and other aspects of development, in response to the characteristics of materials and diversification. In recent years widely used in national defense, aerospace, medical, steam locomotives, electrical and electronic industry, 3D printing technology has gradually become an important high-tech industries [6].

1.2 Turbine blades

Aquarium water quality control is an important work of farming, whether the quality of water or not affect the bait status of fish [7]. The general public aquarium for the adjustment of water quality will use the submersible motor, put it in the tank can create similar to the role of circulating water to control the oxygen content, temperature, turbidity, etc., and make the vibration effect of inhibition of algae in the waters of the breeding, to keep the water flow within the aquarium smooth [8]. And the most important part of the process of water circulation is the turbine blade. According to the discussion of the turbine blade force under the water, the analysis of the turbine blade strength indicates that the width of the turbine blade affects the force area of the turbine blade. The pressure is equal to the pressure multiplied by the pressure area, and the magnitude of the force applied to the turbine blade rotation is proportional to the pressure area of the turbine blade when the hydrostatic condition is constant [9].

The two-dimensional fluid pressure effects of the blade cross-sectional area of the turbine blades are related to the optimum performance achievable by the turbine blades. For the smooth rotation of the blades, the turbine blades are usually designed with existing airfoils [10]. When designing the thickness and chord length of the turbine blades, the structural strength of the turbine blades must be considered. In the preliminary stage of the turbine blade, the 3D computer model structure should be built quickly. Therefore, the rule of thumb is usually used according to the two- Based on the formula of fluid dynamics and logarithmic helix, the turbine blade blade size was constructed by using Inventor 2016 simulation software to establish the turbine blade geometry model and the engineering elasticity of the turbine blade. The parameters and strength parameters are as follows:

Table 1. ABS material engineering elastic parameters and strength parameters.

Name	Parameters
Material	ABS plastic
Number of model grids	10,000
Material density	1.06 g/cm^3
Yield strength	20 MPa
Ultimate tensile strength	29.6 MPa
Young's modulus	2.24 GPa
Poisson's ratio	0.38 ul
Shear modulus	0.811594 GPa

The force of the turbine blades is measured under the fixed output power of the submersible motor, and the calculated leaf load is used as the condition of the finite element analysis method to analyze the turbine blade. The required structural strength explored the 3D printing of the turbine blade parts and analog data differences to provide the aquarium turbine blade design reference.

1.3 The purpose of research

In order to make the feed flow from the internal flow of the aquarium, the feed is effectively spread out due to the flow. In this study, the turbine blades of the tank are made by 3D printing, and the turbine blades of different thickness are designed by the fluid mechanics calculation method, and the relationship between the stress and strain state of the force area and the blade thickness of the turbine blades in 3D operation is discussed.

2 Research methods

In this study, Inventor 2016 was used to design and draw the model, and the 3D blade turbine blades suitable for the aquarium were drawn with reference to the common turbine blade appearance. The same airfoil, stacking angle, blade number, water pressure and so on were designed parameters, only the same thickness of the turbine blades. The material was set to ABS material, and the stress and strain state of the turbine blade thickness at 2.5mm were simulated. Finally, the turbine blades with thickness of 2mm and 3mm were fabricated by 3D printing technology to explore the use of 3D printing technology. Whether the stress strain of the turbine blade can correspond to the turbine blade of the actual injection, and understand the effect of the stress strain value on the turbine blade produced by 3D printing technology. Fig. 1 and Fig. 2 show the three - dimensional model of the turbine blade at thickness 2mm.

Fig.1. Turbine blade model (2mm)(up).

Fig. 2. Turbine blade model (2mm)(down).

In this paper, the fluid calculation based on the Renault mean (RANS) equation viscous flow theory, that the uniform flow to maintain a certain speed around the propeller is a typical three-dimensional incompressible turbulent fluid. The RANS expression for the entire computational flow field is as follows:

Continuous equation Eq. (1):

$$\frac{\partial_{ui}}{\partial_{xi}} = 0 \tag{1}$$

Momentum equation Eq. (2):

$$\rho \frac{\partial(u_i u_j)}{\partial x_j} = \frac{\partial_p}{\partial x_j} + \frac{\partial}{\partial x_j} \left[\mu \left(\frac{\partial u_i}{\partial x_j} + \frac{\partial u_j}{\partial x_i} \right) \right] - \frac{\partial(\rho \overline{u'_i u'_j})}{\partial x_j} + \rho g_i \tag{2}$$

The fluid calculation domain is set to two parts of the fixed domain and the rotating domain to ensure the calculation in the normal state. The design of the 3D modeling geometry is set as stationary stationary objects such as water and aquarium body, and the turbine blades are set to rotation. The simulation set of turbine blade surface is defined as non-slippable wall, turbine blade placed in the aquarium depth and blade size according to the size of the aquarium. The simulation set the turbine blade thickness of 2.5mm, the simulation set material for the ABS Material.

3 Results and discussion

3.1 Effective stress and strain analysis of turbine blade - force 5N

Fig. 3 shows effective stress of turbine blades on 5N, and Fig. 4 shows effective strain of turbine blades on 5N. It is understood from the figure that the effective strain is evenly distributed at the leaf edge of the turbine blade with a minimum effective strain of 0.005 and a maximum effective strain of 5.97. The minimum effective stress is 0.0001 MPa and the maximum effective stress is 0.1436 MPa.

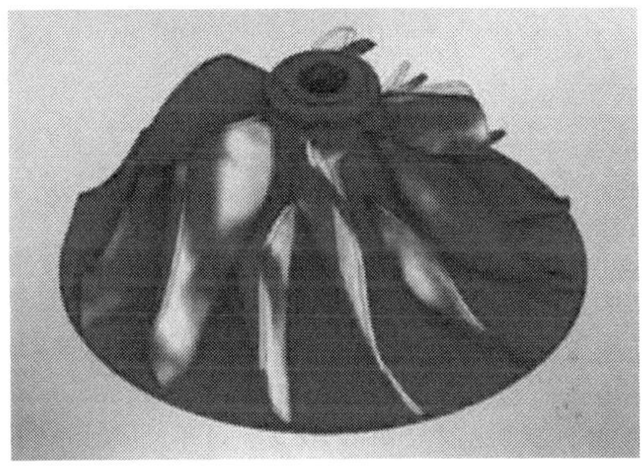

Fig. 3. Effective stress of turbine blades on 5N.

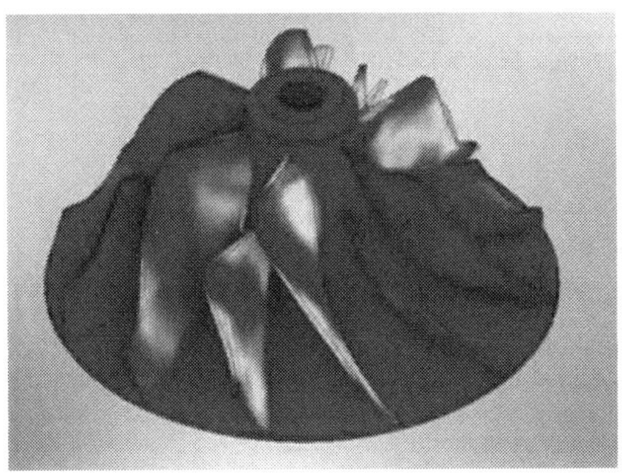

Fig. 4. Effective strain of turbine blades on 5N.

3.2 Effective stress and strain analysis of turbine blade - force 10N

Fig. 5 shows efective stress of turbine blades on 10N, and Fig. 6 shows effective strain of turbine blades on 10N. It is understood from the figure that the effective strain is evenly distributed at the leaf edge of the turbine blade with a minimum effective strain of 0.006 and a maximum effective strain of 7.17. The minimum effective stress is 0.0001 MPa and the maximum effective stress is 0.1725 MPa.

DOI:10.1051/matecconf/201712300008

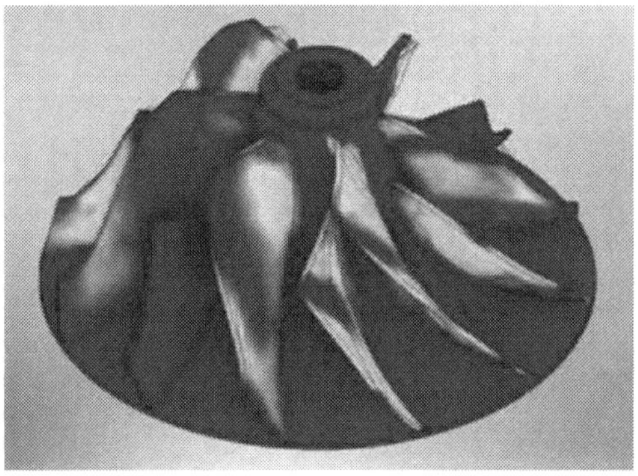

Fig. 5. Efective stress of turbine blades on 10N.

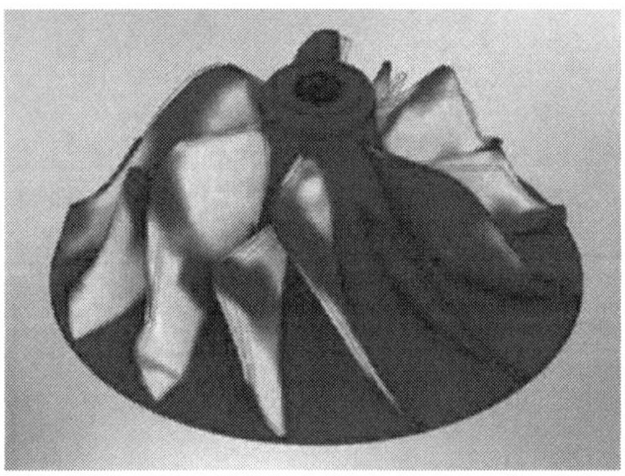

Fig. 6. Effective strain of turbine blades on 10N.

3.3 Effective stress and strain analysis of turbine blades - force 15N

Fig. 7 shows efective stress of turbine blades on 15N, and Fig. 8 shows effective strain of turbine blades on 15N. It is understood from the figure that the effective strain is evenly distributed at the leaf edge of the turbine blade with a minimum effective strain of 0.013 and a maximum effective strain of 10.66. Its effective stress quality minimum 0.0003MPa, the maximum effective stress value of 0.256MPa.

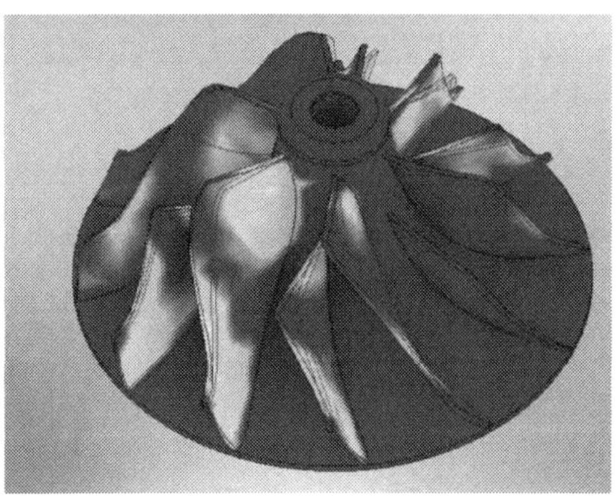

Fig.7 Effective stress of turbine blades on 15N.

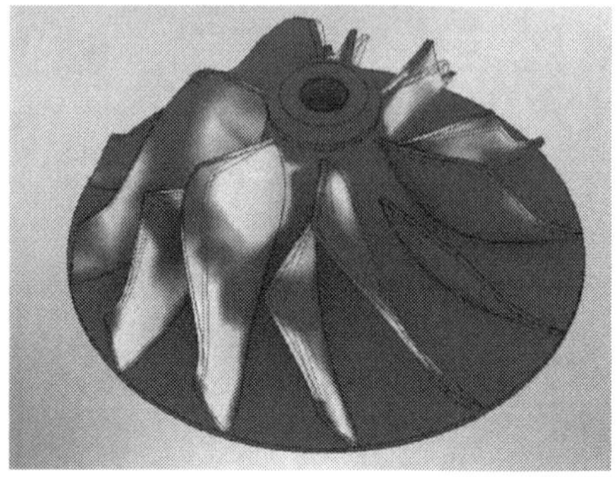

Fig.8. Effective strain of turbine blades on 15N

3.4 Effective stress and strain analysis of turbine blade - force 20N.

Fig. 9 shows efective stress of turbine blades on 20N, and Fig. 10 shows effective strain of turbine blades on 20N. It is understood from the figure that the effective strain is evenly distributed at the leaf edge of the turbine blade with a minimum effective strain of 0.096 and a maximum effective strain of 11.52. The minimum effective stress is 0.0004MPa, the maximum effective stress is 0.278MPa.

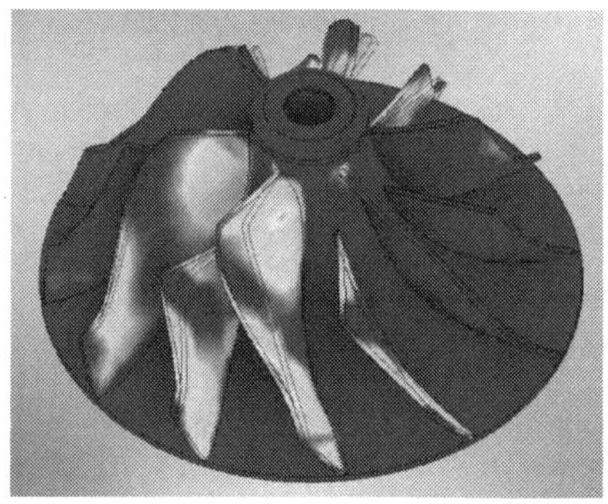

Fig. 9. Effective stress of turbine blades on 20N.

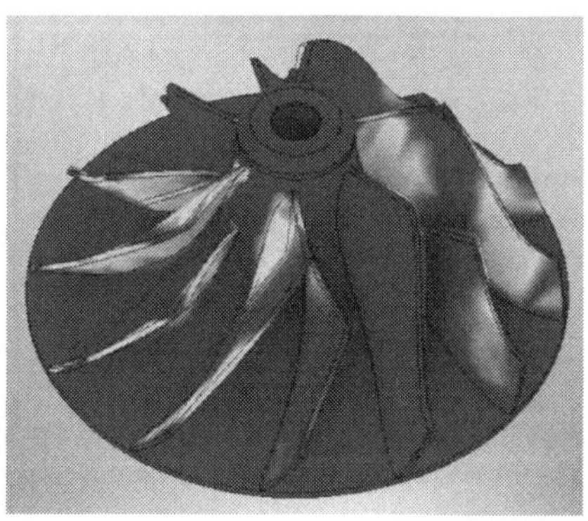

Fig.10. Effective strain of turbine blades on 20N.

4 Conclusions

In this study, the model of the turbine blade is actually made by 3D printing technology. It is found that the turbine blade with different blade thickness can be consistent with the simulation. Through the above analysis, it is found that the turbine blade with the thickness of 2mm is in the actual operation. After 10 hours, there was a serious damage to the tube wall. The turbine blades with a blade thickness of 3 mm were able to operate normally after 24 hours of operation. It indicated that the stress and strain energy of the turbine blades produced by 3D printing technology could be simulated with the simulated injection turbine blades correspond to their effective stress and effective strain. The turbine blades made by 3D printing technology produce breakage of material under the condition of blade thickness of 2mm. The turbine blades made by 3D printing technology can be used in the case of blade thickness of 3mm consistent with the simulation results, and in the actual operation of the process has a good strength.

References

1. P.K. Yuen, Embedding objects during 3D printing to add new functionalities. Biomicro-fluidics,**10**(4), 1-10 (2016)
2. W. Lee, S. Jeon, D. Kim, 3d-printed micromixer with helical blades for highviscosity fluids. Department of Chemical Engineering, [online] Available:http://www.rsc.org/images/LOC/2015/PDFs/Papers/1253_T.320e. (2015)
3. Y.S. Huang, Process and application of metal powder. Science Development Monthly, **523**,10-14 (2016)
4. Iñigo Flores Ituarte, Eric Coatanea, Mika Salmi, Jukka Tuomi, Jouni Partanen, Additive manufacturing in production: a study case applying technical requirements, Physics Procedia, **78**, 357-366 (2015)
5. T. Seno, Y. Ohtake, Y. Kikuchi, N. Saito, H. Suzuki, Y. Nagai, 3D scanning based mold correction for planar and cylindrical parts in aluminum die casting, Journal of Computational Design and Engineering, **2**(2), 96–104 (2015)
6. W. Lin, W. Liu, X. Liang, Development and prospecting application of metal powder process technology and equipment for curing, Taichung City: Metal Industry Research and Development Center, (2014)
7. M. Li, M. Hong, J. Chen, Q. You, X. Huang, Automatic feeding and water quality analysis of indigenous type indoor circulation aquaculture system, Journal of Agriculture and Forestry, **49** (4), 85-98 (2000)
8. G. Peng, How to create a viable ecological pool. Conservation of Nature, **35** (9), 6-10 (2001)
9. Y.R. Jiang, Yi Ru, Design and analysis of small wind turbine blade, Wind Power Monthly, 2(2006) From:http://www.twtia.org.tw/upload/111/2006661713205148
10. J. Tangler, J. David, Kocurek, Wind turbine post-stall airfoil performance characteristics guidelines for blade-element momentum methods, NREL/CP-500-36900 Oct. 2004

A histogram statistical method for the detection of localized faults in deep groove ball bearing

Shang-Chih Lin[*] and *Yennun* Huang

Research Center for Information Technology Innovation, Academia Sinica, 128 Academia Road, Section 2, Nankang, Taipei, Taiwan

Abstract. This study aims to use the histogram statistical method to establish a deep groove ball bearing fault diagnosis strategy. First, statistical indicators are used to excavate the fault characteristics buried in the vibration signal, and use the histogram to define the characteristic area for fault diagnosis. The results show that the indicators 1, 3, 6 have better statistical differences. Based on this, the accuracy of pattern recognition for all test data is 100 %. Finally, the statistical significance of ball damage was significant, and the results showed high correlation (56~73 %). The correlation between inner race damage model was 49~57 % and healthy model was 52 %. As the inner race damage and health model in the statistical sense, there are some similar, so there is a relatively high correlation. In the future research work, it will be committed to mining more representative indicators to enhance the relevance of abnormal characteristics.

1 Introduction

In the 21st century depends on human technological breakthroughs, gave birth to the history of the fourth industrial revolution. There will be a large number of mechanical equipment towards the development of automation, means that machinery must maintain long-term and uninterrupted operation, in order to maintain active production behavior. However, in the precision machining industry, if the machine can not effectively detect the abnormal situation, long-term will deepen the seriousness of mechanical failure, and finally lead to production line shut down, shortened equipment life and operator safety threats and other issues. In addition, the machine tool is one of the main core facilities of precision machinery, and its closely related industries such as aerospace, shipbuilding, automobile, factory, generator. Of course, there are many factors that affect the normal operation of the machine, such as materials, process, thermal deformation, cutting tools, rotating machinery, processing conditions.

In the course of research on rotating machines, Chang et al. [1-10] used in-depth analysis of rotating machinery, using different techniques [1], such as electrical detection, vibration detection and partial discharge detection. However, the characteristics of the vibration signal are analyzed by the axis trajectory technique, and the characteristic parameters are extracted by using the fractal theory. Finally, the neural network is used to identify the fault [2-3]. In addition, the use of electrical detection method [4], partial discharge detection method [5] to analyze the impact of rotating machinery failure on the severity of the operating state, and the use of partial discharge detection technology [6-7] diagnosis of high voltage motor stator coil insulation. Further mixing the different signals, such as electrical signals combined with vibration signals [8-9], electrical signals combined with partial discharge signals [10], to develop a more robust expert decision-making system.

However, when the rotating machinery gradually produce abnormal state, often the abnormal current, vibration, temperature, noise and other phenomena, affecting the machine tool quality and personnel safety issues. Antonio et al. [11] used the neural network-like algorithm to predict the surface roughness of the process. Palani et al. [12] applied the machine vision technique to detect the surface condition, and studied the roughness model for the cutting parameters and surface roughness images to find out the relationship between the roughness and the design to ensure that the design of the processing conditions met the quality requirements. Hazza et al. [13] used SVM, EML and ANN and other prediction methods for signal analysis, to find the relationship between the vibration surface roughness, the predicted value and the actual measurement error. Dahbi et al. [14] analyzed the processing conditions, such as cutting speed, feed rate, depth of cut and tool nose radius, so as to learn from the surface roughness of the neural network, from the forecast results can be learned very well, the error range within 10 %. Srinivasa Pai et al. [15] considered the critical effect of tool radius in processing, and then used multi-layer sensing neural network to predict surface roughness. The results show

[*] Corresponding author: sclinciti@citi.sinica.edu.tw

that the predicted value has a roughness parameter close to the actual measurement.

In recent years, Lin et al. [16] discussed the fuzzy diagnosis strategy of rotating electrical machinery in the event of inner ring damage by statistic method. In order to analyze the fault diagnosis technology more deeply, this study aims to use the histogram statistical method to establish a deep groove ball bearing fault diagnosis strategy. Using the statistical indicators to excavate the fault characteristics of the vibration signal, and use the histogram to define the characteristic area of the model, and then achieve the purpose of effective fault diagnosis, detailed method described in Section 2, the Section 3 for the results and analysis, the Section 4 is the conclusion.

2 Methodology

In the analysis method, mainly the use of statistical methods for data analysis, in order to master the deep groove ball bearing state. First of all, the use of statistical indicators to dig out the abnormal characteristics of the data, and by the three-dimensional distribution determine the key indicators, and finally use the histogram statistical method for abnormal diagnosis, detailed process shown in Fig. 1.

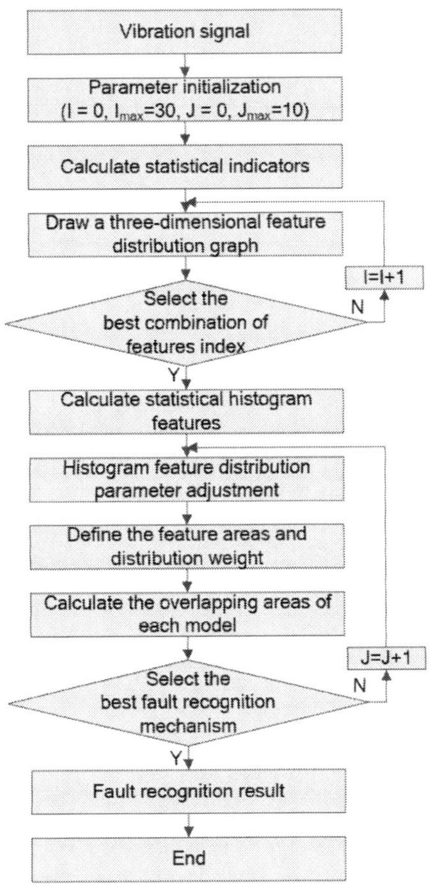

Fig. 1. The proposed method flow chart.

2.1 Feature indicators

In the evaluation of the characteristic index, nine statistical methods were used for the analysis. The main functions used are (1)-(9), such as standard deviation, rms value, peak value, crest factor, kurtosis, skewness, upper bound value of histogram, lower bound value of histogram and mean value.

1) Standard deviation

From the standard deviation index can be viewed vibration intensity distribution in the proportion of the average around, when the values are quite consistent, the standard deviation is relatively small.

$$\text{Standard } \textit{Deviation} = \sqrt{1/N \sum_{i=1}^{N} (X_i - \overline{X})^2} \tag{1}$$

where X is the data, i is the number of data, N is the total of data.

2) Root mean square value

From the RMS value indicators, it can be learned that with the mechanical aging, the value will gradually rise, and when the occurrence of mechanical anomalies, the vibration change is more obvious.

$$\textit{RMS Value} = \sqrt{1/N \sum_{i=1}^{N} (X_i)^2} \tag{2}$$

where X is the data, i is the number of data, N is the total of data.

3) Peak value

From the peak value reflects the intensity of the vibration signal, when the abnormal situation occurs, the reaction is often strong vibration intensity.

$$\textit{Peak Value} = \max(|X_i|) \tag{3}$$

where X is the data, i is the number of data.

4) Crest factor

From the crest factor, we can see the ratio of the peak and the RMS value, and observe the change of the vibration waveform.

$$\textit{Crest Factor} = \max(|X_i|) \Big/ \sqrt{1/N \sum_{i=1}^{N} (X_i)^2} \tag{4}$$

where X is the data, i is the number of data, N is the total of data.

5) Kurtosis

From the kurtosis index to observe the vibration pulse characteristics, the normal operation showed a gentle Gaussian distribution, and in the object damage, it presents a high peak phenomenon.

$$\textit{Kurtosis} = 1/N \sum_{i=1}^{N} \left(X_i - \overline{X} \Big/ \sqrt{1/N \sum_{i=1}^{N} (X_i - \overline{X})^2} \right) \tag{5}$$

where X is the data, i is the number of data, N is the total of data.

6) Skewness

From the skewness index, observe the value on the left or right side of the mean value, so that the signal is positive or negative skew, but if the value is zero, it means that the values are distributed on both sides of the mean.

$$Skewness = 1/N \sum_{i=1}^{N} \left(X_i - \bar{X} \Big/ \sqrt{1/N \ \sum_{i=1}^{N} (X_i)^2} \right)^3 \quad (6)$$

where X is the data, i is the number of data, N is the total of data.

7) Upper bound value of histogram

From the upper bound value of histogram indicators to observe the distribution of the situation, according to the needs of the analysis to dismantle the different distance of data for vibration intensity analysis.

$$UB = \max X_i + 0.5(\max X_i - \min(X_i)/N - 1) \quad (7)$$

where X is the data, i is the number of data, N is the total of data.

8) Lower bound value of histogram

From the upper bound value of histogram indicators to observe the distribution of the situation, according to the needs of the analysis to dismantle the different distance of data for vibration intensity analysis.

$$LB = mixX_i - 0.5(\max X_i - \min(X_i)/N - 1) \quad (8)$$

where X is the data, i is the number of data, N is the total of data.

9) Mean value

From the peak value reflects the intensity of the vibration signal, when the abnormal situation occurs, the reaction is often strong vibration intensity.

$$Mean \ \ Value = average(|X_i|) \quad (9)$$

where X is the data, i is the number of data.

When the vibration signal is calculated by the characteristic index, then through the three-dimensional feature distribution to determine the applicability of indicators, and a total of 30 combinations such as 1-2-3,1-2-4,1-2-5,1-2-6, 1-2-7, 1-2-8, 1-2-9, ... , 1-8-9.

2.2 Histogram

In the multiple index analysis, the use of histograms to calculate the characteristics of the distribution area, and two-dimensional graphics were presented statistical samples and measures, as shown in Fig. 2. At the same time, in order to make the statistical samples of each experimental model have significant difference, the fuzzy calculation is carried out for the characteristic area, so as to improve the performance of fault identification.

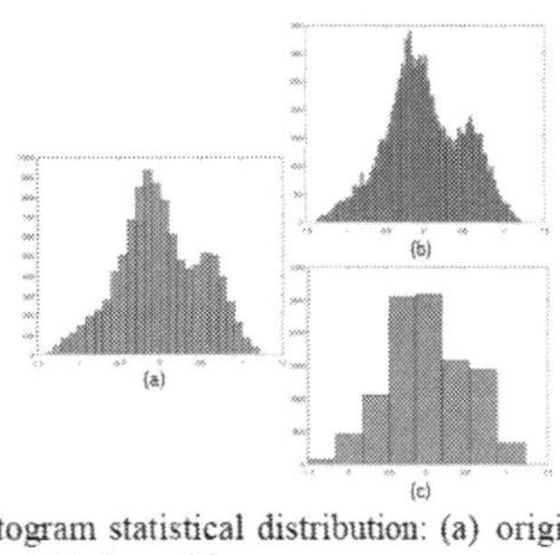

Fig. 2. Histogram statistical distribution: (a) original bins, (b) more bins and (c) fewer bins.

Then use (10) to calculate the weight of each feature distribution area, to infer the possible fault type.

$$A(x)\_w_y = y_i / \left(i^{\max} \cdot z^{\max}\right) \quad (10)$$

where A is the experimental model, x is the category, the range is 1, 2, 3, respectively, health, inner race damage and ball damage model. W is the weight, y is the characteristic distribution area, the range is 1, 2, ... , f, f is the fuzzy parameter. i is the amount of data, the number is 12. z is the best indicator, the number is 3.

The third section will detail the characteristics of the experimental model and its fault recognition performance.

3 Result and Analysis

In the experimental analysis section, the first use of data (number of twelve) to establish a statistical model, and the use of data (number of nine) for performance testing. The following will be discussed in two stages, namely, the selection of the best characteristic index (Section 3.1) and fault diagnosis (Section 3.2).

3.1 The select of the best feature index

This section is the characteristic index of the analysis experimental data, and the average is the criterion of the evaluation index. The health model and the ball damage model have similarities in the indicators 4, 5 and 8; while the health model and the inner race damage model have similarities in the index 8; the inner race damage model and the ball damage model are similar in the indicators 8 and 9, the detailed information is shown in Table 1.

Table 1. The statistical relationship between the index and the model.

Index	Fault type		
	H	I	B
1	0.0633	0.2996	0.1471
2	0.0644	0.2996	0.1472
3	0.2423	1.5276	0.5415
4	3.7611	5.0992	3.6788
5	2.9342	5.5637	2.8305
6	-0.1838	0.0901	0.0265
7	0.2423	1.5276	0.5416
8	0.0000	0.0000	0.0000
9	0.0118	0.0046	0.0046

After analysis and comparison, the selection of the most significant differences in the indicators, respectively, indicators 1, 3 and 6. In addition, the three-dimensional spatial distribution characteristics shown in Figs. 3 and 4. The diagnostic analysis results are described in Section 3.2.

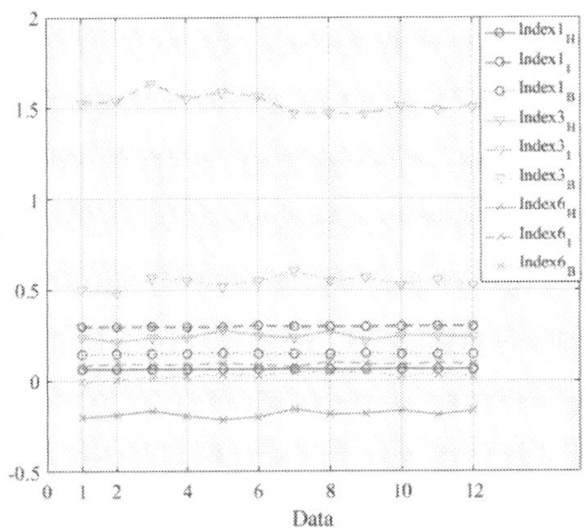

Fig. 3. Comparison of the best feature index of each model.

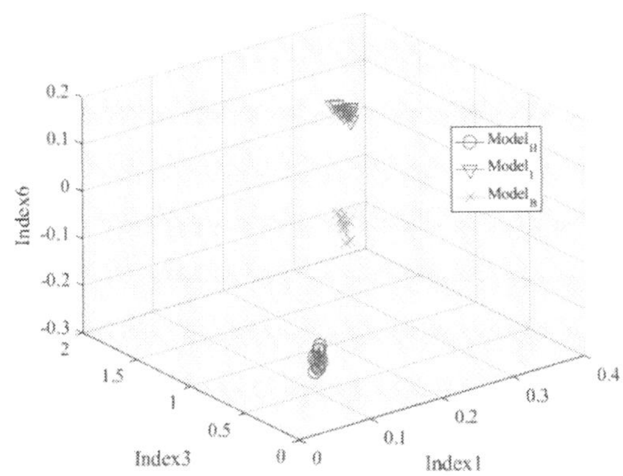

Fig. 4. Three-dimensional characteristic distribution of each experimental model.

3.2 Fault diagnosis

In the fault diagnosis section, the decision analysis is based on the characteristic parameters of Section 3.1.

However, in Section 3.2.1, a histogram feature distribution is used to determine an ideal feature model.

Then compare the differences between the data and the various feature models in Sections 3.2.2 to 3.2.4; Finally, we discuss the relevance of each model in Section 3.2.5 to verify the validity of the proposed method.

3.2.1 The select of the best histogram feature distribution

This section is intended to explore the use of histograms to analyze the performance of statistical indicators and to define valid feature models from them. Mainly it is focused on the selection of the histogram of the feature distance, in order to obtain the greatest difference between the various experimental models. However, in order to obtain the ideal feature model, the bin value of the histogram with different degrees of fuzz is tested, which can effectively define the characteristic interval of the statistical index. And when the bin value is more, that has a more feature interval.

At the same time, there is a direct and positive effect on the difference of the model, but it will increase the load of the computer; on the other hand, fewer bin values are less accurate for the interpretation of feature metrics and reduce the differences between the various feature models, but the computational load is lighter. In this case, the performance of different bin values is tested, as shown in Fig. 5.

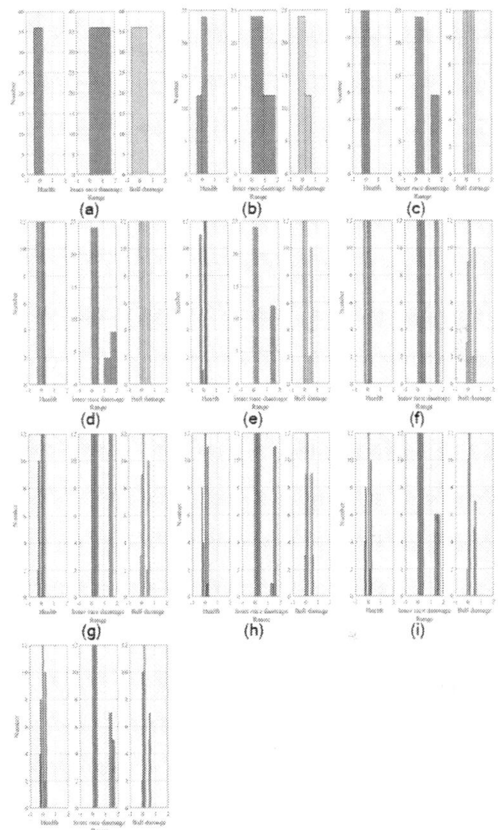

Fig. 5. Histogram feature distribution parameter adjustment: (a) first to (j) tenth.

It is clear from Fig. 5 that the difference in each feature interval is significant. When the bin value is small, the feature area distribution is highly overlapped, which means that each model has high similarity, as shown in Fig. 5 (a); however, if the bin value is more, the feature area is divided more detailed, as shown in Fig. 5 (j).

In order to determine the characteristic interval that is more suitable for each model, we first calculate the overlapping regions of the model with the intersection characteristics (gray in Table 2), then the minimum overlap range as the parameters of each experimental model, as shown in Table 2.

Table 2. Simplified feature model and overlapping area distribution.

No.	Feature area			Overlapping area
	Health	Inner race damage	Ball damage	
1	[-0.40 0.28]	[0.00 1.64]	[-0.60 0.61]	[-0.40 0.61]
2	[-0.40 0.40]	[0.00 1.80]	[-0.30 0.70]	[-0.30 0.70]
3	[-0.30 0.30]	[0.00 0.60] [1.20 1.80]	[-0.20 0.70]	[-0.20 0.60]
4	[-0.30 -0.15] [0.00 0.30]	[0.00 0.50] [1.00 2.00]	[-0.10 0.26] [0.44 0.62]	[0.00 0.50]
5	[-0.27 -0.05] [0.06 0.28]	[0.00 0.33] [1.32 1.65]	[-0.10 0.20] [0.35 0.65]	[-0.10 -0.05] [0.00 0.28]
6	[-0.24 -0.15] [0.03 0.12] [0.21 0.30]	[0.00 0.56] [1.40 1.68]	[-0.10 0.26] [0.38 0.62]	[0.00 0.56]
7	[-0.28 -0.12] [0.04 0.12] [0.20 0.28]	[0.00 0.48] [1.44 1.68]	[-0.08 0.22] [0.42 0.62]	[0.00 0.48]
8	[-0.24 -0.11] [0.02 0.09] [0.15 0.28]	[0.00 0.42] [1.26 1.68]	[-0.07 0.20] [0.47 0.65]	[0.00 0.28]
9	[-0.25 -0.13] [0.05 0.10] [0.16 0.28]	[0.00 0.38] [1.33 1.71]	[-0.06 0.16] [0.46 0.61]	[0.00 0.28]
10	[-0.24 -0.14] [0.02 0.07] [0.18 0.28]	[0.00 0.34] [1.36 1.70]	[-0.06 0.07] [0.14 0.21] [0.48 0.61]	[0.00 0.28]

It is known from Table 2 that each model has a high degree of overlap in the feature areas of numbers 1 and 2.

In addition, to further consider the impact of the distribution of weight, the health model in the number 10 features and weight distribution is most clear.

However, from the sixth fuzzy adjustment to obtain a stable result, and in order to reduce the complexity of modeling and computing resources, so it is the basis for failure decision; and so on, determine the characteristics of the inner race damage and ball damage model, respectively, for the number 10 and 6.

3.2.2 Diagnostic results based on health characteristics

In the health model section, the feature distribution regions are defined by histograms, [-0.24 -0.15], [0.03 0.12] and [0.21 0.30], respectively.

After the data were analyzed by statistical index and histogram, the health model test data (H1, H4, H7) fully accorded with the characteristic area of the health model; the inner race damage model test data (I2, I5, I8) have only a few of them that meet the health characteristics; ball damage model test data (B3, B6, B9) are almost non-compliant with health characteristics, as shown in Table 3.

Table 3. Comparison of test data with health model.

Test data	Feature area & Distribution weight		
	[-0.24 -0.15]	[0.03 0.12]	[0.21 0.30]
	0.33	0.33	0.33
H1	-0.20	0.06	0.24
I2	-	0.08	0.30
B3	-	-	-
H4	-0.17	0.06	0.24
I5	-	0.09	0.30
B6	-	-	-
H7	-0.20	0.06	0.25
I8	-	0.09	-
B9	-	0.03	-

According to the results in Table 3, calculate the weight of the test data. In the diagnostic results, the test data were considered similar to the health status of 52 % (health test data), 24~37 % (inner race damage test data), 0~22 % (ball damage test data), as shown in Fig. 6.

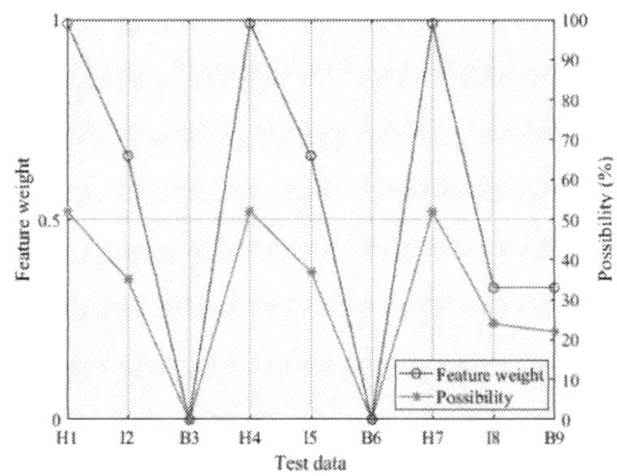

Fig. 6. The similarity coefficients and diagnostic results in healthy feature model.

3.2.3 Diagnostic results based on inner race damage characteristics

In the inner race damage model section, the feature distribution regions are defined by histograms, [0.00 0.17], [0.17 0.34], [1.36 1.53] and [1.53 1.70], respectively.

After the data were analyzed by statistical index and histogram, the health model test data (H1, H4, H7) have only a few of them that meet the inner race damage characteristics; the inner race damage model test data (I2, I5, I8) are very consistent with the inner race damage characteristics; ball damage model test data (B3, B6, B9) are almost non-compliant with inner race damage characteristics, as shown in Table 4.

Table 4. Comparison of test data with inner-race damage model.

Test data	Feature area & Distribution weight			
	[0.00 0.17]	[0.17 0.34]	[1.36 1.53]	[1.53 1.70]
	0.33	0.33	0.22	0.11
H1	0.06	0.24	-	-
I2	0.08	0.30	1.53	1.53
B3	0.14	-	-	-
H4	0.06	0.24		-
I5	0.09	0.30	1.51	-
B6	0.02, 0.15	-	-	-
H7	0.06	0.25	-	-
I8	0.09	0.31	-	1.56
B9	0.03, 0.15	-	-	-

According to the results in Table 4, calculate the weight of the test data. In the diagnostic results, the test data were considered similar to the inner race damage status of 35 % (health test data), 49~57 % (inner race damage test data), 22~33 % (ball damage test data), as shown in Fig. 7.

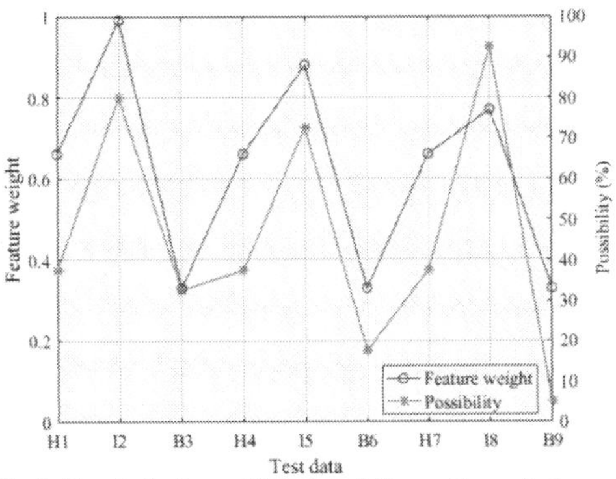

Fig. 7. The similarity coefficients and diagnostic results in inner race damage feature model.

3.2.4 Diagnostic results based on ball damage characteristics

In the ball damage model section, the feature distribution regions are defined by histograms, [-0.07 0.02], [0.02 0.11], [0.11 0.20], [0.47 0.56] and [0.56 0.65], respectively.

Then, the results of the analysis show that the health model test data (H1, H4, H7) and the inner ring damage model test data (I2, I5, I8) do not meet the ball damage characteristics; ball damage model test data (B3, B6, B9) conforms to the characteristic area properties, as shown in Table 5.

Table 5. Comparison of test data with ball damage model.

Test data	Feature area & Distribution weight				
	[-0.07 0.02]	[0.02 0.11]	[0.11 0.20]	[0.47 0.56]	[0.56 0.65]
	0.08	0.25	0.33	0.25	0.08
H1	-	0.06	-	-	-
I2	-	0.08	-	-	-
B3	-0.01	-	0.14	0.50	-
H4	-	0.06	-	-	-
I5	-	0.09	-	-	-
B6	0.02	0.02	0.15	0.52	-
H7	-	0.06	-	-	-
I8	-	0.09	-	-	-
B9	-	0.03	0.15	0.55	-

According to the results in Table 5, calculate the weight of the test data. In the diagnostic results, the test data were considered similar to the ball damage status of 13 % (health test data), 13~19 % (inner race damage test data), 56~73 % (ball damage test data), as shown in Fig. 8.

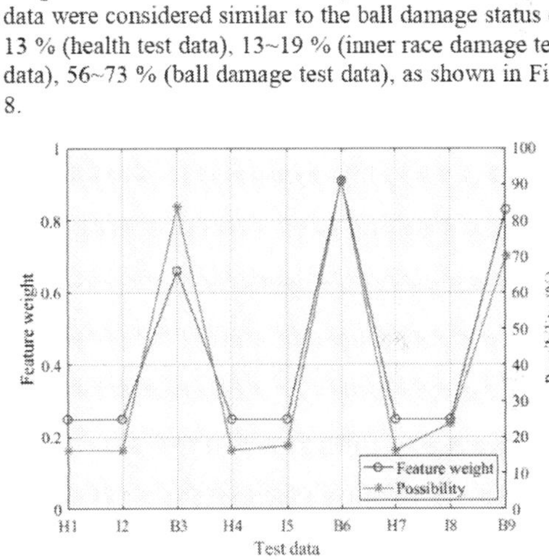

Fig. 8. The similarity coefficients and diagnostic results in ball damage feature model.

3.2.5 Evaluate pattern recognition performance

This section will be based on the analysis of the previous three sections to assess its performance, through the analysis results that all the test data can get the correct pattern recognition results, in addition to prove that the proposed method is feasible, as shown in Fig. 9.

In Fig. 9, it is known that the health model has some common characteristics with the inner ring damage model, but does not produce a misjudgment, and does not have the characteristics of ball damage; however, in the part of the ball damage model, the correct pattern can be clearly determined because it has distinct characteristics from other models, as shown in Fig. 9.

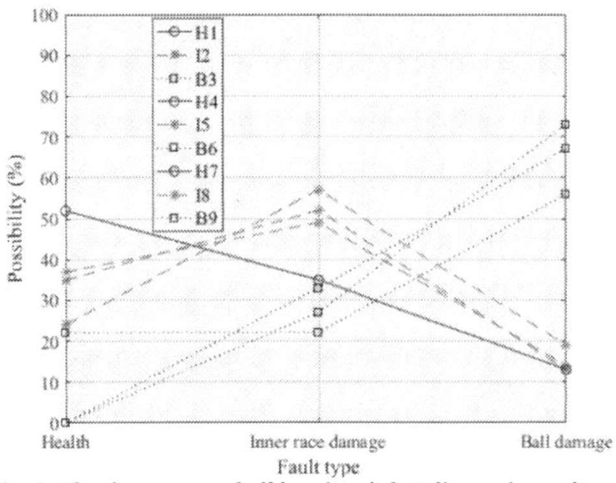

Fig. 9. The deep groove ball bearing defect diagnosis results.

4 Conclusion

In this paper, the statistical features are used to excavate the hidden faults in the vibration signal, and the most suitable indicators are screened for histogram statistics. Finally, the feature regions of each model are used as the basis for fault diagnosis. The results showed that the indexes 1, 3 and 6 had significant differences in characteristics. However, all test data have 100 % fault resolution accuracy. On the other hand, due to the significant difference in the statistical index of the ball damage model, the evaluation results showed high correlation (56~73 %); while the correlation of the inner race damage model was 49~57 %; the health model correlation was 52 %. As the inner race damage and health model in the statistical sense, there are some similar, so there is a relatively high correlation. In the future study will continue to carry out data mining, to find more significant differences in the characteristics of indicators to strengthen the performance of correlation analysis. We sincerely hope that the research results will have more help to the maintenance staff.

Acknowledgements

The work is supported by the Ministry of Science and Technology, Taiwan, under Grant MOST 103-2221-E-001-028-MY3.

References

1. H. C. Chang, C. Cheng, Y. T. Lin, J. F. Chen, C. C. Kuo, H. C. Chen, Y. H. Lin, Y. M. Jheng, S. C. Lin, C. F. Hsieh, and C. Y. Lin, "Smart Grid Development of On-line Monitoring and Diagnosis Methods for High Voltage Motor," *Monthly Journal of Taipower's Engineering*, **822**, pp. 94-115, (2017)
2. H. C. Chang, S. C. Lin, C. C. Kuo, C. Y. Lin, and C. F. Hsieh, "Using Neural Network based on the Shaft Orbit Feature for Online Rotating Machinery Fault Diagnosis," *IEEE International Conference on System Science and Engineering*, pp. 1-4, (2016)
3. H. C. Chang, C. C. Kuo, S. C. Lin, and C. Y. Lin, "Application of fractal theory to the rotating machinery of the vibration orbit feature extraction," *The 37th Symposium on Electrical Power Engineering*, pp. 1-6, (2016)
4. H. C. Chang, S. C. Lin, C. C. Kuo, and C. F. Hsieh, "Induction Motor Diagnostic System Based on Electrical Detection Method and Fuzzy Algorithm," *International Journal of Fuzzy Systems*, **18**, 5, pp. 732-740, (2016)
5. H. C. Chang, S. C. Lin, C. C. Kuo, and C. Y. Lin, "Fuzzy Theory-Based Partial Discharge Technique for Operating State Diagnosis of High-Voltage Motor," *International Journal of Fuzzy Systems*, **18**, 6, pp. 1092-1103, (2016)
6. H. C. Chang, S. C. Lin, C. C. Kuo, and C. Y. Lin, "Application of Artificial Neural Network to the Partial Discharge Recognition of High-Voltage Motor," *Proceedings of 2016 National Symposium on System Science and Engineering*, pp. 1-4, (2016)
7. H. C. Chang, C. C. Kuo, S. C. Lin, C. Y. Lin, and B. R. Chen, "Partial Discharge Pattern Recognition on the Stator Winding Insulation of High Voltage Motors Using Artificial Neural Network," *The 37th Symposium on Electrical Power Engineering*, pp. 1-6, (2016)
8. H. C. Chang, Y. T. Lin, C. Cheng, C. C. Kuo, S. C. Lin, Y. M. Jheng, C. F. Hsieh, and C. Y. Lin, "Fuzzy-based Fault Diagnosis System for Induction Motor," *The 37th Symposium on Electrical Power Engineering*, pp. 1-6, (2016)
9. H. C. Chang, S. C. Lin, C. C. Kuo, C. F. Hsieh, and C. Y. Lin, "Fuzzy Theory-Based Hybrid Decision-Making System for Rotating Electrical Machinery Fault Diagnosis," *International Conference on Fuzzy Theory and Its Applications*, pp. 1-6, (2016)
10. H. C. Chang, S. C. Lin, C. C. Kuo, C. F. Hsieh, and C. Y. Lin, "A Two-Stage Rotating Machinery Condition Monitoring and Diagnostic System," *International Conference on Engineering and Natural Science–Summer Session*, pp. 136-154, (2016)
11. A. J. Vallejo, M. M. Ruben, R. M. Ricardo, and G. C. Luis, "Online Prediction of Surface Roughness in Peripheral Milling Processes," *European Control Conference*, pp. 3695-3700, (2009)
12. S. Palani and Y. Kesavanarayana, "Prediction of surfaceroughness in end milling process by machine vision using neuro fuzzy network," *International Conference on Science Engineering and Management Research*, pp. 1-5, (2014)
13. M. H. F. A. Hazza, E. Y. T. Adesta, and M. F. Amin, "Seder Using Soft Computing Methods as an Effective Tool in Predicting Surface Roughness," *International Conference on Advanced Computer Science Applications and Technologies*, pp. 9-13, (2015)
14. S. Dahbi, L. Ezzine, and H. E. Moussami, "Modeling of Surface Roughness in Turning Process by using Artificial Neural Networks," *International*

DOI:10.1051/matecconf/201712300009

Conference on Logistics Operations Management, pp. 1-6, (2016)

15. N. Fang, P. S. Pai, and N. Edwards, "Multidimensional Signal Processing and Modeling with Neural Networks in Metal Machining_Cutting Forces, Vibrations, and Surface Roughness," *IEEE International Conference on Communication Software and Networks*, pp. 77-80, (2016)
16. S. C. Lin and Y. N. Huang, "Application of Statistical Methods to Excavate the Abnormal State Characteristics of Rotating Machinery," *Proceedings of 2017 National Symposium on System Science and Engineering*, pp. 112-116, (2017)

Applying virtual reality to remote control of mobile robot

Chin-Shan Chen[1,*] and *Ching-Wen* Lui[2]

[1]National Pingtung University of Science and Technology, No. 1, Shuefu Road, Neipu, Pingtung,91201, TAIWAN R.O.C.
[2]Van Guard Mold Enterprise CO. LTD., No.11, Dazhan Ln , Yanchao Dist., Kaohsiung City, 82444, TAIWAN R.O.C.

Abstract. The purpose of this research is based on virtual reality to assisted pick and place tasks. Virtual reality can be utilized to control remote robot for pick and place element. The operator monitored and controlled the situation information of working site by Human Machine Interface. Therefore, we worked in harsh or dangerous environments that thing can be avoided. The procedure to operate mobile robot in virtual reality describes as follow: An experiment site with really experimental equipment is first established. Then, the experimental equipment and scene modeling are input to virtual reality for establishing a environment similar to the reality. Finally, the remote mobile robot is controlled to operate pick and place tasks through wireless communication by the object operation in virtual reality. The robot consists of a movable robot platform and robotic arm. The virtual reality is constructed by EON software; the Human Machine Interface is established by Visual Basic. The wireless connection is equipped the wireless Bluetooth, which is set the PC and PLC controller. With experimental tests to verify the robot in virtual reality and the wireless remote control, the robot could be operated and controlled to successfully complete pick and place tasks in reality by Human Machine Interface.

1 Introduction

Virtual Reality (VR) has been widely applied to various industries with the development of science and technology, such as education, entertainment, military affairs, space science, industrial design, and manufacturing, etc. Playing a important role in such domains, VR integrates the machinery and becomes the development of human machine interface (HMI). VR is often applied to aided instruction, training, or simulation of manufacturing process and restoring the real scenes. In addition to excluding the restrictions of location and time, it could reduce costs and avoid damage for people and machine.

The popularity of automation has resulted in many factory largely replacing manpower with robots. Mobile robots are expected to enhance the ability of the users in executing complex work in a remote environment [1,2]. However, wrong operation would still occur when robots are set or operated by manpower that danger and machinery breakdown could be caused. For this reason, training with VR replacing the real operation training could reduce the costs resulted from accident and machinery breakdown.

Furthermore, the development costs for a real prototype is very high, and the pick and place process could not be repeatedly applied that pick and place presents equal importance to the design and manufacturing process [3,4]. For pick and place tasks, the aid of reality could reduce errors and quickly completing the work. Two issues are required for simulating manufacturing systems with computers, namely simulation of non-continuous events and Virtual Reality software[5]. During the remote operation, the awareness of the remote environment is primary for an operator. With remote presentation, an operator could receive the information from the remote environment that favorable object modeling is required for simulating the object behaviors in Virtual Reality.

This study expects to applied the interactive VR to having the users rapidly and clearly understand the environment and the operation, implementing the simulation of robot operation in the actual environment, and operate and control the mobile robot in the reality through wireless remote monitoring and control so as to successfully complete the pick and place tasks by a robot.

2 Experimental system structure

The experimental system structure for controlling mobile robot proceeding pick and place with the integration of VR and remote control is explained as follows.

The hardware for the system contains robot arm, mobile robot, wireless Bluetooth, PC, and Programmable Logic Controller (PLC). The model for the robot arm with two degrees of freedom is mechanism for pick and place tasks. Besides, design an electromagnet for pick and place tagger. The model for the mobile robot platform is mobile car. The model for the wireless Bluetooth is HL-MD08P-PLC-C1(v2.1+EDR). PC is used for the control, which is equipped with Intel Corei5-4670 CPU 3.40GHz processing unit, Windows 7

* Corresponding author: chencs@mail.npust.edu.tw

DOI:10.1051/matecconf/201712300010

operating system, 4G memory (RAM), and Intel HD Graphics 4600 NVIDIA GeForce GT 610 graphics card. Fig. 1 Shows that software for the system contains Pro/ENGINEER Wildfire 2.0, 3D Studio Max 7, EON Studio 5.2 (EON Studio is a development tools in VR which supplier is EON Reality company.), Visual Basic 6.0 and GX Developer 8. Physical property is included in EON, functions to present the physical properties of an object, such as physical collision, friction, and gravity effect, so as to vivify the objects in VR. The solid graphic module of robot arm and mobile robot are constructed with Pro/ENGINEER software and transferred into 3D Studio Max software for being edited. Then, transferred into .eoz file with EON Trans software for VR being edited in EON Studio. This HMI of the remote control programming software in this study is designed by Visual Basic. The control program for mobile robot and arm is wrote by GX Developer. A PC is used for connecting with Wireless Bluetooth.

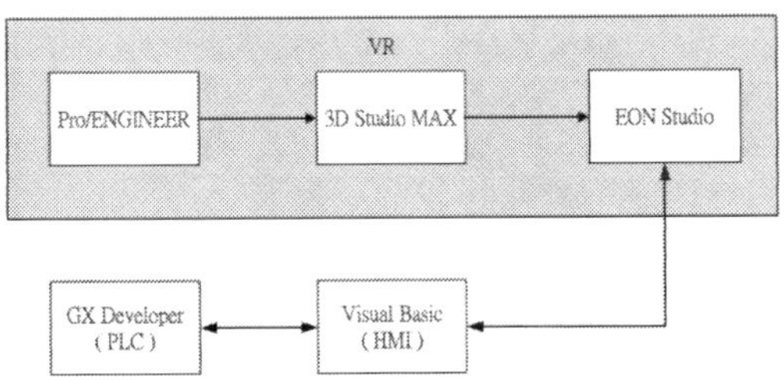

Fig. 1. Software system.

3 Virtual reality

VR is defined as the science integrating human and information. In other words, it is the 3D and interactive environment created by a computer, standing for a real world or an imagined virtual environment.

For Virtual Reality, 3D is an essential object for simulating a real world. Being a virtual environment constructed by computer graphics, Virtual Reality allows the users experiencing the real world [6] that they could observe an object according their demands, rather than being restricted in the defined objects. With the example of stereoscopic puzzles, learners could observe the positions of puzzles by changing the angle of view in Virtual Reality. It is considered being closer to problem-solving in the real world, without being restricted in certain solution.

The Robot Association started studying the applications of Virtual Reality in 1990 and regarded the great potential of Virtual Reality for Human-Machine Interface [5]. The applications of Virtual Reality have expanded the remote monitoring domain for human beings [7,8]. With the effects of computer computing speed and network transmission rate, the early research could merely transmitted limited information and uncontinuous frames. To overcome the delay problem, robots and the 3D interface for the work environment were utilized for remote operation, providing an operator with real scenes for replacing the delayed images [9]. Considering the applications of robotic arms, 2D plane displays could not satisfy the high-degree of freedom operates of robot arms. 3D graphs therefore is applied for displaying various information to better correspond to an operator's demands [10]. The renewal and update of robot hardware allow real-time addition and change in Virtual Reality. Having 3D virtual scenes to present the on-site information, present problem of insufficient bandwidth could be solved, as 3D virtual scenes offer several angles of view, which could better provide an operator with operation information, comparing to a camera shooting from fixed angles [11].

Virtual Reality is also well developed in other fields, like medical rehabilitation[12,13], simulation of surgery, driving simulation of large vehicles, virtual manufacturing process of various products and materials, virtual operation of arms in distinct military affairs, virtual training of a soldier, and virtual cities [14], and virtual training process of education [15].

To have identical motions of virtual objects and real objects, the translation and rotation of a real object should be understood. The control of virtual object is also described as follows. Fig. 2 shows the coordinates and the rotation of virtual objects in various angles, including Roll, Pitch, and Heading. In consideration of the control being the key in dynamic mechanical design, virtual simulation should present the same concept[16]. For this reason, the action and overall moving procedure of a real object should be actually realized in advance so as to establish the control of virtual objects in EON.

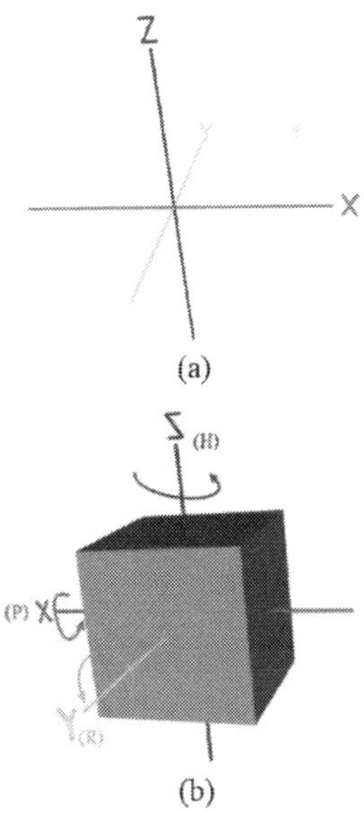

Fig. 2. (a) Definition the coordinates of EON; (b) Definition the Roll, Pitch, and Heading of EON.

With the example of translation control of a mobile robot in VR, the explanations of the translation control is as follows. To finish the translation of mobile robot is setting by the place node which in simulation tree of EON as in Fig. 3. First, open the node of place in EON. Second, key in the parameters for forward and click icon of the type coordinates. Then, the switch of communication for HMI is setting in the 'InEvent' as in Fig. 4. Finally, we adjusted the value of the parameters to match the mobile robot in reality. Such judgments could control the direction of mobile robot; besides, the parameters for forward, backward, left turn, right turn, and distance movement are set, which are respectively operated by click on the push button graphic 「FOR」, 「REV」, 「TURN L」, and 「TURN R」 and 「STOP」 on the HMI as in Fig. 5. Fig. 6 shows the HMI with EON in start status. The adjustment of damping parameter could change the speed of module motion.

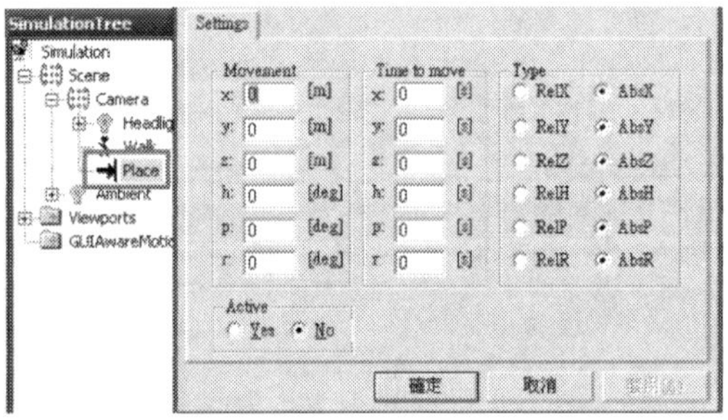

Fig. 3. Setting the node of place in EON.

(a)

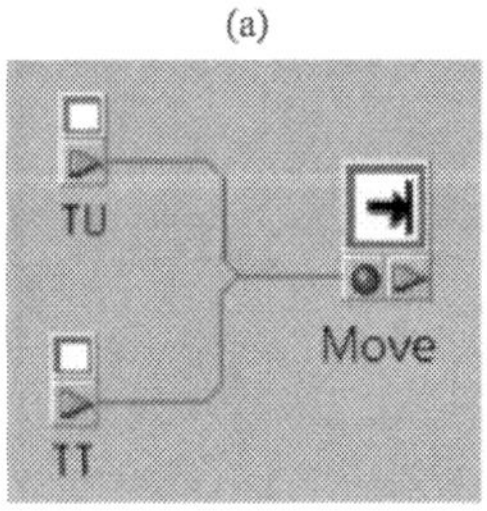

(b)

Fig.4. (a) Create an external communication event; (b) Setting the node of switch in EON.

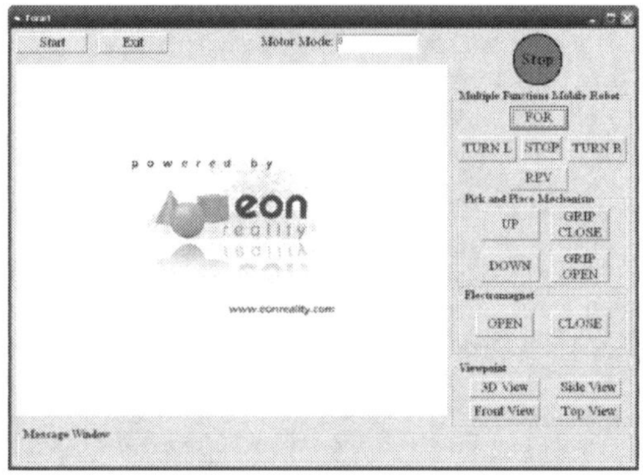

Fig. 5. HMI with EON.

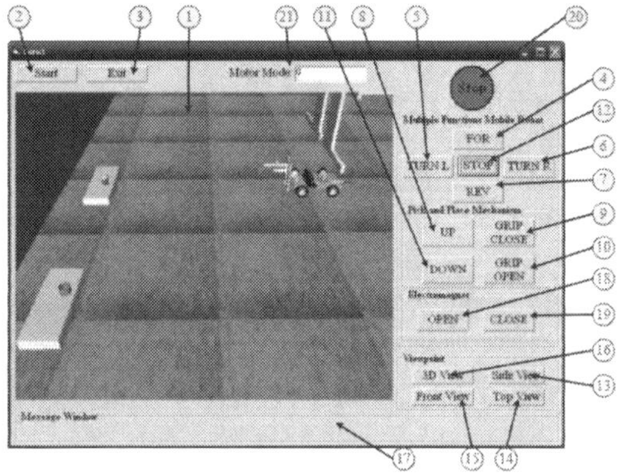

Fig. 6. HMI with EON in start status.

4 Virtual reality integrated mobile robot control

The virtual scenes are first input to the module in EON, simulation is used for setting the control parameters of objects, and mobile robot and communication framework are further installed. The robot test is then proceeded in VR, as the following procedure.

4.1 Control procedure

An indoor solid experiment is first setup, which includes mobile robot, robot arm, work platform, and communication framework. A similar scene is also established in VR to verify the synchronous operation of the robot. In the experimental scene, a mobile robot equipped with robot arm and two platforms are established; No. 2 platform is equipped with object and No. 3 platform is equipped a hole with round shape. The robot is controlled to walk a straight line from No. 1 to No.2 platform, takings the object and turning 90 degree to No. 3 platform, place the object in the hole with the same shape, and return to the start point, Fig. 7.

DOI:10.1051/matecconf/201712300010

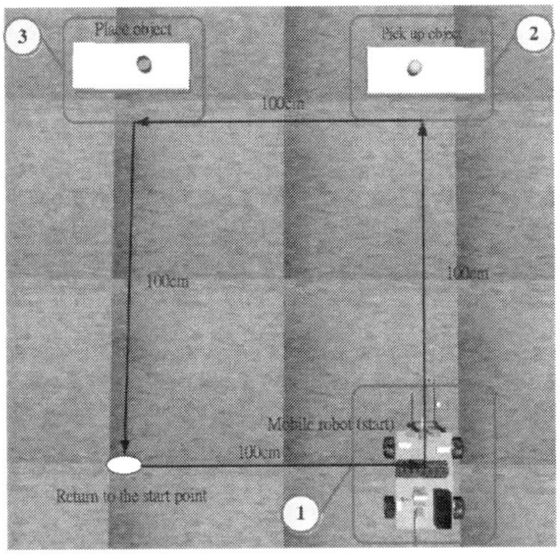

Fig. 7. Control procedure.

4.2 Connection of robot

The connection is proceeded with Bluetooth; then, the robot motion, which is originally controlled by push button under PLC based, is called out the connection function in windows by switch node in EON. The synchronous connection control could be operated through VR scenes. Fig. 8 shows a HMI, VR relationship with reality of mobile robotic control system. The wireless communication control utilizes the wireless transmission specifications and standards of IEEE 802.15.1 bandwidth.

4.3 Operate and control of mobile robot

The mobile robot motion mentioned in section 3 is controlled by click on the push button graphic 「FOR」, 「REV」, 「TURN L」, and 「TURN R」 and 「STOP」 on the HMI. The control of robotic arm for pick and place mechanism is similar to it of wheels. Robotic arm, on the other hand, are divided and controlled by 「UP」 for moving up and 「DOWN」 for moving down for pick and place mechanism, 「GRIP CLOSE」 for operating close and 「GRIP OPEN」 for operating open for gripper, 「OPEN」 for generating the magnetic force and 「CLOSE」 for not to generate the magnetic force an electromagnet, as shown in Fig. 9. We can set the different viewpoint in the human-machine interface. The user can select the different viewpoint to watch a mobile robot in VR, Such as, 「3D View」, 「Side View」, and 「Front View」.

The mobile robot and arm, gripper could simultaneously move without any effects. However, simultaneously moving forward and backward at the same part would not appear any motions.

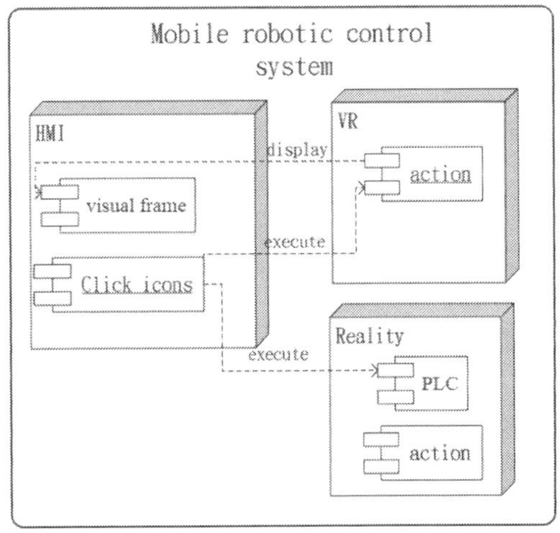

Fig. 8. Mobile robotic control system.

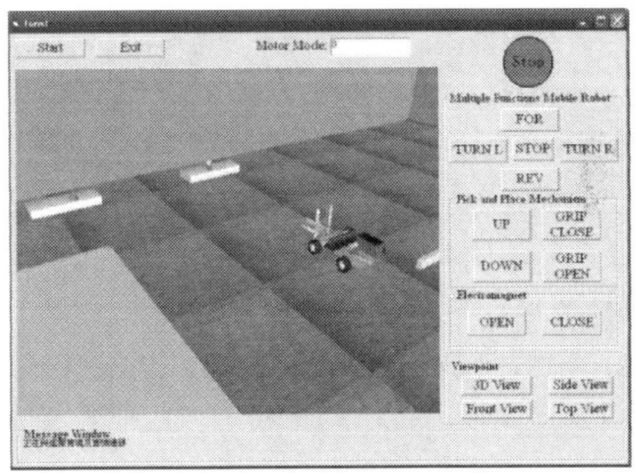

Fig. 9. To operate and control the mobile robot.

4.4 Experimental results and discussions

The mobile robot is first controlled to move to No. 2 work platform, and the clamp of robotic arms with pick and place mechanism is then controlled to the top of the plug-in for grabbing the plug-in by opening the clamp of robotic arms and operating the clamp to the proper position. The mobile robot is further moved to No. 3 work platform for the assembly of plugging the plug-in to the plug with the same shape, and then returned to the start point.

With the experimental test, the mobile robot forward speed appears 0.5804m/sec in reality, while the virtual mobile robot forward speed being adjusted the speed parameter reveals 77.84m/sec, about 0.5804m/sec, and the mobile robot backward speed appears 0.6046m/sec in reality, while the virtual mobile robot backward speed being adjusted the speed parameter reveals 81.09m/sec, about 0.6046m/sec. The robot arm moves upward speed about 0.0227m/sec in reality, while the virtual robot arm forward speed being adjusted the speed parameter reveals 0.00789m/sec, about 0.0227m/sec, and the robot arm backward speed appears 0.0241m/sec in reality, while the virtual mobile robot backward speed being adjusted the speed parameter reveals 0.00845m/sec, about 0.0241m/sec. The detailed data are shown in Table 1.

The follow-up adjustment of speed parameter reduces the operation errors between VR and the real robot down to the least so that the movement and assembly of the robot in VR and the solid robot are almost synchronous. Furthermore, the wireless remote control assists in successfully controlling the robot in reality to complete the assembly, as shown in Fig. 10.

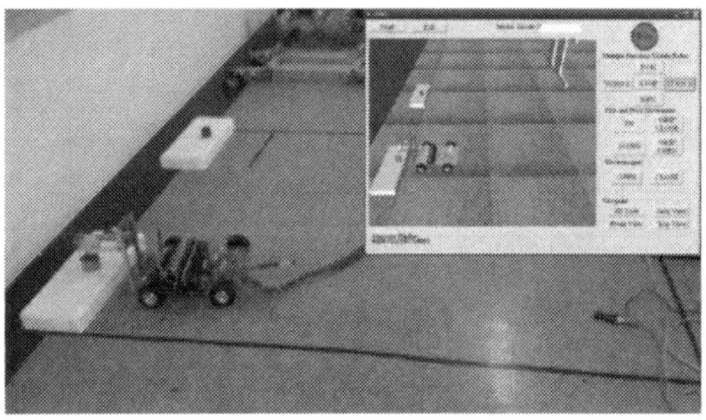

Fig. 10. Experimental procedure.

Table 1. Mobile robot speed parameter relative speed in virtual and reality.

	Virtual reality robot speed (m/sec)	Reality mobile robot speed (m/sec)
mobile robot forward	77.84	0.5804
mobile robot backward	81.09	0.6046
Pick-up	0.00793	0.0227
Pick-down	0.00845	0.0241
Gripper close	22.0327(deg/sec)	0.3845(rad/sec)
Gripper open	21.7468(deg/sec)	0.3796(rad/sec)

5 Conclusions

EON is utilized for real-time connection control of solid robot to automatically complete the pick and place tasks in VR. The past research mostly reduced the errors by improving the transmission speed after the camera transmitting the on-site images. In this study, VR, replacing a camera, revises the speed parameter of the robot for synchronism between virtual and real frames. Moreover, simulation tree in EON is applied to presenting the real physical properties of an object, such as the moving speed and collision of a mobile robot and arm. When such physical properties are wrongly or not set, the simulation in VR would be largely reduced. For example, when the robot touches the work platform without stopping or even passing it through, the plug-in assembly would not be completed.

In the real operation, it requires several attempts and corrections to grab the plug-in and successfully complete the assembly. VR allows freely changing the angles of view to make the grab and assembly be more easy. A new function to show the information of coordinate, moving distance, and rotation angle of the robot or other objects in VR frames could have the users more clearly understand the situation in the field. Or, including a virtual guidance could assist the users more easily operation the process, which are considered as a different research domain.

The fraction that this search needs to be strengthened which explained as follows: To raise the robot feedback information of HMI, that can be provided for user to make references. Because of many bugs happening while testing, Bluetooth v2.1+EDR is used for VR with PLC may be not so suitable.

References

1. F.C. Hsu, VR-Based Teleoperation System for Industrial Robots, Master's thesis, Department of Mechanical Engineering, Chung-Yuan Christian University (2000)
2. Y.H. Huang, Applying Virtual Guidance for Robot Teleoperation, Master's thesis, Department of Electrical and Control, National Chiao Tung University (2007)
3. S.W. Chuang, Virtual Assembly Modeling, Master's thesis, Department of Mechanical Engineering, National Central University (1997)
4. T.J. Ke, Personal Computer Assembly by Virtual Reality Technique, Master's thesis, Department of Computer Science, National Chiao Tung University (1998)
5. E.A. Orady, T.A. Osman, *21st International Conference on Computers and industrial Engineering*, **33**, pp. 87-90, (1997)
6. T.A. Chen, Development of Robotic Gripper for Telepresence in a Telerobotic System, Master's thesis, Department of Electrical and Control Engineering, National Chiao Tung University (2003)
7. H.B. Chen, Controller Design and Virtual Reality of Tatung TTUP3D Parallel Robot, Master's thesis, Department of Mechanical Engineering, Tatung University (2001)
8. E. Freund, J. RoBmann, *Proceedings of the 2001 IEEE International Conference on Robotics and Automation*, pp. 21-26, (2001)
9. X. Xie, L. Sun, Z. Du, Z. Zong, *Proceedings of the 6th World Congress on Intelligent Control and Automation*, pp. 21- 23, (2006)
10. Y.C. Lin, Virtual Assembly Modeling, Master's thesis, Department of Mechanical Engineering, National Central University (2002)

11. D.Y. Chiou, Object-oriented System Modeling for Web-besed Telerobot Systems with a Real Implication, Master's thesis, Institute of Industrial Engineering, National Taiwan University (2000)
12. Imre Cikajlo, *Journal of Medical and Biological Engineering*, **34**, 4, pp. 319-326, (2014)
13. C.W. Yin, N.Y. Sien, L.A. Ying, S.F. Chung and D.T. Leng, *Clinical Rehabilitation*, **28**, 11, pp. 1107-1114, (2014)
14. Junji Nomura and Kazuya Sawada, *Annual Reviews in Control*, **25**, pp. 99-109, (2001)
15. H. Brenton, J. Hernandez, F. Bello, P. Strutton, S. Purkayastha, T. Firth, and A. Darzi, *Computers & Education*, **49**, 1, pp. 32-53, (2007)
16. A. Jönsson, J. Wall, G. Broman, *International Journal of Machine Tools & Manufacture*, **45**, pp. 795-801, (2005)

Effects of particle shape and temperature on compaction of copper powder at micro scale

Chao-Cheng Chang* and *Ming-Ru* Wu

Department of Mold and Die Engineering, National Kaohsiung University of Applied Sciences, Kaohsiung, 80778, Taiwan

Abstract. This study investigated the effects of particle shape and temperature on the compaction of copper powder at micro scale. Copper powder particles were compressed inside a cylindrical die cavity with 2 mm diameter to form compacts with about 3 mm height. Two kinds of particle shapes, spherical and dendritic, and two forming temperatures, room temperature and 400°C, were considered in the experiments. Some of the produced compacts were further sintered at 600°C. The study also used simple upsetting tests to investigate the characteristics of the deformation of the compacts under compressive stresses. The results showed that the compacts produced at room temperature demonstrated brittle deformations. However, by increasing the forming temperature to 400°C, ductile deformations have been observed on the compacts of dendritic particles. Furthermore, the sintering treatment resulted in increases in dimensions, decreases in relative density and hardness, and an increase in ductility. It also led to pore growths which have been seen on scanning-electron microscope images. These phenomena were most significant in the dendritic powder compacts which were produced at 400°C and treated by the sintering process.

1 Introduction

Micro forming processes have been developed in recent years with the objective of fabricating micro metal parts available for the mass production in the fields of electronics products, life science, and transport industry. To rapidly fulfil the need in markets and reduce the lead time of the micro metal parts, the powder metallurgy (PM) process which can produce parts at net or near-net shape is widely interested in academia and industry.

The PM process basically consists of powder production, blending, compaction, sintering, and finishing operations. By using different methods of compressing and shaping, PM has been developed into several processes including cold isostatic pressing (CIP), hot isostatic pressing (HIP), and powder injection molding (PIM). PM has not only been used in industrial products but also in medical applications, such as porous metals for use in orthopaedic applications [1]. On the other hand, the direct laser sintering process of metallic powder, which does not require a compaction of powder, has also developed for rapid tooling [2] and producing metallic microstructures [3]. Moreover, it is possible to use a direct powder rolling process for manufacturing titanium strips [4].

Similar to PIM, micro powder injection molding, which can produce the micro-components of metal, ceramic and hard materials, is also an attractive process for the mass production of complex micro-components with features at the micrometer or sub-micrometer scale [5, 6]. The size and shape of powder particle, which may be described in terms of aspect ratio or shape factor, greatly influence the operations of PM and the quality of the product [7]. The shape factor, which is the ratio of surface area of the particle to its volume, affects the contact between particles that has an important effect on their behavior in processing them. For the same pressure of compaction, the coarse powder particles result in a higher density of the compact than the fine powder particles [8]. In addition to the effects of shape and size, temperature affects the process of PM, especially playing an important role on the bonding of the particles. It is possible to vary compaction temperature for producing a PM product with a graded microstructure which has a nearly homogeneous solid at the high-temperature side and clearly porous structure at the low-temperature side [9]. Furthermore, an increase in the sintering temperature may result in faster densification and sintering as demonstrated in the PIM process of tungsten carbide components in which the relative density and hardness increase dramatically at a certain sintering temperature [10]. Recently, Bai et.al [11] proposed a direct forging process of powdered nickel-based superalloys. In the proposed process, encapsulated and vacuumed powder particles were heated up to a forming temperature and forged directly at high speed to the final shape. The process provides an attractive approach for producing near net-shape fine grained products at low cost and with high productivity.

In this study, copper powders were compressed inside a cylindrical die cavity to form compacts with 2 mm diameter. Two kinds of particle shapes in copper powders, dendritic and spherical shapes, were considered. To investigate the effect of temperature, experiments

* Corresponding author: ccchang@kuas.edu.tw

were conducted on two forming conditions which were at room temperature and in vacuum at an elevated temperature. Some of the produced compacts were further treated by sintering. Furthermore, the study used simple upsetting tests to investigate the characteristics of the deformation of the compacts under compressive stresses. The dimension, relative density, hardness, and microstructure of the produced compacts were investigated.

2 Experimental procedures

In this study, copper powder particles were compressed inside the die cavity with 2 mm diameter to form micro cylindrical compacts as illustrated by Fig. 1. Two shapes of copper particles, spherical and dendritic shapes, and two forming temperatures, room temperature and 400°C, were considered in experiments. Some of the produced compacts were further treated by sintering. Therefore, there were 8 process conditions according to various particle shapes and temperature, and considering with or without sintering to produce compacts. The compacts were subsequently compressed by the flat dies in simple upsetting tests (Fig. 2) for obtaining the stress-strain curves which can provide information about the deformation of the compacts under compressive stresses. The study also examined the properties of the compacts including dimension, relative density, hardness, and microstructure. Fig. 3 displays the flowchart of the study.

To prevent the oxidation of copper powder in the compacting process at an elevated temperature, a vacuum chamber for experiments was developed. Fig. 4 shows the schematic diagram of the experimental setup in which the powder particles were heated by the die placed inside the vacuum chamber. The material of the vacuum chamber was stainless steel and the die components (Fig. 5) were mainly made of tool steel (JIS SKD 61). The developed vacuum chamber was installed on an electro press with a capacity of 50 kN and a positioning accuracy of 10 μm, as shown in Fig. 6.

The study used spherical powder particles with sizes ranging from 14 to 25 μm and dendritic powder particles with a mean size of 45 μm in experiments to investigate the effect of particle shape. Fig. 7 shows scanning-electron microscope (SEM) images of the copper powder particles. By carefully controlling the weight of the powder and feeding them into the die cavity, the powder compacts with 80 ± 1 mg and 2 mm diameter were produced by a punch speed of 0.1 mm/s and a maximum load of 5.88 kN (1872 MPa). In the experiments conducted at 400°C, the powder particles were directly fed into the die cavity and heated with a rate of 40°C per minute inside the vacuum chamber with a condition of 2.6×10^{-2} Torr before the compaction process was carried out. The height of the produced powder compact was about 3 mm and varied with the shape of particle and forming temperature. Furthermore, some of the powder compacts produced at room temperature and 400°C were subsequently treated by sintering in vacuum at 600°C for 1 hour and cooled down to room temperature inside the furnace. The powder compacts obtained from different processing conditions were compressed by two flat dies at a die speed of 0.1 mm/s to perform simple upsetting tests. To reduce the friction at the contact interfaces between the powder compact and dies in the tests, molybdenum disulfide (MoS_2) based lubricants were applied. Fig. 8 presents some of dendritic copper powder compacts and the deformed compacts from simple upsetting tests.

The micrographs of the fabricated powder compacts were taken using an SEM to reveal the microstructures. By measuring the dimensions of the powder compact and its weight, the volume was calculated and thus relative density can be estimated. Furthermore, hardness numbers across the section of the powder compact were determined using micro Vickers hardness.

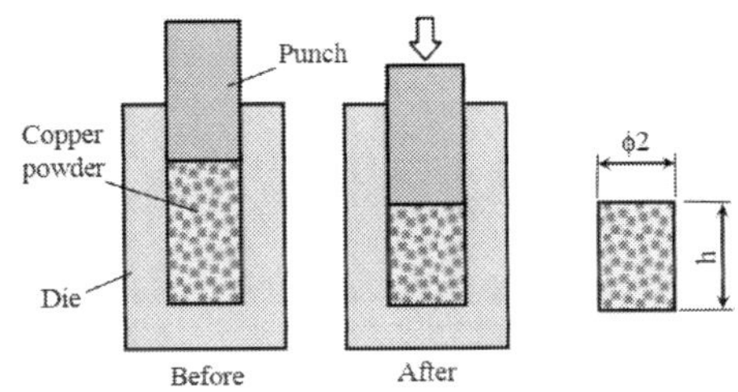

Fig. 1. Schematic illustration of powder compaction and dimensions of powder compact.

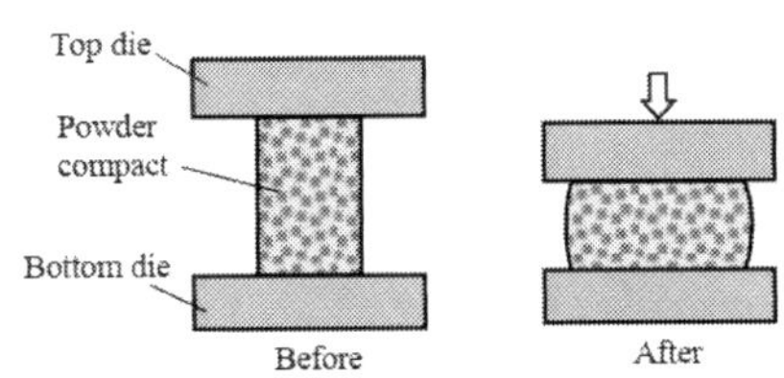

Fig. 2. Schematic illustration of simple upsetting test.

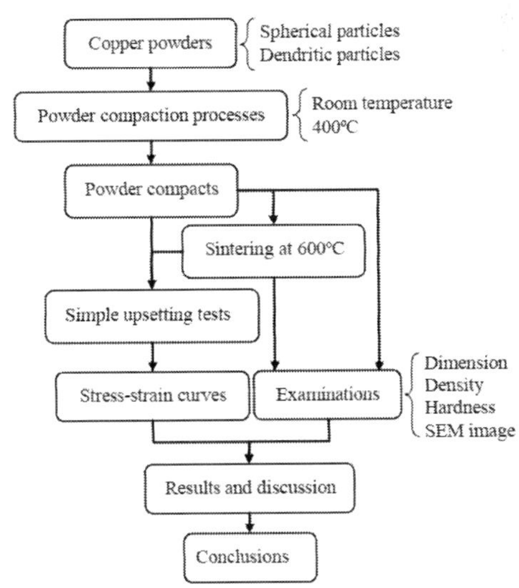

Fig. 3. Flowchart of the study.

DOI:10.1051/matecconf/201712300011

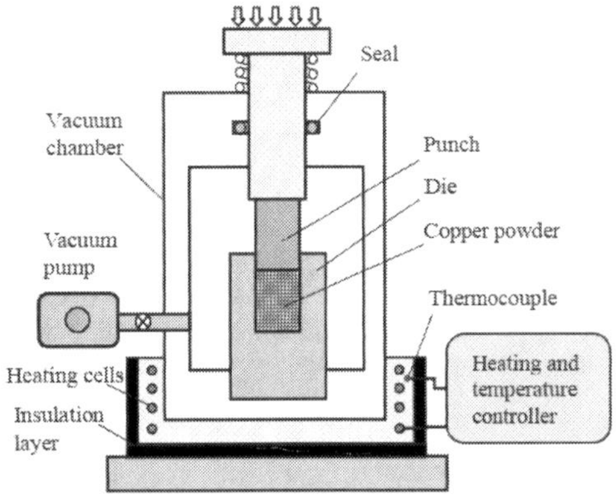

Fig. 4. Schematic diagram of the experimental setup.

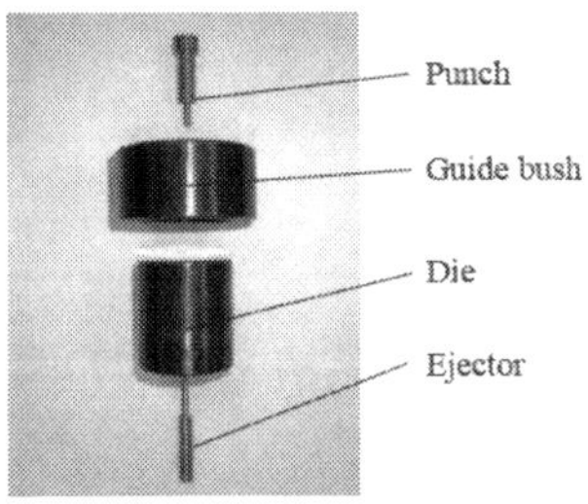

Fig. 5. The die components for experiments.

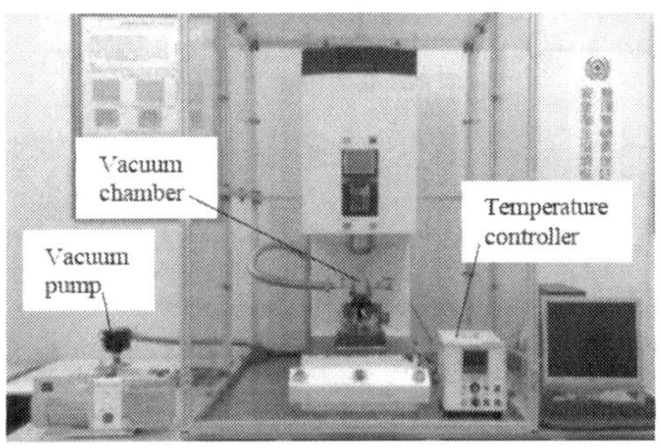

Fig. 6. The electro press installed with the vacuum chamber.

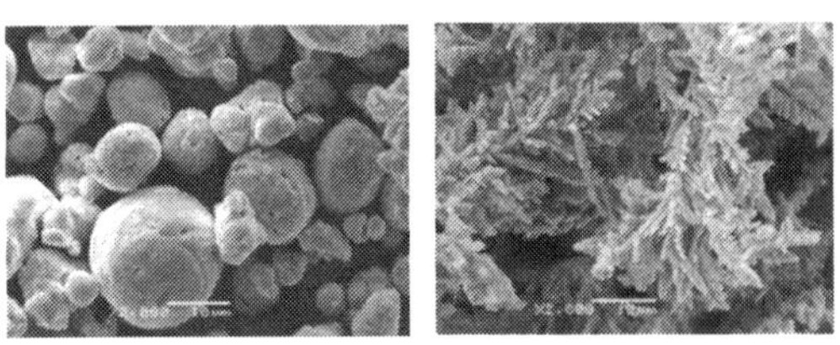

(a) Spherical powder (b) Dendritic powder

Fig. 7. SEM images of copper powder particles.

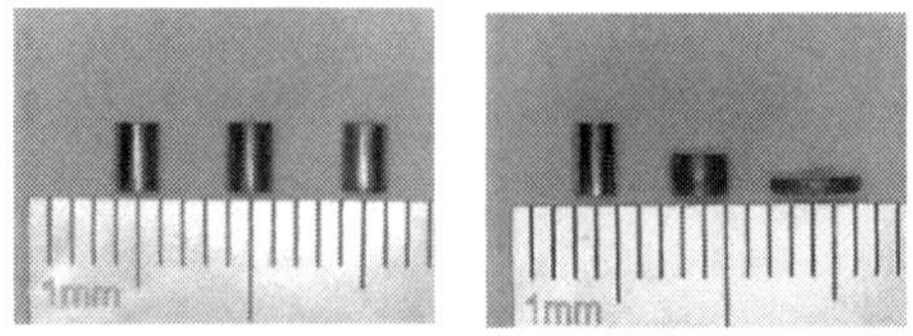

(a) Powder compacts (b) Deformed powder compacts

Fig. 8. The powder compacts and deformed powder compacts from simple upsetting tests.

3 Results and discussion

3.1 Dimensions and relative densities

The diameters and heights of the produced powder compacts (Fig. 9) were measured to understand the variations in dimensions and used to calculate the volumes for estimating the densities. Tables 1 and 2 list the dimensions of spherical and dendritic powder compacts, respectively. These data were average values of the measurements from 12 powder compacts. For the same weight and compressive pressure, the compacts of dendritic particles had smaller heights than those of spherical particles on the condition without sintering. The phenomena occurred in the compaction processes at room temperature as well as 400°C. The results implied that dendritic particles led to a higher relative density. Furthermore, after the spherical powder compacts produced at room temperature were sintered, their average diameter and height slightly increased by 0.5% and 0.6%, respectively, as summarised in Table 3. Raising the forming temperature to 400°C, the increase in the dimensions became greater. The sintering treatment had more considerable effects on the expansion of the dendritic powder compacts than those of the spherical powder compacts. The increase of the diameter and height raised by 5.2% and 10.4% respectively, in the powder compact produced at 400°C and treated by sintering.

In this study, the relative density was the ratio of the density of the copper powder compact to the density of solid copper. By using the measured weights and dimensions of the compacts, their densities and relative densities were calculated. Fig. 10 shows that the relative densities were about 93% in the compacts of spherical particles produced at both room temperature and 400°C. The effect of forming temperature on the relative density was not significant. The phenomena also occurred in the compacts of dendritic particles but the relative densities were higher and increased to about 97%. After the compacts were treated by the sintering process, the relative densities decreased. In the cases of room temperature, the relative densities of sintered compacts of spherical and dendritic particles were about 92% and 87%, respectively. The relative densities dramatically decreased in the cases of 400°C. In particularly, the sintered compact of dendritic particles had the lowest relative density about 78%.

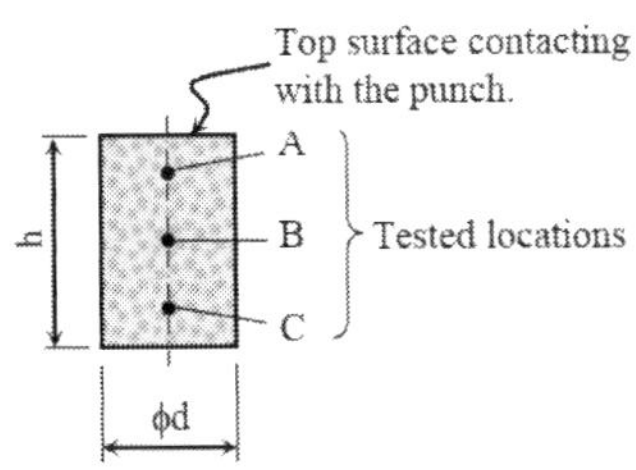

Fig. 9. Illustration of the dimensions and locations on the section of the powder compact for hardness tests and SEM images.

Table 1. Dimensions of spherical powder compacts.

Particle shape	Spherical			
Compaction temperature	Room temperature		400°C	
Treatment	Without sintering	Sintered	Without sintering	Sintered
d (mm)	2.001	2.019	2.003	2.072
h (mm)	3.002	3.031	2.989	3.091

Table 2. Dimensions of dendritic powder compacts.

Particle shape	Dendritic			
Compaction temperature	Room temperature		400°C	
Treatment	Without sintering	Sintered	Without sintering	Sintered
d (mm)	2.009	2.052	2.002	2.103
h (mm)	2.940	3.192	2.934	3.207

Table 3. Increase of dimensions after sintering.

Particle shape	Spherical		Dendritic	
Compaction temperature	Room temp.	400°C	Room temp.	400°C
Increase of d (%)	0.5	3.4	2.2	5.2
Increase of h (%)	0.6	3.5	8.2	10.4

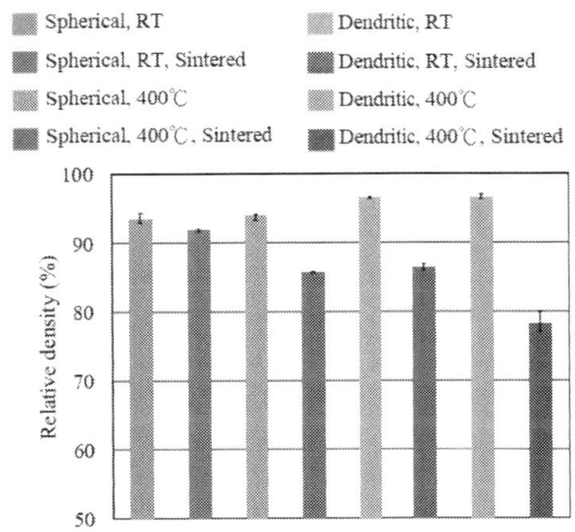

Fig. 10. Relative densities.

3.2 Hardness

Hardness tests were performed at three locations on the section of the powder compact as illustrated by Fig. 9. To correctly observe the hardness variations in the powder compacts, the surfaces contacting with the punch were marked after the compaction processes. Fig. 11 demonstrates that the dendritic particles resulted in greater hardness numbers ranging from about 130 (at location C) to 142 HV (at location A) in the cases of room temperature, and about 141 to 152 HV in the cases of 400°C. For the compacts made of spherical particles, the hardness numbers varied from about 106 to 123 HV. However, the effect of particle shape on the hardness was different in the sintered compacts. The hardness numbers dropped from high values to a range from about 63 to 87 HV in the cases of spherical particles. The effect of the sintering process was more significant in the cases of dendritic particles in which the hardness number decreased to 64 HV at location A for the process at room temperature, and to 46 HV at location C for the process at 400°C. Moreover, the highest and lowest numbers appeared at locations A and C, respectively, for all process conditions. The hardness number slightly decreased from the upper to lower locations that probably indicated the density variation in the powder compacts.

3.3 Micrographs

The SEM images of the spherical powder compacts produced at room temperature (Fig. 12) clearly revealed the deformations and boundaries of the particles in the case without the sintering treatment. However, in the sintered compacts, most of the boundaries have disappeared and pores have developed that could be caused by the bonding of the particles in the sintering process. The location A had less quantity of pores than the location C. This indicated that the quantity of pores slightly increases with the locations away from the top surface which contacted with the punch in the compaction process. In the cases produced at 400°C as presented in Fig. 13, the deformations and boundaries of particles were also visible. Furthermore, the effects of sintering on the bonding of particles and the development of pores were similar to those in the cases of room temperature. Comparing Figs. 12 and 13 shows that an increase in the compaction temperature to 400°C could improve the bonding of the particles that leads to the vanish of the boundaries of the particles.

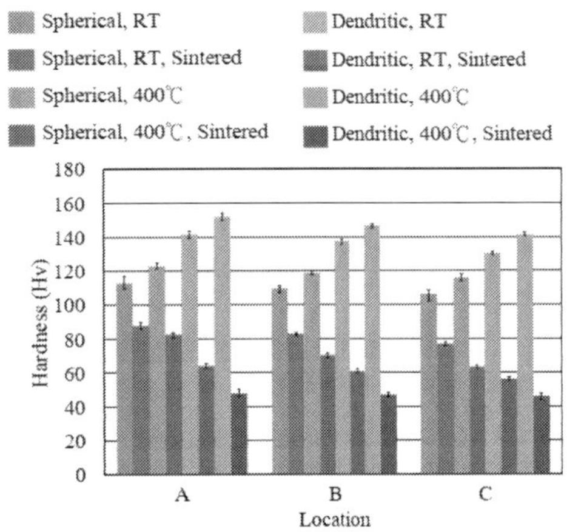

Fig. 11. Hardness.

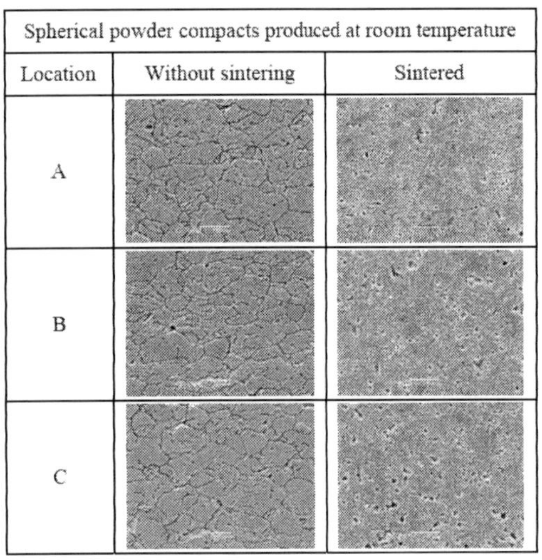

Fig. 12. SEM images of spherical powder compacts produced at room temperature.

The micrographs of the dendritic powder compacts obtained at room temperature displayed smaller boundaries of particles as presented in Fig. 14. After the sintering process, the bonding of the particles has occurred. Moreover, comparing Figs. 12 and 14 shows that the sizes of the pores in the dendritic powder compacts are smaller than those in the spherical powder compacts. This difference could be caused by an easy fracture and deformation of dendritic particles that leads to smaller sizes of particles in the compaction process. Furthermore, Fig. 15 displays that the bonding of particles has also occurred in the powder compacts produced at 400°C on the conditions without as well as with sintering. Some pores can also be clearly found and appear greater sizes in the cases with the sintering process. These trends could lead to lower densities and are in line with the previously discussed variations of the relative densities.

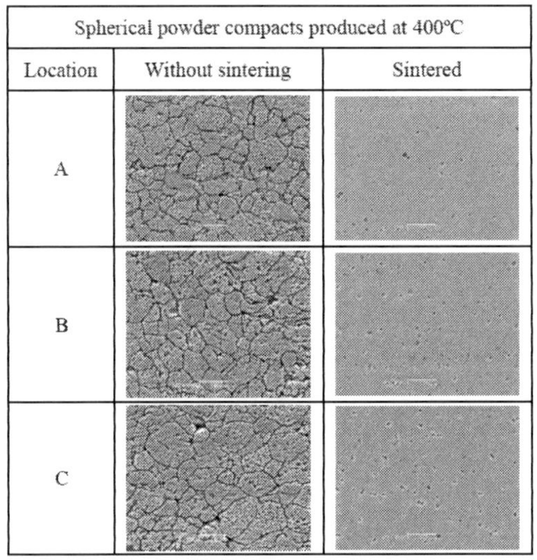

Fig. 13. SEM images of spherical powder compacts produced at 400°C.

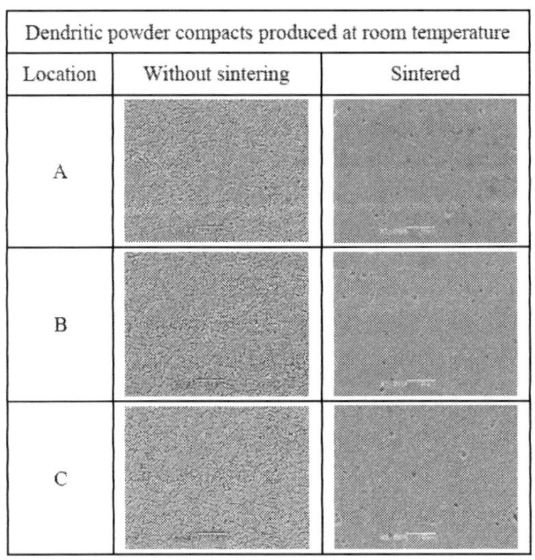

Fig. 14. SEM images of dendritic powder compacts produced at room temperature.

Dendritic powder compacts produced at 400°C		
Location	Without sintering	Sintered
A	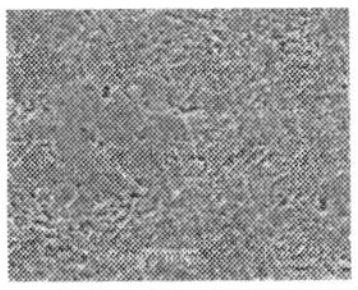	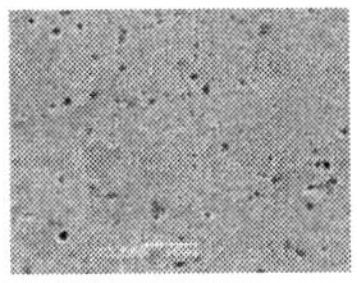
B	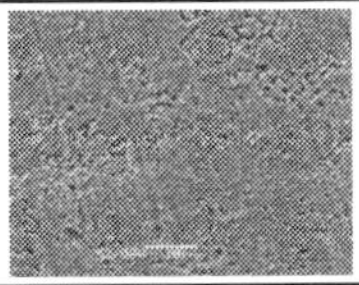	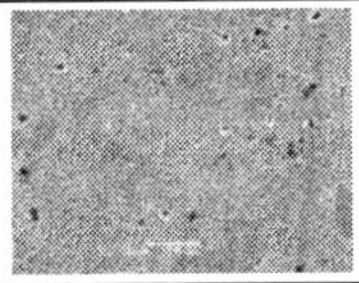
C	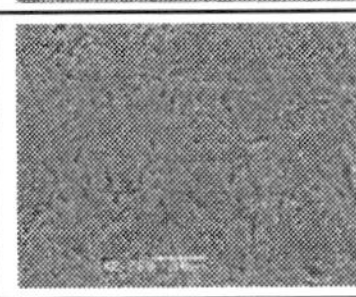	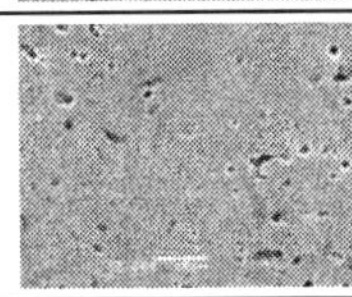

Fig. 15. SEM images of dendritic powder compacts produced at 400°C.

3.4 Stress-strain curves

The true stress-true strain curves of powder compacts are presented in Fig. 16. The copper powder compacts produced at room temperature from both spherical and dendritic particles demonstrated elastic deformations to a maximum compressive stress about 400 MPa and sudden fractures. The phenomena were also observed in the compacts of the spherical particles produced at 400°C although a slightly greater strength of 450 MPa was found. These results were probably caused by weak bonding of the particles that commonly appears in a green compact from a PM process. However, in the cases of dendritic particles, an increase of the compaction temperature to 400°C improved the ductility of the powder compact. Its true stress-true strain curve showed little strain-hardening and had a stress about 450 MPa at a strain of 0.8.

The sintering process significantly influenced the true stress-true strain curves of the powder compacts, and causes an increase in the ductility and great strain-hardening effects, as displayed in Fig. 16. The strength was higher in the cases of spherical particles than those of dendritic particles. It can also be seen that the dendritic powder compact formed at 400°C and treated by the sintering process resulted in the lowest strength among all tests. Moreover, the true stress-true strain curves of the sintered dendritic copper powder compacts were close to the curve of an annealed solid copper.

4 Conclusions

The study investigated the effects of particle shape and temperature on the compaction of copper powder at micro scale. Copper powder particles were compressed inside a cylindrical die cavity to form powder compacts. Two kinds of particle shapes and two forming temperatures were considered in the experiments. Some of the produced compacts were further treated by the sintering process. The study also used simple upsetting tests to investigate the characteristics of deformation of the powder compacts under compressive stresses. The results showed that the powder compacts produced at room temperature demonstrated brittle deformations. However, by increasing the forming temperature to 400°C, ductile deformations have been observed on the compacts of dendritic particles. Furthermore, the sintering treatment resulted in increases in dimensions, decreases in relative density and hardness, an increase in ductility, and pore growths. The phenomena were most significant in the dendritic powder compacts produced at 400°C and treated by the sintering process.

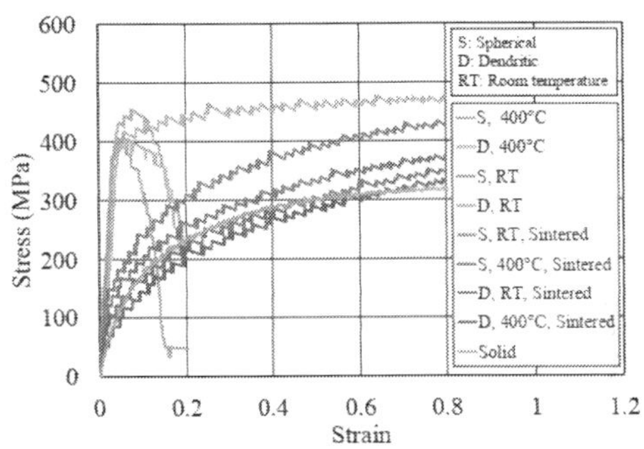

Fig. 16. True stress-true strain curves.

The authors gratefully acknowledge the financial support from the Ministry of Science and Technology, Taiwan (Grant No. MOST 105-2221-E-151-018).

References

1. G. Ryan, A. Pandit, D. P. Apatsidis. Biomaterials **27**(13), 2651 (2006)
2. A. Simchi, F. Petzoldt, H. Pohl, *J. Mater. Process. Technol.*, **141**(3), 319 (2003)
3. Y. Kathuria, *Surf. Coat. Tech.*, **116** 643 (1999)
4. S. Chikosha, T. Shabalala, H. Chikwanda, *Powder Technology*, **264**, 310 (2014)
5. R. Zauner, *Microelectron. Eng.*, **83**(4) 1442 (2006)
6. S. Heng, M. R. Raza, N. Muhamad, A. B. Sulong, A. Fayyaz, *Int. J. Refract Metals Hard Mater.*, **45**, 189 (2014)
7. K. Nishiyabu. Micro metal powder injection molding In Wang, J., ed. Some critical issues for injection molding (InTech, 2012).
8. S. Kalpakjian, S. R. Schmid, *Manufacturing engineering and technology* (SI 7e Ed., Pearson Singapore, 2013)
9. B. Kieback, A. Neubrand, H. Riedel, *Mat. Sci. Eng. A-Struct.*, **362**(1), 81 (2003)
10. A. Fayyaz, N. Muhamad, A. B. Sulong, J. Rajabi, Y. N. Wong, *J. Mater. Process. Technol.*, **214**(7), 1436 (2014)
11. Q. Bai, J. Lin, J. Jiang, T. Dean, J. Zou, G. Tian, *Mat. Sci. Eng. A-Struct*, **621**, 68 (2015)

Applying dynamic mold temperature control to cosmetic package design

Shih-Wen Hsiao[1], *Chih-Huang* Yen[1,*], and *Chu-Hsuan* Lee[1]

[1] National Cheng Kung University, Industrial Design Department, 701, Tainan, Taiwan

Abstract. Owing to the fashion trend and the market needs, this study developed the eco-cushion compact. Through the product design and the advanced process technology, many issues have improved, for instance, the inconvenience of transportation, the lack of multiuse capability, the increase of costs, and the low yield rate. The eco-cushion compact developed in this study was high quality, low cost, and meets the requirements of the eco market. The study aimed at developing a reusable container. Dynamic mold temperature control was introduced in the injection modeling process. The innovation in the product was its multi-functional formula invention, eco-product design, one-piece powder case design, and multifunctional design in the big powder case, mold flow and development of dynamic mold temperature control. Finally, through 3D drawing and modeling, and computer assistance for mold flow and verification to develop and produce models. During the manufacturing process, in order to solve the problems of tightness and warping, development and manufacture of dynamic mold temperature control were introduced. This decreased the injection cycle and residual stress, and deformation of the products has reduced to less than 0.2 mm, and the air tightness increased. In addition, air leakage was less than 2% and the injection cycle decreased to at least 10%.The results of the study can be extended and applied on the future design on cosmetic package and an alternative can be proposed to solve the problems of air tightness and warping. In this study, dynamic mold temperature control is considered as a design with high price-performance ratio, which can be adopted on industrial application for practical benefit and improvement.

1 Introduction

In a prosperous economy and aging society, the aspects of using cosmetic spread rapidly. Skin aging is an inevitable biological phenomenon of human life. Advancing age brings changes to all components of the integumentary system with consequent signs on the skin. Skin aging is mainly due to intrinsic (chronologic) and extrinsic aging (photo-aging). Photo-aging is a consequence of exposure to ultraviolet radiations [8]. People use cosmetic to secure the aging problem, and global cosmetic markets are increasing annually.

Figure 1 presents the data announced by international market research resource, Research and Market [5] indicating that the turnover of 2014 in the global cosmetic market was 46 billion US dollars in 2014 and it will grow to 67 billion US dollars in 2020. The annual growth rate is around 6.4%... Table 1 presents cosmetic categorization. Liquid foundation is a product combing water and power applied to the face to create an even color to the complexion, and to look pretty. Moreover, it is also a base layer for more complex cosmetics. The product was first introduced in 1937, and since then it has been a favor for female customers. However, the traditional package of liquid foundation resulted in the inconvenience of transportation and usage. In 2008, Korean skincare brand Amorepacific produced Cushion Compact [4] to solve the inconvenience. Thus customers can carry it and reapply makeup any time. The design principle was to inject liquid foundation to the sponge (cushion compact) [7] with 800 thousand air holes. To obtain the liquid foundation, customers simply press the pat. In each pressure, air will be driven into the liquid foundation. Thus it will be full of air and lighter than the normal foundation. Moreover, latex puff does not absorb water, thus, it will not attach too much powder. Lightly press the foundation onto your skin, the skin can absorb the powder, and to cover the flaws. During the developing process, the public did not accept the product widely. Turnover in the first year was only 3.8 billion Korean won (KRW). However, with the success in Korean drama in 2013, along with celebrity endorsement, subsidiaries of Amorepacific, Sulwhasoo, and Laneige sold cushion compact every 1.2 second. In 2015, cushion compacts from Amorepacific have been sold over 500 million. Among all, the biggest market is China, Taiwan, and Hong Kong. Turnovers in other compact cushion related products have risen for about 140% compared to other related products [3]. In the huge market opportunities, many cosmetic manufacturers were

* Corresponding author: jordan_yan@hotmail.com

attracted for research and development. Owing to the fashion trend and market demand, this study aimed to develop co-cushion compact.

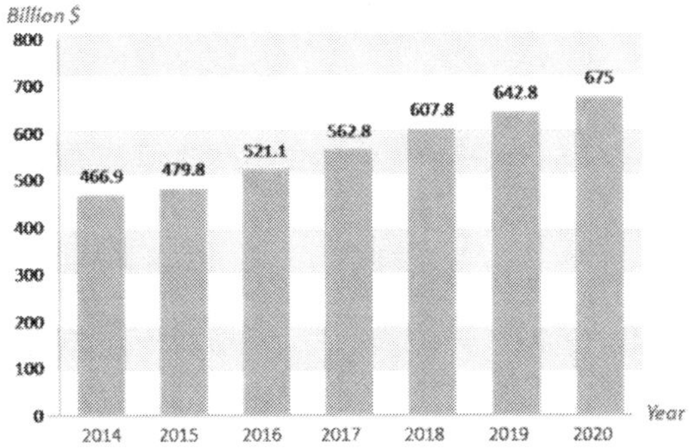

Fig. 1. Prediction for Global Cosmetic Market between 2014 to 2020.

Table 1. Cosmetic Categorization.

	Item	Details
Categorization	Cosmetics in General	Foundation, pressed powder, lipsticks, body makeup
	Cosmetics for eyes	Brow makeup, eye shadow, eyelid makeup, mascara, eye makeup, makeup remover
	Cosmetics for lips	Lip balm, lip gloss, lipstick, lip liner

1.1 Research motivation

Traditional liquid foundation underwent problem analysis. Figure 2 presents the problems of the traditional liquid foundation. Descriptions are illustrated in the followings.

1.1.1 The inconvenience of transportation

Liquid foundation in general belongs to liquid cream category. Bottles or soft tubes are normally used in packaging. To obtain the cosmetics placing in the bottles, customers simply press the lid. However, it is very likely to over squeeze due to the difficulty in controlling the proper amount of liquid. Moreover, similar problems can occur to soft tubes as well. If customers do not tighten the lid, the liquid foundation will be excreted. In terms of applying makeup onto your face, people normally collect the product in the palm of their hands and dip the finger into the liquid and smear on their face. However, this method will not provide a balanced complexion, and it will produce an uneven finish. Furthermore, skin infections may occur if hands are not properly sanitized.

1.1.2 Cost increase in containers

Traditional liquid foundation added some chemical synthetic for achieving specific functions such as protecting against burning, or as rejuvenation. Plastic containers are not compatible and thus corrosion might occur. Therefore, materials such as glass containers are normally used. However, the production of glass containers is high cost, and the container itself is heavier, and easily breakable. These lead to the increase cost in manufacturing.

1.1.3 Environmental pollution issues

Global cosmetic market grows stably every year. Regarding the ingredients and the function, the design of the package is also an important factor that consumers take into consideration when purchasing cosmetic products. Most of the consumers throw away the containers once they finish the products, which bring damage to the environment. According to the statistics calculated by the Environmental Protection Agency of USA, a quarter of landfill waste is made up of the disposable package. Thus, how to design an eco friendly package, which is reusable, is an issue that manufacturers need to face. At the same time, this is the future trend for eco cosmetics [9].

1.1.4 Inconvenience in storage

The amount of the liquid foundation is at least 30 ml. Therefore, for customers who use it rarely or those who use a small amount of foundation per time, liquid foundation remains a lot before it is expired or it will be dried. Thus, it is a waste.

Fig. 2. Problem analysis of liquid foundation in the market.

Problem analysis in industrial development

(1) An appeal to safe and natural products: People pay more attention on being healthy and nurturing life. Thus, the needs for organic cosmetic markets came out. When purchasing and using cosmetics, customers bear in mind products that are healthy, safe and efficient. Products that are organic, with a few chemicals or chemical-free are the ones that customers are looking for.

(2) Need in product diversity: Cosmetics plays an important role in cosmetic industry. Cosmetics used to target female customers, and now it has caught the

attention towards male customers. Along with the cosmetic technology and the invention of new cosmetics, cosmetic product is now approaching diverse and multiple functions. In terms of the applying time difference, there is makeup for morning, and night. Regarding the change of functions, there is makeup for sun protection, maintenance or it is water resistant. Considering the applying time, natural makeup, light makeup, outgoing makeup and so on, cosmetic is now with diverse function.

(3) Need in specific function: In the busy lifestyle, people still think highly of how they look. Cosmetics do not simply help make the skin looks better, but also provide maintenance and protection such as sun block, whitening, moisture holding, anti-aging to accord with people's need.

1.2 Research purpose

Based on the above problem analysis, this study invented an eco-cushion compact. Table 2 illustrated the proposed alternatives and research purpose.

Table 2. Alternatives and research purpose.

Item	Alternative	Research purpose
The development of cushion compact package	1. Small amount: 15g 2. Single thermoform-sealed packaging 3. Along with powder box, cushion compact can be filled in directly	1. This can solve the problems of products being expired, or dried. 2. Customers only open the package when applying. This ensures the product quality. This is convenience and hygienic.
The development of powder case	1. It can be replaceable, and reusable. 2. It can stand along, thus it is easy for transportation. 3. It is a one-pieced design, thus there is no post-process assembly. 4. The application of dynamic mold temperature control and systematic development.	1. It is consistent with eco design. 2. It provides multiple functions for customers. 3. It reduces production cost, and increases competitiveness. 4. It increases the yield rate.
The development of big powder case	1. It can be reusable. 2. It is elegant, fashionable and rectangular shape. 3. It has an interlocking structure. 4. The cushion is an air permeable design.	1. It is consistent with eco design. 2. It can be widely applied. 3. The powder box is less likely to fall. 4. It prevents cushion from stagnant smells and bacterial growth

2 Research structure

The study was based on the existing developmental process. In addition to investigating and developing a natural multifunctional formula, the aim is to provide an eco-friendly and reusable products. The technology of mold flow analysis and dynamic mold temperature control were used to develop eco-cushion compact related products. Thus, it reduced production cost and increased the yield rate.

Figure 3 presents the flow chart of the study. The development of formula and the sample production will be described in stage 1. Mold design and production will be illustrated in stage 2. Injection will be presented in stage 3.

2.1 Sample production and development

The initial design for cushion compact was a rectangular case. The product set included the development of cushion compact case, powder case (including powder), and big powder case (including cushion and related items). The internal and external design arrangement was shown in Figure 4. The design aimed at an eco friendly and reusable design. That features are illustrated as followings.

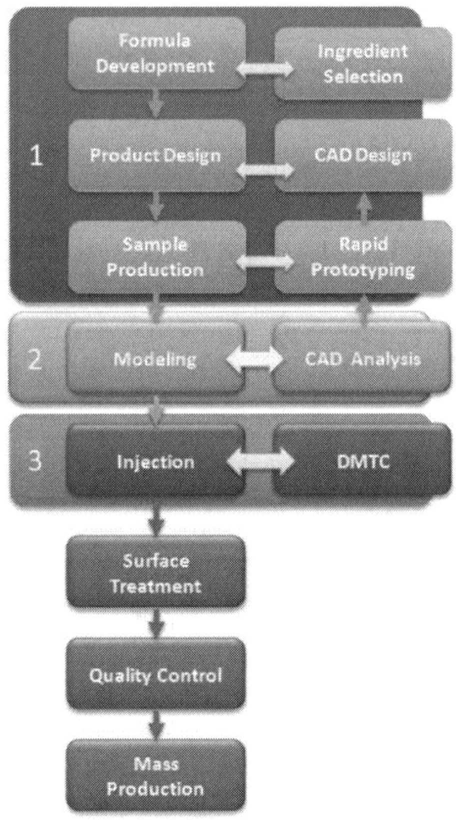

Fig. 3. Research structure and flow chart.

2.1.1 Cushion compact case

Cushion sponge is used to pad the powder liquid. It is a single thermoform-sealed packaging with small amount. Thus, customer can use it when they are going to use it. This ensured the product from exposing to the air and resulting in oxidation before it is used. Hence, it is fresh, and with stable quality. For products with bigger volume, it avoids product from going off due to the expiry date and provide an option for customers who use the products rarely or simply with a small amount. Figure 4(a) presents the case design and details.

2.1.2 Powder case

The design of the power case covered the shortcoming of traditional design. In the traditional design, it was a two-set device, and customer could not carry it outside. Now the container has changed to single package, which can stand along to carry outside. It is also reusable. In addition to providing customers multiple functions it accords with environmental need. It is a one-pieced design, thus, only one mold is required. Moreover, there is no post-process assembly, which reduces production cost and increases production. Figure 4(b) presents the case design and details.

2.1.3 Powder case

Big powder case: It is designed in an elegant rectangular shape. It is separated from the powder case to provide a clean shell. Moreover, it is reusable. The interlocking design makes it easy to remove and reseal. Furthermore, air permeable design of the cushion prevents cushion from stagnant smells and bacterial growth. Figure 4(c) presents the case design and details.

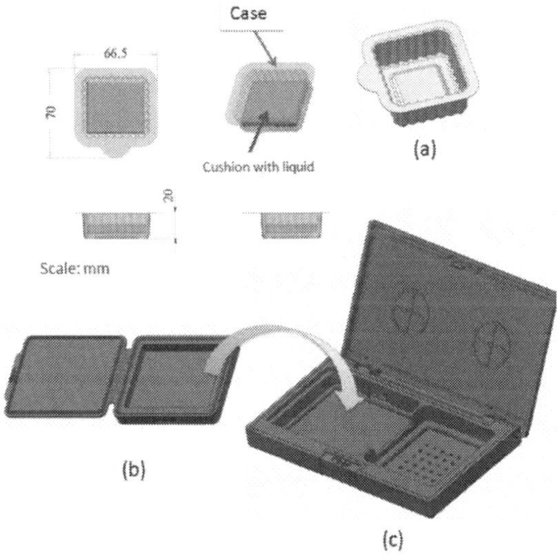

Fig. 4. The internal and external design arrangement.

2.2 Mold design and production

In the initial stage of developing mold for the products, mold flow was used for feasibility analysis of injection as shown in Figure 5. Through mold flow, optimal flow and cooling design [6] were observed. Thus, the maximum contraction controlled and maintained at 0.22 mm. Hence, it ensured the air tightness of the products, and avoided spillage and drying. It also saved time used in testing, and reduced the developing time and production cost. Furthermore, it increased the product quality.

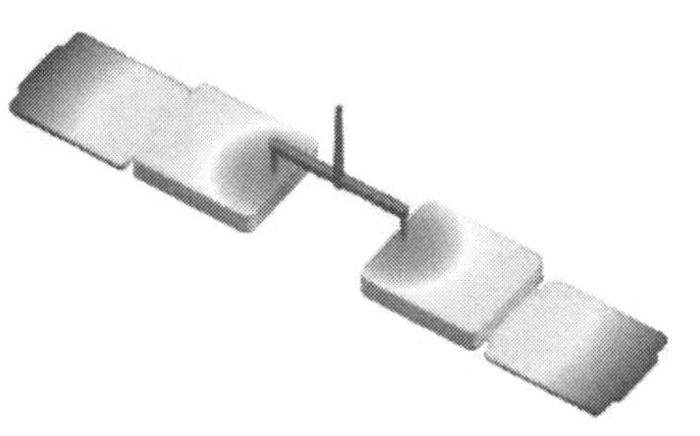

Fig. 5. Mold flow for the powder case

2.3 Injection

Dynamic mold temperature control was introduced [1, 2], plug-in mold temperature control system was designed as shown in Figure 6. Instant heating and cooling processes shortens the time for the mold to increase temperature and cooling. It shortens the injection period. Moreover, it reduces the residual stress and warping problem after the injection. This improves the surface brightness of the product, increases the product quality, accords energy saving and fulfills the market need in the green process.

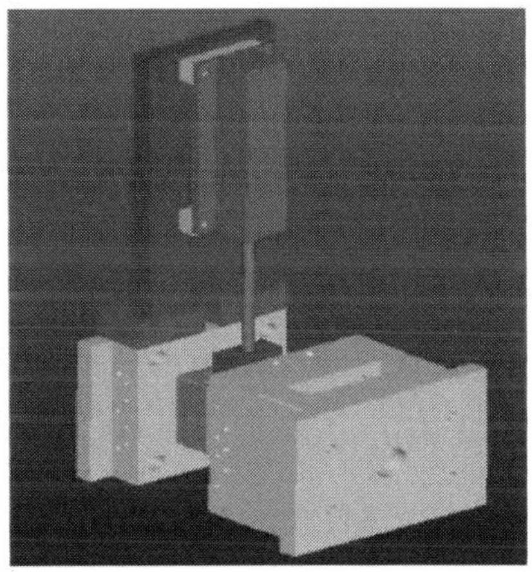

Fig. 6. Design of the dynamic mold temperature control.

3 Research method

3.1. Eco product design

Cushion compact is in the single thermoform-sealed packaging. Customer simply fills it to the powder case when applying the product. When the customer runs out of the cushion compact, they can easily placed it with a new one. Thus powder case can be reusable. The powder case is a single set design, the customer simply withdraws the case when applying, and the big powder case is less likely to be dirty. Power case is easy for transportation. Both the powder case and the big power case are reusable, which accords with eco design.

3.2 One-pieced design of the powder case

In the market, most of the powder cases are designed with upper and lower cases. During injection, two molds are required in manufacture along with manual assembly to complete the products. In this study, the powder case is a one-pieced design. Thus, only one mold is needed, and there is no post-process assembly, which reduces production cost and increases production.

3.3 Design of the multiple functions of the big powder case

Unlike the traditional circular design, an elegant rectangular shape was chosen to conform to customers' feel need among different countries. The internal arrangement and the clip were designed for customers to take out and utilize easily. The placement for cushion had ventilation holes, which can prevent cushion from stagnant smells and bacterial growth.

3.4 The introduction of mold flow

Before the development of mold, the assistant of computer analysis can be used to observe online the status of plastic in mold cavity. Thus, mold, product, and injection criterion can be modified, and the optimal design can be obtained. This saved the time used in testing, and reduced the developing time and production cost. Furthermore, it increased the product quality.

3.5 The establishment and introduction of dynamic mold temperature control

In the current cosmetic industry of package development, traditional injection molding is still the main stream. This method results in the difference between thickness and thinness of the products. Moreover, products are likely to be warped after the injection. Problems such as insufficient in jet flow and surface brightness also exist. Through the innovative mold temperature control technology, dynamic mold temperature control is applied to study. Figure 7 presents the advantages of the process. Before filling, the surface temperature of the mold is heated to glass transition temperature. This increased the fluidity of melt in the mold, decreased the injection pressure, and cutting force during the melt filling process. Furthermore, it reduced the residual stress and warping problem and increased the stability of the case. It avoided spillage and drying and at the same time, it shortened the injection period and increased the surface brightness of the product, the yield rate. This is a green technology.

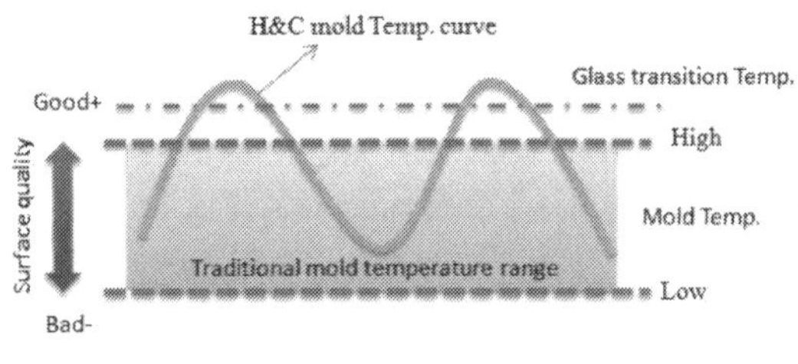

Fig. 7. H&C mold temperature curve and glass transition temperature.

3.5.1 Dynamic mold temperature control, DMTC

Dynamic mold temperature control has advantages such as improving product quality and shortening injection period. This technology includes "internal mold heating temperature controlling technology" and "mold surface heating temperature controlling technology". There are various ways of heating. After comparison between disadvantages and advantages, it can be seen that mold surface heating temperature controlling technology has higher heating efficiency and the ability to save energy.

Induction Heating Temperature Control, IHTC, adopted different types of induction coils for rapid heating on mold surface. This eliminates defects from the product such as surface stitch line and contraction. The feature of the technology is superficial heat of the mold surface, which is about 0.1mm. It resulted in rapid change of the temperature on the mold, which is about 20-40 degree Celsius. It was used to approach mold surface for heating. After designing, the system is able to heat both single-sided and double-sided mold. Infrared halogen lamp can operate vertical move, which is presented in Figure 8. Structure and related design of the temperature controlling set and injection machine is also shown in the Figure 8.

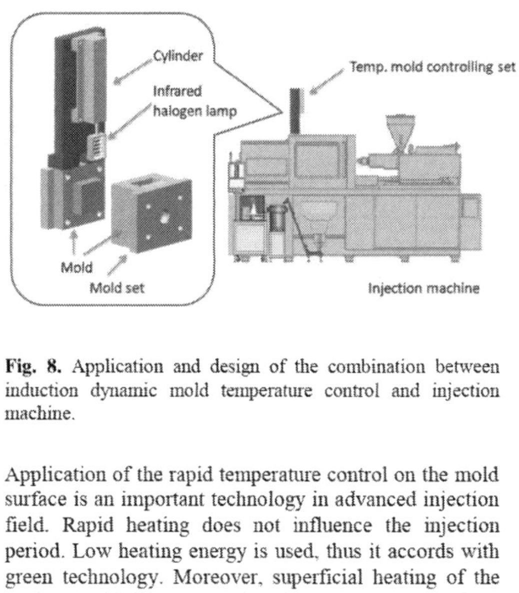

Fig. 8. Application and design of the combination between induction dynamic mold temperature control and injection machine.

Application of the rapid temperature control on the mold surface is an important technology in advanced injection field. Rapid heating does not influence the injection period. Low heating energy is used, thus it accords with green technology. Moreover, superficial heating of the surface mold does not lead to thermal expansion of the overall mold. Hence, it won't influence its usable life. On the other hand, in terms of mold cooling, to achieve the goal of rapid cooling, surface heating provided low heat energy to the mold. Thus, cooling process is faster. The freedom of the future temperature control technology can be increased efficiently.

4 Results and discussion

Results that were referred to green manufacture are described as follow.

4.1 Energy efficiency increased

It reduced 10% of the injection period, and yield rate of the product was over 95%. Moreover, it reduced the cost of manpower and saved mechanical energy. Temperature control of the traditional injection mold had strong influence towards product injection period and surface quality. Surface mold with high temperature is good quality; however it is very likely to be warped, as shown in Figure 9 (a). Introduction of dynamic mold temperature control increased the fluidity of melt in the mold and the heating efficiency. Moreover, it decreased the injection pressure, 10% of the injection period and increased the product size stability. In addition, it improved the appearance, and increased the yield rate to over 95%, as shown in Figure 9 (b).

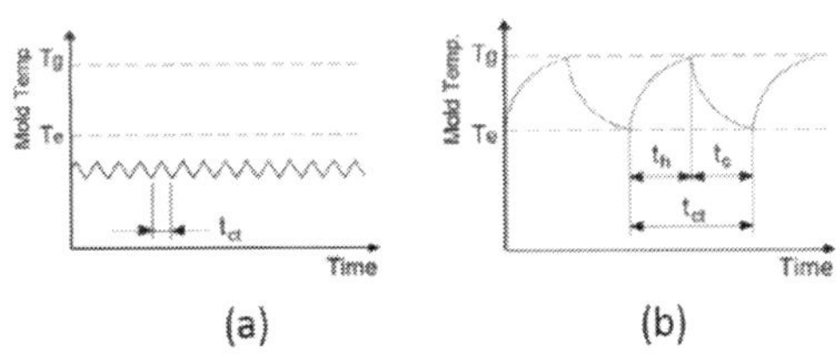

Fig. 9. Comparison of injection period between traditional injection and dynamic mold temperature control.

4.2. Material reduced

It reduced over 20% of the plastic material and over 40% of the metal material for the mold. The traditional powder case includes two set of mold as shown in Figure 10 (a). After manual assembly, post-process assembly is required for cosmetic filling process. In this study, the powder case is a one-pieced design, as shown in Figure 10 (b). Thus, only one mold is needed, and manual assembly is not needed, which reduces materials, mold cost and manual cost

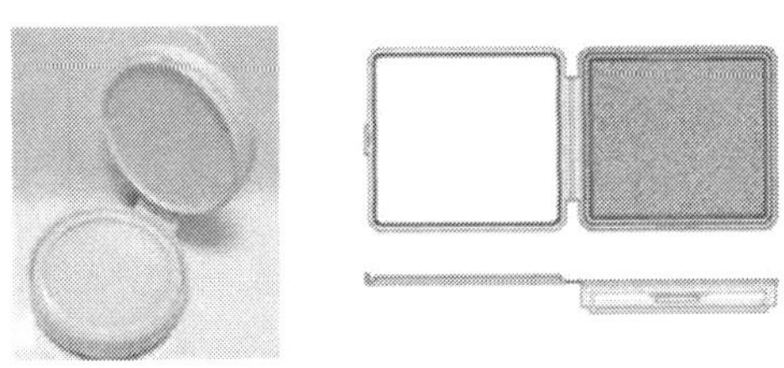

Fig. 10. Comparison of injection period between traditional injection and dynamic mold temperature control

Table 3 illustrates concrete results of the injection; meanwhile current technology and industrial condition are discussed.

Table 3. Advantage and improvement of the injection under DMTC

	Before	After
Technology Situation	Thickness and thinness difference of the powder case products and imbalance control of the mold temperature influence product injection quality. Long injection period leads to wrapped products for over 0.05 mm. Worse air tightness, air leakage is over 8%. Powder is very likely to spill or dry.	The development of dynamic mold temperature control reduced injection period, and residual stress. Warping problem is reduced to less than 0.2 mm. The increase of the air tightness leads to less than 2% of the air leakage. Injection period is decreased at least 10%.
Industrial Situation	Traditional injection cannot satisfy the rapid change of the industry. Considering manual cost and labor needs, most of the factories are set in China, or Vietnams. There is not much difference in products compared to its competitors. Hence, price cannot be raised to create higher income.	Dynamic mold temperature control is now developing in the country. Introducing the new technology is beneficial to future product competiveness and industrial development.

Acknowledgement

The authors are grateful to the Ministry of Economic Affairs of Taiwan for supporting this research under CITD, Grant No. E10500006011-274.

References

1. D. Yao, B. Kim, Development of rapid heating and cooling system for injection molding applications, *Polymer Engineering and Science*, **42**(12), pp. 2471-2481, (2002)
2. P. C. Chang, S. J. Huang, Simulation of infrared rapid surface heating for injection molding, *International Journal of Heat and Mass Transfer*, pp. 3846-3854 (2006)
3. http://www.harpersbazaar.com/beauty/makeup/g7327/best-cushion-compacts- summer-skin/
4. http://www.glamour.com/story/how-to-use-cushion-compact-foundation
5. http://www.researchandmarkets.com/reports/4040196/global-organic-cosmetics-market-forecast-and
6. D. Pierick, R. Noller, The effect of processing conditions on shrinkage, *Society of Plastics Engineers*, ANTEC, pp. 252-258, (1991)
7. Huh, J. W., Choi, K. H., & Jue, M. S. The accuracy of ultrasonography on the location of lipomas in the

forehead, *Program book (old book collection)*, **67**(2), pp. 443-444 (2015)

8. Tundis, R., Loizzo, M. R., Bonesi, M., & Menichini, F. Potential role of natural compounds against skin aging. *Current medicinal chemistry*, **22**(12), pp. 1515-1538 (2015)
9. Philippe, M., Didillon, B., & Gilbert, L. Industrial commitment to green and sustainable chemistry: using renewable materials & developing eco-friendly processes and ingredients in cosmetics. *Green Chemistry*, **14**(4), pp. 952-956 (2012)

CAD-based intelligent robot system integrated with 3D scanning for shoe roughing and cementing

Cheng-Chang Chiu[1,*], *Wen-Teng* Wang[2], and *Wan-Shan* Yin[1]

[1]Mechatronics Section, Metal Industries Research Development Centre,
[2]Taiwan National Formosa University, Taiwan

Abstract. Roughing and cementing are very essential to the process of bonding shoe uppers and the corresponding soles; however, for shoes with complicated design, such as sport shoes, roughing and cementing greatly relied on manual operation. Recently, shoe industry is progressing to 3D design, thus 3D model of the shoe upper and sole will be created before launching into mass production. Taking advantage of the 3D model, this study developed a plug-in program on Rhino 3D CAD platform, which realized the complicated roughing and cementing route planning to be performed by the plug-in program, integrated with real-time 3D scanning information to compensate the planned route, and then converted to working trajectory of robot arm to implement roughing and cementing. The proposed 3D CAD-based intelligent robot arm system integrated with 3D scanning for shoe roughing and cementing is realized and proved to be feasible.

1 Introduction

Roughing and cementing are very essential to the process of bonding shoe uppers and the corresponding soles, especially for the cold cement construction of shoe manufacturing. The purpose of roughing before cementing is to help breaking the smooth surface for better adhesion and thus reaching the desired bonding strength. The upper and the sole are usually designed to be in complicated freeform surface, which causes shoe manufacturers in the past required and relied greatly on manual roughing and cementing. In the recent years, because of the severe increase of personnel cost, awareness of labor right, and frequent happened strikes in Southeast Asia, those Taiwan shoe manufacturers who allocated factories in oversea started striving for transforming to automation manufacturing, and thus the automatic roughing and cementing process would be the first step toward automation. Cold cement shoes can be divided into two types: cup sole which has tall side wall to cradle the upper of the shoes and flat sole without side wall, as shown in Fig. 1. These two types have very different sole design, and the work trajectories on shoe sole for roughing and cementing are differed; thus the automation for these two different sole types need to be separated designed. Flat sole is attached directly to the bottom of the slip-lasted shoe upper, hence the area for roughing and cementing will be the whole shoe bottom; as for cup sole, the area will be extended to the margin of an upper. The side roughing will be presented on the die line of the sole as the yellow curve line shown in Fig. 1(b).

Fig. 1. Two sole types: (a) flat sole [1], and (b) cup sole [2].

Chuanyu Wu et al. developed a 5-axis apparatus where a 3D scanner and a work tool for roughing were fixed on the main axis respectively as well as a upper was positioned on the machine bed [3]. Yuzhen Jin et al. focused on the characteristics of shoe upper scanning and designed an infrared lasers 3D scanner specifically for shoe upper 3D measurement [4]. Chuanyu Wu integrated the achievements of [3] and [4] and further proposed a method to generate the work trajectory based on the obtained scanned shoe sole information. Francesco Jatta et al. designed a 6-DOF manipulator integrated with CAD/CAM to realize automatic bottom and side roughing and cementing of shoe uppers [6]. The above mentioned approaches that generated roughing working trajectory from CAD model of the upper and did not take the error caused by the previous process into consideration, thus only applicable to the situation of small error. Zhongxu Hu et al. combined a 3D scanner to a mechanical arm and roughing the surface via high pressure grit blasting [7], which was consisted of a rotary table to position the upper, a 3D scanner fixed on the linear slide mechanism, and and automatic manipulated blasting nozzle.

Wen-Teng Wang et al. propose an algorithm to generate and control the trajectory of roughing and cementing for an intelligent roughing and glue spraying robot arm with 3D visualization. The proposed

* Corresponding author: chang@mail.mirdc.org.tw

DOI:10.1051/matecconf/201712300013

algorithm was adoptable to different shoe uppers by taking advantage of the 3D visualization to obtain the data of the differences of different shoe shapes, and calculated the couture area of the shoe bottoms and working trajectories to drive the robot arm to complete the process automatically. Due to the features of every shoe upper is considered, the automatic roughing and cementing processes can fulfill different characteristics and quality requirements of different shoe types.

Different from the 2D modeling of the traditional footwear industry, owing to the current development of 3D CAD system, the application of 3D modeling in shoe manufacturing has been widely adopted, such as some international big brands like Nike, Adidas, UA and etc. Therefore, today's shoe factory has already had 3D shoe model before launching into production, which enables the automatic generation of roughing and cementing working trajectory to cup soles with high side walls. Based on the existing 3D models, this paper took advantage of the open characteristics of Rhino 3D CAD platform to develop a plug-in program of 3D graphic interface for generating the working trajectory of roughing and cementing. Further combined with the 3D visualization scanner to retrieve the real information of each work piece to compensate and turn the route, and then drive the robot arm to execute the processes of roughing and cementing.

2 Experiment and methodology

In order to equip the auto roughing and cementing system with broaden applications and more friendly operations, this paper introduced the Rhino 3D CAD platform and developed a plug-in program to plan and generate the working trajectory after the information of 3D shoe models is retrieved. Meanwhile, combined with the real-time 3D visualization scanner and algorithm of route conversion, the robot arm is auto-driven to complete the processes of roughing and cementing. The experiment set-up of the proposed design is presented in Fig. 2. The end effector of the robot arm is installed with clamp to fetch a lasted upper. After the lasted upper is clamped, the lasted upper is fixed in vertical or certain inclination angle to IR laser of the 3D scanner, moved horizontally at a stable speed, and 3D scanner is activated to retrieve the information of the upper. Followed with the calculation of roughing and cementing trajectory, the format of the robot arm working path is converted and loaded to the controller of the robot arm to drive the robot arm and the lasted upper to compete the roughing and cementing with planned route.

Fig. 2. Experiment set-up of the proposed design [8].

Figure 3 demonstrates the flow-chat of the proposed design, and the processes are elaborated as follows.

1. Activate Rhino 3D and load the plug-in roughing and cementing route planning program.
2. Start up the plug-in program and retrieve the 3D model of IGES format.
3. Plan the working trajectory with the curve rendering function of Rhino 3D.
4. Set the working trajectory and posture with plug-in program.
5. Output the working trajectory and posture.
6. Activate the modification program.
7. Read the working trajectory and posture.
8. Read the information of 3D visual scanner.
9. Perform the calculation of route compensation.
10. Output the route of robot arm.

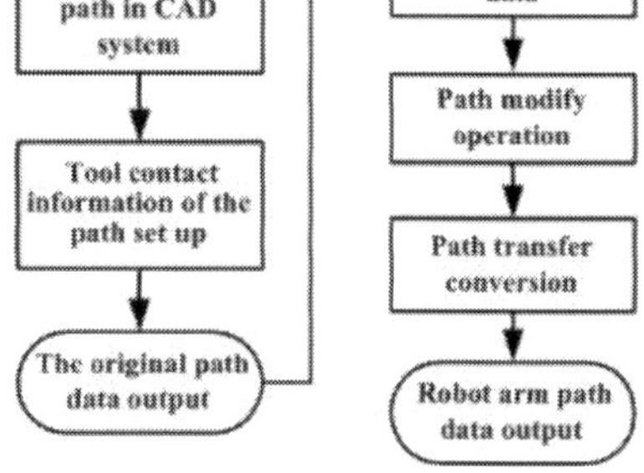

Fig. 3. Flow-chat of the proposed design.

3 Trajectory planning by 3D graphical interface of CAD platform

Due to the high side wall, the roughing and cementing trajectory of a cup sole is not available directly from the shape of upper. In order to enable the auto roughing and cementing of cup soles, the 3D design model is adopted and imported to the proposed plug-in program to do the working trajectory planning. Rhino 3D possesses strong function in free curve model and surface rendering and is in very reason price, which has made it popular to shoe manufacturers. Besides, Rhino.Net SDK helps developers to extend Rhino's functionality using Rhino plug-ins. Rhino.Net SDK supports VB.NET, C#, C++. NET, Delphi.NET and other programming languages to develop plug-ins. Due to Rhino.Net SDK itself uses C++ to package the category library, C++.Net would be the most direct and compatible option for plug-in development; thus, this study uses VC++2010 for Rhino.Net SDK plug-in development.

3.1 Operation and development of plug-in programs

1. Load the plug-ins to Rhino 3D.
2. Read the IGES file for the shoe 3D model.
3. Use the Rhino 3D curve rendering feature.
4. Generate the contact points of the roughing and cementing trajectory.
5. Set curve segmentations of the roughing and cementing posture.
6. Graphically set vectors of the sub-section for the roughing and cementing posture
7. Input the values of vectors of the sub-section for the roughing and cementing posture.
8. Output the trajectory of the contact points and posture.

3.2 Background design of roughing and cementing plug-in

In order to bring the reality of live operation and direct interaction with the hardware to users, this study designed to import the 3D model that is build according to the real size and position of the hardware to be the background of the plug-in, as shown in Fig. 4.

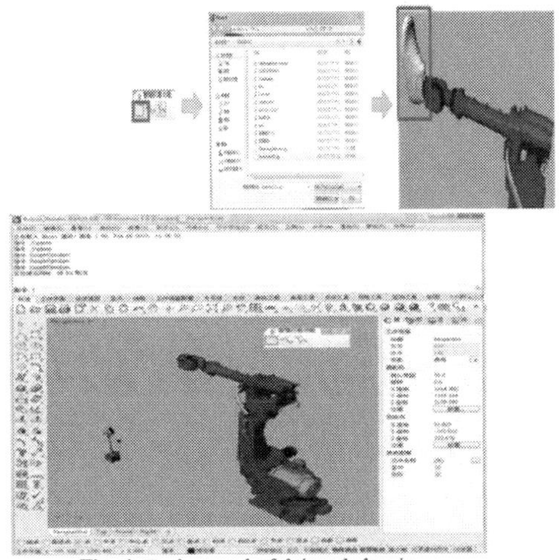

Fig. 4. Background of deigned plug-in.

3.3 Functionality development of plug-in

The functions of the plug-in program are developed by using VC++2010 to refer to the API functions provided by Rhino.Net SDK, and the function algorithm is introduced into the plug-in program. Table 1 shows the main API functions referenced from Rhino.Net SDK in this study.

Table 1. Main API functions used by plug-in program

<table>
<tr><th>Function</th><th>Corresponding Functions</th></tr>
<tr><td>Read IGES files</td><td>1. //define model's format in IGES
const wchar_t* szFilter = L"IGES Files (*.igs;*.iges)|*.igs; *.iges||";
2. //show dialog to choose a model
CFileDialog dialog(TRUE, L"igs", 0, dwFlags, szFilter, pParentWnd, 0, TRUE);
3. //script an imported model
script.Format(L"_-Import \"%s\" _Enter _Enter _Enter", filename);</td></tr>
<tr><td>Generate contact points for roughing and cementing</td><td>1. //choose a desired curve
CRhinoGetObject go;
go.SetCommandPrompt(L"Select curve");
2. //get the curve's length
CRhinoGetNumber gn;
gn.SetCommandPrompt(L"Length from start");
gn.SetLowerLimit(0.0, TRUE);
gn.SetUpperLimit(crv_length, TRUE);
gn.GetNumber();
if(gn.CommandResult() != success)
return gn.CommandResult();
double length = fabs(gn.Number());
3. //get the curve's length
ON_3dPoint pt = crv->PointAt(length);
context.m_doc.AddPointObject(pt);</td></tr>
<tr><td>Rhino Command prompt</td><td>gn.SetCommandPrompt(L"Do you want to modify the vector?(Yes:0/No:1)");</td></tr>
</table>

3.4 Function of designed plug-in

The process of the roughing and cementing trajectory planning by plug-in program is explained as follows.

1. Read the IGES format file of the 3D model. IGES is the standard exchange format of geometric information for CAD/CAM system, thus 3D model is format in IGES in this paper. Figure 5 shows the operation procedure and results of the plug-in.

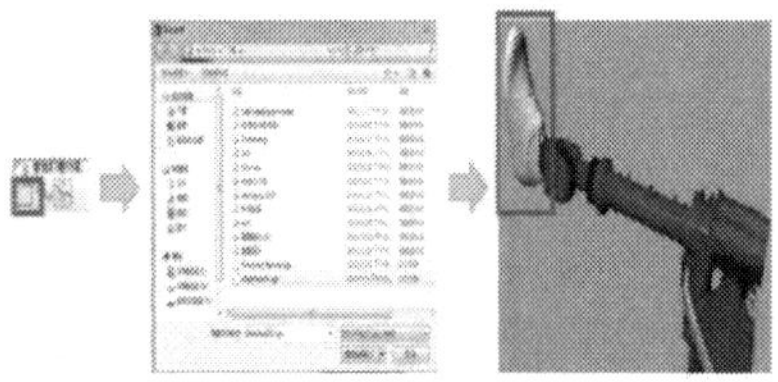

Fig. 5. Plug-in program reads 3D shoe model.

2. Using Rhino 3D graphic interface to render the contours of roughing and cementing area in the existed 3D model. For flat-bottom shoe type, the planned contour would be the max peripheral profile of the bottom. For the cup sole shoe type, the planned contour would be the die line of the 3D shoe sole model which mapped on the upper. Figure 6 is the planned contour for the cup sole type shoe.

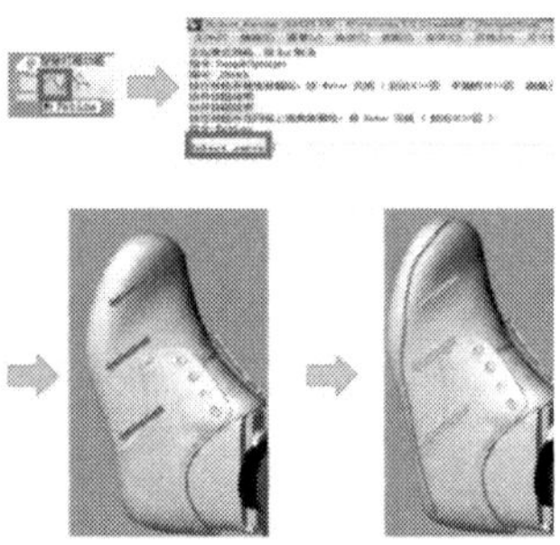

Fig. 6. Contour planning for cup sole shoe type.

3. Vector setting for the segments of the roughing and cementing contour posture. Plug-in program is employed to select the segmentation points on the contour, as shown in Fig. 7.

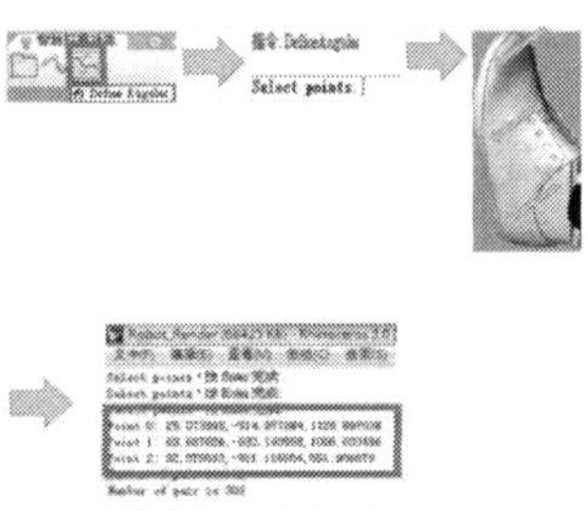

Fig. 7. Segmentation point selection.

4. Set the vectors of the contact posture of each segment. The contact angle of roughing on the segments of the upper may be different; the plug-in can be further developed to individually set the information for the specific segment. Figure 8 shows the vector setting of the segmented contours.

5. Output the contact points and the corresponding vectors for the roughing and cementing posture.

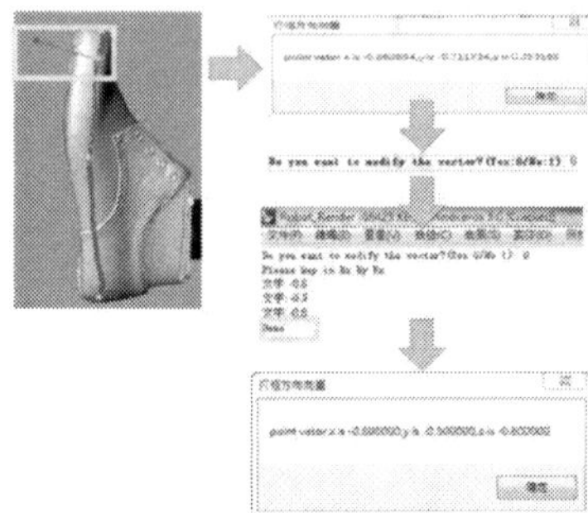

Fig. 8. Vectors setting of each segment of contour.

4 3D Visualization

In practice, there existed errors in the previous process of upper manufacturing; therefore, the trajectory planning requires modification and compensation with the 3D geometry information of the upper retrieved by the 3D scanner. When the scanning is completed, the 3D coordinator data originated at the 3D scanner is obtained and then real-time transformed to origin of the coordinator to the end effector of the robot arm.

4.1 Application of 3D visualization

The proposed system is developed to fulfill the requirement of small quantity, customization, and fast production of footwear manufacturing; hence, the system platform is designed to be equipped with the 3D visualization system Gocator 2340 to scan the lasted upper clamped by the robot arm. The 3D scanner is fixed on the platform; the robot arm clamps the lasted upper and move toward the scanner and then move horizontally in stable velocity for scanning the lasted upper, shown as Fig. 3.

4.2 3D scanning of cup sole shoe type

In order to retrieve the 3D geometry information of the cup sole shoe type, the lasted upper would be scanned from three different angles. Figure 9 (a) presents the three scanned postures of the cup sole lasted upper, respectively are right, bottom, and left. After the geometric information of these three postures is obtained, the synthesis diagram is shown as Fig. 9 (b) and saved for later trajectory compensation.

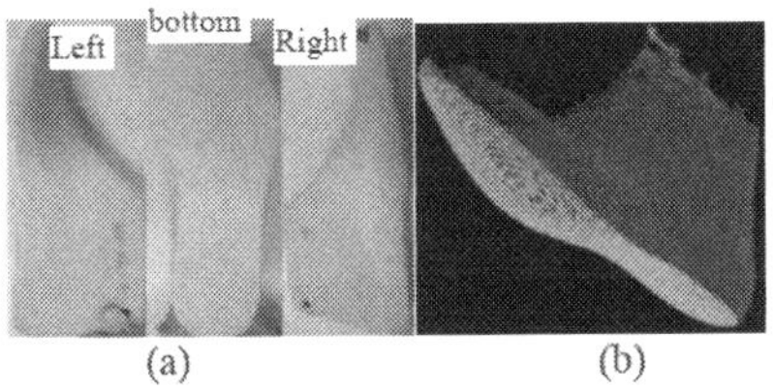

Fig. 9. Scan Postures of cup sole shoe type: (a) Three Postures, and (b) 3D Fitted Plot.

5 Trajectory compensation and modification

Sport shoes is the one that require the highest functionality among all shoe types; hence, the materials for the upper, except leather or fabric, would design to use foam to rise the comfortability of wearing, or add the supported hard structure, such as stiffener in back parts, to create additional support. Due to the combination of different materials in upper manufacturing, it causes differences even to the same size of the same shoe type.

Therefore, the precious working trajectory of shoes like these cannot be obtained directly and only with the 3D shoe upper model; thus, the assistance of 3D visualization system is required to retrieve the real geometry information of every shoe. Further, via the algorithm to calculated and then compensate the differences of each shoe upper, and then the robot arm is possible to correctly execute the process of roughing and cementing.

The compensation process is descripted as follows.

1. Read the above stated scanned data of two sides scanning of the lasted upper.
2. Construct the data of the lasted upper to KD tree format.
3. Search 3 nearest scanned points of each standard contour point in KD tree.
4. Project each standard contour point to the surface constructed by the 3 nearest scanned points, and thus the compensated contact points of the roughing and cementing trajectory are obtained.
5. Fit the compensated points to NURBS curves, smooth the curves, and generate the new contact points.
6. Calculate the origin coordinator of the end effector with Rx, Ry, and Rz according to the transformation matrix of the robot end effector.
7. Output the JBI file for the robot arm trajectory.

5.1 KD tree data structure

KD Tree is a space-partitioning data structure for organizing points in a k-dimensional space. KD tree structure is characterized to fast search the nearest neighbor in space, which is often applied to search the nearest adjacent point to the specific point group. As shown in Fig. 10, for the 2-dimensional point data, KD Tree would separate the points into two sides. The points greater than the root are in the right side, and each node will be separated in the next level of dimension till all the levels of dimensions used and the classification is completed [9].

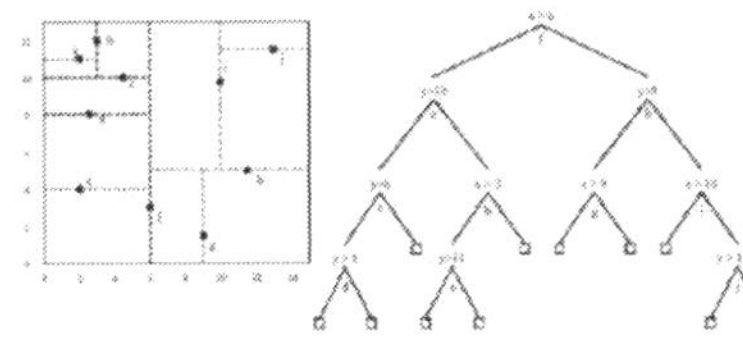

Fig. 10. 2-Dimension KD Tree [10].

5.2 Contact Point Compensation on 3D Fitted Plot

Because there are errors existed between the contact point of the 3D scanned data and the existed 3D model, this study compensated the errors by projecting the contact points at its nearest scanned point. Via the above mentioned KD tree format, on the scanned data of the lasted upper, three nearest neighbor points to the original contact point are searched and constructed a surface; and then the new compensated contact point is calculated as shown in Fig. 11. Figure 12 presents the result that all the contact points of the roughing and cementing contour are compensated.

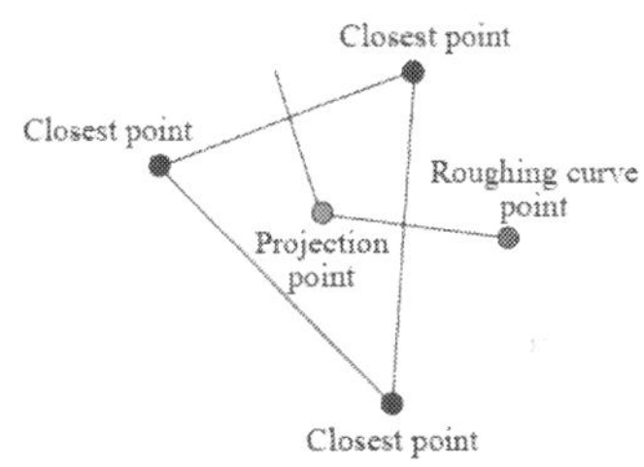

Fig. 11. Diagram of Contact Point Projected to Surface Constructed by Three Neighbor Points.

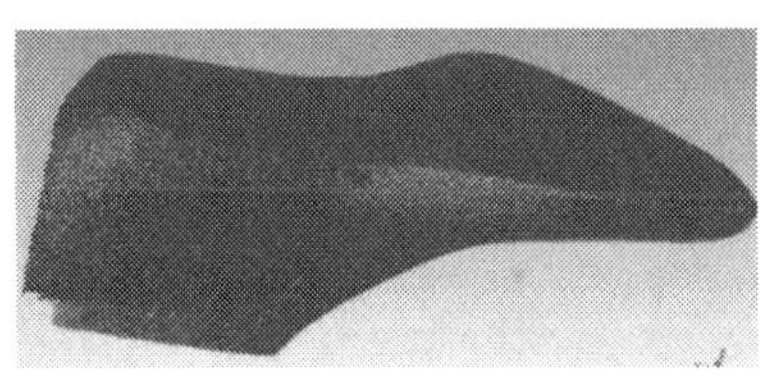

Fig. 12. New contact points after compensation.

5.3 Fitting and smoothing of NURBS curve

Figure 13 shows the contact points of the contour line after compensation, which presents to be a fluctuant line. If these contact points are directly converted to the working trajectory of robot arm that might cause vibration to the robot arm because of an uneven route.

Therefore, before the calculation of trajectory conversion, smoothing is required. In this study, the non-smooth contact points were fitted to Non-uniform Rational B-spline (NURBS) curve, and then the contact points were rearranged in accordance with the fitted curve to generated new contact points after smoothing.

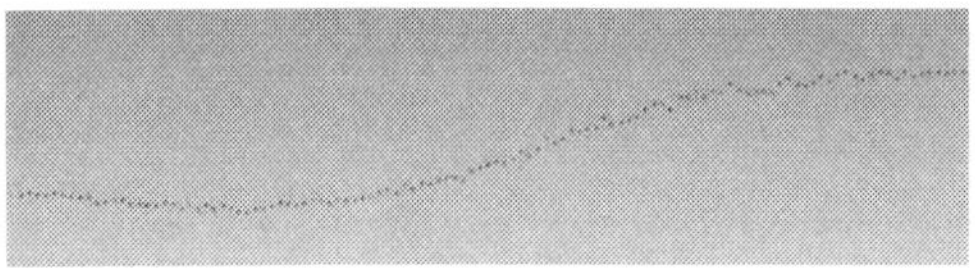

Fig. 13. Enlarged Segment of Compensated contact Points Before Smoothing.

The mathematic definition of Non-uniform Rational B-spline (NURBS) is descripted as Eqs (1) to (4), where n+1 is the number of control factor, k number of the contact points, $N_{i,k}$ the basic function of B-spline, P_i control points, and u the parameter value.

$$C(u) = \sum_{i=0}^{n} N_{i,k} P_i \qquad (1)$$

$$N_{i,0}(u) = \begin{cases} 1, & if\ u_i \le u < u_{i+1} \\ 0, & otherwise \end{cases} \qquad (2)$$

$$N_{i,p}(u) = \frac{u - u_i}{u_{i+p} - u_i} N_{i,p-1}(u) + \frac{u_{i+p+1} - u}{u_{i+p+1} - u_{i+1}} N_{i+1,p-1}(u) \qquad (3)$$

The curve fitting is processed by matching to a $C(u_j)$ NURBS curve to make the contact point q_j and the curve to have the smallest error e calculated as Eq. (4).

$$e = \sum_{j=0}^{m} [c(\bar{u}_j) - q_j]^2 = \sum_{j=0}^{n} \left[\sum_{i=0}^{n} N_{i,k} p_i - q_j \right]^2 \qquad (4)$$

That is, if n and k are set to be constant, the control point of P_i curve is obtained by calculating the extremum of $min\ (e) = 0$. And then the fitted curve is used to smooth u and re-distribute the contact points, thus the new smoothed contact points are acquired shown as Fig. 14.

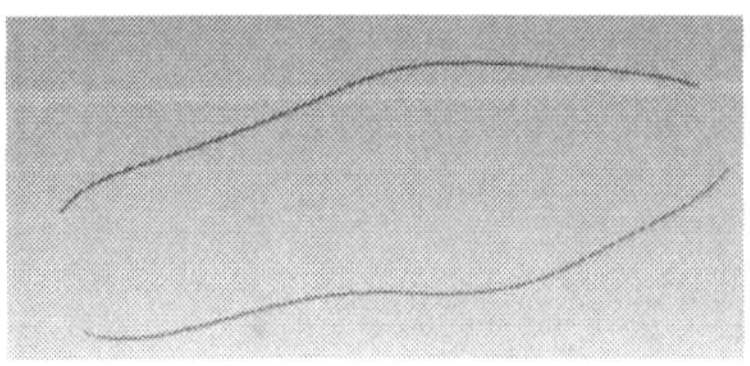

Fig. 14. Smoothed Contact Points for Roughing and Cementing.

6 Robot arm

The robot arm employed in this study is Yaskawa MH6, which is an 6-axis articulated manipulator, portable wight 6 kg, and repeatbility ±0.08mm. The research system setup is shown as Fig. 2.

7 Experimental result and discussion

In this study, the path is designed by the plug-in program based on Rhino CAD platform, and its data is generated automatically as shown in Figure 15. Each contact point data contain position components(x, y, z) and orientation components (Rx, Ry, Rz) arranged in a row. The six values represent x, y, z, Rx, Ry and Rz from left to right. In addition, the units of position and orientation are in [mm] and [degree] respectively.

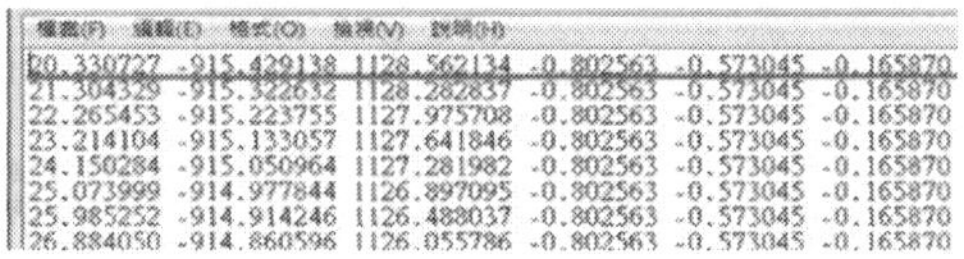

X	Y	Z	Rx	Ry	Rz
20.330727	-915.429138	1128.562134	-0.802563	-0.573045	-0.165870
21.304329	-915.322632	1128.282837	-0.802563	-0.573045	-0.165870
22.265453	-915.223755	1127.975708	-0.802563	-0.573045	-0.165870
23.214104	-915.133057	1127.641846	-0.802563	-0.573045	-0.165870
24.150284	-915.050964	1127.281982	-0.802563	-0.573045	-0.165870
25.073999	-914.977844	1126.897095	-0.802563	-0.573045	-0.165870
25.985252	-914.914246	1126.488037	-0.802563	-0.573045	-0.165870
26.884050	-914.860596	1126.055786	-0.802563	-0.573045	-0.165870

Fig. 15. Format of Contact Points.

Four Snapshots in Figure 16 show the proposed system implementing the processes of roughing and cementing for a cup sole shoe type after the integration of hardware and software. The white blush is used as roughing tool. According to a path, the blush can contact the shoe well along the curve during 1st snapshot→ 2nd snapshot and 3rd snapshot→ 4th snapshot. However, the shoe runs an unexpected path and pushes the blush suddenly during 2nd and 3rd snapshots, when the robot changes its orientation to achieve the desired contact angle. Thus, it should be interpolated some transition points to escape collisions.

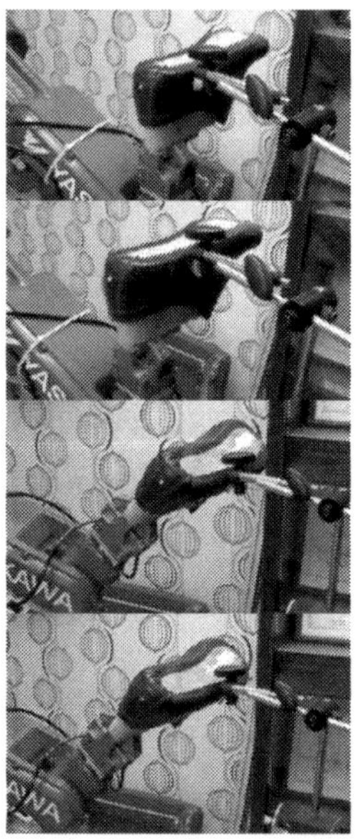

Fig. 16. Testing of Proposed System Implementing Roughing and Cementing on Cup Sole Shoe Type.

In the plug-in program, it is assumed that the shoe contact with a roughing tool at a sharp point, and the contact force is omitted. In the experiment, the contact is a plane, and it has an obvious friction force. Therefore, the blush is damaged over time.

8 Conclusion

Designated to the current shoe industry situation of customization, small quantity, and variation of shoe types, in this study, the automatic robot arm is developed, which is capable to automatically compensate errors between 3D model and each lasted upper and able to implement the processes of roughing and cementing on the cup sole shoe type. The developed plug-in program for Rhino 3D CAD platform enables the contour planning for roughing and cementing on 3D graphic interface, the calculation of the compensated contact points with the real-time 3D scanned information, and generation of working trajectory of Yaskawa robot arm after smoothing and coordinator conversion. Thus the employed robot arm are successfully driven to implements the roughing and cementing. The proposed system is potential to further development, such as equipping force sensors on the robot arm to acquire the value of the force applied on roughing shoe uppers for advanced adjustment of the roughing position for the purpose of higher accuracy on error compensation.

References

1. http://www.adidas.com/us/originals-shoes
2. http://www.nike-taiwan.com.tw/
3. Chuanyu Wu, Yong Ni, Qinchuan Li, Yuzhen Jin, Implementation of 5-DOF apparatus used for adhesive spray and roughing of shoe upper, *Proceedings of the IEEE International Conference on Automation and Logistics*, Qingdao, China, pp. 1408-1413, September 200.
4. Yuzhen Jin, Lei Zhang, Chuanyu Wu, Zuchao Zhu, Detection of 3D curve for shoe sole spraying based on laser triangulation measurement, *Proceedings of the IEEE International Conference on Automation and Logistics*, Shenyang, China, pp. 865-868, 2009
5. Chuanyu Wu, Xiaojie Mao, Xiangjun Shi, Lei Zhang, Methods of generating robot spraying trajectory based on shoe sole information, Intelligent systems and applications, *ISA 2009. International Workshop*, pp. 1-4, 2009
6. Francesco Jatta, Lorenzo Zanoni, Irene Fassi, S. Negri, A roughing/cementing robotic cell for custom made shoe manufacture, *Int. J. Computer Integrated Manufacturing*, **17**(7), pp. 645-652, 2004
7. Z. Hu, C. Marshall, R. Bicker, P. Taylor, Automatic surface roughing with 3D machine vision and cooperative robot control, *Robotics and Autonomous Systems*, **55**(7), pp. 552-560, 2007
8. W. T. Wang, T. Y. Lee, C. M. Yang, W. S. Yin, Research of 3D vision intelligent robot system for shoe roughing and cementing, *Proceedings of 2015 CSME*, 2015/12
9. T. H. Chu, H. Kuo, W. Su, Cloud point PCL Learning, Beijing University of Aeronautics and Astronautics Press, 2012/10
10. https://en.wikipedia.org/wiki/K-d_tree

Warpage behavior analysis in package processes of embedded copper substrates

Yeong-Maw Hwang[1], *Tsang-Hung* Ou[2], and *Wei-Hung* Su[1,*]

[1]National Sun Yat-Sen University, Department of Mechanical and Electro-Mechanical Engineering, 80424 Gushan Dist, Taiwan
[2]Advanced Semiconductor Engineering, 81170 Nanzi Dist Kaifa Rd. No.25, Taiwan

Abstract. With the advance of the semiconductor industry and in response to the demands of ultra-thin products, packaging technology has been continuously developed. Thermal bonding process of copper pillar flip chip packages is a new bonding process in packaging technology, especially for substrates with embedded copper trace. During the packaging process, the substrate usually warps because of the heating process. In this paper, a finite element software ANSYS is used to model the embedded copper trace substrate and simulate the thermal and deformation behaviors of the substrate during the heating package process. A fixed geometric configuration equivalent to the real structure is duplicated to make the simulation of the warpage behavior of the substrate feasible. An empirical formula for predicting the warpage displacements is also established.

1 Introduction

Thermal bonding of copper pillar flip chip packages is a relatively new bonding process in packaging technology field, especially for a substrate with an embedded copper trace. During the heating package process of an embedded copper trace, the substrate warps at high temperatures as shown in Fig.1 [1]. These warpage causes a gap between the substrate and die, which will lead to welding failure.

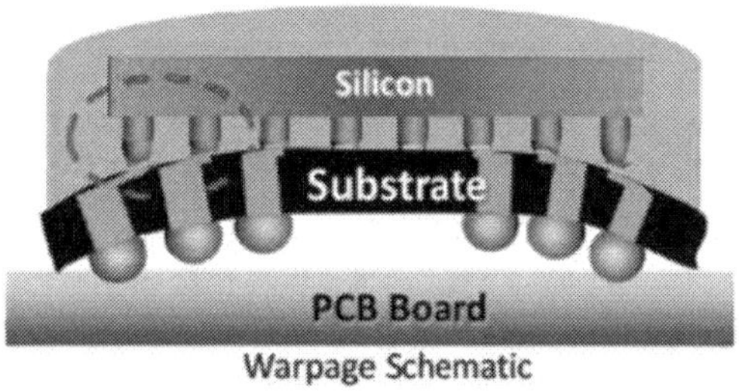

Fig. 1. Warpage schematic during heating package process of an embedded copper trace.

Kim et al. [2] used an anisotropic shell model, considering their viscoelastic properties, to simulate the warpage behavior of a high-density multilayer printed circuit board for solid-state disk drive, with homogenized copper patterns. Both the maximum warpage and the residual warpage of the full microelectronic package were predicted.

McCaslin et al. [3] developed a methodology to predict the warpage of a particular substrate. The methodology accounted for both the trace pattern planar density and planar orientation in material property calculations for each layer of a multilayer substrate. Their results using the developed methodology agreed with the experimental data.

Tsai et al. [4] used shadow moiré method to measure the warpage of flip-chip PBGA packages subject to thermal loading. They used the finite element method and Suhir's theory [5] to analyze the thermal deformations of the package and discuss the mechanism.

The substrate geometry discussed in the literature was relatively simple, this study uses a Finite Element Analysis software ANSYS to simulate the embedded copper trace substrate and a model for analysis of thermal behavior at high temperatures is established [6]. An empirical reference formula from the finite element analysis results is established to predict product's warpage performances with its structure parameters.

In order to achieve this goal, there are two approaches; one is to simulate all substrate with its actual structure. The advantage of this method is the simulation results can reflect the warpage of the whole substrate. However, the actual structure of the substrate is very complicated and asymmetric as shown in Fig. 2, and it needs a lot of resources and spent a lot of time to finish one simulation. So another approach of simplifying the actual structure is used it to replace the complicated structure of the substrate. The entire structure is assumed to be homogeneous. The most important thing is that only a small part at the center of substrate is simulated, and then curve fitting is used to predict entire substrate's warpage performances [7].

* Corresponding author: M043020067@student.nsysu.edu.tw

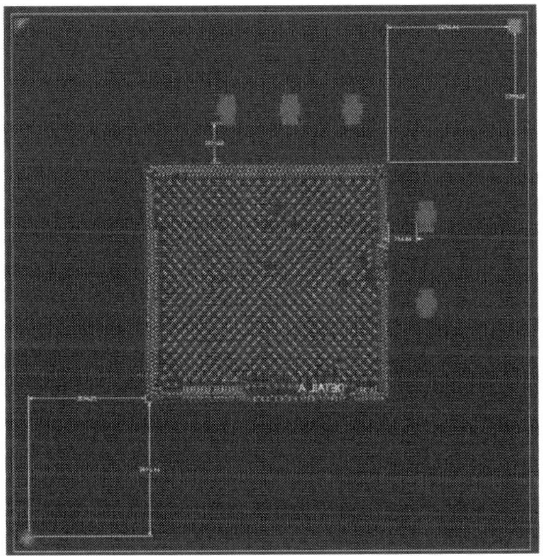

Fig. 2. Top view of substrate (red frame is the center of substrate).

2 Finite element modelling

2.1 Material properties

This substrate is composed of three kinds of materials: copper (Cu), polypropylene (PP) and solder resist (SR). The material properties are shown in Table 1. The coefficients are assumed to change linearly between the two temperatures.

Table 1. Material properties used in finite element simulations.

Material	Young's modulus (GPa)	CTE(ppm/℃)	Poisson's ratio
Cu	E_c=110	α_C=16.5	0.34
PP	E_{PP}=18(T=25℃) E_{PP}=11(T=250℃)	α_{PPX}=α_{PPY}=12 α_{PPZ}=27	0.18
SR	E_{SR}=6.1(T=25℃) E_{SR}=0.19 (T=250℃)	α_{SR}=45(T=25℃) α_{SR}=130(T=250℃)	0.37

2.2 Geometric structure of substrate

The substrate structure can be roughly divided into five layers: top SR, M1, PP, M2 and bottom SR, as shown in Fig. 3.

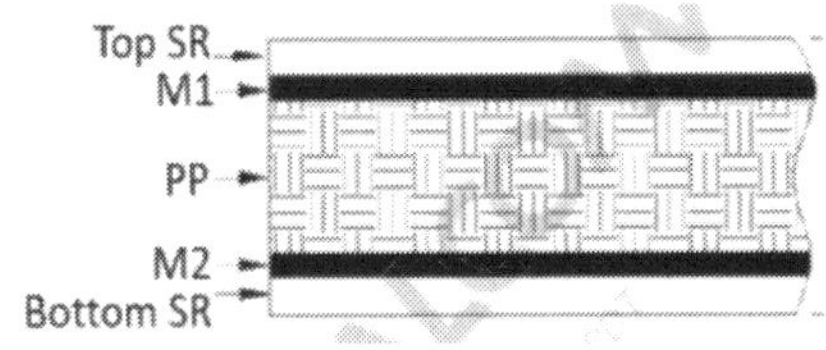

Fig. 3. Cross section schematic diagram of substrate.

Top SR and bottom SR are solder resist, M1 and M2 are copper, and PP is Polypropylene. The configurations of each part are shown in Figs. 4-8.

The top SR and bottom SR have the same structures as real structure, but M1 and M2 are based on the copper area occupation ratio. Assume a square pad has a pitch of L_{M1}=L_{M2}=100μm, according to pitch and copper area occupation ratio, W_{M1} and W_{M2} for M1 and M2 can be determined.

As the title, this is an embedded copper substrate, therefore, copper should be embedded into PP, so clearly there are many notches corresponds to M1 and M2 as shown in Fig. 8.

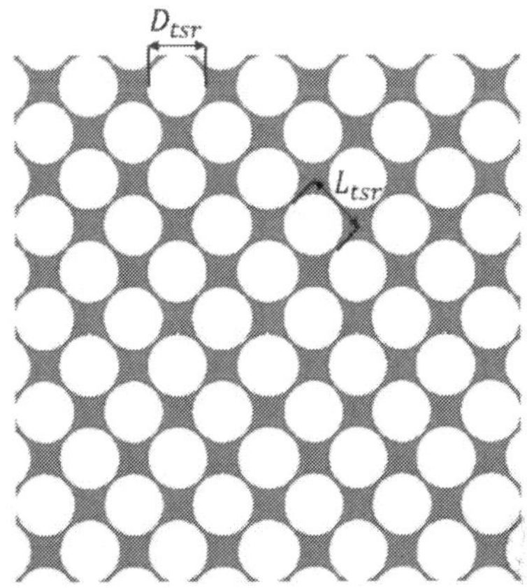

Fig. 4. Top view of top SR.

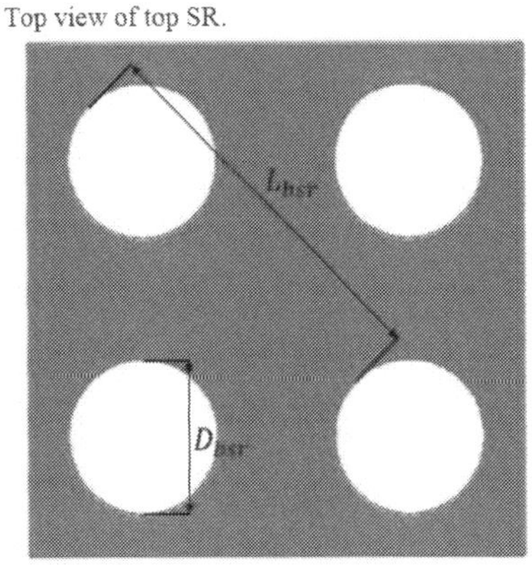

Fig. 5. Top view of bottom SR.

DOI:10.1051/matecconf/201712300014

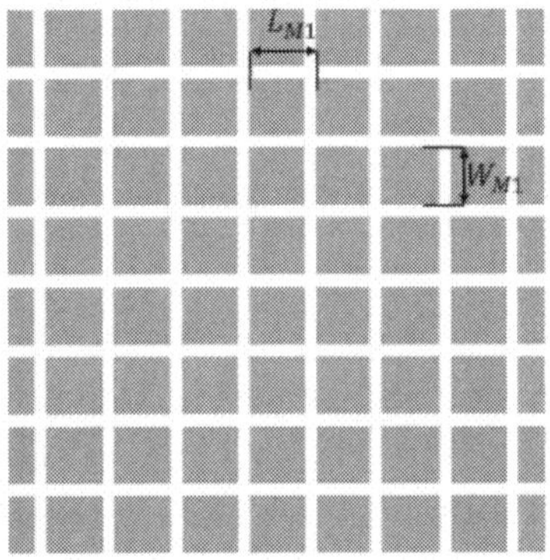

Fig. 6. Top view of M1.

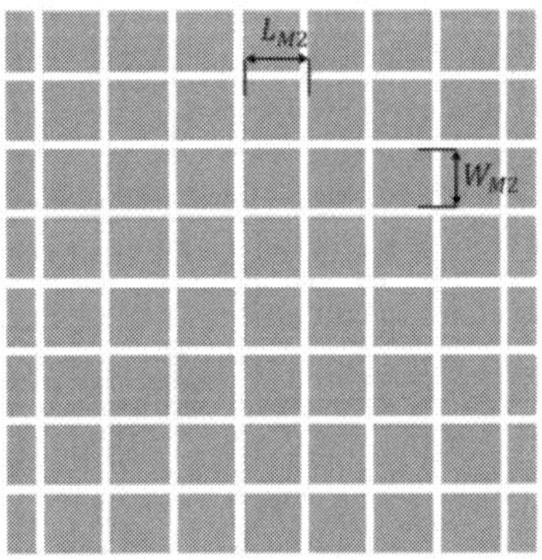

Fig. 7. Top view of M2.

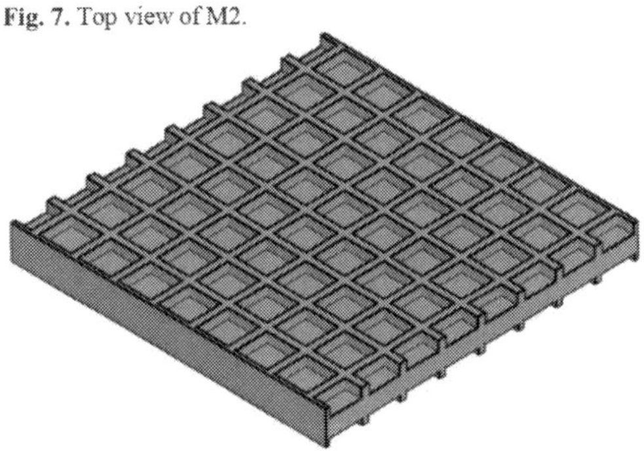

Fig. 8. Stereogram of PP.

2.3 Mesh setting and boundary condition

2.3.1 Mesh setting

The model is auto meshed by tetrahedron 10 node element, and set the longest element edge length as 7(μm). The mesh configuration is shown in Fig. 9.

Because this model is selected from the center of substrate, it has one-eight symmetry structure as shown in Fig. 10. In order to save resource and time, a symmetry boundary condition on the symmetry surface is set, and the node on the bottom of the substrate center is fixed, Gravity in all elements is not considered.

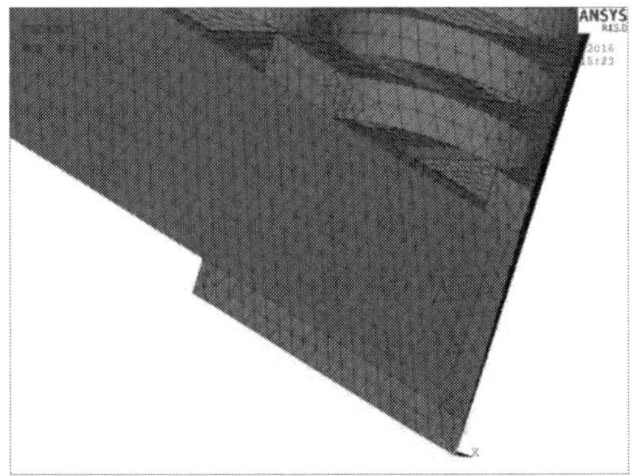

Fig. 9. Mesh configuration.

2.3.2 Boundary conditions

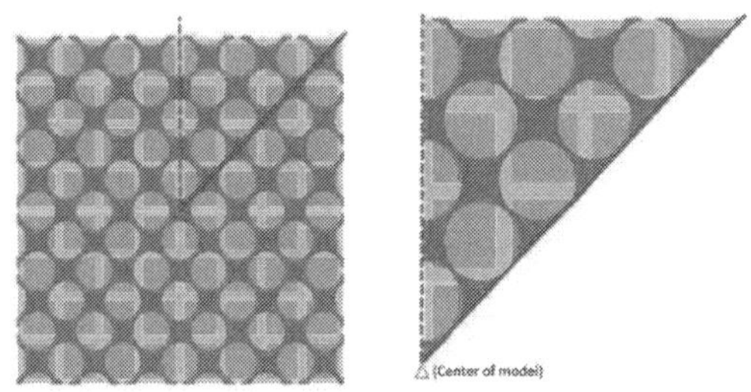

Fig. 10. Boundary condition.

The last boundary condition is temperature. In order to analyse substrate's warpage behavior during package process, the initial temperature is assumed at room temperature (25℃). The temperature are given as following: 50,100,150,183,217,245,260℃.

3 Results and discussion

The displacements at the thickness direction on the diagonal line are measured as shown in Fig. 11. Using these results, curve fitting and expansion to entire substrate scale are established.

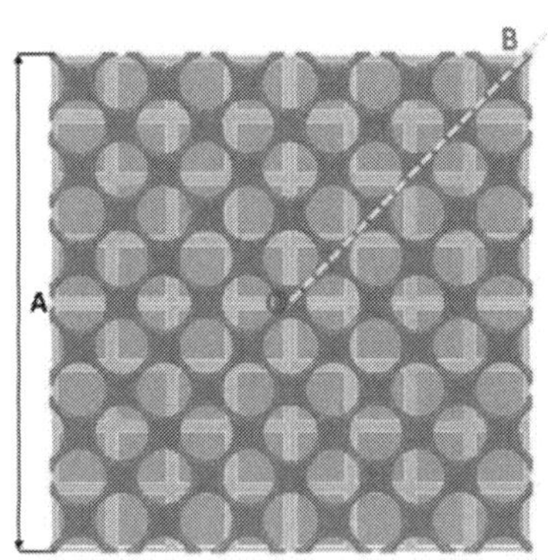

Fig. 11. Select displacement of thickness direction on diagonal line OB.

The displacement distributions on the diagonal line OB (shown in Fig. 11.) for all temperatures are shown in Fig. 12., diagonal line OB is set started from O end to B (distance 0 to 2000). Based on these displacement values, quadratic equations are used for curve fitting as shown in Table 2.

The empirical equations are expanded to predict the entire substrate deformation and the warpage behavior as shown in Fig. 13 [8].

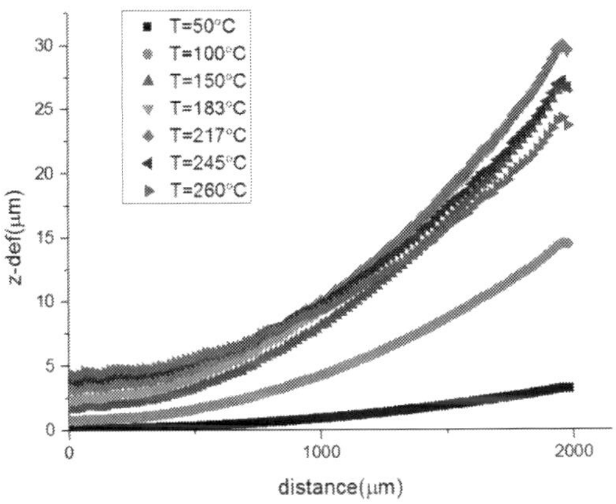

Fig. 12. Z(thickness) direction displacements for all temperatures.

Table 2. Empirical equations.

	Empirical Equations
T=50℃	$Z=8.07\times10^{-7}x^2-2.6\times10^{-5}x+0.2$
T=100℃	$Z=3.58\times10^{-6}x^2-1.2\times10^{-4}x+0.8$
T=150℃	$Z=6.50\times10^{-6}x^2-2.3\times10^{-4}x+1.8$
T=183℃	$Z=7.10\times10^{-6}x^2-2.6\times10^{-4}x+2.5$
T=217℃	$Z=6.94\times10^{-6}x^2-2.5\times10^{-4}x+3.3$
T=245℃	$Z=5.99\times10^{-6}x^2-2.2\times10^{-4}x+4.0$
T=260℃	$Z=5.10\times10^{-6}x^2-1.8\times10^{-4}x+4.4$

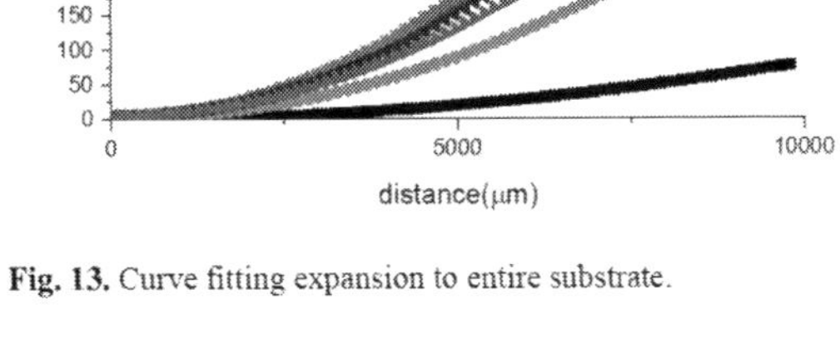

Fig. 13. Curve fitting expansion to entire substrate.

Shadow morié method [9] is dependent on two diagonal lines of the substrate, and adds the highest and lowest displacement values on the two diagonals to represent warpage magnitude, as shown in Fig. 14. Each temperature has its own warpage value.

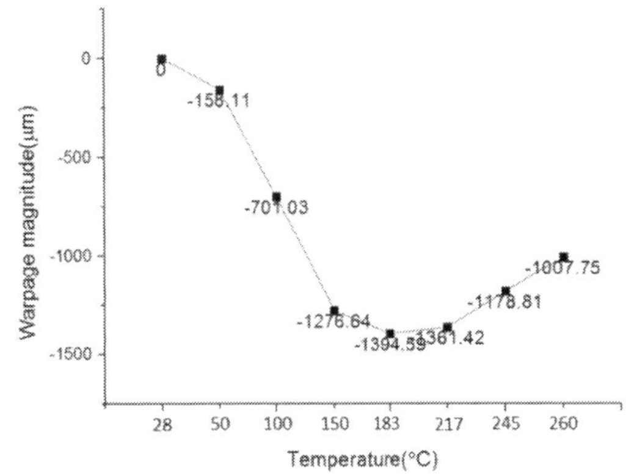

Fig. 14. Transfer deformation into shadow moiré values.

4 Taguchi method

Taguchi method is also called Quality Engineering. It is used to decrease the experimental numbers. In this method, orthogonal arrays is selected according to factor number and level number, and discuss the influence of each control factor on quality characteristic, this method is simple and efficient [10].

During the analysis process of Taguchi methods, quality characteristic is the most important index, Taguchi method divides quality characteristic into the following three types as shown in Table 3.

Table 3. Quality characteristic types of Taguchi method.

Quality characteristic types	
Smaller-the better	Y=0
Large-the better	Y=∞
Nominal-the better	Y=m

In order to find out the main influencing factor for warpage behavior, Taguchi method is used in the finite element analysis [11, 12].

The final goal of this study is to reduce the warpage behavior as much as possible, so the maximum deformation value in thickness direction(Y) of the substrate was selected as the quality characteristic, type of Smaller-the better quality characteristic is adopted in this study.

Using the Taguchi method, the influences of the copper thickness and area occupation ratio on the thermal deformation are discussed. Four geometric variables related to copper thickness and area occupation ratio were selected as control factors as shown in Table 4, where H_{M1} and H_{M2} are the copper trace thicknesses of M1 and M2 layers, ρ_{M1} and ρ_{M2} are the copper area occupation ratio of the M1 and M2 layers. $L_9(3^4)$ orthogonal array shown in Table 5. The corresponding simulation plan for the nine groups is shown in Table 6.

Table 4. Control factors and levels.

Factor	Level 1	Level 2 (Original)	Level3
$A(H_{M1})$	13μm	18μm	23μm
$B(H_{M2})$	10μm	15μm	20μm
$C(\rho_{M1})$	62%	67%	72%
$D(\rho_{M2})$	70%	75%	80%

Table 5. $L_9(3^4)$ Orthogonal array.

Simulation \ Factor	A	B	C	D
1	1	1	1	1
2	1	2	2	2
3	1	3	3	3
4	2	1	2	3
5	2	2	3	1
6	2	3	1	2
7	3	1	3	2
8	3	2	1	3
9	3	3	2	1

Table 6. $L_9(3^4)$ Orthogonal array of simulation plan.

Simulation \ Factor	H_{M1} (μm)	H_{M2} (μm)	ρ_{M1}	ρ_{M2}
1	13	10	62%	70%
2	13	15	67%	75%
3	13	20	72%	80%
4	18	10	67%	80%
5	18	15	72%	70%
6	18	20	62%	75%
7	23	10	72%	75%
8	23	15	62%	80%
9	23	20	67%	70%

Through finite element, deformation value in thickness direction from simulation and the effect of each factor are shown as Table 7.

Table 7. Maximum deformation values for orthogonal array and the effect of each factor.

Simulation \ Factor	A (μm)	B (μm)	C	D	Y (μm)
1	13	10	62%	70%	34.03
2	13	15	67%	75%	32.95
3	13	20	72%	80%	32.18
4	18	10	67%	80%	28.06
5	18	15	72%	70%	26.13
6	18	20	62%	75%	35.67
7	23	10	72%	75%	22.09
8	23	15	62%	80%	31.69
9	23	20	67%	70%	28.95
Level 1	33.05	28.06	33.80	29.70	
Level 2	29.95	30.26	29.99	30.24	
Level 3	27.58	32.27	26.80	30.65	
Effect	-5.47	4.20	-6.99	0.94	

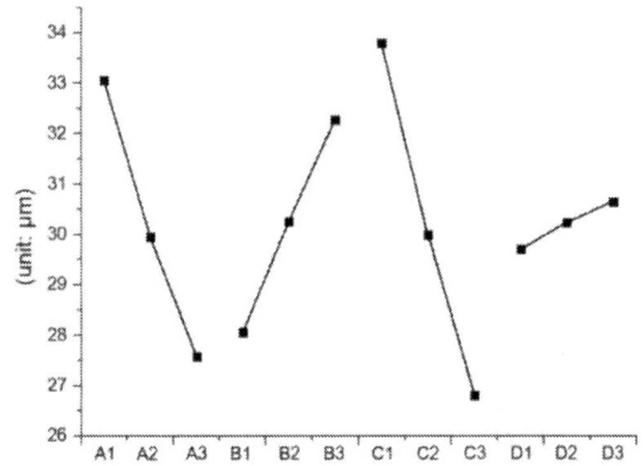

Fig. 15. Reaction diagram of factors to quality characteristic.

Reaction diagram of factors to quality characteristic is shown in Fig. 15, factor $C(\rho_{M1})$, copper area occupation ratio of M1, is the most effective factor in thermal deformation. Factor $D(\rho_{M2})$, copper area occupation ratio of M2, is the smallest effective factor in thermal deformation. The order of the factors is given in Eq (1) as following:

$$C(\rho_{M1}) > A(H_{M1}) > B(H_{M2}) > D(\rho_{M2}) \quad (1)$$

From Fig. 15, the best combinations for the smallest warpage behavior is A3=23μm, B1=10μm, C3=72%, D1=70%. From the tendency, it can be said that the increase of the thickness and copper area occupation ratio of M1 is better, the decrease of the thickness and copper area occupation ratio of M2 is better for decreasing the thermal warpage behavior of substrate.

5 Conclusions

In this paper, a finite element modelling was proposed to simulate the thermal warpage behavior of a substrate with embedded copper traces. To establish a simpler layout, an area occupation ratio is used to construct the complicated configurations of copper traces M1 and M2 layers. This concept can get a symmetrical configuration for M1 and M2. It can also simplify the complicated geometry and save a lot of simulation time. Empirical equations were also established to predict the deformation of the whole substrate.

Taguchi method was used to analyze the influence of copper traces M1 and M2. In this method, four geometric variables are H_{M1}, H_{M2}, ρ_{M1}, ρ_{M2}, analysis results for the influence in thermal deformation of four geometry factors are shown as Eq (1).

In the future, this finite element modelling will be modified. Different area occupation ratio will be set to make this model closer to the actual situation. Other parameters, such as the occupation ratio of top SR and bottom SR or the thickness of top SR and bottom SR will be optimized to get a smaller warpage of the substrate.

References

1. M. Koide, K. Fukuzono, H. Yoshimura, IEEE, *56th Electronic Components and Technology Conference*, 1869 (2006)
2. D.H. Kim, S.J. Joo, D.O. Kwak, IEEE, *IEEE Transactions on Components, Packaging and Manufacturing Technology*, **6**, 1667 (2016)
3. L.O. McCaslin, S. Yoon, H. Kim, IEEE, *IEEE Transactions on Advanced Packaging*, **32**, 740 (2009)
4. M.Y. Tsai, H.Y. Chang, M. Pecht, IEEE, *IEEE Transactions on Device and Materials Reliability*, **9**, 419 (2009)
5. M.Y. Tsai, C.H. Hsu, C.N. Han, ASME, *Journal of Electronic Packaging*, **126**, 115 (2004)
6. S.H. Chao, C.P. Hung, M. Chen, Y. Lee, J. Huang, G. Kao and D.B. Luh, Microelectronics Reliability, **57**, 101 (2015)
7. S.Y. Chiou, eThesys, *Finite Element Analyses of Copper Pillar Flip-Chip Package Processes*, (2015)
8. C. Zhu, H. Li, G. Xu and l. Luo, Microelectronics Reliability, **55**, 418 (2014)
9. Y.L. Guo, eThesys, *Thermo-Mechanical Deformation and Stress Analysis of Flip-Chip Ball Grid Array*, (2003)
10. H.H. Li, Gau Lih Book Co., Ltd., *Principles and Practices of Quality Design*, (2000)
11. T.S. Lan, M.C. Chiu, and L.J. Yeh, Science Alert, *Information Technology Journal*, **7**, 299 (2008)
12. H. Huh, J.H. Heo, H.W. Lee, Int. J. Mach. Tools Manuf, **43**, 345 (2003)

Research on energy recycle percentage of servo motor

Yen-Ming Tseng[1] , *Li-Shan* Chen[2, *], *Jeng-Shyang* Pan[3], and *Jing Huang*[4]

[1,3] Key Laboratory of Big Data Mining and Application/School of Information Science and Engineering, Fujian University of Technology, Fuzhou City, 350118, Fujian, China
[2] School of Management, Fujian University of Technology, Minhou, Fuzhou City, 350118, Fujian, China
[4] School of Information Science and Engineering, Fujian University of Technology, Fuzhou City, 350118, Fujian, China

Abstract. The energy recycle circuit fed back the energy to batteries is one of the important means to extend the travel distance of action vehicle for an electric vehicle or hybrid vehicle by driven batteries. In order to achieve energy recycle must be use, such as the motor characteristics, the principle of motor controller and design of circuit to reach the energy recycle of motor in action vehicles. Focus on the 24V 8A 150W servo motor and its shaft coupling with a condition by magnetic brake to simulate as the analog load and adjust the brake to 0.2 Newton-meter. According to the characteristics of the motor, the principle of the motor controller and the design and energy recovery of the circuit system for discussion, that include drive control circuit of motor and based on the former to design the circuit of energy recycle circuit of motor and to feed back the energy to the driving batteries. According to the characteristics of the motor, the principle of the motor controller and the design and energy recovery of the circuit system for discussion, that include drive control circuit of motor and based on the former to design the circuit of energy recycle circuit of motor and to feed back the energy to the driving batteries. By mean of the architecture of energy recycle circuit to test the power recovery which divide into without load and without energy collector, without load and with energy collector, with load and without energy collector and with load and energy collector four cases under these conditions to investigate the power recovery analysis. For the statistics of energy, the average energy recovery power percentage by without load and without energy collector is 9.7%, the average energy recovery power percentage by without load and with energy collector is 20.5%, the average energy recovery power percentage by with load and without energy collector is 1.82%, and the average energy recovery power percentage by with load and with energy collector is 3.1%.

1 Introduction

From Electricity power source transfer to Electromagnetic energy and Mechanical movement is factions of motors that is already exist in daily life of and in industrial factories as well as active traffic vehicles. Braking or deceleration in the operation or rotation status of the motor, but which can produce energy or called electricity power by this operation and to transfer another form that called heat to diverge to air. It is mean to waste the energy that made low effective and not useful well. And form the view of useful, it can to collect this energy to energy collector or and storage reuse or feedback to source of force. In recent years, the relevant literature and this article are closely related to the following: (1) Energy storage for the interface of energy recovery [1, 2]. (2) Energy recovery using super capacitors and neural networks [3, 4]. (3) Electric vehicle in driving torque and braking energy feedback charge control [5]. (4) In the train on the automatic driving when the gearshift and brake cause slow down status made the momentum transfer electricity type to the power regeneration of energy management system [6, 7]. (5) Using of ultra-capacitance with experimental characteristics for the regulating voltage and providing the inrush current to improve the driving torque of motor [8]. (6) Used in the train on the automatic driving and braking to energy recycle to DC/AC power system [9, 10]. Design and implementing considerations of a novel rapid charger for lead-acid batteries are using a single-stage power converter with an energy recovery cell [11]. In this paper, that focus at the servo motor and the specification is DC24V, 8A, 150W; 1350RPM which base on the load characteristics of motor, the principle of the motor controller of the circuit system and to reach the energy recycle by designing recovery circuit. Any motor is controlled by the drive circuit to control the speed, torque and power of motor which by modifying the motor control drive circuit without changing its drive system architecture that to design control recovery circuit system and calculate the energy recovery power at specific stage. Because the choice of motor size and power, are more suitable for small appliances and children's electric vehicles, such as the use of a wide range of applications,

* Corresponding author: sun56@ms8.hinet.net

so as the energy recovery power and the motor itself to compare the characteristics of the selected motor specifications both of no load and loaded energy recycling statistics. How to use the energy generated by the principal of generator as a research goal, and can be used for electric vehicles and electric bicycles to develop applications. It is to study on the architecture of drive and control circuits and according to the characteristics of the motor speed control and power generation by motor speed, the possess the energy collector or not is chosen to design the structure of the control energy recycle circuit system, furthermore, actual measurement of the recycling voltage, current and power of the motor energy recovery. The energy recovery power in different environments or different types of motors is different.

2 Design the energy recycle of motor

2.1 Structure of the energy recycles of motor

When brakes is active and the inertia movement of the electric vehicle in the driving mode that controller will cut off the power supply let the batteries no more output to drive the motor but in this case, the motor is converted the effect motor to generator made lots of electricity or energy to support the energy recycle, if not support the architecture to energy recycle that electricity or energy will consumption become waste in lines. The role change from motor to generator that produce the electricity or energy and store in energy collector or storage to regular the voltage via DC to DC boost feedback to batteries to complete the function of energy recycle and achieve the benefits. Fig. 1 is the motor energy recovery process block diagram by battery which is the energy source and through the controller to drive the motor for speed or torque control, at the same time the motor deceleration of speed and cut off the supply power but based on the inertia law the motor still reverse and the magnetic field over cross the coil to produce the electricity made the motor become generator. Due to the electricity output by motor which not stable and the waveform like lighting in very short time interval appear to catch and regulate voltage in energy storage. It is through the chargeback controller to step up voltage and energy feedback. The main function can be let the wasted to recycle to source become a part of the power source that can improve the battery efficiency and extended the endurance.

2.2 The current flow of the motor in the energy recycles

Such as the motor operate in multi-quadrant and base on the equivalent circuit that current flow can decide the mode of motor. When the braking is active and the terminal voltage polarity of reminding unchanged and the motor reverse direction at the same status that only one parameter such as direction of current flow in contrast between in motor mode and generator mode. As shown in Fig. 2 in general motor mode which current I_m and path for solid line. When the brake active that breaking control make the switch box change the function of controller to cut off the power and he motor function into a generator and with the controller anti-current, energy recovery I_c path for the dotted line feedback to the battery.

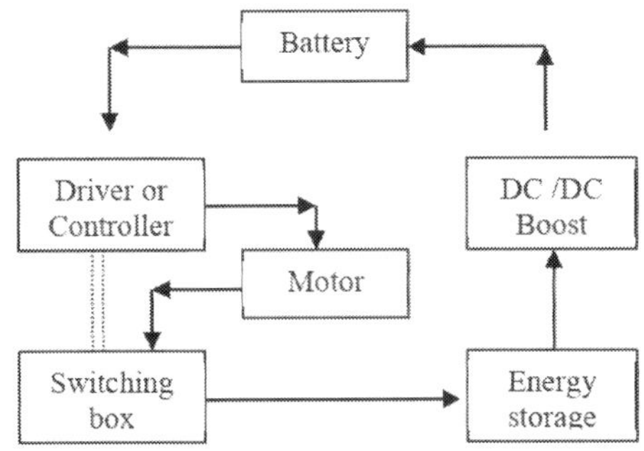

Fig.1. Motor energy recovery process block diagram.

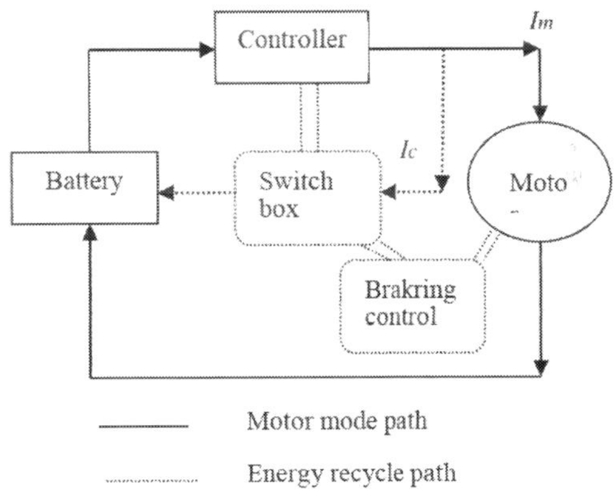

Fig.2. Energy recovery control diagram.

The motor is divided into both modes, one for drives the vehicle in motor mode and another for energy recycle for generator mode. According to equivalent circuit of motor analysis which without change the structure of the motor it is only need to change the input and output energy. As shown in Fig. 3, the equivalent circuit and current of the motor flow diagram. When used for the motor which V_{in} for the power input polarity and T for the mechanical torque output direction as shown in the current direction for the figure is the same direction and the motor for the power consumption.

When used as a generator which V_{in} for the power input polarity and T for the mechanical torque output direction is the same as the motor mode, V_{in} for the power output that the current direction is the opposite direction and the generator for the power supply. In the direction of the generator and the motor current will be a difference of 180 degrees, which is the most important point of energy recovery point of view, the general energy recovery to DC permanent magnet motor recovery more than then on-permanent magnet motor.

DOI:10.1051/matecconf/201712300015

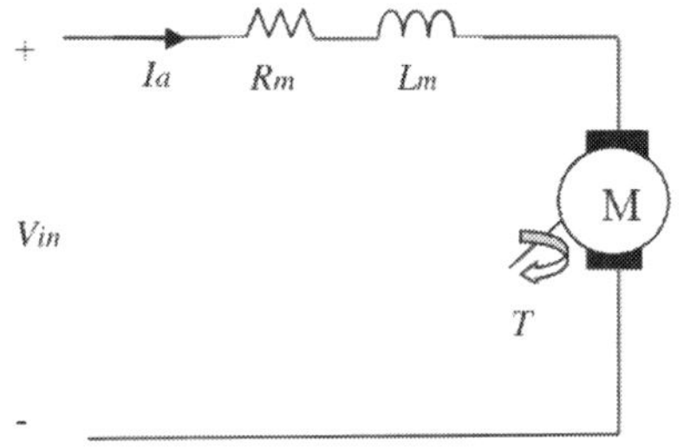

Rm: internal resistance of motor
Lm: Self inductance of motor

Fig.3. Equivalent circuit and current flow diagram of motor.

2.3 Pulse width modulation (PWM) control drive circuit

By using the servo motor as the research object to design a simple pulse width modulation(PWM) technology to drive the circuit and regulate the voltage and current feed in the motor and base on the voltage changing and current changing at the same time to varying the input power to drive the motor such as Fig. 4.

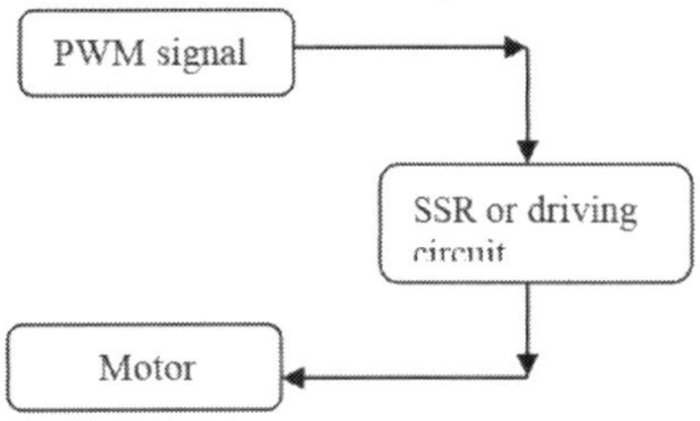

Fig.4. Motor speed control by PWM.

The period $T = 1/f$ is also fixed by the signal f which is set by the PWM controller to output and the period T = drive controller on time (T_{on}) + drive controller cut off time (T_{off}) , As shown in Fig. 5. By changing the Ton and T_{off} time to change SSR conduction status and time , in other words, that determines the input power into the motor time and changing the voltage across the motor.
General DC motor relationship between speed and voltage as show in equation (1), when $K\Phi$ fixed, the voltage and speed is proportional. The torque of motor is based on the formula (2), when $K'\Phi$ fixed, the current is proportional to the torque.

In general, when the voltage changes made the current also changes and reflecting the motor speed and torque also will change.

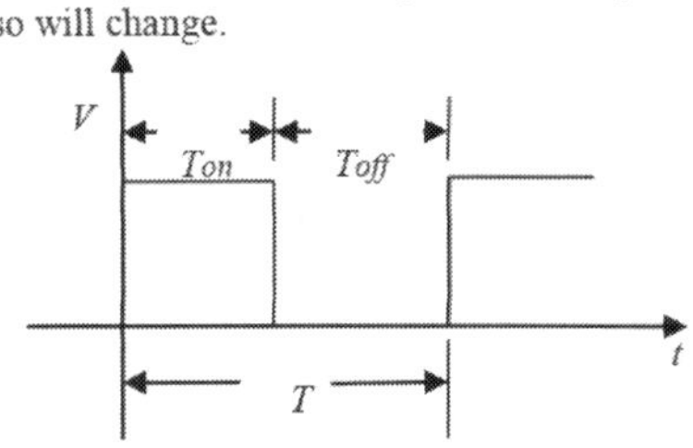

Fig.5. PWM conduction signal.

$$E = K\Phi N \quad (1)$$

$$T = K'\Phi I_a \quad (2)$$

where

E: Terminal voltage of motor
K: Voltage coefficient
Φ: Magnetic flux
N: Rotating speed
T: Torque of motor
K': troque coefficient
Ia: Armature current

3 Measurement and analysis of motor energy recycle

In order to understand the energy recovery different that divide into four cases to discuss, one for without load and without energy collector, two for without load and with energy collector, three for with load and without energy collector, and the final case for with load and with energy collector to compare the difference.

3.1 Energy recovery waveform

Fig. 6 shows the servo motor in energy recovery status which voltage, current and recovery power waveform. It is clear that the recovered voltage curve and the current curve are in the relationship between the time period as the abscissa, in which the voltage curve approaches the linear and slightly declining trend and closely related to the length of the time period. Because the circuit design has a boost circuit, the voltage boost to the battery voltage higher than the back after the rechargeable battery, because the chargeback voltage is higher than the battery voltage and will be charged to power. The current curve with the wave form of the waveform, because the booster has a minimum voltage to a level before the booster can only boost before the charge. The voltage of the discharge voltage and the current has a pulse wave shape, and the voltage has a decreasing trend. The maximum value of the boost voltage and the maximum current and the instantaneous effective power are shown in the figure. The chargeback period depends on the current waveform till the current to zero that mean no energy back to charge to battery which no matter with the voltage.

Fig. 6 shows the maximum voltage of energy recycle status is 28.8 *V*, the maximum current value is 463 *mA* of the energy back to the battery and the power of the battery back to the battery is 192 *mW*.

3.2 Energy recovery of servo motor

In the servo motor energy recovery situation, (1) without load and without energy collector that energy recycling curve as voltage, current and power as Fig. 7, (2) without load and with energy collector that energy recycling curve as voltage, current and power as Fig. 8, (3) with load and without energy collector that energy recycling curve as voltage, current and power as Fig. 9, and (4) with load and with energy collector that energy recycling curve as voltage, current and power as Fig. 10. Comparison of these four conditions to discuss the curve of the trend in the case of no load voltage curve is smoother, the voltage is proportional to the motor speed, and the current curve is irregular. In the case of loaded current curve is more smooth and fixed, the voltage curve is irregular.

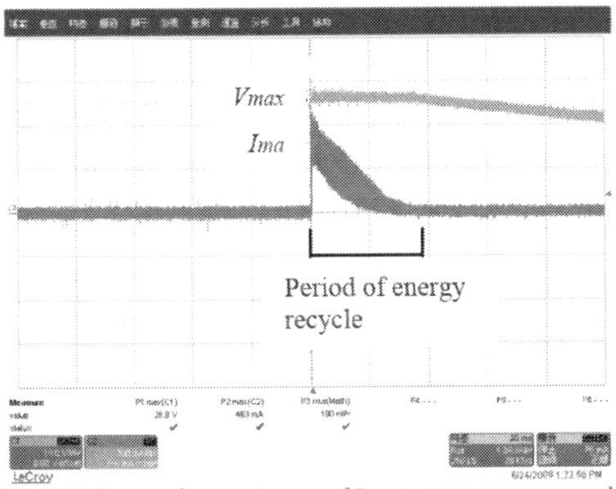

Fig.6. Voltage and current curve of Servo motor energy recycle.

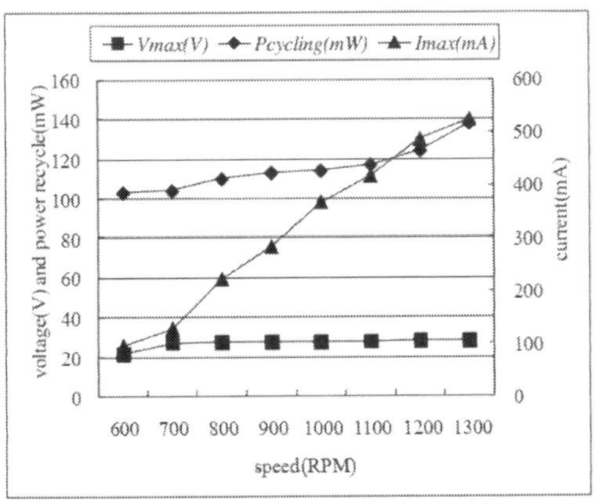

Fig.7. Energy recycle curve of without load and without energy collector of servo motor.

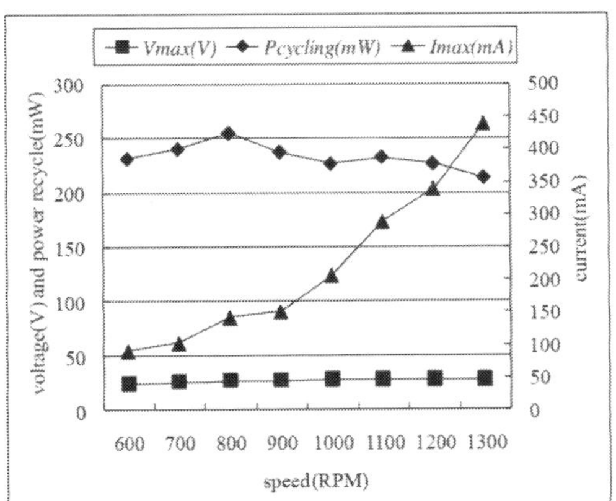

Fig.8. Energy recycle curve of without load and with energy collector of servo motor.

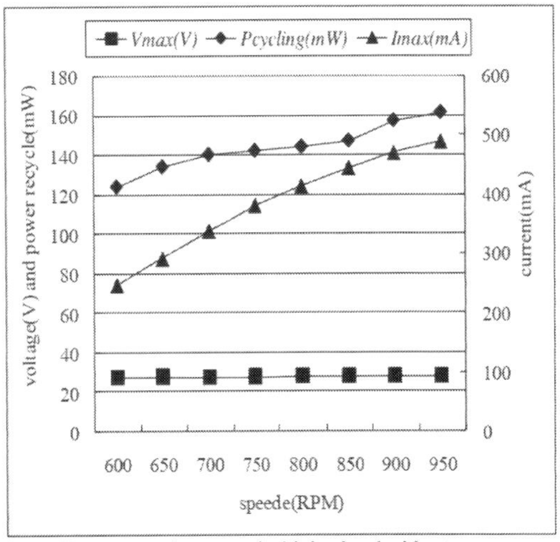

Fig.9. Energy recycle curve of with load and without energy collector of servo motor.

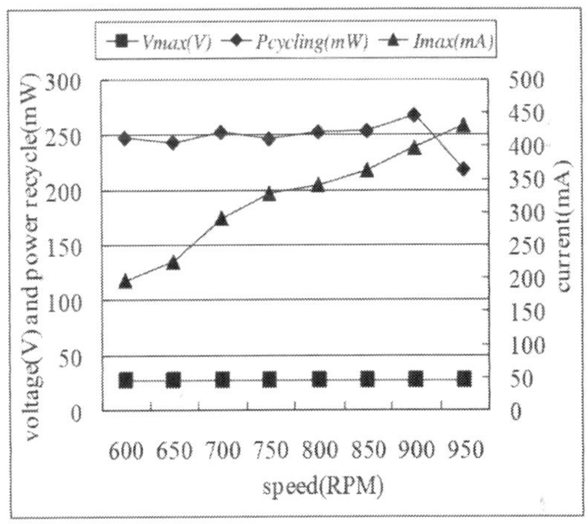

Fig.10. Energy recycle curve of with load and with energy collector of servo motor.

3.3 Comparison of Energy Recovery of with load and without load of servo motor

Comparison of the energy recycle of servo motor with or without energy collector and load as shown in table1, with the same speed, there is an average energy recovery of 251*mW* with load and with energy collector more than the without load and without energy collector 120*mW*, and with load and with energy collector carrying the average power is without load and without energy collector more than 2 times.

Fig. 12 shows the comparison of the energy recovery power curve of the servo motor. At any speed, the recycling power of with load and with energy collector is higher than that of the without load and without energy collector.

Table 1. Comparison of the energy recycles of the servo motor

speed (RPM)	Without load and without energy collector (mW)	With load and with energy collector (mW)
600	103	247
700	141	243
800	138	252
900	114	246
1000	104	252
1100	113	253
1200	124	267
average	120	251

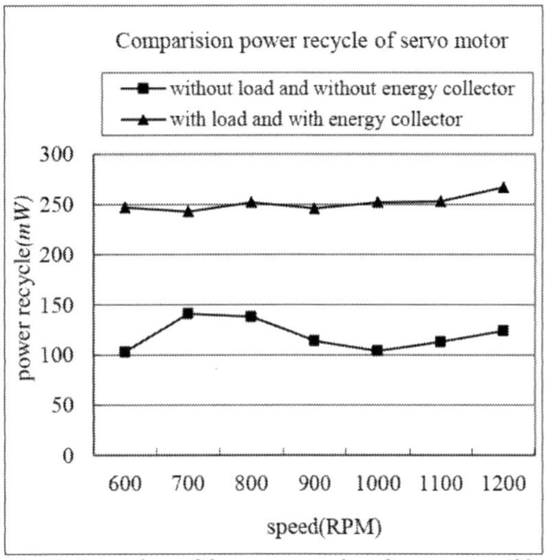

Fig.11. Comparison of the energy recycles of servo motor with or without energy collector and load.

3.4 Energy recovery rate

For the battery-driven power of the action of the vehicle, because the battery capacity can be used after the completion of the charge is determined, in accordance with the traditional use of methods, such as the reservoir in general, if no input, the reservoir will step by step pay and one day run out. For the battery-driven power of the action of the vehicle, because the battery capacity can be used after the completion of the charge is determined, in accordance with the traditional use of methods, such as the reservoir in general, if no input, the reservoir will step by step pay and one day run out. This article will use the energy consumption of the battery can be used so that the battery can increase the capacity to use and can lengthen its battery life, if used in mobile vehicles on its mileage will be lengthened with substantial application benefits.

Of course, the amount of recovery power, the capacity of the battery can be used and can lengthen its battery life; the recovery rate is also related to investment in economic benefits. The power recycle rate is calculated as shown in equations (3) and (4).

$$P_{in} = V \times I \times T \tag{3}$$

$$PRR = \frac{P_{recycling}}{P_{in}} \times 100\% \tag{4}$$

Where

P_{in} : input power
V : input voltage
I : input current
T : peroid of power recycle
$P_{recycling}$: power recycle of motor
PRR : power recycle rate

Table 2 is power recycling of servo motor under without load and without energy collector, it can be seen that the maximum recovery power rate is 14.9% and the average energy recovery rate is 9.7%.

Table 2. Power recycling of servo motor under without load and without energy collector.

$P_{in}(mW)$	RPM	$V_{max}(V)$	I_{max} (mA)	$P_{cycling}$ (mW)	PRR (%)
0.78	600	21.3	96	103	14.9
0.79	700	26.9	128	104	13.2
0.88	800	27.2	222	110	12.6
1.34	900	27.5	283	113	8.5
1.46	1000	27.5	368	114	7.8
1.62	1100	27.6	418	117	7.2
1.75	1200	27.8	487	124	7.1
1.98	1300	28.1	525	138	7.0
average(%)			9.8		

Table 3 is power recycling of servo motor under without load and with energy collector that the maximum recovery power rate is 33.59% and the average energy recovery rate is 10.7%.

Table 4 is power recycling of servo motor under with load and without energy collector that the maximum recovery power rate is 2.7% and the average energy recovery rate is 1.8%.

Table 5 is power recycling of servo motor under with load and with energy collector that the maximum recovery power rate is 4.0% and the average energy recovery rate is 3.1%.

Table 3. Power recycling of servo motor under without load and with energy collector.

$P_{in}(mW)$	RPM	$V_{max}(V)$	I_{max} (mA)	$P_{cycling}$ (mW)	PRR (%)
0.69	600	23.4	90	231	33.5
0.79	700	25.3	102	240	30.5
0.88	800	26.9	141	255	29.1
1.34	900	26.9	150	237	17.7
1.46	1000	27.5	206	226	15.5
1.62	1100	27.2	288	232	14.3
1.75	1200	27.8	338	226	12.9
1.98	1300	27.5	438	213	10.7
average(%)			20.5		

Table 4. Power recycling of servo motor under with load and without energy collector.

$P_{in}(mW)$	RPM	$V_{max}(V)$	I_{max} (mA)	$P_{cycling}$ (mW)	PRR (%)
6.22	600	27.2	247	124	2.7
6.64	650	27.5	291	134	2.4
7.52	700	27.4	338	140	2.1
8.05	750	27.5	381	142	1.8
8.54	800	27.7	413	144	1.7
9.38	850	27.8	444	147	1.4
10.03	900	27.8	469	157	1.3
11.15	950	27.8	488	161	1.3
Average(%)			1.8		

Table 5. Power recycling of servo motor under with load and with energy collector.

$P_{in}(mW)$	RPM	$V_{max}(V)$	I_{max} (mA)	$P_{cycling}$ (mW)	PRR (%)
6.1	600	27.5	196	247	4.0
6.6	650	27.5	224	243	3.7
7.5	700	27.5	291	252	3.3
8.1	750	27.8	328	246	3.1
8.5	800	27.8	341	252	3.0
9.0	850	27.5	363	253	2.8
9.2	900	27.8	397	258	2.9
10.4	950	27.8	431	235	2.1
average(%)			3.1		

Compare four cases of power recycle of servo motor. (1) Without load and without energy collector of average energy recycle is 9.7%. (2) Without load and with energy collector of average energy recycle is 20.5%. (3) With load and without energy collector of average energy recycle is 1.8%. (4) With load and with energy collector of average energy recycle is 3.1%.

4 Conclusions

Due to the electricity output by motor which not stable and the waveform in very short period appear to catch and regulate voltage in energy storage. It is through the chargeback controller to step up voltage and energy to battery. The recycle rate is inversely proportional to the speed and the maximum recycling current. If recycle rate is reduced which let the speed increasing and the maximum value of the recycling current becomes larger. And the recycling period is greatly reduced will made power recycle rate smaller.

In the case of energy recovery power; no matter how in load or without load that is a better energy recovery by the structure of with energy collector. From the view of recycle energy rate, the structure of the energy collector is better result.

References

1. M. Ortúzar, J. Moreno, and J. Dixon, "Ultracapacitor-based auxiliary energy system for an electric vehicle: implementation and evaluation," *IEEE Trans. Ind. Electron*, **54**(4), 2147-2156, (2007)
2. Zhongyue Zou, Junyi Cao, Binggang Cao, Wen Chen. " Evaluation strategy of regenerative braking energy for supercapacitor vehicle," *Isa Transactions*, **55**, 234-240, (2015)
3. Shuai Lu, K. A. Corzine, and M. Ferdowsi, "A new battery/ultracapacitor energy storage system design and its motor drive integration for hybrid electric vehicles," *IEEE Trans. Veh. Technology*. **56**(4), 1516-1523, (2007)
4. J. Moreno, M. E. Ortúzar and J. W. Dixon, "Energy-management system for a hybrid electric vehicle, using ultracapacitors and neural networks," *IEEE Trans. Ind. Electron*, **53**(3), 614-623, (2006)
5. Chaofeng Pan, Liao Chen, Long Chen, Haobin Jiang, Zhongxing Li, Shaohua Wang. "Research on motor rotational speed measurement in regenerative braking system of electric vehicle," *Mechanical Systems and Signal Processing*, **66**, 829-839, (2016)
6. Xiaohong Nian, Fei Peng, Hang Zhang, "Regenerative braking system of electric vehicle driven by brushless DC Motor," *IEEE Transactions on Industrial Electronics*, **61**, 5798-5808, (2014)
7. Zhang Junzhi, Li Yutong, Lv Chen, Yuan Ye, "New regenerative braking control strategy for rear-driven electrified minivans," *Energy Conversion and Management*, **82**, 135-145, (2014)

8. S Lu, K/ Corzine, M Ferdowsi, " A New Battery/Ultracapacitor Energy Storage System Design and Its Motor Drive Integration for Hybrid Electric Vehicles, " *IEEE Transactions on Vehicular Technology*, **56**(4), 1516-1523, (2007)
9. J. Wang, B. Taylor, Z. G. Sun, and D. Howe, "Experimental characterization of a super capacitor-based electrical torque-boost system for downsized ICE vehicles," *IEEE Trans. Vet. Technology*, **56**(6), 3674-3681, (2007)
10. Y. P. Yang and T. H. Hu. "A new energy management system of directly-driven electric vehicle with electronic gearshift and regenerative braking," *in Proc., Amer. Contr. Conf.*, 4419-4424, (2007)
11. HM Huang, HJ Chiu, MH Tseng, LW Lin. "Design and implementation of a novel rapid charger with energy recovery," *IEEE International Conference on Industrial Technology*, 1300-1303, (2005)

MATEC Web of Conferences **123**, 00016 (2017) DOI:10.1051/matecconf/201712300016
ICPMMT 2017

A compact low insertion loss bandpass filter based on meandered self-coupled ring resonator

Ching-Chien Cheng [1,*], *Kong-Xin* Cheng [1], *Huang-Kuang* Kung[1], *Chin-Yu* Wang[1], and *Yng-Huey* Jeng[2]

[1]Institute of Mechatronics Engineering, Cheng Shiu University, Kaohsiung City, Taiwan
[2]Department of Electronic Engineering, National Formosa University, Yunlin, Taiwan

Abstract. The traditional structure of dual mode bandpass filter is formed by a circular or rectangular resonator, which causes excessive circuit areas to be occupied. A 5.2 GHz bandpass filter, based on the triple-branch self-coupled ring resonator, is designed to achieve most essential features such as low insertion loss, compact size, and wide stopband. The coupling of each coupled branch introduces the even- and odd-mode perturbation, which produce sharp bandpass and wide stopband responses due to the generation of band-edge transmission zeros. The simulation and measurement in this paper are respectively verified by using IE3D electromagnetic simulator and Agilent's HP8722C network analyzer. Experimental results show that the filter has 19.2% bandwidth centered at 5.2 GHz, 1.2 dB insertion loss, 25 dB stopband rejection from 6 to 9 GHz. Moreover, the circuit size in this paper has been down to 9×6 mm^2 and can be applied to communication of microwave applications.

1 Introduction

The component of passive circuit, such as the filter occupying the largest percentage of communication system structure, is more important to preserve high performance and reduce the electronic circuit size. Many researches of dual-mode bandpass filter have been proposed in papers [1-9].

The bandpass filters based on the dual-mode ring resonator was first analyzed by Wolff [1]. Some papers proposed dual mode ring resonator bandpass filters by using step-impedance or ring resonators with different perturbation schemes [2-4]. The perturbation element is a plus to or cut from the resonator, and it is also viewed as capacitive or inductive [5]. One of the methods is the investigation of designing a compact, low insertion loss and sharp rejection by cascaded dual mode ring resonator [6]. Furthermore, dual-mode microstrip filters are highly attractive for radio frequency systems because their sizes are compact and easy to fabricate [7]. By adding open stubs symmetrically on ring resonator, the dual mode filter is realized [8].

However, there is a novel distributed perturbation by self-coupled segments of the ring resonator, which brings the easier implementation scheme [9]. In this paper, the ring resonator is based on the theory of a self-coupled, simple bandpass filter with low insertion loss and the compact size designed, which showed the easy structure and narrow bandwidth with two transmission zeros near the passband. This filter of 5.2 GHz band is designed. The measurement results show good consensus with simulation.

*Corresponding author: chching@kmvs.km.edu.tw

2 Filter design

The traditional structure of dual mode bandpass filter is shown in Fig. 1. Generally, the filter structure is formed by a circular or rectangular resonator, which causes excessive circuit areas to be occupied. To make up for the disadvantage, the ring resonator was bended in this study to achieve the goal of saving the filter size.

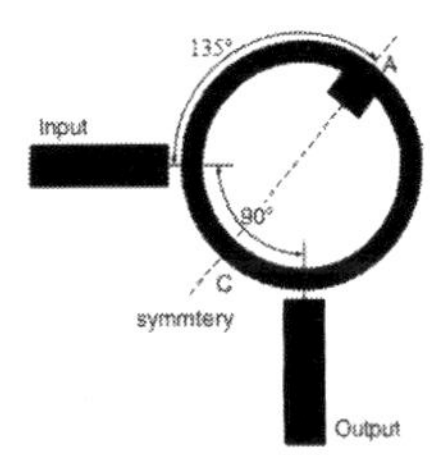

Fig. 1. Structure of traditional dual mode bandpass filter.

2.1 Schematic of proposed bandpass filter

Unlike those studies in the past using the perturbation approach, the ring resonator with coupling method is applied to cause the even and odd mode to be separated, and to achieve resonance effects. Thus, the filter offers an advantage of a simpler circuit layout and is easier to be implemented. The schematic types of the proposed self-coupled bandpass filter with input and output ports are shown in Fig. 2. The proposed bandpass filter is composed of self-coupled ring resonator with parallel-coupled input and output networks. The self-coupled ring includes a central coupled branch and two other coupled branches, connected at both ends of the central coupled branch. An intuitive way to understand this self-coupled ring resonator is the meandering of the conventional circular ring resonator such that the ring size can be considerably miniaturized. Different from the lumped perturbation of the conventional ring resonator, the distributed coupling of the proposed ring resonator provides the mode perturbation between even and odd modes.

2.2 Resonance conditions

In Fig. 2, the layout discontinuity effect of both coupled sections are neglected, with the structure of the ring resonator symmetric to line $\overline{AB}$, and the even-odd mode theory applied to analyze the resonance mode characteristics. Fig. 3(a) and (b) show that the equivalent even-mode, odd-mode circuits, the even-mode and odd-mode resonance conditions can be derived from equation (1)-(6) and ignore the coupling support effect on both sides of the bent section. Two coupled sections at ends have the same parameters: the electrical length of θ_1, the coupling coefficient of c_1, and the characteristic impedance Z_1. The central coupled section has the electrical length of θ_2, the coupling coefficient of c_2, and the characteristic impedance of Z_2.

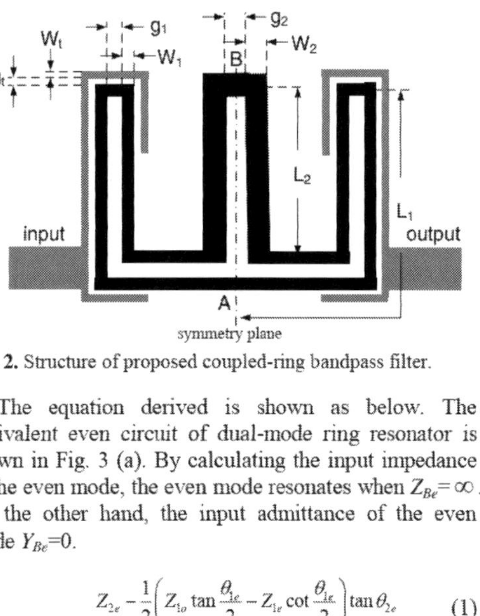

Fig. 2. Structure of proposed coupled-ring bandpass filter.

The equation derived is shown as below. The equivalent even circuit of dual-mode ring resonator is shown in Fig. 3 (a). By calculating the input impedance of the even mode, the even mode resonates when $Z_{Be}=\infty$. On the other hand, the input admittance of the even mode $Y_{Be}=0$.

$$Y_{Be}=jY_{2e}\frac{Z_{2e}-\frac{1}{2}\left(Z_{1o}\tan\frac{\theta_{1e}}{2}-Z_{1e}\cot\frac{\theta_{1e}}{2}\right)\tan\theta_{2e}}{\frac{1}{2}\left(Z_{1o}\tan\frac{\theta_{1e}}{2}-Z_{1e}\cot\frac{\theta_{1e}}{2}\right)+Z_{2e}\tan\theta_{2e}}=0 \quad (1)$$

Then

$$Z_{2e}-\frac{1}{2}\left(Z_{1o}\tan\frac{\theta_{1e}}{2}-Z_{1e}\cot\frac{\theta_{1e}}{2}\right)\tan\theta_{2e}=0 \quad (2)$$

The even mode resonance condition becomes

$$\tan\theta_{2e}=\frac{2\sqrt{\frac{1+C_2}{1-C_2}}}{Z\left(\sqrt{\frac{1-C_1}{1+C_1}}\tan\frac{\theta_{1e}}{2}-\sqrt{\frac{1+C_1}{1-C_1}}\cot\frac{\theta_{1e}}{2}\right)} \quad (3)$$

Where

$$C_1=\frac{Z_{1e}-Z_{1o}}{Z_{1e}+Z_{1o}},C_2=\frac{Z_{2e}-Z_{2o}}{Z_{2e}+Z_{2o}},Z_1=\sqrt{Z_{1e}Z_{1o}},Z_2=\sqrt{Z_{2e}Z_{2o}}$$

$$Z=\frac{Z_1}{Z_2}$$

$$\frac{\theta_{2e}}{\theta_2}=\frac{f_{re}}{f_{norm}}=\frac{\theta_{1e}}{\theta_1}\ ;\theta_1+\theta_2=\pi$$

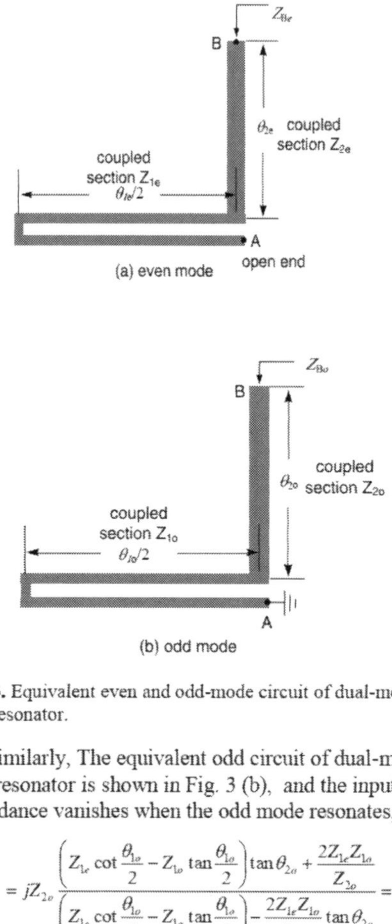

Fig. 3. Equivalent even and odd-mode circuit of dual-mode ring resonator.

Similarly, The equivalent odd circuit of dual-mode ring resonator is shown in Fig. 3 (b), and the input impedance vanishes when the odd mode resonates,

$$Z_{Bo}=jZ_{2o}\frac{\left(Z_{1e}\cot\frac{\theta_{1o}}{2}-Z_{1o}\tan\frac{\theta_{1o}}{2}\right)\tan\theta_{2o}+\frac{2Z_{1e}Z_{1o}}{Z_{2o}}}{\left(Z_{1e}\cot\frac{\theta_{1o}}{2}-Z_{1o}\tan\frac{\theta_{1o}}{2}\right)-\frac{2Z_{1e}Z_{1o}}{Z_{2o}}\tan\theta_{2o}}=0 \quad (4)$$

Then

$$\left(Z_{1e}\cot\frac{\theta_{1o}}{2}-Z_{1o}\tan\frac{\theta_{1o}}{2}\right)\tan\theta_{2o}+\frac{2Z_{1e}Z_{1o}}{Z_{2o}}=0 \quad (5)$$

The odd mode resonance condition is obtained as

 DOI:10.1051/matecconf/201712300016

$$\tan\theta_{2o} = \frac{2Z\sqrt{\frac{1+C_2}{1-C_2}}}{\left(\sqrt{\frac{1-C_1}{1+C_1}}\tan\frac{\theta_{1o}}{2} - \sqrt{\frac{1+C_1}{1-C_1}}\cot\frac{\theta_{1o}}{2}\right)} \quad (6)$$

Where

$$C_1 = \frac{Z_{1e}-Z_{1o}}{Z_{1e}+Z_{1o}}, C_2 = \frac{Z_{2e}-Z_{2o}}{Z_{2e}+Z_{2o}}, Z_1 = \sqrt{Z_{1e}Z_{1o}}, Z_2 = \sqrt{Z_{2e}Z_{2o}}$$

$$Z = \frac{Z_1}{Z_2}$$

$$\frac{\theta_{2o}}{\theta_2} = \frac{f_{ro}}{f_{norm}} = \frac{\theta_{1o}}{\theta_1};\ \theta_1 + \theta_2 = \pi$$

Under the even-odd mode resonant frequency f_{re} and f_{ro}, the θ_{1e}, θ_{2e}, θ_{1o} and θ_{2o} are the electrical lengths corresponding to the physical lengths ℓ_1 and ℓ_2 respectively. Thus, the resonance frequency condition is obtained as $f_r = \sqrt{f_{re}\cdot f_{ro}}$.

Since the complicated form of equation (1)-(6) is difficult to obtain the analytic solution, we use the graphic expression for analyzing the effect of the even-mode and odd-mode resonant frequencies with the impedance ratio and coupling coefficients.

2.2.1 Impedance ratio effect Z

A case about c_1=0.2 and c_2=0.2 is plotted in Fig. 4 and it shows the dependence of mode resonant frequencies with respect to the impedance ratio of Z with several characteristics as below.

1) When impedance ratio Z = 1, the even- and odd-mode resonant frequency are not separated.
2) When impedance ratio Z < 1, the odd-mode resonant is lower than the even-mode and has the inductive perturbation feature.
3) When impedance ratio Z > 1, the odd-mode resonant frequency is higher than the even-mode and the type is in a capacitive perturbation state.
4) The geometric average resonant frequency, $f_{r,SCR} = \sqrt{f_{re}f_{ro}}$ is apathetical to Z.

Therefore, these arguments are determined to support the operating band and bandwidth of the filter in this paper.

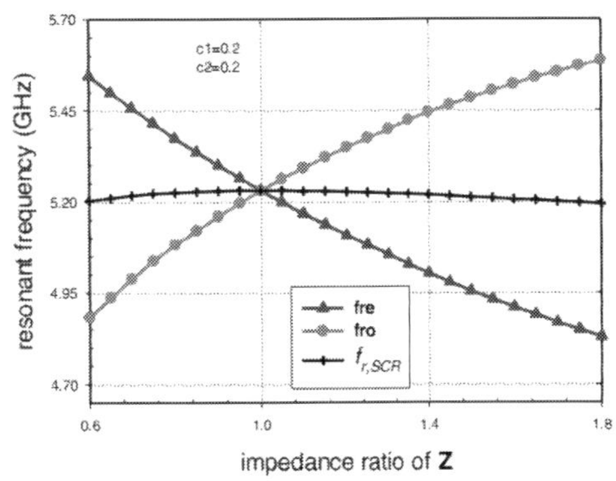

Fig. 4. Resonant frequencies with respect to the impedance ratio of Z.

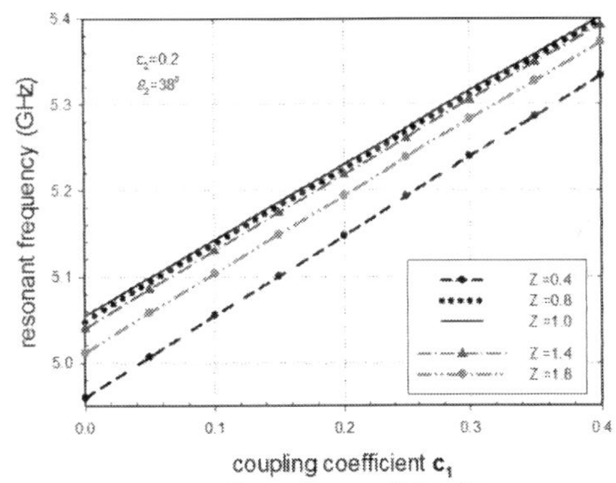

(a) coupling coefficient C_1

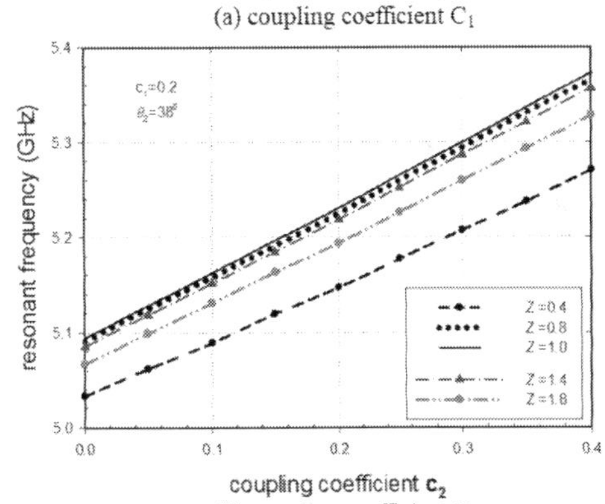

(b) coupling coefficient C_2

Fig. 5. Resonant frequencies with respect to the coupling coefficient.

2.2.2 Characteristics of Coupling Coefficient

Fig. 5 (a) and (b) are the relative resonant frequency between coupling coefficient C_1 and C_2 for different impedance ratios under the $\theta_2 = 38^\circ$. There are several characteristics as below.

1) When the coupling coefficient increases, the resonant frequency will increase at any impedance ratio.

2) When Z < 1, the resonant frequency will increase following the incremental impedance ratio in the any coupling coefficient.
3) When Z > 1, the the resonant frequency will decrease following the incremental impedance ratio in the any coupling coefficient.
4) When Z = 1, it shows the highest resonant frequency.

2.3 Transmission zeros analysis

The main advantage of the ring resonator filter is generating two transmission zeros near the passband. Because there are two transmission paths between the input and output ports, therefore when the current of the two paths is offset each other at the output, the transmission zeroes on both sides of the passband are generated. In this way, Y-parameter matrix method (Admittance Matrix Method) is used to facilitate the analysis of transmission zero generation.

Ignoring the discontinuous effect of the circuit branch bending, as shown in Fig. 6, the entire resonator can be disassembled into a parallel coupling line with electrical length $\theta_1/2$, impedance Z_1 and a parallel coupling line with electrical length θ_2, impedance Z_2. Here $\theta_c = \theta_{IO}/2$, $\theta_a = \theta_1/2 - \theta_c$ and θ_{IO} as the length of the electrical input between the input and the output port are defined, and $\theta_1 + \theta_2 = \pi$.

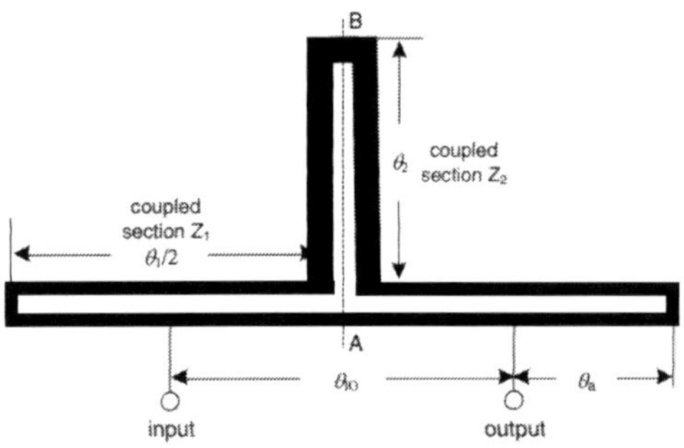

Fig. 6. Self-coupling ring resonator two-way transmission path equivalent circuit.

In Fig. 7, the equations (7) to (12) are the formulas for the parallel coupling microstrip Y parameter of the four ports and the two ports. Further, the formulas (7) to (12) are converted into the following formulas (13) to (18) using the relationship between the coupling coefficient and the impedance.

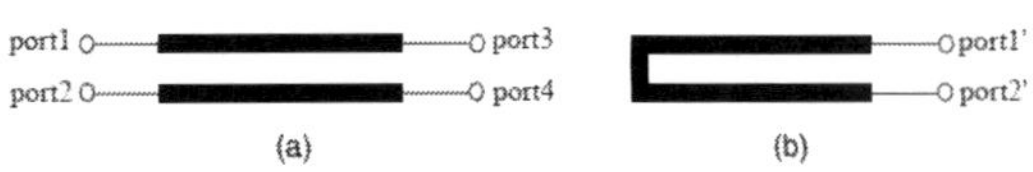

Fig. 7. parallel coupling transmission line diagram (a) four port (b) two port.

$$\begin{bmatrix} Y_{11} & Y_{12} & Y_{13} & Y_{14} \\ Y_{21} & Y_{22} & Y_{23} & Y_{24} \\ Y_{31} & Y_{32} & Y_{33} & Y_{34} \\ Y_{41} & Y_{42} & Y_{43} & Y_{44} \end{bmatrix} = \begin{bmatrix} Y_{11} & Y_{12} & Y_{13} & Y_{14} \\ Y_{12} & Y_{11} & Y_{14} & Y_{13} \\ Y_{13} & Y_{14} & Y_{11} & Y_{12} \\ Y_{14} & Y_{13} & Y_{12} & Y_{11} \end{bmatrix}$$

$$Y_{11} = (-\frac{j}{2})[Y_e + Y_o]\cot\theta \tag{7}$$

$$Y_{12} = (-\frac{j}{2})[Y_e - Y_o]\cot\theta \tag{8}$$

$$Y_{13} = (-\frac{j}{2})[Y_e - Y_o]\csc\theta \tag{9}$$

$$Y_{14} = (-\frac{j}{2})[Y_e + Y_o]\csc\theta \tag{10}$$

$$\begin{bmatrix} Y_{11}' & Y_{12}' \\ Y_{21}' & Y_{22}' \end{bmatrix} = \begin{bmatrix} Y_{11}' & Y_{12}' \\ Y_{12}' & Y_{11}' \end{bmatrix}$$

$$Y_{11}' = \frac{j}{2}[Y_e \tan\theta - Y_o \cot\theta] \tag{11}$$

$$Y_{12}' = \frac{j}{2}[Y_e \tan\theta + Y_o \cot\theta] \tag{12}$$

$$Y_{11} = Y\frac{-j}{\sqrt{1-c^2}}\cot\theta \tag{13}$$

$$Y_{12} = Y\frac{-jc}{\sqrt{1-c^2}}\cot\theta \tag{14}$$

$$Y_{13} = Y\frac{j}{\sqrt{1-c^2}}\csc\theta \tag{15}$$

$$Y_{14} = Y\frac{-jc}{\sqrt{1-c^2}}\csc\theta \tag{16}$$

$$Y_{11}' = Y\frac{-j}{\sqrt{1-c^2}}\cot\theta \tag{17}$$

$$Y_{12}' = Y\frac{-j}{\sqrt{1-c^2}\sin 2\theta}(c + \cos 2\theta) \tag{18}$$

By using the formulas (13) to (18), we can solve the position of transmission zero in Fig. 6. The transitive formula is shown in equation (19), and when Y_{21} is zero, the transmission zero is generated.

$$Y_{21} = Q_d + \frac{2S_dT_d(G_d + P_u) - (H_d + Q_u)(S_d^2 + T_d^2)}{(H_d + Q_u)^2 - (G_d + P_u)^2} \tag{19}$$

And within the formula (19),

DOI:10.1051/matecconf/201712300016

$$P_u = -\frac{jY_2}{\sqrt{(1-c_2^2)}\sin 2\theta_2}(c_2+\cos 2\theta_2),$$

$$Q_u = \frac{jY_2}{\sqrt{(1-c_2^2)}\sin 2\theta_2}(1+c_2\cos 2\theta_2),$$

$$S_d = Q_{ac} - \frac{S_{ac}G_{ac}}{2H_{ac}},\ T_d = -\frac{S_{ac}G_{ac}}{2H_{ac}},$$

$$G_d = T_{ac} - \frac{G_{ac}^2}{2H_{ac}},\ H_d = -\frac{G_{ac}^2}{2H_{ac}},\ Q_d = -\frac{Q_{ac}^2}{2H_{ac}},$$

$$Q_{ac} = T_c - S_c\frac{(Q_a+Q_c)}{P_a+P_c},$$

$$S_{ac} = S_c - T_c\frac{(Q_a+Q_c)}{P_a+P_c},$$

$$T_{ac} = P_c - \frac{S_c^2}{P_a+P_c},$$

$$G_{ac} = Q_c - \frac{S_cT_c}{P_a+P_c},\ H_{ac} = P_c - \frac{T_c^2}{P_a+P_c},$$

$$P_c = Y_1\frac{-j}{\sqrt{1-c_1^2}}\cot\theta_c,\quad Q_c = Y_1\frac{jc_1}{\sqrt{1-c_1^2}}\cot\theta_c,$$

$$S_c = Y_1\frac{j}{\sqrt{1-c_1^2}}\csc\theta_c,\quad T_2 = Y_1\frac{-jc_1}{\sqrt{1-c_1^2}}\csc\theta_c.$$

The complicated form of (19) makes it difficult to obtain the analytic solutions in calculation. Then we turn to the graphic expression of Y_{21} for analyzing the effects of the impedance ratio Z and input-output separation length θ_{IO} [9].

2.3.1 Effect of the impedance ratio Z

Fig. 8 is the coupling coefficient with C_1 and C_2 being equal to 0.2, $\theta_1 = 140^\circ$, $\theta_2 = 40^\circ$ and Z ratio being 0.5, 1.0, 1.8 in the normalization of the frequency of the Y_{21} relationship diagram when the angle between the input and output is 80 degree. As shown in the Fig. 8, it can be found that there is no transmission zero on both sides of the fundamental frequency passband when the inductive coupling Z < 1. When Z > 1 is capacitively coupled, two transmission zeros are generated on both sides of the passband and a transmission zero occurs near the second passband. When Z = 1, the transmission zero will occur near the base band, destroying the passband characteristics.

2.3.2 Effect of angle between input and output ports

Fig.9 is the normalized frequency of Y_{21} relationship diagram showing when the coupling coefficient C_1 and C_2 are equal to 0.2, the Z ratio is 1.8, with the angle between input port and output being 60°, 70°, 80°, 90°, 100°. It can be seen from the Fig. 9 that when the angle between the input and output port is 70° ~ 80°, the passband on both sides of the transmission zero points shows the most symmetry.

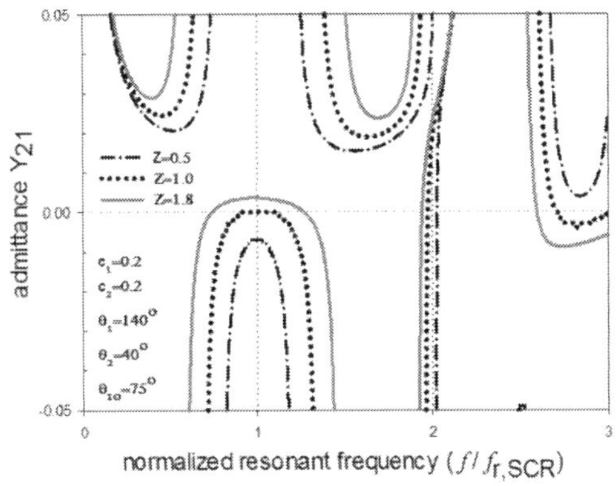

Fig. 8. Z ratio of the normalization frequency of Y_{21} relationship diagram.

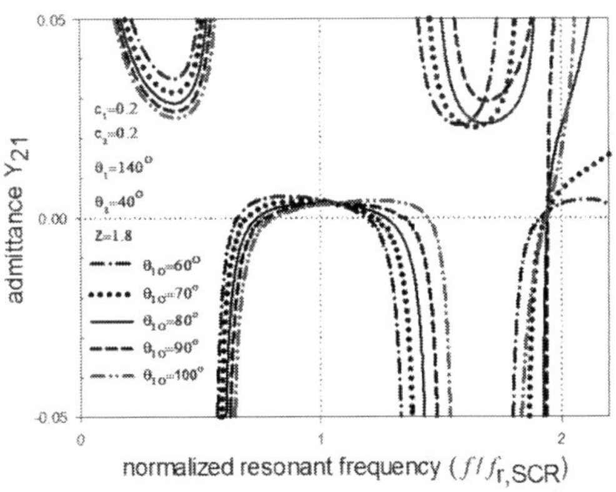

Fig. 9. shows the Y_{21} relationship diagram difference between the input and output ports at the normalized frequency.

3 Experimential results

Following the theoretical analysis in section 2, in this paper, the coupling coefficients C_1 and C_2 are all used at 0.2, the electrical length $\theta_1 = 142^\circ$, $\theta_2 = 38^\circ$, the characteristic impedance Z_1 is 109Ω, Z_2 is 61.5Ω, and the impedance ratio Z =1.77. Moreover, the relative dimensions of the resonator is equivalent to W_2 = 1.2 mm, W_1 = 0.3 mm, W_f=0.2 mm, L_1=8.5 mm, L_2=4.5 mm, g_2 = 0.6 mm, g_1 = 0.6 mm, and g_f=0.2 mm.

This filter is fabricated on the substrate with a relative dielectric constant of 3.38, with loss tangent being 0.0025 and thickness being 0.762 mm. The actual photograph of 5.2GHz self-coupled bandpass filter is shown in Fig. 10. The size of self-coupled bandpass filter is 9 x 6 mm^2. On the other hand, the circuit size is relatively 54% of the size to the traditional uniform ring case.

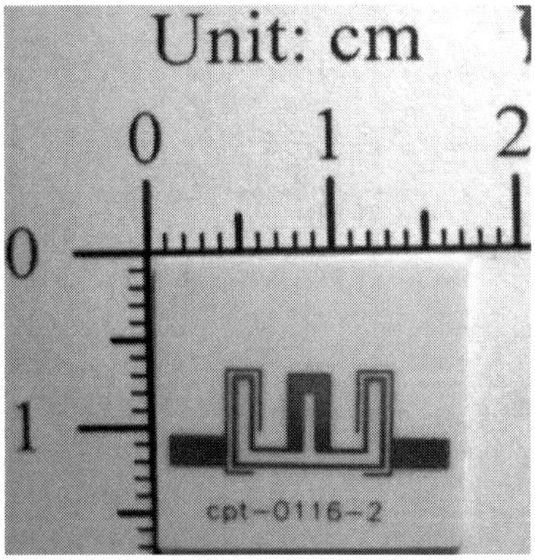

Fig. 10. Photograph of 5.2 GHz bandpass filter.

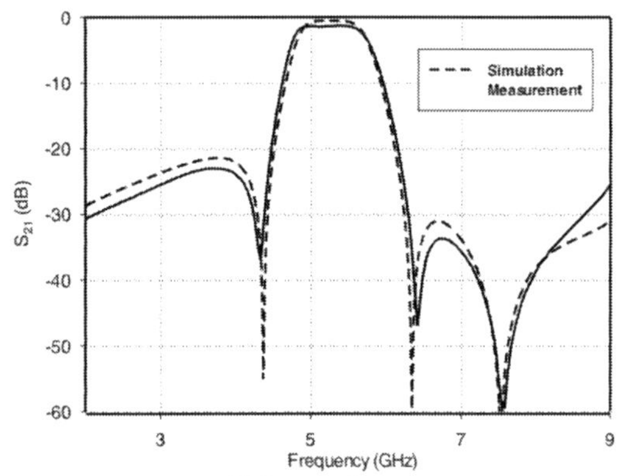

(a) Insertion loss

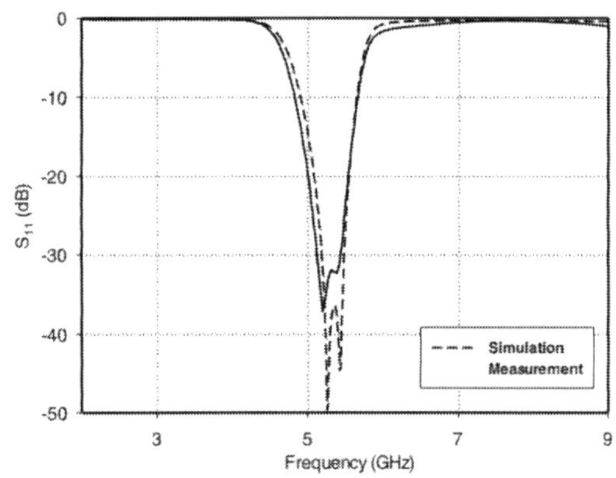

(b) return loss

Fig. 11. shows the measurement and simulation results of bandpass filter.

The simulation and measurement are respectively verified by using IE3D electromagnetic simulator and Agilent's HP8722C network analyzer [10]. Fig. 11 shows the simulated and measured results. Finally, there is very good consensus between the measured and simulated results in this paper that within 5-5.4 GHz, the measurement return loss is greater than 22 dB and measurement insertion loss is less than 1.2 dB and this condition improves the response of passband. The measured 3-dB bandwidth is about 19.2% at the center frequency of 5.2 GHz. The filter has three transmission zeros close to the passband at 4.39 GHz, 6.42 GHz and 7.58 GHz, which produces sharp bandpass and wide stopband response.

4 Conclusion

A 5.2 GHz dual-mode bandpass filter has been proposed in this study by using self-coupled ring resonator for small size, low insertion loss, wide stopband and it is easy to implement. The proposed filter size is 9 x 6 mm^2, three transmission zeros are close to the passband at 4.39 GHz, 6.42 GHz and 7.58 GHz, and the 3-dB bandwidth is about 19.2% from 4.75 GHz to 5.73 GHz, and insertion loss is less than 1.2 dB at center frequency 5.2 GHz. A good consensus has been reached in this study by simulation and measurement showing highly valuable application of proposed bandpass filter in wireless communications.

Acknowledgments

We would like to express our sincere heartfelt thanks to our supervisor, Mr. Sheng-Fuh Chang, professor of Department of Electrical Engineering, National Chung Cheng University, Chiayi, Taiwan, for his invaluable advice and constant help on the paper.

References

1. I. Wolff, N. Knoppik, Microstrip ring resonator and dispersion measurements on microstrip lines, *Electron. Lett.*, **7** (1971)
2. I. Wolff, Microstrip bandpass filter using degenerate modes of a microstrip ring resonator, *Electron. Lett.*, **8** (1972)
3. M. Matsuo, H. Yabuki, M. Makimoto, Dual-mode stepped-impedance ring resonator for bandpass filter applications, *IEEE Trans. Microwave Theory Tech.*, **49** (2001)
4. S. Luo, L. Zhu, A novel dual-mode dual-band bandpass filter based on a single ring resonator, *IEEE Microw. Wireless Compon. Lett.*, **19** (2009)
5. T.W. Lin, J.T. Kuo, S.J. Chung, Dual-Mode Ring Resonator Bandpass Filter With Asymmetric Inductive Coupling and Its Miniaturization, *IEEE Trans. Microwave Theory Tech.*, **60** (2012)
6. L.H. Hsieh, K. Chang, Compact, low insertion-loss, sharp-rejection, and wide-band microstrip bandpass filters, *IEEE Trans. Microwave Theory Tech.*, **51** (2003)

7. L. Athukorala, D. Budimir, Compact Dual-Mode Open Loop Microstrip Resonators and Filters, *IEEE Microw. Wireless Compon. Lett.*, **19** (2009)
8. W. Feng, X. Gao, W. Che, Q. Xue, Bandpass Filter Loaded With Open Stubs Using Dual-mode Ring Resonator, *IEEE Microw. Wireless Compon. Lett.*, **25** (2015)
9. Y.H. Jeng, S.F. Chang, Y.M. Chen, Y.J. Hung, A novel self-coupled dual-mode ring resonator and its application to bandpass filters, *IEEE Trans. Microwave Theory Tech.*, **54** (2006)
10. IE3D V10.1, *ZELAND SOFTWARE*, INC., CA, U.S.A.

A study of small sized quartz tuning fork using finite element analysis

Gwo-Chung Tsai

Department of Mechanical and Electro-Mechanical Engineering, National I-Lan University, I-Lan, Taiwan, R.O.C

Abstract. This is a preliminary study to small sized quartz tuning fork operating at 32.768 kHz for several types of supporting structure. Tine length and tine width can be calculated by Euler-Bernoulli equation (Exact Solution), but the base shall be analyzed by FEM (Approximation Solution) due to flexibility of base design. Finite Element Method (FEM) software ANSYS is used to simulate quartz tuning fork to extract electric response from harmonic analysis. Several types of supporting structure are to compare to modified tuning fork with notch by simulation.

1 Introduction

The design of quartz tuning fork was early treated as U-shape model, and analysis method has been derived from Transfer Matrix Method by J.A. Kusters and...et al. in 70s. The model (Fig. 1) was divided into several segments, and each segment was assumed to operate flexure and length-extensional mode then to extract characteristic frequency of system [1]. The base of quartz tuning fork in the above method didn't include rotary inertia effects. In 90s, Hideaki Itoh and...et al. also took quartz tuning fork to be U-shape model, and the right half section similar to L-shape was composed of three bars then to calculate the characteristic frequency of system by Sezawa's approximation [2]. But in the base, the frequency equation was considered as flexure modes. Based on the assumption of Euler-Bernoulli equation, the equation should be suitable for long bar not for short bar (Tuning Fork Base). In 80 s, the frequency equation of the flexure modes including rotary inertia effect was introduced by Shih S. Chuang [3], and the frequency equation including rotary inertia effect is suitable for small ratio of tine length to tine width and for higher frequency tuning fork. The quartz tuning fork using overtone flexure modes was developed by Shih S. Chuang.

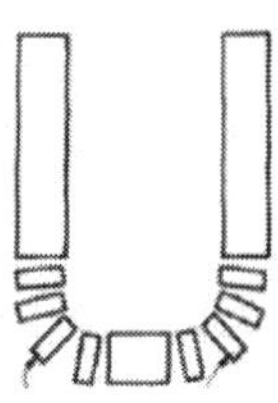

Fig. 1. U-shape tuning fork model [1].

In order to improve the electric characteristic of quartz tuning fork, Yoshiro Tomikawa[4][5] brought up ideas of modified base-width and arm corners cut off (Fig. 2). Those ideas were used in the real products indeed to enhance better electric characteristic, such as a higher Quality factor and lower resistance...etc. In 2003, Fumitaka Kitamura [6] patented a new base shape–notch which is located between tines and mounting area (Fig. 3) in order to get well electric characteristic in miniaturization of quartz tuning fork unit.

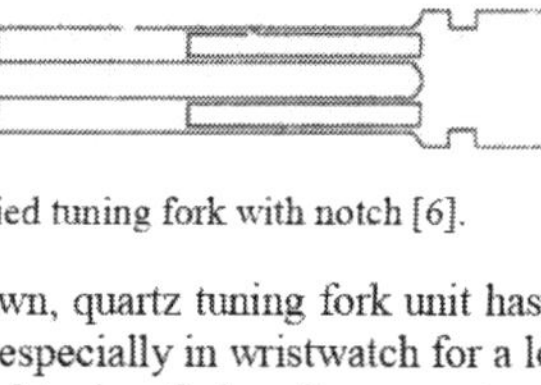

Fig. 2. Modified tuning fork with arm corners cut off and base-width [5].

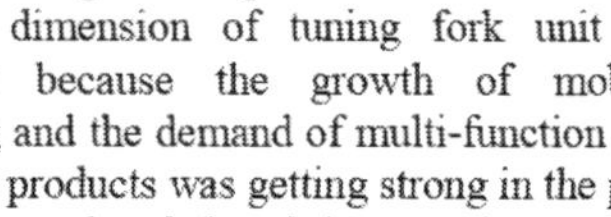

Fig. 3. Modified tuning fork with notch [6].

As known, quartz tuning fork unit has been used in many fields especially in wristwatch for a long time. The dimension of tuning fork unit was not most important issue due to capacious space of wristwatch. However the issue of dimension of tuning fork unit was more important because the growth of mobile phone marketing and the demand of multi-function of consume electronic products was getting strong in the past decade. The quartz tuning fork unit is not only required to meet the tight space but also required to hold the well electric characteristic. Nowadays, the popular packaged dimension of tuning fork unit which used in mobile phone is 4.1x1.5 mm^2 and 3.2x1.5 mm^2. Recently, the smallest quartz tuning fork unit 2.0x1.2 mm^2 has been developed by competitors. Those quartz tuning fork units

* Corresponding author: gctsai@niu.edu.tw

are much small than before. mm^2 has been developed by competitors. Those quartz tuning fork units are much small than before.

For design of small sized quartz tuning fork unit, both tine length and base length shall be decreased to meet tight space. As there is a relationship between length of tine and width of tine governed by Euler-Bernoulli equation[7] operating at 32.768 kHz, the decrease of tine length is limited. So, we are focusing on the analyses of base shape to get proper shape and dimension of base in order to meet requirement of tight space and good electric characteristic. Quartz tuning fork have been widely used in sensor, AFM and gyroscope [8-11].

In this paper, the design of tuning fork is to first extract tine width and tine length from Euler-Bernoulli equation with clamped-free boundary condition, then uses FEM software (ANSYS, Inc.) to simulate quartz tuning fork. The package dimension is 3.2x1.5 mm^2 with inner dimension 2.45x0.8 mm^2. Several types of supporting structure of quartz tuning fork unit shall be demonstrated to compare to modified tuning fork with notch (Fig.3).

2 Euler-Bernoulli equations

The Euler-Bernoulli equation [7] and its frequency equation with clamped-free boundary condition is given by

$$\frac{\partial^4 y}{\partial x^4} + \frac{1}{a^2}\frac{\partial^2 y}{\partial t^2} = 0,\ where\ a^2 = \frac{EI}{\rho A} \tag{1}$$

$$Cos\beta|Cos\beta| + 1 = 0 \tag{2}$$

E is constant derived from anisotropic elastic constants, and it depends on the cutting angle. β is wave number and l is beam length. The first root, which means fundamental mode, of (2) is 1.875. The frequency equation function of tine width and tine length is given by

$$f = k\frac{w}{l^2} \tag{3}$$

where f is frequency (Hz), k is constant in a specified cutting angle, w is tine width and l is tine length.

3 Improvement by modified base

As decrease of base length, the electric characteristic of quartz tuning fork is getting poor. It is better that ratio of base length to tine length is larger than 40 % [6]. According to this condition, of which ratio of base length to tine length is equal to 40 %, and formula (3) which operates at 32.768 kHz, tine width, tine length and base length can be tabularized as Table 1.

Table 1. Dimensions of specimens.

Tine width	Tine length	Total length
0.05	1.165	1.63
0.06	1.276	1.79
0.07	1.378	1.93
0.08	1.473	2.06
0.09	1.563	2.19
0.1	1.647	2.31
0.11	1.728	2.42
0.12	1.805	2.53
0.13	1.878	2.63
0.14	1.949	2.73

Unit : mm

As ordinary wafer thickness using in quartz tuning fork is 0.1 mm and tine thickness larger than tine width may cause unwanted vibrating displacement which occurs in not main vibration direction [12], the tine width small than 0.09 mm to be not discussed this time. The tine width large than 0.12 mm is also to be not discussed due to total length over inner space. Tine width 0.1 mm is picked up for case study.

The model of modified tuning fork with base-width (Fig. 3) is to verify that it is better ratio of base length to tine length is larger than 40 % by simulation, the result (Fig. 4) shows the electric characteristic is descent once the ratio is small than 40 %.

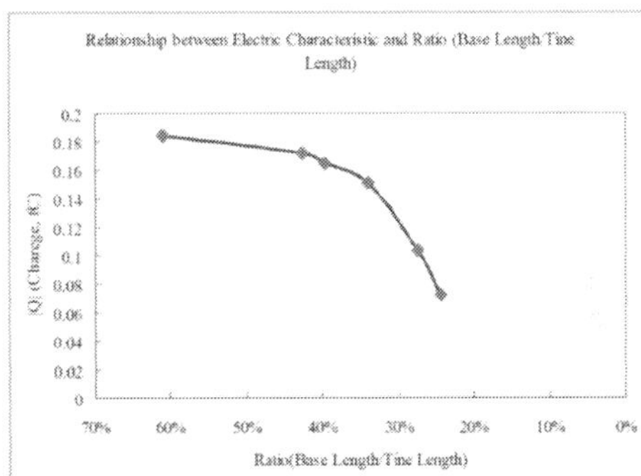

Fig. 4. Relationship between electric characteristic and ratio (Base length/Tine length).

In order to overcome poor electric characteristic due to ratio less than 40 %, Fumitaka Kitamura[6] patented a new base shape "notch" which is located between tines and mounting areas. The improvement caused from notch was also verified under ratio (Base length/Tine length) 40 % by simulation as shown in Fig. 5. It is obvious that modified tuning fork with notch has better electric characteristic as vibration energy doesn't leak to mounting areas.

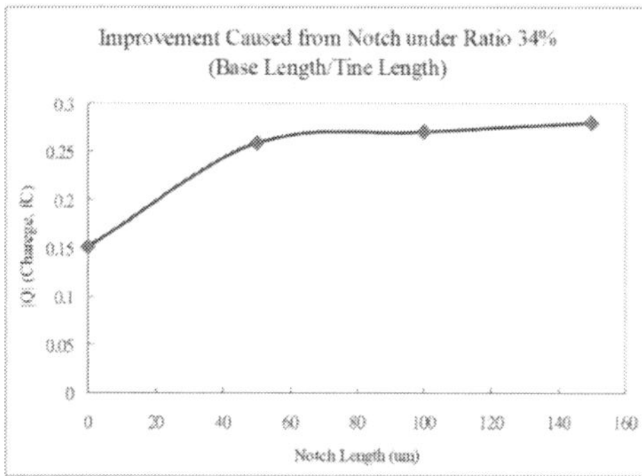

Fig. 5. Improvement resulted from notch under ratio 34 % (Base length/Tine length).

4 Simulation

The improvement by modified base comes out and idea. If one can clearly understand the mechanism of supporting structure for vibration energy trapping, small sized quartz tuning fork unit can be realized. Based on the idea [8], several types of supporting structure are built to compare to modified tuning fork with notch. Equations should be centred and should be numbered with the number on the right-hand side.

4.1 Procedure, conditions and modelling

Tine length and tine width calculated by Euler-Bernoulli equation at $+2^{\circ}$ X-cut (Operating at 32.768 kHz) are 1.64mm and 0.1 mm respectively. The thickness of tuning fork model is 0.1 mm. The finite element analysis software ANSYS is used to simulate quartz tuning fork resonator.

4.1.1 Procedure and conditions

Modal analysis is first to extract the characteristic frequency then to proceed with harmonic analysis for electric response (Charge per node, |Q|, fC).

a. Average number of element is 4000.
b. Average number of node is 22400.
b. Damping ratio is 0.0001.
c. The comparison is based on electric response of each model calculated by harmonic analysis.

4.1.2 Model

The model material is pure quartz crystal without electrode, and voltage as boundary condition applies on specified area on tines. There are five kinds of model regarding various base lengths to be analyzed (Table 2, Fig.6 ~ Fig.10)

Table 2. Dimensions of different notch.

Type	Base Length(μ m)		
Notch	560		
A	250	450	610
B	250	450	610
C	250	450	610
D	250	450	610

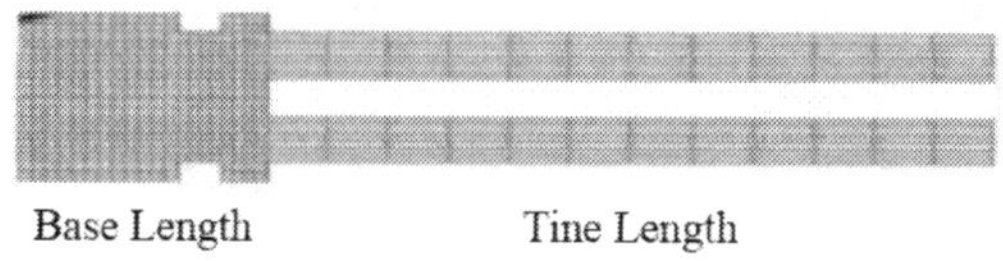

Fig. 6. Modified tuning fork with notch.

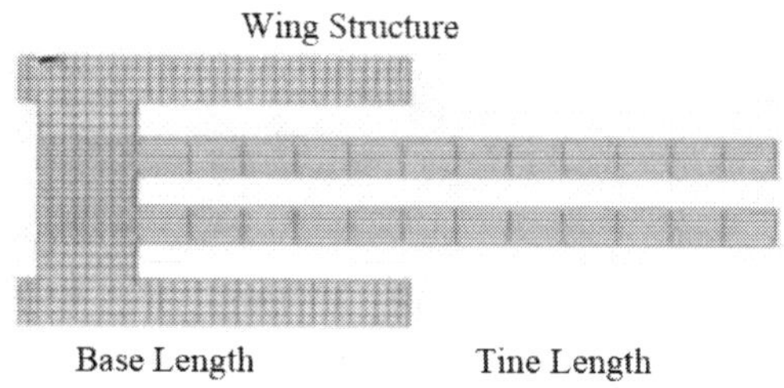

Fig. 7. Tuning fork type A.

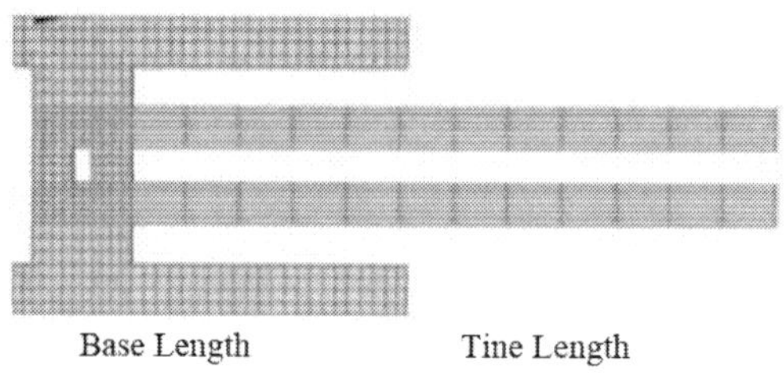

Fig. 8. Tuning fork type B.

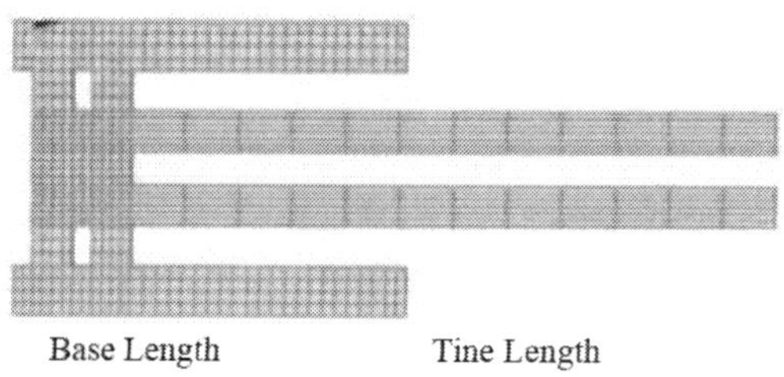

Fig. 9. Tuning fork type C.

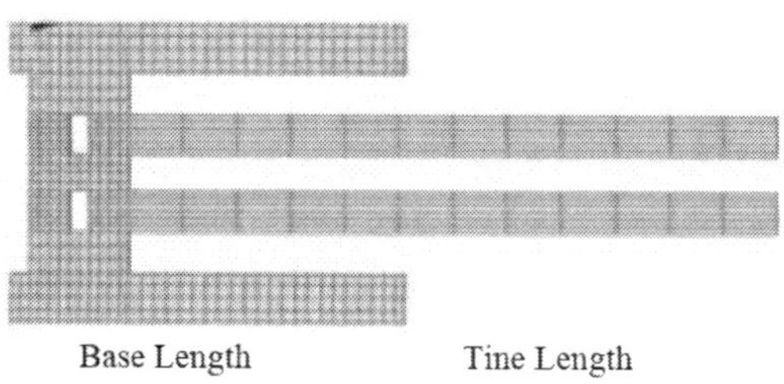

Fig. 10. Tuning fork type D.

4.2 Results

The electric response (|Q|, Charge, fC) of type A ~ D was calculated by harmonic analysis as follows (Fig. 11). Fig. 11 shows it is evident that electric response of model is better when base length is equal to 250 um. In case of base 250 um, type C is clearly better than other types, and location of hole shall be an important factor when base length is short. When base length is equal to 450 um or 610 um (Long base length), location of hole is not the important factor, but type B, C and D still have better electric response than type A.

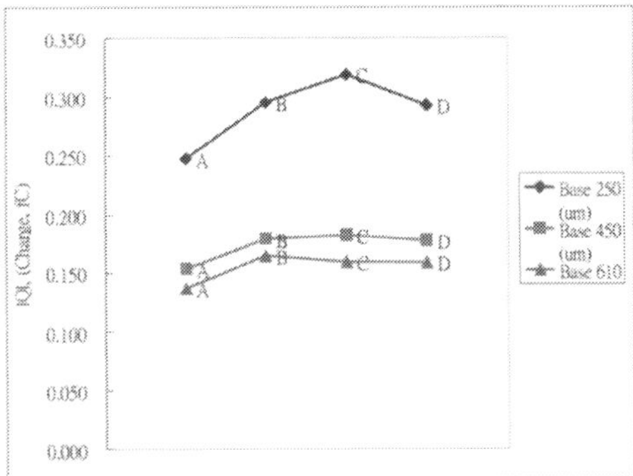

Fig. 11. Electric response.

Fig. 12 is that the result of type A ~ D compares to modified tuning fork with notch. It is obvious that the type B, C and D with base length 250 um and "hole" are better than modified tuning fork with notch. The total length of type A, B, C and D with base length 250 um is 1890 um, it is shorter than modified tuning fork with notch (Total length: 2200 um).

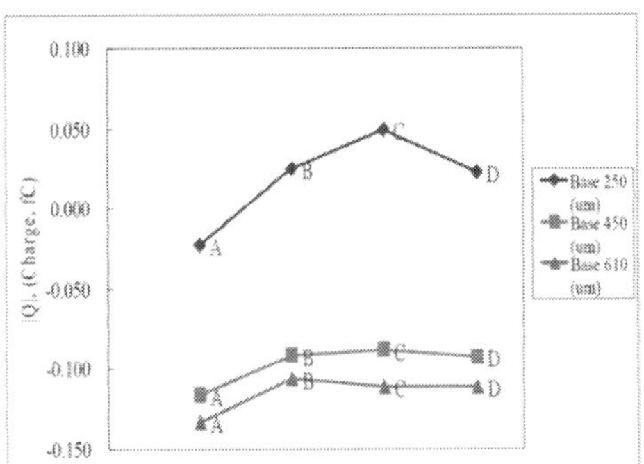

Fig. 12. Analytical results.

Fig. 13 shows that electric response is descent no matter what type is when base length is increasing, and it has an opposite trend to Fig. 4. However, type with hole really has better electric response than type without hole.

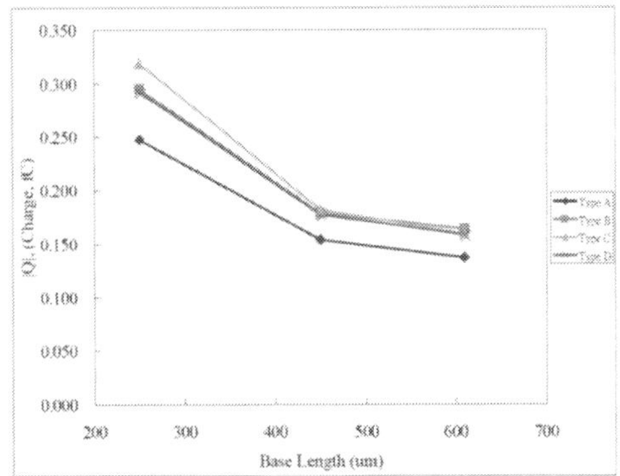

Fig. 13. Electric response.

5 Conclusion

In order to meet the restriction of tight space and well electric characteristic, several types of supporting structure of quartz tuning fork were discussed. The simulation result shows new supporting structure has a better electric response than modified tuning fork with notch, and hole located in base can effectively increase electric response. The author thinks hole may play a role of reflection mechanism to avoid vibration energy leakage to mounting areas, like beveling of AT-cut crystal. However, large hole may decrease stiffness of base. The balance between electric response and base stiffness needs to be confirmed.

Compare to modified tuning fork with notch, electric response is descent as base length of new supporting structure is increasing. This shall be a good info to miniaturization.

Next is to build models with electrode so to approach real device. The more complicated phenomenon can be involved in model to verify this view point. Experiment will be executed also.

References

1. J.A. Kusters, C.A. Adams, and H.E. Karrer and R.W. Ward, "Analytical and experimental investigations of 32kHz quartz tuning forks", *Proc. 30th Ann. Frequency Control Symp.*, 175-183(1976)
2. Hideaki Itoh and Takashi Matsumoto, "An analysis of frequency of a quartz crystal tuning fork by sezawa's approximation – The effect of clamped position of its base", *Joint Meeting EFTF – IEEE IFCS*,172-179(1999)
3. Shih S. Chuang, "Quartz tuning fork crystal using overtone flexure modes", *Proc. 35th Ann. Frequency Control Symp.*, 6-15(1981).
4. Yoshiro Tomikawa, Sumio Sugawara, and Masashi Konno , "Finite element analysis of displacement at base portion of a quartz crystal tuning fork", *IEEE Transactions on Sonics and Ultrasonics*, **26**(3), 23-29(1979)

5. Yoshiro Tomikawa, Shigeru Oyama, and Masashi Konno, "A quartz crystal tuning fork with modified basewidth for a high quality factor: finite element analysis and experiments", *IEEE Transactions on Sonics and Ultrasonics*, **29**(6),124-129(1982)
6. Fumitaka Kitamura, Junichiro Sakata and Hideo Tanaya, "US Patent 6587009", 2003
7. Karl F. Graff, "Wave Motion in Elastic Solids", Dover Publications, Inc., New York
8. Roger Oria, Jorge Otero, Laura González, Luis Botaya, Manuel Carmona and Manel Puig-Vidal," Finite element analysis of electrically excited quartz tuning fork devices," *Sensors*, **13**, 7156-7169 (2013)
9. M.R. Abdel-Rahman, B.A. Albassam, M.A. Alduraibi, " Driveless gyroscope response of a quatz piezoelctric vibratory tuning folk, " *Acta Physica Polonica A*, **127**, 1352-1354 (2015)
10. Jorge Otero, Laura Gonzalez and Manel Puig-Vidal , "Nanocharacterization of soft biological samples in shear mode with quartz tuning fork probes, " *Sensors*, **12**, 4803-4819 (2012)
11. Anatoliy A. Kostereva and Frank K. Tittel, "Applications of quartz tuning forks in spectroscopic gas sensing," *Review of Scientific Instruments*, **76**, 043105-1,043105-9 (2005)
12. Hideo Tanaya, "US Patent 20070024163", 2007

Piercing and surface-crack defects in combination extrusion of fasteners

Heng-Sheng Lin[1,*], *Chien-Yu* Lee[1], *Wen-Shun* Li[1], and *Cheng-Tsung* Wu[1]

[1] Department of Mold and Die Engineering, National Kaohsiung University of Applied Sciences, Kaohsiung 807, Taiwan

Abstract. Piercing or surface-crack defect occurs in combination-extruding the support pin. There has been so far insufficient knowledge in avoiding such forming defects in forming the similar products. In this paper, a combination-extrusion experiment, with various combinations of forward and backward extrusion ratios, was conducted to observe the occurrence of the defects. The result shows that the defects occur at low level of both forward extrusion ratio and backward extrusion ratio used in the combination-extrusion process. The low forming level causes the onset of insufficient plastic metal-flow, which is similar to that of inhomogeneous deformation in the unidirectional extrusion or drawing process. A forming limit diagram was constructed and can be used in the defect prediction for the future development of similar products.

1 Introduction

The extrusion process has been commonly used in forming fasteners. Such forming operation includes forward, backward, and combined forward-backward extrusion. The forming defects with the forward or backward extrusion processes, such as Chevron cracks, have been investigated extensively [1-6]. However, the investigations on the forming defects with combination extrusion, such as piercing or surface-crack defects, are still scarce. Lee et al. [7] report the occurrence of a lapping defect caused by the dead metal in forming a piston-ring with respective forward and backward extrusion. The lapping defect is avoided by controlling the material flow and eliminating the dead metal zone with a novel die design with combination extrusion.

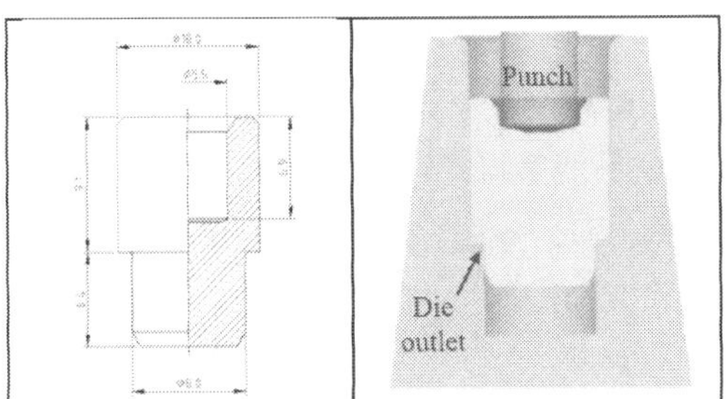

(a) (b)

Fig. 1. (a) Dimensions of support pin; (b) combination-extrusion of support pin.

The demand of the current work arises from the forming design of a support pin, as shown in Fig. 1. The occurrence of the piercing or surface-crack defects, as shown in Fig. 2, renders the combination extrusion as an unreliable operation. The surface cracks may appear sporadically, and be recognized as the forerunner of piercing to happen. Similar approach has been reported by Yoon et al. [8], in their hot bi-directional extrusion with magnesium alloy. There has been so far insufficient knowledge in avoiding such forming defects in forming the similar products.

(a) (b)

Fig. 2. Forming defects of combination extrusion: (a) surface crack; (b) piercing.

* Corresponding author: hslin@kuas.edu.tw

In this paper, a combination-extrusion experiment with various combinations of forward and backward extrusion ratios was conducted to observe the occurrence of the defects. A complete forming limit diagram was also constructed and can be used in the defect prediction for the future development of the similar products. Finite element analysis for effective strain and ductile fracture was performed.

2 Methods

2.1 Experiment

Annealed ASTM A36 structural steel bar was used as the workpiece material for the combination-extrusion experiment. The Young's modulus was 200 GPa, and the Poissson ratio was 0.26. The workpiece was 25 mm in diameter and 25 mm in height, whose size was 2.5 times of the support pin as shown in Fig. 1. The flow-stress of workpiece is shown in Fig. 3. The combination extrusion was conducted on a hydraulic press. The punching speed was 10 mm/s.

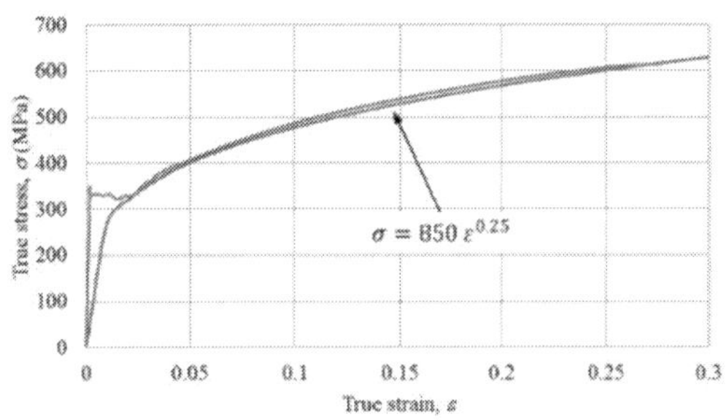

Fig. 3. Flow stress of workpiece.

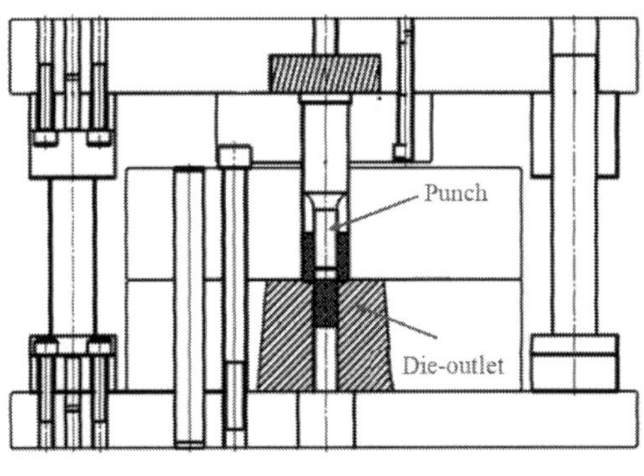

Fig. 4. Die-set configuration of combination extrusion.

Various combinations of forward extrusion ratios and backward extrusion ratios were tested in the experiment. The design of the die set is show in Fig. 4. The forward extrusion ratio is defined as the cross-sectional area of the original workpiece divided by that of the downward extruded part from the die-outlet. The backward extrusion ratio is defined as the cross-sectional area of the original workpiece divided by that of the upward extruded part from the punch. For instance, the forward and backward extrusion ratios in forming the support pin shown in Fig. 1 are

$$\frac{\frac{\pi}{4}10^2}{\frac{\pi}{4}8^2} = 1.56 \quad (1)$$

and

$$\frac{\frac{\pi}{4}10^2}{\frac{\pi}{4}(10^2-5.5^2)} = 1.43 \quad (2)$$

, respectively. In the combination-extrusion experiment, three dies of different outlet diameters and three punches of different diameters were used. This will provide various combinations of forward and backward extrusion ratios. The dimensions and design parameters used in the combination-extrusion experiment are shown in Table 1.

Table 1. Dimensions and design parameters used in combination-extrusion experiment.

Billet height, mm	25.0
Billet diameter, mm	25.0
Die-outlet diameter, mm (Forward extrusion ratios are in parentheses)	12.50 (4.00) 17.68 (2.00) 21.65 (1.33)
Punch diameter, mm (Backward extrusion ratios are in parentheses)	12.50 (1.33) 17.68 (2.00) 21.65 (4.00)

2.2 Finite element simulation

Finite element software DEFORM 2D was used for the analysis. The workpiece was assumed to be elasto-plastic and the punch and die were assumed to be rigid. The flow rule is

$$\sigma = 850\varepsilon^{0.25} \quad (3)$$

which was obtained from curve fitting of the tensile test. 15,000 elements were used in the simulation and the friction factor was 0.12. The effects of strain rate and temperature were neglected. The normalized Cockcroft and Latham damage criterion was used in determining the level of ductile fracture, which is defined as

$$C = \int_0^{\bar{\varepsilon}_f} \frac{\sigma^*}{\bar{\sigma}} d\bar{\varepsilon} \quad (4)$$

where C is the critical damage value, σ^* is the maximum tensile stress, $\bar{\sigma}$ is the effective stress, $\bar{\varepsilon}$ is the effective strain, and $\bar{\varepsilon}_f$ is the critical effective strain upon material fracture, respectively. The corresponding damage value upon fracture was 1.3, which was obtained from matching the tensile test for the ASTM A36 structural steel bar of 12.5 mm in diameter.

DOI:10.1051/matecconf/201712300018

3 Results and discussion

3.1 Experimental result

Fig. 5 shows the experimental result of the combination extrusion. The corresponding forming limit diagram plotted against the forward and backward extrusion ratios is shown in Fig. 6. In order to explicitly compare the experimental result with the forming limit diagram, the corresponding forward and backward extrusion ratios are reversed on both the horizontal and vertical labels of Fig. 6, respectively. The patterns of flow line formed with the combination-extrusion are similar to those of Fig. 2, in which the piercing cracks are crossing instead of parallel to the metal flow. This indicates that the forming defect was not caused by the discrepancy of metal flow and the dead metal blocked by the die corner.

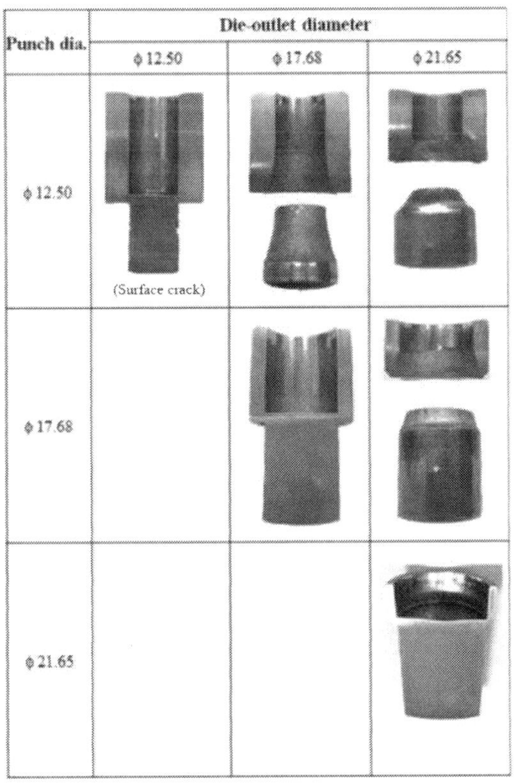

Fig. 5. Cross-section of combination-extrusion experiment.

The experiment shows that satisfactory combination extrusion can be obtained when the die-outlet diameter was small and the punch diameter was large. This translates to a large forward extrusion ratio and a large backward extrusion ratio used during the combination extrusion. Surface-crack defect occurred at a small die-outlet diameter (12.5 mm) and a small punch diameter (12.5 mm), i.e. a large forward extrusion ratio and a small backward extrusion ratio. Piercing defect occurred when the die-outlet diameter was large and the punch diameter was small, i.e. a small forward extrusion ratio and a small backward extrusion ratio, respectively.

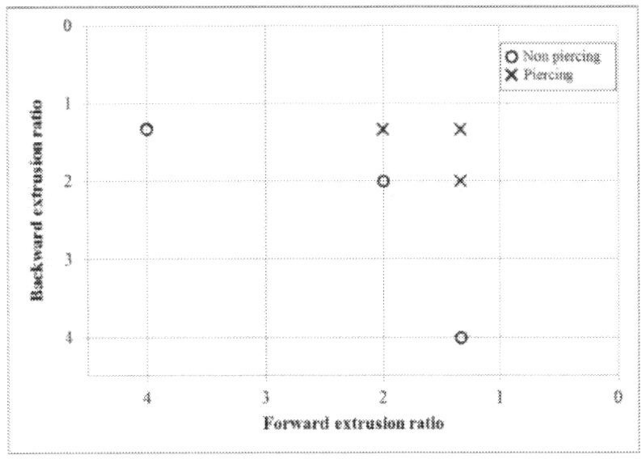

Fig. 6. Forming limit diagram of combination-extrusion experiment.

The result shown in Figs. 5 and 6 is similar to that of Yoon et al. [8]. Contrary to the conventional wisdom, the piercing defect does not necessarily occur when the die-outlet diameter is close to that of punch. Moreover, the piercing defect does not tend to occur when the consumption of distortion energy is high, i.e. high level of extrusion ratio. Both phenomena indicate that the forming defects occurred in combination extrusion are attributed to insufficient deformation level, which is also a primary factor causing the strain inhomogeneity of workpiece [9] in the forming processes. The low forming level also is an indication that the hydrostatic pressure is low, and the tendency of crack initiation in the workpiece is not suppressed.

3.2 Simulation result

Fig. 7 shows the distribution of effective strain obtained from the FEM analysis corresponding to the conditions used in the combination extrusion of Fig. 5. The strain appears concentrated along the workpiece's forward-extruded outer surface and the backward-extruded inner surface when the diameter of the die-outlet is equivalent to that of the punch. This phenomenon might lead to the speculation that piercing or surface defect would occur under such condition attributed to the concentrated deformation. However, the experimental result shown in Fig. 5 reveals that piercing does not occur under such condition. On the contrary, piercing occurs at both low forward and low backward extrusion ratios where the effective strain appears mild and scattered. In this analysis, the distribution of effective strain does not explain well on the forming defects of combination extrusion.

A similar trend is found in the distribution of the normalized Cockcroft and Latham damage values as shown in Fig. 8. The damage values also appear concentrated along the workpiece's forward-extruded outer surface and the backward-extruded inner surface when the diameter of the die-outlet is equivalent to that of the punch. However, for the piercing conditions when both forward and backward extrusion ratios are low, the

DOI:10.1051/matecconf/201712300018

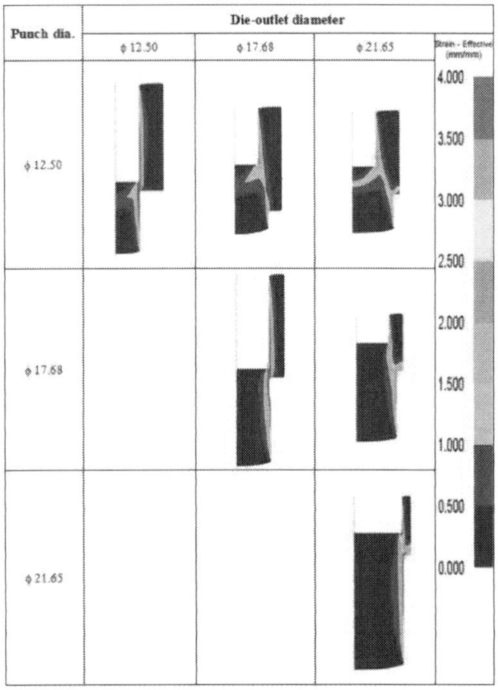

Fig. 7. Distribution of effective strain.

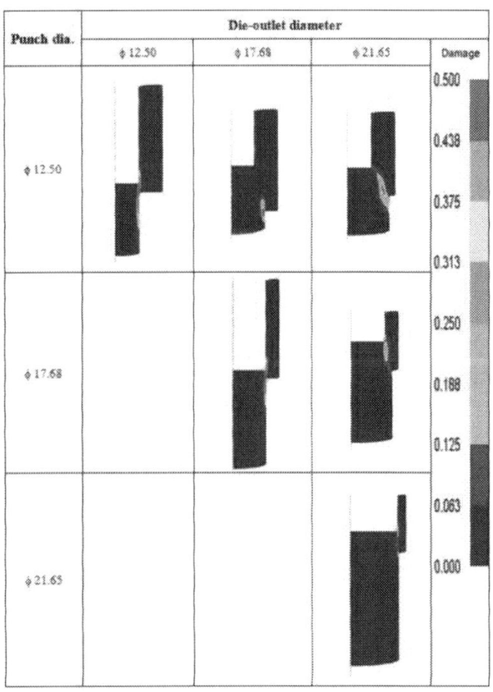

Fig. 8. Distribution of the normalized Cockcroft and Latham damage values.

damage values appear smaller and less concentrated. This is no surprise because the damage values are obtained by the integration of the maximum tensile stress and the effective strain. And the effective strain affects the damage values significantly. The damage values were less than 1.3, the critical damage value upon fracture obtained from matching the tensile test. The analysis of damage values does not predict the occurrence of cracks during the combination extrusion. Therefore, the normalized Cockcroft and Latham damage criterion does not apply well in explaining the forming defects of combination extrusion.

3.3 Improvement of forming defect for support pin

3.3.1 Modification of the pre-forming stage

For the support pin shown in Fig. 1, the forward extrusion ratio and the backward extrusion ratio are 1.56 and 1.43, respectively. The forming level falls on the border of the piercing and the non-piercing conditions, according to the forming limit diagram of the combination-extrusion experiment shown in Fig. 6. This indicates that possibility of applying the combination extrusion and avoiding the piercing defect, if some modification measures are made.

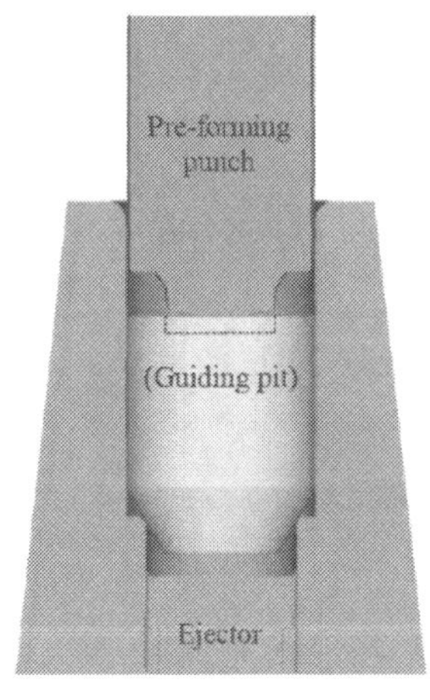

Fig. 9. Die-set configuration of pre-forming stage.

The simple way to avoid the forming defect for the support pin is to increase the corner angles at the die outlet and the punch. But this measure would affect the usability of the product. A quick solution to avoid the forming defect in support pin is to increase the penetration depth of the punch at the pre-forming stage, as shown in Fig. 9. According to the field practice, the purpose of the pre-forming is to provide a shallow circular pit on the back of the workpiece as the guiding pit for the subsequent combination extrusion with the slim punch, as shown in Fig. 1(b). Without the pre-forming stage, the slim combination-extrusion punch breaks easily because the eccentric buckling might occur from the heavy extrusion load. Therefore, by increasing the pre-forming punch stroke slightly further, the bottom side of the workpiece will contact with the ejector and the tendency for backward extrusion will be enhanced. This measure at the same time can significantly distribute the load in the stage of combination extrusion to the pre-forming stage, and the occurrence of piercing defect can be avoided.

3.3.2 Decomposition of the combination extrusion

Another robust way to avoid the forming defects in combination extrusion is to decompose the operation into respective forward extrusion and backward extrusion. Although the hydrostatic pressure in the workpiece during the unidirectional extrusion operation is usually very high, the tendency for the crack initiation is suppressed. Punch and die life maybe decreased because of the larger hydrostatic pressure generated in the unidirectional extrusion. The decomposition also requires an extra forming stage and additive tool cost as well as a larger forming press.

Fig. 10 shows the flow lines of the cross-section formed with the unidirectional extrusion. There are neither cracks formed on the corner of the forward-extruded outer surface nor on the corner of the backward-extruded inner surface. Normally the forward extrusion followed by backward extrusion is favored because of the convenience of punch ejection. Otherwise, after the completion of the preliminary backward extrusion, the punch gets wrapped by the workpiece during the following forward extrusion. This causes the risk of breakage on the punch.

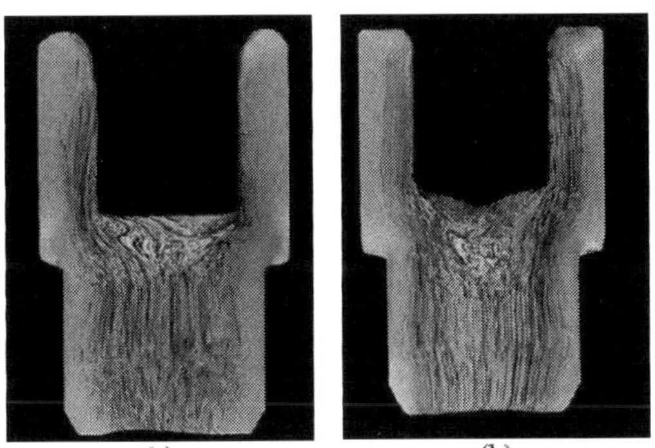

(a) (b)

Fig. 10. Flow line images of the cross-section formed with unidirectional extrusions: (a) forward extrusion followed by backward extrusion; (b) backward extrusion followed by forward extrusion.

4 Conclusion

In this paper, a combination-extrusion experiment with various combinations of forward and backward extrusion ratios was conducted to observe the occurrence of forming defects. A forming limit diagram was also constructed and can be used in the defect prediction for the future development of the similar products. Finite element analysis for effective strain and ductile fracture was performed.

(1) The experiment shows that satisfactory combination extrusion can be obtained when the die-outlet diameter was small and the punch diameter was large. This translates to a large forward extrusion ratio and a large backward extrusion ratio used during the combination extrusion. Surface-crack defect occurred at a small die-outlet diameter (12.5 mm) and a small punch diameter (12.5 mm), i.e. a large forward extrusion ratio and a small backward extrusion ratio. Piercing defect occurred when the die-outlet diameter was large and the punch diameter was small, i.e. a small forward extrusion ratio and a small backward extrusion ratio, respectively. Contrary to the conventional wisdom, the piercing defect does not necessarily occur when the die-outlet diameter is close to that of punch. Moreover, the piercing defect does not tend to occur when the consumption of distortion energy is high, i.e. high level of extrusion ratio. Both phenomena indicate that the forming defects occurred in combination extrusion are attributed to insufficient deformation level which is also a primary factor causing the strain inhomogeneity of workpiece in the forming processes.

(2) The distribution of effective strain appears concentrated along the workpiece's forward-extruded outer surface and the backward-extruded inner surface when the diameter of the die-outlet is equivalent to that of the punch. This phenomenon might lead to the speculation that piercing or surface defects would occur attributed to the concentrated deformation. However, the experiment reveals that piercing does not occur under such condition. On the contrary, piercing occurs at both low forward and low backward extrusion ratios where the effective strain appears mild and scattered. In this analysis, the distribution of effective strain does not explain well on the forming defects of combination extrusion. A similar trend is found in the distribution of the normalized Cockcroft and Latham damage values. The damage values also appear concentrated along the workpiece's forward-extruded outer surface and the backward-extruded inner surface when the diameter of the die-outlet is equivalent to that of the punch. However, for the piercing condition when both forward and backward extrusion ratios are low, the damage values appear smaller and less concentrated. The damage values were less than 1.3, the critical damage value upon fracture obtained from matching the tensile test. The analysis of damage values does not predict the occurrence of cracks during the combination extrusion.

(3) A quick solution to improve the forming defect in support pin is to increase the penetration depth of the punch at the pre-forming stage. The bottom side of the workpiece will contact with the ejector and the tendency for backward extrusion will be enhanced. This measure at the same time can significantly distribute the load in the stage of combination

extrusion to the pre-forming stage, and the occurrence of piercing defect can be avoided. Another robust way to avoid the forming defects in combination extrusion is to decompose the operation into separate forward extrusion and backward extrusion. Although the hydrostatic pressure in the workpiece during the unidirectional extrusion operation is usually very high, the tendency for the crack initiation is suppressed.

References

1. M. Terhorst, A. Feuerhack, D. Trauth, *Int J Mater Form*, 9(4), 449 (2016)
2. P.O. Bouchard, L. Bourgeon, S. Fayolle, K. Mocellin, *Int J Mater Form*, **4**(3), 299 (2011)
3. K. Komori, *Int J Mech Sci*, **45**,141 (2003)
4. N.V. Reddy, P.M. Dixit, G.K. Lal, *Int J Mach Tool Manufact*, **40**, 95 (2000)
5. P.J. McAllen, P. Phelan, *J Mater Process Technol*, **183**, 210 (2007)
6. K. Saanouni, J.F. Mariage, A. Cherouat, P. Lestriez, *Comput Struct*, **82**, 2309 (2004)
7. D.J. Lee, D.J. Kim, B.M. Kim, *J Mater Process Technol*, **139**, 422 (2003)
8. D.J. Yoon, S.J. Lim, H.G. Jeong, E.Z. Kim, C.D. Cho, *J Mater Process Technol*, **201**, 179 (2008)
9. H.S. Lin, Y.C. Hsu, C.C. Ke, *Int J Adv Manuf Technol*, **81**, 53 (2015)

 DOI:10.1051/matecconf/201712300019

Experimental and FEM analysis of three carbon steel characterization under quasi-static strain rate for bumper beam application

Dawit Bogale Alemayehu[1,*], *Song-Jeng* Huang[2], and *Ermias Geberekidan* Koricho[3]
[1]Department of Mechanical Engineering, Bahir Dar University, Bahir Dar City, Ethiopia
[1,2]Department of Mechanical Engineering, National Taiwan University of Science and Technology, Taipei, 10607, Taiwan
[3]Department of Mechanical Engineering, Georgia Southern University, Statesboro, GA 30458, USA

Abstract. This paper investigates the mechanical behavior of three selected steel materials which are considered to be the bulk material of front most bumper beam of a vehicle that is suddenly loaded in the quasi-static range. Thirty-six constant strain rate uniaxial tension tests were performed. The test was performed on a HUALONG electro-hydraulic universal testing machine at four strain rates ($3.33\times10^{-3}, 3.33\times10^{-2}, 3.33\times10^{-1}, 3.33s^{-1}$). The FEM which is ABAQUS/CAE is used to simulate the bumper subsystem using the three selected steel. The outcome shows that ultimate tensile strength (UTS) increase with an increase in strain rate and high alloy steel (HAS) material has the maximum mean UTS. The FEM in the post -processing stage gives the minimum displacement and maximum strain energy for HAS material when compared to the other two materials. Finally, from both experimental and ABAQUS explicit analysis the result shows HAS material is better suit for the bumper beam application.

1 Introduction

Steel is a material used to build the foundations of society. It is an iron-based material containing low amount of carbon and alloying elements that can be made into thousands of compounds with exciting properties to meet a wide range of engineering application. Steel is truly a versatile material. The value of steel produced annually is estimated over 400 billion USD. Thus steel finds its way into the world economy. Nowadays, need of steels is soar up due to newly introduced massive construction projects in civil, mechanical, naval, aeronautical engineering and in other engineering fields. Therefore, it is necessary for designers to understand the dynamic behavior of steel materials.

Bumper beams are designed to with stand the high speed impact from the opposite vehicle or rigid wall during collision and the material to be selected should have a property which can absorb shock load or store strain energy during deformation, this special property is achieved by selecting traditional or reference steel materials with reasonable carbon content.

Recently, considerable attention has been given to the use of steel materials in construction sectors to understand their mechanical and thermo-mechanical behavior under different circumstances. This initiation mainly came from collapse of the two world trade center buildings which were attacked in Sep, 2001 by terrorist. Most of the literatures have concentrated on high–strain rate due to airplane impact and fire impact on the building materials [1].

The study concerns the effect of strain rates caused by dynamically loaded with impacts in the range of ($3.3*10^{-3}$ to $3.3s^{-1}$) or quasi-static and fire impacts. But the thermal effect (at elevated temperature) is not considered here because the tensile test has been takes place at room temperature. Therefore, the final aim is to increase the strength of this steel products to overcome slow speed impact by absorption of energy through deformation modeling (characterizing) it using numerical and experimental work. And also descriptions of permanent plastic deformations that are designed deliberately in a wide range of industrial processes like automobile industry for this paper, or occur accidentally under undesired overloads. The ABAQUS/CAE software (finite element method) is used to perform reliable simulations of impact experiments that produce plastic deformation in the selected steel specimen from the steel industries in Ethiopia. The numerical model accounts for the actual experimental setup used in the experiments, including testing of the selected items at different strain rate and tensile deformation using universal testing machine performed at room temperature. For the application part the form the car safety part the bumper subsystem is modeled using CATIA/V5 software and exported to ABAQUS/CAE in order to study the effect of strain rate during slow speed impact. Energy absorption from ABAQUS/CAE is compared observed and analyzed as well as the displacement.

In their work, introduce the mechanical properties and microstructures of AISI 4340 high strength alloy steel under different tempering conditions are investigated [2]. They also believe that an understanding of the mechanical properties of metals during deformation over a wide range of loading conditions is of considerable importance for a number of engineering applications. It has been reported that the superiority of super austenitic stainless steel, 254 SMO (S31254), lies in its high strength, good weld ability and great resistance to stress corrosion and pitting because of its higher content of alloying chromium and molybdenum compared to general stainless steel. Furthermore, the

* Corresponding author: bogale.sintayehu@gmail.com

purpose of this paper is to investigate the effect of strain rate on the mechanical properties and microstructure variation of the super-austenitic stainless steel 254 SMO [3]. The development of physically-based dislocation models of a mechanical response of materials under different strain rates and temperatures is an important problem of modern mechanics. If the microstructure is not affected much by changes in the strain rate and temperature, the models can be employed for predicting mechanical properties of pure metals and alloys under uniaxial loading conditions where in the specimen experiences isotropic homogeneous deformation up to the point of failure solely due to the motion of uniformly distributed defects (dislocations) [4].

One of the defining features of impacts that occur at velocities sufficiently large to cause inelastic (and particularly plastic) deformations is that most of these deformations occur at high strain-rates. These deformations may also lead to large strains and high temperatures. Unfortunately, we do not understand the high-strain-rate behavior of many materials (often defined as the dependence σ_f (ε, ˙ε, T) of the flow stress on the strain, strain rate and temperature), and this is particularly true at high strains and high temperatures. A number of experimental techniques have therefore been developed to measure the properties of materials at high strain-rates. In this section we focus on those experimental techniques that develop controlled high rates of deformation in the bulk of the specimen, rather than those in which high-strain-rates are developed just behind a propagating wave front [6]. The primary experimental techniques associated with the measurement of the rate dependent properties of materials are described in Fig 1 (note that the stress states developed within the various techniques are not necessarily identical). An excellent and relatively recent review of these methods is presented by [7]. For the purposes of this discussion, strain rates above 102 s^{-1} are classified as high strain rates, strain rates above 104 s^{-1} are called very high strain rates, and strain rates above 106 s^{-1} are called ultra-high strain rates. Conventionally, strain rates at or below 10^{-3} s^{-1} are considered to represent quasi-static deformations, and strain rates below 10^{-6} s^{-1} are considered to be in the creep domain. Quasi static experiments are typically accomplished through a variety of servo hydraulic machines, and ASTM standards exist for most of these experiments. Most servo-hydraulic ma-chines are unable to develop strain rates larger than 100 s^{-1} repeatedly, but some specialized servo-hydraulic machines can achieve strain rates of 101 s^{-1}. The primary approach to testing in this strain rate range uses drop towers or drop weight machines [8], and great care must be exercised in interpreting the data because of the coupling between impact-induced wave propagation and machine vibrations. More recently, consideration was given to the impact of the rate of straining of a particular material or component on its performance. Since steel is a strain rate sensitive material, its yield strength increases as the loading rate increases. This provides further benefits in its ability to sustain and absorb higher loads and higher input energy, such as in the case of deformation of a bumper or other structural component. Again, this is not a new discovery but it was only through the introduction of the advanced vehicle concepts phase of the ULSAB (Ultra-Light Steel Auto Body) development that this benefit of steel began to be introduced in structural design of automobile components. Considerable effort was then expended in various laboratories around the world to generate tensile data at straining rates ranging from quasi-static (10-3 s^{-1}) to 103 s^{-1} for many of the above steel grades [9].

2 A Bumper system and components

A bumper is a structure attached to or integrated with the front and rear ends of a motor vehicle, to absorb impact in a minor collision, ideally protecting occupants and minimizing repair costs. The bumper components are shown in the Fig. 1.

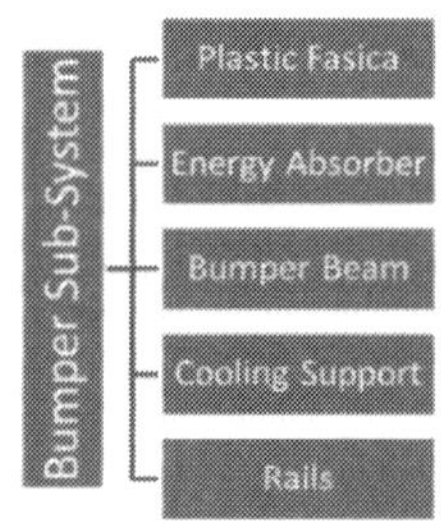

Fig. 1. Automotive bumper system and components.

3 Methodology

3.1 Experimental method

If a material is to be used as part of an engineering structure that will be subjected to a load, it is important to know that the material is strong enough and rigid enough to withstand the loads that it will experience in service.

3.1.1 Description of the experiment

This experiment is done to get the property of the sample materials which are three in number (Low carbon Steel, High Alloy Steel, Medium Carbon Steel). The tension test is done at Addis Ababa Institute of Technology department of mechanical engineering in mechanical workshop. The sample material from the bulk material of the bumper is taken to be tested in our laboratory for tensile strength. The samples are prepared according to ASTM E8 standards in order to achieve the experimental objective. The tensile testing is done using the servo electrohydraulic universal testing machine with data acquisition in order to have the X-Y data information during the testing of the specimen. The experiment is done by applying load from the servo hydraulic universal testing machine by applying tension uniaxial load on one side of the specimen and the other side is constrained to be fixed. The measurement is taken on the gauge section in which its length is 50mm, 2mm thick and 12.5mm width.

3.2 Experimental design

The experimental design is shown in Table 1, the three X's each represents the number of samples to be tested for tensile loading for each strain. For instance, MS material is tested three times (XXX) for 3.333*10-3 s-1. It describes the three important parameters such as, the Replica, Factor, and Variable. Here the replica is three for each sample and most importantly the total samples of this testing for the three factors are 36.

Table 1. Matrix of experimental to be conducted under room temperature.

Strain rates (s^{-1})	3.33×10^{-3}	3.33×10^{-2}	3.33×10^{-1}	3.333
Crosshead(mm/min)	10	100	1000	10000
MS	XXX	XXX	XXX	XXX
HAS	XXX	XXX	XXX	XXX
LCS	XXX	XXX	XXX	XXX

MS=Mild Steel, **HAS**=High Alloy Steel, **LCS**=Low Carbon Steel

3.2.1 Materials used

The materials under the test are three types in quantity and all are low carbon steel materials. These materials indicated by alphabet A(HAS), B(LCS) and C(MS) are shown in the Fig.2.

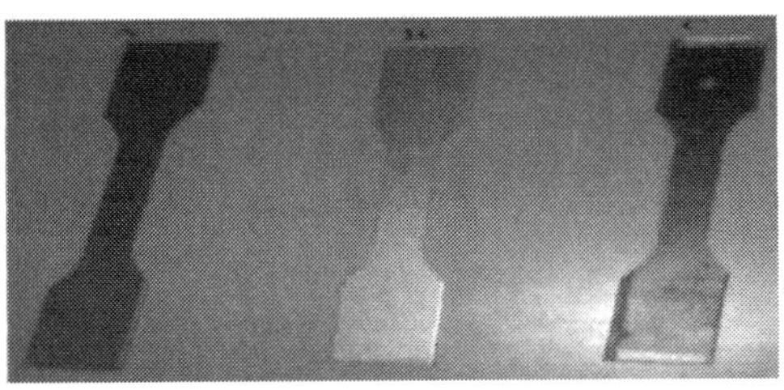

Fig. 2. Specimens "dog bone" used for tensile testing.

3.3 Finite element modelling

3.3.1 Brief introduction to ABAQUS/CAE model

The simulation work was carried out using commercial software called ABAQUS/CAE which is an engineering tool for programming that is used to solve various degrees of engineering problems ranging from linear to non-linear problems that are complex. ABAQUS/CAE enables models to be solved as quickly as possible by simply importing the geometry under investigation with the right physical and material properties associated to it, meshing it, loading and also by applying the boundary conditions to the material to be modelled [11]. The reference bumper subsystem model was taken from one of the complete FEM car models available in the NHTSA web site and imported into ABAQUS environment.

The 3D part and meshed part that represents the Specimens "dog bone" which is simulated for tension load using ABAQUS/CAE are shown in Fig.3.

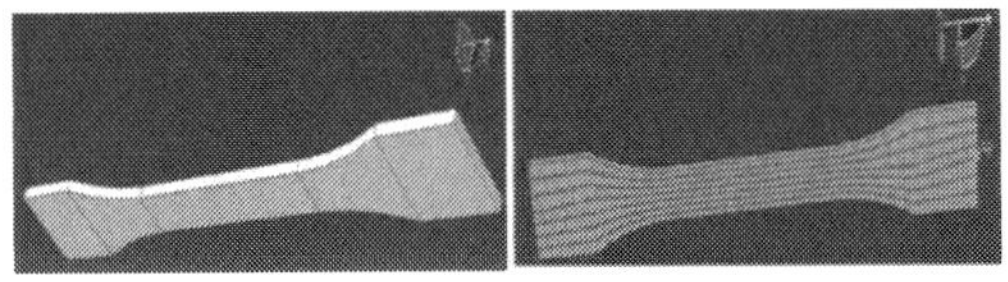

Fig.3. Extruded in 2mm thickness specimen model on ABAQUS/CAE and mesh part.

Assembled car bumper system built part by part by CATIA V5 available in the NHTSA web site is shown in Fig.4 to the left. And this assembled as one part and then saved as STEP* or IGES file extension to be imported to ABAQUS/CAE software for meshing it as desired shown in Fig.4 to the right.

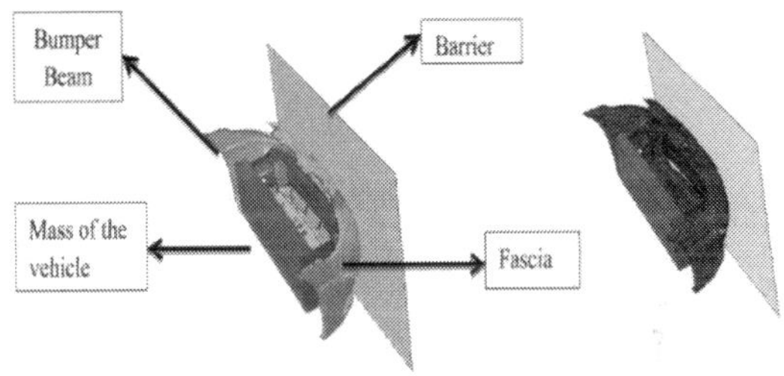

Fig.4. Assembly of the parts and in its initial position and its mesh.

4 Result and discussion

4.1 Results from universal Testing Machine with different Speed

Fig.5 shows two different results depending on the loading rate or the speed. The three materials to the right is with the same speed so that the resulting displacement is almost the same. At high strain rate the displacement is less and almost at quasi-static strain rate displacement is high.

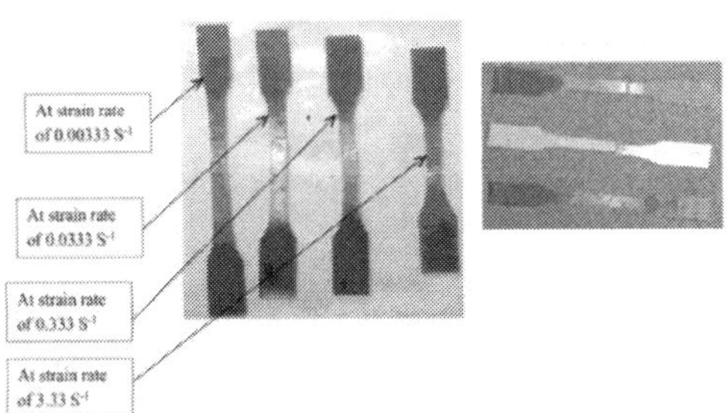

Fig.5. HAS specimen after fracture for different speeds and all samples.

Fig. 6 shows true stress vs. true strain at four different strain rates for the three materials. From this figure, the flow stress is increased with .the increase of strain rate.

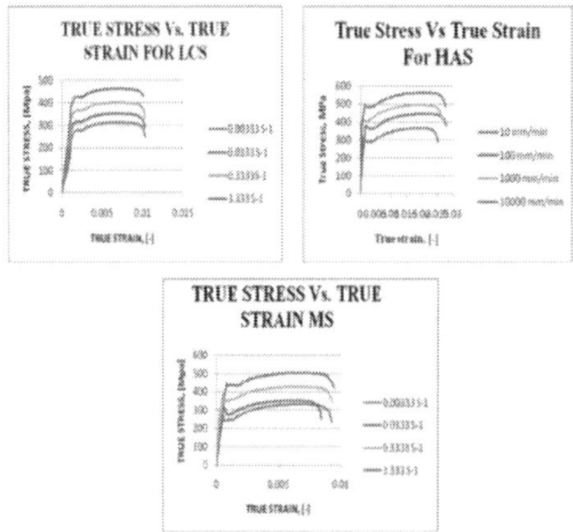

Fig.6. True stress vs. true strain at four different strain rates for the three materials.

From Fig.7, it's indicated that the relation between true stress vs true strain which are the experimental results from the tensile test with different cross head speed of 100, 1000, and 10000 mm/min. From all the true stress vs true strain curves stated below HAS material has the maximum true ultimate strength as well as maximum yield strength for all strain rate used during the tensile test.

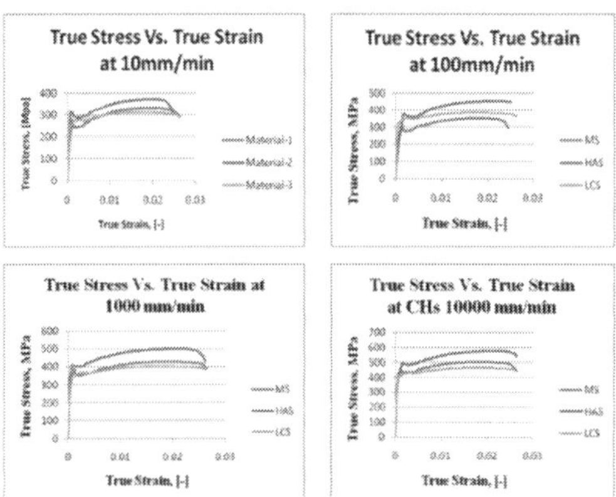

Fig.7. True stress vs true strain for the three materials at 0.00333 s^{-1}, 0.0333 s^{-1}, 0.333 s^{-1}, and 3.33 s^{-1} strain rates.

In Fig.8, three materials are compared considering the mean ultimate tensile strength (UTS) for the stated strain rate under study. Among all the three materials HAS shows better improvement in its ultimate tensile strength.

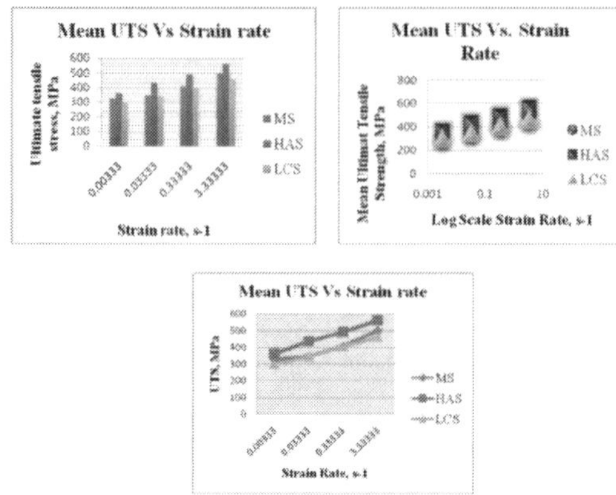

Fig.8. Mean UTS vs. strain rate for the three materials.

4.2 Results from FEM

4.2.1 FEM/ABAQUS/CAE/–explicit analysis result

Results from visualization module for the sample specimen under study without tensile loading from universal testing machine (to the left: no stress) and increasing application of tensile load (to the right: with highly stressed in the gauge region) are indicated in Fig.9.

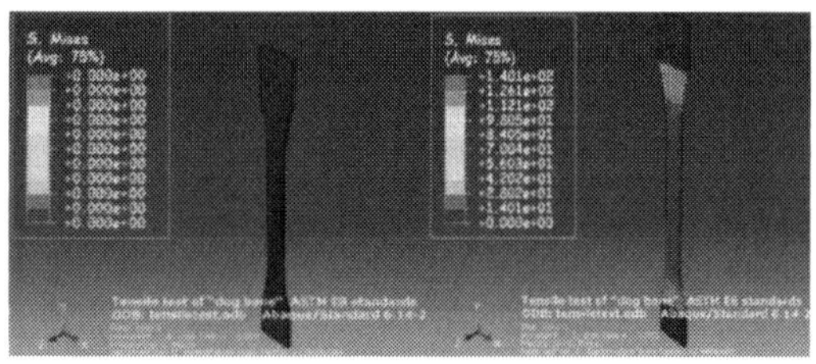

Fig.9. The displacement counter for sample specimen.

Fig.10 shows the velocity magnitude and displacement magnitude for the bumper system which is meshed in ABAQUS/CAE and analyezed for low speed impact. As it is expected in the real scenario and literature results the car will have decelerated after collision and the velocity magnitude will be lower.

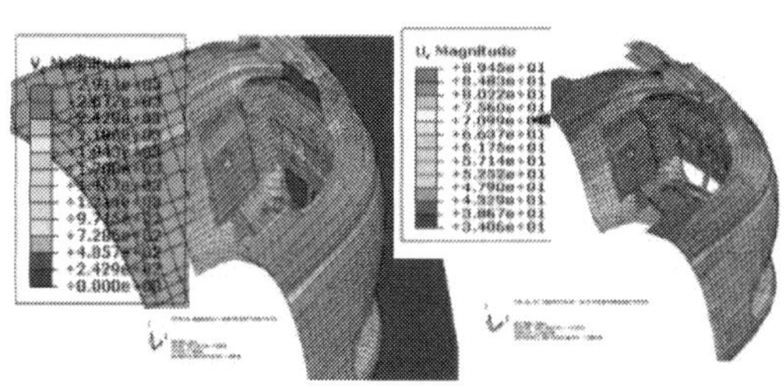

Fig.10. Displacement counter for the whole bumper energy absorber system.

4.2.2 The Bumper subsystem ABAQUS/CAE result

The results are from the ABAQUS/CAE for the overall displacement of the bumper system after low velocity impact for LCS which higher when compared to the other two sample materials with lower energy storing capability shown in Fig. 11.

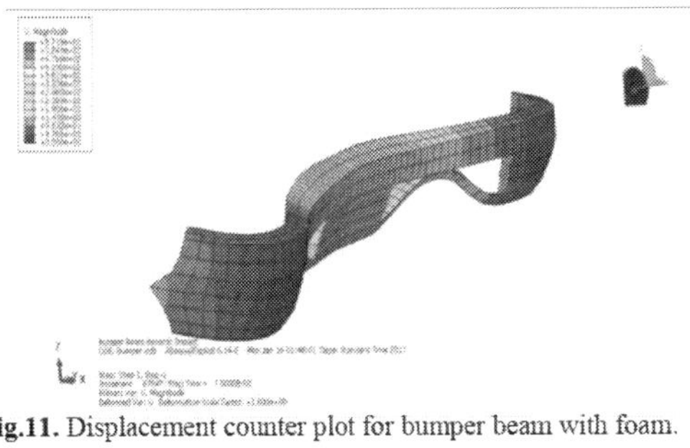

Fig.11. Displacement counter plot for bumper beam with foam.

4.2.3 The strain-energy plot from the ABAQUS/CAE visualization

Fig.12 shows that the displacement and the strain energy stored in the bumper beam during deformation after collision or impact takes place. As it is observed from both diagrams from all the three sample materials considered in this paper high alloy steel (HAS) show lower deformation and higher strain energy during straining of the bumper under low speed impact which occurred during indoor test.

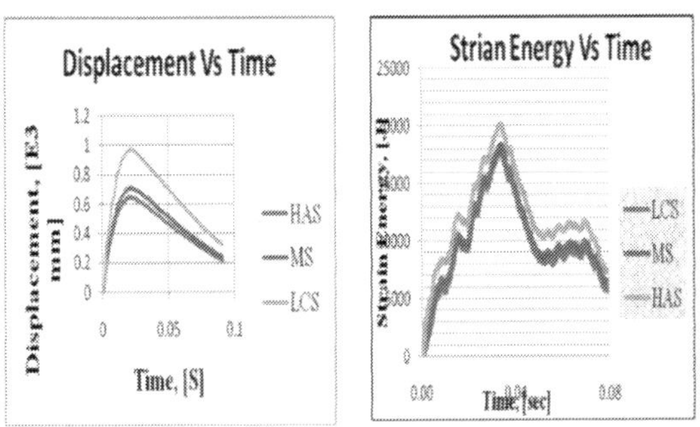

Fig.12. Strain energy vs time for the three sample steel materials.

5 Conclusion

In this study three different materials-namely low carbon steels (LCS, HAS, and MS) have been evaluated for the bumper beam, to understand the effect of strain rate on a crash behaviour of automobile during low speed frontal crash test performed by 36-sample specimen test in the experiment for tension at different strain rate and the data used as input to numerical simulation with ABAQUS code.

To study the steel material for bumper beam, two different factors are considered; strain rate and material type. When the materials are tested under different strain rate for their mechanical behaviours such as yield stress and ultimate tensile strength it is observed that both the yield stress and mean UTS of the material shows positive response with the strain rate; as the strain rate increases the mean yield stress and the mean UTS increases. And material two shows 20% increase in its UTS and 14% of its yield strength when compared to the other two steel material. Hence from the experimental result HAS improves the crash worthiness and shows better mechanical responses for such application relative to the other two materials under investigation.

On the other hand, from the FEM the ABAQUS/CAE result the displacement and strain energy are plotted against the time during low speed impact of the bumper subsystem for the three low carbon steel. HAS absorbs more energy and shows low deformation or has the minimum displacement when it applied with dynamic loading and from this for this quasi-static strain rate **HAS** has best mechanical behaviour as compared to the other two steel materials for this low speed impact.

References

1. American Iron and Steel Institute, Steel bumper systems for passenger cars and light trucks report. *American Iron and Steel Institute* (2006)
2. American Iron and Steel Institute, Steel bumper system for passengers vehicles and light tracks Report. Southfield, Michigan 48075: (2006)
3. Chia, C.W. et al., Inverse effect of strain rate on mechanical behavior and phase transformation of super austenitic stainless steel, Scripta Materialman, 17-20 (2007)
4. Coffey, C.S. & Davos, V.F., Drop weight impact machines: a review of recent progress, In Ramesh, K.T., ed. JANNAF Propulsion Systems Hazards Subcommittee Meeting, (1986)
5. Council, A.C., Automotive Plastic. [Online] (2) Available at: http://www.plasticscar.com/bumpers (Accessed 29 January 2013)
6. Field, J.E., Review of experimental techniques for high rate deformation and shock studies, *Int. J. Imp. Eng.*, **30**,725–775(2004)
7. M. Vural, D. Rittel & G. Ravichandran, Large strain mechanical behavior of 1018 cold-rolled steel over a wide range of strain rates, *Metallurgical and Materials Transactions*, **34A**(A), 2873-2885(2003)
8. Ravi Yatnalkar, B.E. , Experimental investigation of plastic deformation of Ti-6Al-4V under various loading conditions, Masters. Ohio: The Ohio State University The Ohio State University (2010)
9. Simulia, D.S., *Abaqus Analysis User's Manual*. 1st ed. Chicago: Simulia Software company (2010)
10. Springer, High *strain rate and impact experiments. in K.T. Ramesh, ed. Springer Handbook of*

DOI:10.1051/matecconf/201712300019

Experimental Solid Mechanics. New York: W.N. Sharpe, Jr **3**,1-29 (2008)

11. Woei, S.L. & Tzay, T.S., Mechanical properties and microstructure of AISI 4340 high strength alloy steel under tempering condition. ELSEVIER, 198-199 (1997)

An artificial flexible robot arm based on pneumatic muscle actuators

Jyh-Chyang Renn* *and Bo-Han* Li
Department of Mechanical Engineering, National Yunlin University of Science and Technology, No. 123, University Road Sec. 3, Douliou, 640 Yunlin, Taiwan, R.O.C.

Abstract. The purpose of this paper is to develop a novel human-friendly artificial flexible robot arm using four parallel-connected pneumatic muscle actuators (PMAs). The PMA is a flexible silicone rubber actuator which has some behaviors nearest to the real biological muscle including translational and rotational motions. An inverse kinematic model for the motion control is also developed. Finally, from experiment results, it is proved that not only the axial contraction control of a single PMA but also the attitude control of the whole pneumatic flexible robot arm using PID controller are satisfactory.

1 Introduction

The pneumatic muscle actuator (PMA) is a soft, deformable fluid actuator able to contract and to exert a tension force when supplied with a given air pressure. One advantage of the PMA is its ability to simulate the real human muscle behaviors and motions. Therefore, it has been widely discussed and researched in the past several years. On the other hand, due to its soft surface, the PMA can work with human, contact the human body and handle fragile objects. Consequently, such an actuator already finds many applications in the field of medical engineering, medical welfare system. For examples, the upper limb muscle actuator [1], power-assist wear [2], bio-mimetic actuators [3], rehabilitation robot [4, 5], biceps and triceps actuator [6], and so on. In this paper, a novel human-friendly artificial flexible robot arm using four parallel-connected PMAs is developed. Thanks to the high power-to-weight ratio of the PMA, it is expected that such an artificial arm can help strengthen the force output of a normal human arm or assist some handicapped person. Regarding the control of the artificial arm, one significant achievement of this study is the utilization of only two pneumatic proportional directional control valves to control all four parallel PMAs simultaneously. Thus, the hardware structure can be simplified while the attitude control of the proposed artificial flexible robot arm is still achieved successfully. In the following, two different structures of PMA are firstly outlined.

2 Translational and rotational PMA

Figure 1 shows the basic translational type of PMA. It consists of silicon rubber tube covered with a thin fiberglass layer. When the compressed air is supplied to the PMA, it contracts in the axial direction as shown in Fig. 1. The contraction length reaches 10 mm corresponding to maximal inlet pressure of 6 bar. Another commonly used PMA is the rotational type as shown in Fig. 2. Compared to scheme shown in Fig. 1, a thin metal sheet is inserted between the silicon rubber tube and fiberglass surface. Since the metal sheet is not deformable, the PMA rotates when the compressed air is supplied as shown in Fig. 2. The rotational angle reaches 30^0 when the inlet pressure is the maximal 6 bar.

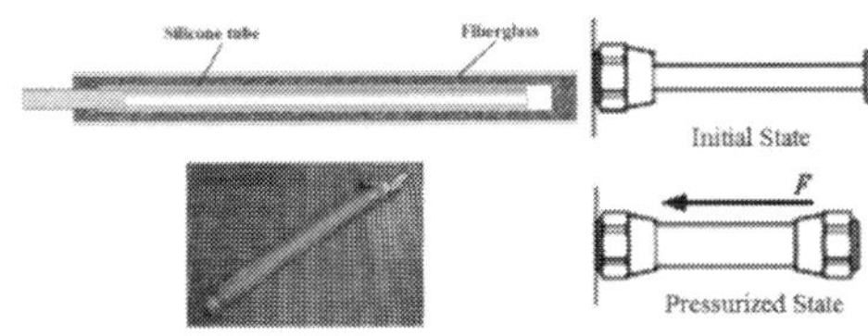

Fig. 1. Translational PMA.

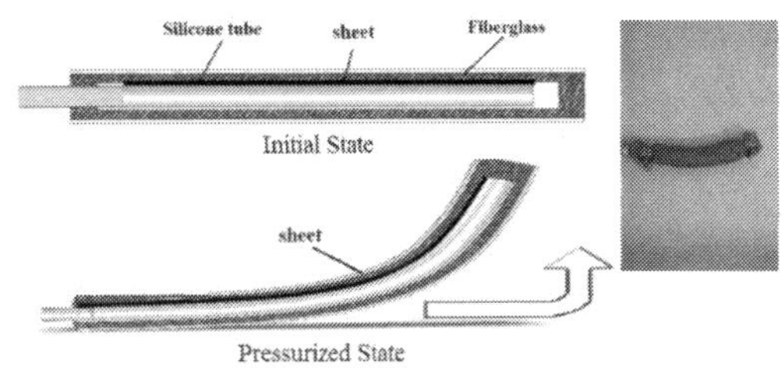

Fig. 2. Rotational PMA.

3 Structure of artificial flexible robot arm

As shown in Fig. 3, the design of proposed artificial flexible robot arm is based on closed and parallel linkage mechanism. It consists of two main rigid bodies. One is not movable and hence is called the base while the other is movable and is called the platform. Four parallel translational PMAs serve as the motion constraints and connect the fixed base with the movable platform. In addition, four universal joints are utilized at the platform end. The attitude controls, including circular, pitch and yaw, can be achieved by controlling the deformations of four translational PMAs.

* Corresponding author: rennjc@yuntech.edu.tw

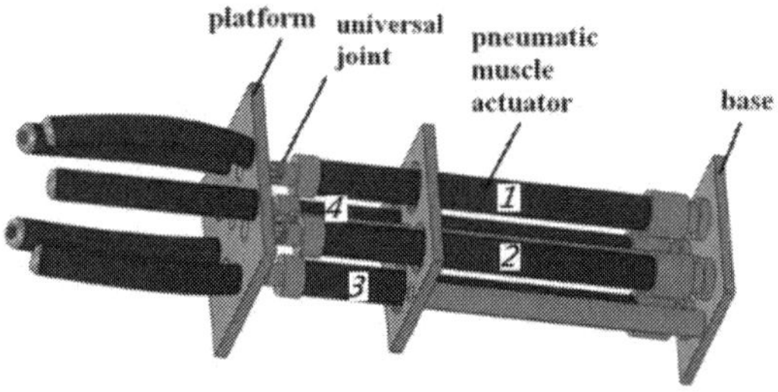

Fig. 3. Scheme of the proposed artificial flexible robot arm.

On the other hand, a soft pneumatic gripper based on four rotational PMAs is also mounted at the other end of the platform to simulate the catch motion of human fingers. Figure 4 shows the real picture of the implemented artificial flexible robot arm mounted on a tripod. Finally, to control the attitude of the movable platform, the contraction lengths of all four PMAs have to be determined by the inverse kinematic model.

Fig. 4. Real picture of the implemented artificial flexible robot arm.

4 Inverse kinematic model

The inverse kinematics that maps Cartesian space into joint space is inevitably necessary for the attitude control [5-7]. Using Cartesian coordinate, the coordinate system {B} for the base and coordinate system {P} for the movable platform can be defined as shown in Fig. 5. Let the matrix ${}^{B}_{P}\mathbf{R}$ be the attitude orientation matrix of platform coordinate relative to base coordinate and the angles of pitch, roll and yaw of the base coordinate are denoted by γ, β and α. Thus, three orientation matrices for the base coordinate can be derived.

$$\mathbf{R}(X_B,\gamma)=\begin{bmatrix}1&0&0\\0&c\gamma&-s\gamma\\0&s\gamma&c\gamma\end{bmatrix},\ \mathbf{R}(Y_B,\beta)=\begin{bmatrix}c\beta&0&s\beta\\0&1&0\\-s\beta&0&c\beta\end{bmatrix} \quad (1)$$

and $\mathbf{R}(Z_B,\alpha)=\begin{bmatrix}c\alpha&-s\alpha&0\\s\alpha&c\alpha&0\\0&0&1\end{bmatrix}$.

where $c\alpha$ denotes $cos(\alpha)$, $s\alpha$ *denotes* $sin(\alpha)$, and so on. Furthermore, the attitude orientation matrix ${}^{B}_{P}\mathbf{R}$ can be calculated by

$$\begin{aligned}{}^{B}_{P}\mathbf{R}&=\mathbf{R}_{Z_B,\alpha}\cdot\mathbf{R}_{Y_B,\beta}\cdot\mathbf{R}_{X_B,\gamma}\\&=\begin{bmatrix}c\alpha&-s\alpha&0\\s\alpha&c\alpha&0\\0&0&1\end{bmatrix}\begin{bmatrix}c\beta&0&s\beta\\0&1&0\\-s\beta&0&c\beta\end{bmatrix}\begin{bmatrix}1&0&0\\0&c\gamma&-s\gamma\\0&s\gamma&c\gamma\end{bmatrix}\\&=\begin{bmatrix}c\alpha\cdot c\beta&c\alpha\cdot s\beta\cdot s\gamma-s\alpha\cdot c\gamma&c\alpha\cdot s\beta\cdot c\gamma+s\alpha\cdot s\gamma\\s\alpha\cdot c\beta&s\alpha\cdot s\beta\cdot s\gamma+c\alpha\cdot c\gamma&s\alpha\cdot s\beta\cdot c\gamma-c\alpha\cdot s\gamma\\-s\beta&c\beta\cdot s\gamma&c\beta\cdot c\gamma\end{bmatrix}\\&=\begin{bmatrix}r_{11}&r_{12}&r_{13}\\r_{21}&r_{22}&r_{23}\\r_{31}&r_{32}&r_{33}\end{bmatrix}.\end{aligned} \quad (2)$$

From the vector diagram shown in Fig. 6, the vector ${}^{B}\boldsymbol{d}=[x\ \ y\ \ z]^T$ represents the vector of platform center relative to the base coordinate. After some derivations, the required length vector ${}^{B}\boldsymbol{q}_i=[q_{ix}\ \ q_{iy}\ \ q_{iz}]^T$ for every PMA at pressurized states can be obtained as follows.

$${}^{\mathrm{B}}\mathbf{q}_i={}^{\mathrm{B}}\mathbf{x}_i+{}^{\mathrm{B}}\mathbf{p}_i\ . \quad (3)$$

where

$$\begin{aligned}{}^{\mathrm{B}}\mathbf{x}_i&={}^{\mathrm{B}}\mathbf{d}-{}^{\mathrm{B}}\mathbf{b}_i\\&=\begin{bmatrix}x-b_{ix}\\y-b_{iy}\\z-b_{iz}\end{bmatrix}=\begin{bmatrix}x-b_{ix}\\y-b_{iy}\\z\end{bmatrix}=\begin{bmatrix}\bar{x}_i\\\bar{y}_i\\\bar{z}_i\end{bmatrix},\end{aligned} \quad (4)$$

and

$$\begin{aligned}{}^{\mathrm{B}}\mathbf{p}_i&={}^{B}_{P}\mathbf{R}\cdot{}^{\mathrm{P}}\mathbf{p}_i\\&=\begin{bmatrix}r_{11}&r_{12}&r_{13}\\r_{21}&r_{22}&r_{23}\\r_{31}&r_{32}&r_{33}\end{bmatrix}\begin{bmatrix}p_{ix}\\p_{iy}\\p_{iz}\end{bmatrix}=\begin{bmatrix}r_{11}p_{ix}+r_{12}p_{iy}\\r_{21}p_{ix}+r_{22}p_{iy}\\r_{31}p_{ix}+r_{32}p_{iy}\end{bmatrix}=\begin{bmatrix}u_i\\v_i\\w_i\end{bmatrix}.\end{aligned} \quad (5)$$

Finally, the length S_i of ith PMA is actually the scalar length of the vector ${}^{\mathrm{B}}\mathbf{q}_i$, that is

$$S_i=\sqrt{(\bar{x}_i+u_i)^2+(\bar{y}_i+v_i)^2+(\bar{z}_i+w_i)^2}\quad (i=1,2,3,4)\ (6)$$

This is exactly the closed-form solution of the proposed artificial flexible robot arm.

DOI:10.1051/matecconf/201712300020

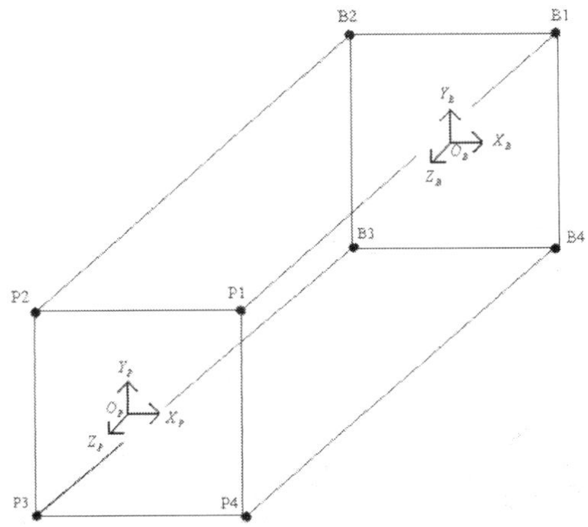

Fig. 5. Coordinate {B} for the base and coordinate {P} for the platform.

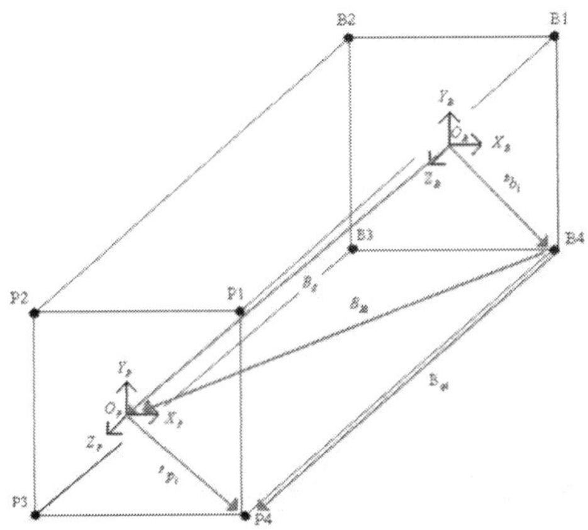

Fig. 6. Vector diagram between two working spaces.

5 Attitude control of the proposed artificial flexible robot arm

The pneumatic circuit diagram as well as the control scheme for the proposed artificial flexible robot arm is shown in Fig. 7. In addition to above-mentioned robot arm, one directional 5/2 valve and two proportional directional valves (FESTO MPYE-5-1/8LF-010-B) are utilized and serve as the main control elements. The 5/2 directional valve controls the motion of soft gripper. When the pressurized air is supplied to the soft gripper, four pneumatic muscle fingers rotate simultaneously and form the catch motion. On the other hand, if the pressurized air in the pneumatic muscle fingers is exhausted to atmosphere, four PMAs will return to their initial states. Thus, the release motion can also be implemented.

Instead of four proportional pressure control valves, only two proportional directional control valves are required to control all four parallel-connected translational PMAs. This is chiefly because one proportional directional control valve provides two outlet ports, denoted by A and B. In addition, these two outlet ports are further connected to two PMAs respectively. Thus, one proportional directional control valve is adequate for the control of two PMAs. As an example, for the pitch control, the upper pair of PMA (1) and (2) as well as the lower pair of PMA (3) and (4) shown in Fig. 3 are sequentially activated. On the other hand, one proportional pressure control valve has only one outlet port. Therefore, four proportional pressure control valves are inevitable for the control of all four PMAs. Disadvantages of such a pressure control scheme, however, are bulky, quite complicated and more expensive. Another important sensor utilized in this study is a gyro (Free-scale MMA7361) that detects and feedbacks the real attitude of the movable platform. The mounting location of this gyro is exactly at the center position of the platform which meets the derivation of forward and inverse kinematic equations. Finally, the closed-loop control block diagram using PID controller is shown in Fig. 8.

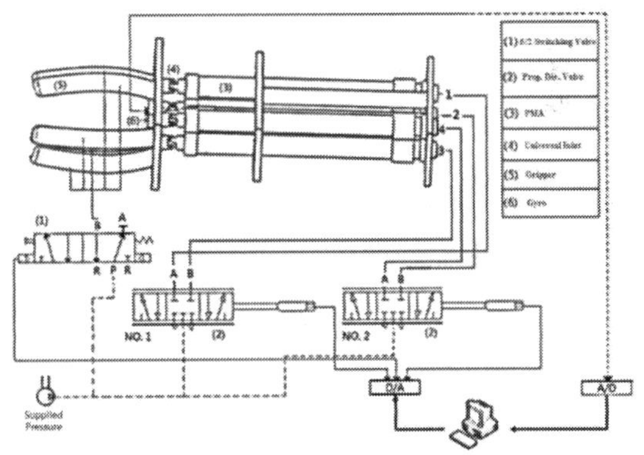

Fig. 7. Pneumatic circuit diagram as well as the control scheme.

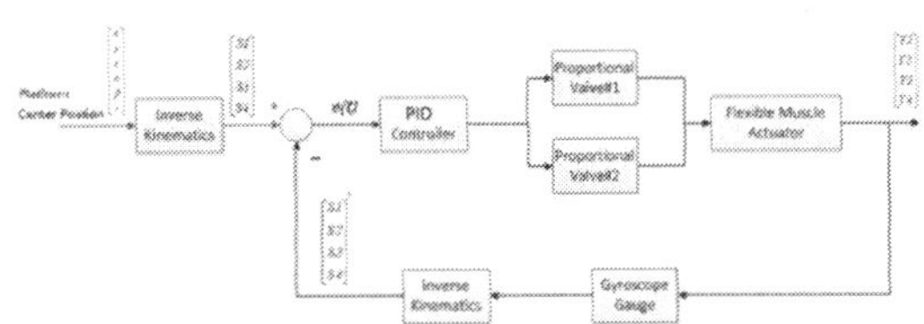

Fig. 8. Closed-loop control block diagram using PID controller.

6 Experimental results of circular tests

Before conducting the circular test, the axial contraction control of one single PMA using PID controller is firstly executed. The initial length of this single PMA is set to be 200 mm. Moreover, five different contraction lengths of PMA from 2 mm to 10 mm are tested. From the experimental results shown in Fig. 9, it is observed that the response is very slow and the position control behavior is quite nonlinear. However, the steady-state error is only around 0.2 mm and is satisfactory. After further frequency response analysis, the bandwidth of the developed PMA is estimated to be 0.24 Hz. For the circular test, the path of the platform's center position using Cartesian coordinate can be defined as

$$X = r\cos(\theta) = r\cos(2\pi f), \quad (7)$$

$$Y = r\sin(\theta) = r\sin(2\pi f). \quad (8)$$

Three different radii are chosen, that is $r = 12$ mm, $r = 9$ mm and $r = 6$ mm. The initial lengths for all four PMAs are also set to be $Z = 200$ mm and the frequency of circular test is set to be $f = 0.01$ Hz. The corresponding results using PID controller are shown in Fig. 10. The maximal error is found to be 0.3 mm which also implies the good and acceptable circular test precision of the proposed artificial flexible robot arm based on PMA.

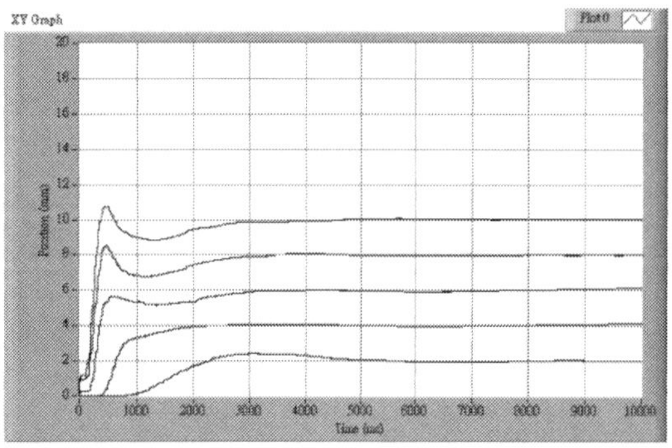

Fig. 9. Five different length contractions control results using PID controller.

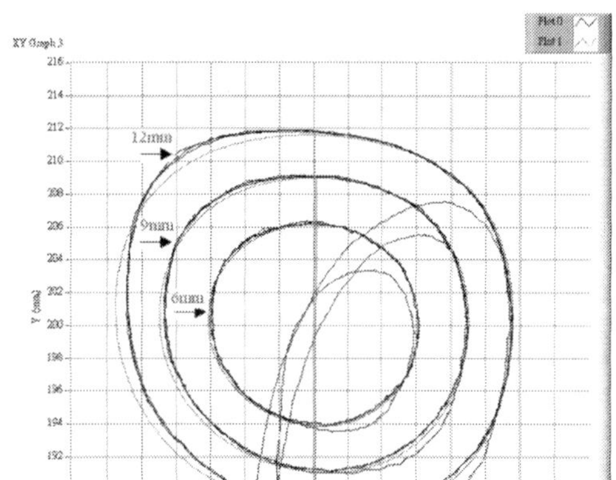

Fig. 10. Three circular test results using PID controller.

Table 1. Specifications of the artificial flexible robot arm.

Items	Spec.
1. Initial length of PMA	200 mm
2. Max. contraction length of PMA	10 mm
3. Max. pitch angle	$\pm 15^{\circ}$
4. Max. yaw angle	$\pm 15^{\circ}$
5. Supplied pressure	5 bar
6. Bandwidth (one PMA)	0.24 Hz

7 Conclusions

In this paper, a novel human-friendly artificial flexible robot arm using four parallel PMAs is successfully developed and realized. Three different motions, including pitch, yaw and circular, can be implemented. Table 1 summarizes some specifications of the proposed artificial flexible robot arm. To get the maximal angle specifications, maximal input (10 V) to the proportional valve is used to control the attitude of robot. On the other hand, to find the bandwidth, the same maximal amplitude input to the proportional valve, but with various frequencies, is employed to control the contraction of PMA. The -3dB frequency of the frequency response is denoted as the bandwidth. In the future, the PMA and the artificial flexible robot arm may find more potential applications in the field of medical engineering or medical welfare system, etc.

Acknowledgment

The financial support of the Ministry of Science and Technology under grant number *MOST- 104- 2221- E- 224- 017* is greatly appreciated.

References

1. M. Aragane, "Power assist wear for upper limb driven by sheet-like pneumatic rubber muscle," *Proceedings of 7th JFPS International Symposium on Fluid Power*, Toyama, Japan, pp. 787-792, 2008
2. T. Noritsugu, D. Sasaki and M. Takaiwa, "Application of pneumatic soft mechanism to a human friendly robot," *Proceedings of 5th JFPS International Symposium on Fluid Power*, Nara, Japan, pp. 61-66, 2002
3. D. G. Caldwell, N. Tsagarakis, and G. A. Medrano-Cerdas, "Bio-mimetic actuators: polymeric pseudo muscle actuators and pneumatic muscle actuators for biological emulation," *Mechatronics*, **10**, pp. 499-530, 2000
4. S. D. Prior and A. S. White, "Measurement and simulation of a pneumatic muscle actuator for a rehabilitation robot," *Simulation Practice and Theory*, pp. 81-117, 1995

5. T. Noritsugu and T. Tanaka, "Application of rubber artificial muscle manipulator as a rehabilitation robot," *IEEE/ ASME Trans. Mechatronics*, **2**, pp. 259-267, 1997
6. J. H. Lilly, "Adaptive Tracking for pneumatic muscle actuators in biceps and triceps configurations," *IEEE Transactions on Neural System and Rehabilitation Engineering*, **11** (3), pp. 333-339, 2003

An automatic energy-saving and thermal monitoring/controlling system for a pond

Ching-Chien Cheng[1], *Min-Chie* Chiu[2,*], *Che-Min* Chiu[3], *Huang-Kuang* Kung[1], *Chin-Yu* Wang[1], and *Zi-Jie* Gao[2]

[1]Institute of Mechatronics Engineering, Cheng Shiu University, Kaohsiung City, Taiwan 83347, R.O.C.
[2]Department of Mechanical and Automation Engineering, Chung Chou University of Science and Technology, Taiwan, R.O.C.
[3]Institute of Biomedical Engineering, National Tsing Hua University, Taiwan, R.O.C.

Abstract. Because of low temperatures and oxygen in cold water, fish will die when cold currents arrive. This will cause tremendous loss of money. In order reduce the cooling of the pond, an automatic thermal detecting and cold-roofing system using a wind-proofing device, heaters, and thermal detectors is proposed. To reduce heat loss due to thermal convection above the pond surface, a motor-driven wind-proofing device automatically controlled by a PLC controller is adopted. Here, the wind-proofing device, thermal detectors, and heating system are connected to the PLC controller. The PLC will also be connected to the PC interface. The temperature thresholds used to trigger the heater and the wind proofing device can be set at the PC interface. Two options for manipulating the heating and the automatic heating can be selected. The related wind-proofing area and the number of heaters will be determined according to the current temperature. Moreover, the PLC can be wirelessly connected to the server PC in the control room. The pond keeper can monitor everything online and control the pond water's temperature. With this, the problem of fish dying in a cold wave can be solved. Consequently, to reduce the electrical exhaust when heating up the pond water, green energy, solar energy and wind energy, is used.

1 Introduction

Primary factors of water quality affecting fish include adequate oxygen, proper temperature, transparency, and limited levels of metabolites. Earlier studies of the water quality of a fish pond in India were conducted by Sewell [1] and Pruthi [2]. During the winter months, plant growth and photosynthesis was limited by low water temperatures. Because of less photosynthesis, the reduction of oxygen caused problems for fish [3, 4]. In order to prevent fish from dying due to low oxygen and temperature when cold currents arrive, several strategies used in reducing the heat loss and increasing the temperature in water were needed and presented in the paper. To decrease heat loss during heat convection above the pond surface, an automatic wind-proofing device linked with a motor as well as a controller is adopted. To heat up the pond's water when the temperature is low, an online temperature monitoring system in conjunction with thermal sensors and heaters is established. Moreover, to save electrical power, green energy (wind and solar) installed around the pond is adopted.

Consequently, to facilitate the operation process, a PC-based interface connected to PLC is built. With this, the temperature thresholds used to trigger the heater and the wind proofing device can be set in the PC's interface.

2 Automatic energy-saving, wind-proofing, and thermal monitoring/controlling system for a pond

The structure of automatic energy-saving, wind-proofing, and a thermal monitoring/controlling system for a pond is depicted in Fig. 1. The related input/output diagram of the PLC connection is shown in Fig. 2. In order to remotely manipulate the energy-saving, wind-proofing, and thermal monitoring/controlling system via the server pc, the auxiliary relays defined in Table 1 have been added into the PLC's ladder program shown in Fig. 3. As indicated in Fig. 3, the auxiliary relays are marked in red in place of the buttons. The server pc can then submit the command to actuate the related functions for the energy-saving, wind-proofing, and thermal monitoring/ controlling system. Moreover, the interface of the server PC programmed by the VB(Virsual Basic) is also illustrated in Fig. 4.

* Corresponding author: minchie.chiu@msa.hinet.net

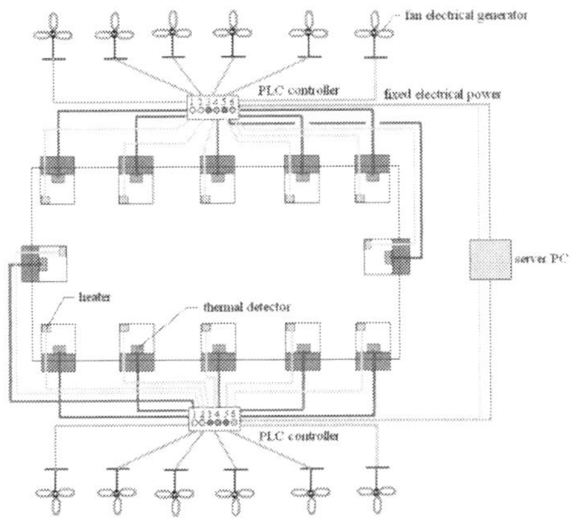

Fig. 1. The structure of automatic energy-saving, wind-proofing, and thermal monitoring/controlling system for a pond.

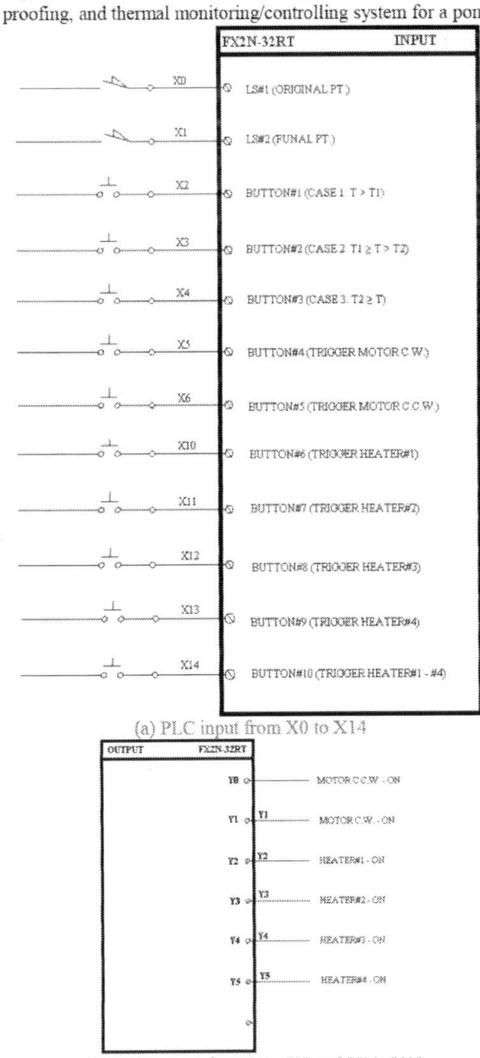

(a) PLC input from X0 to X14

(b) PLC output from Y0 ~Y7 and Y10~Y12

Fig. 2. The related wiring diagram of the PLC connection.

Table 1. The auxiliary relays used to remotely control the action of the PLC program.

Auxiliary Relay Number	Action	PLC Program
M50	MOTOR C.C.W. On/Off	SET ‧ RST M50
M100	MOTOR C.W. On/Off	SET ‧ RST M100
M115	CASE 1 TESTING On/Off	SET ‧ RST M115
M116	CASE 2 TESTING On/Off	SET ‧ RST M116
M117	CASE 3 TESTING On/Off	SET ‧ RST M117
M201	HEATER#1 On/Off	SET ‧ RST M201
M202	HEATER#2 On/Off	SET ‧ RST M202
M203	HEATER#3 On/Off	SET ‧ RST M203
M204	HEATER#4 On/Off	SET ‧ RST M204

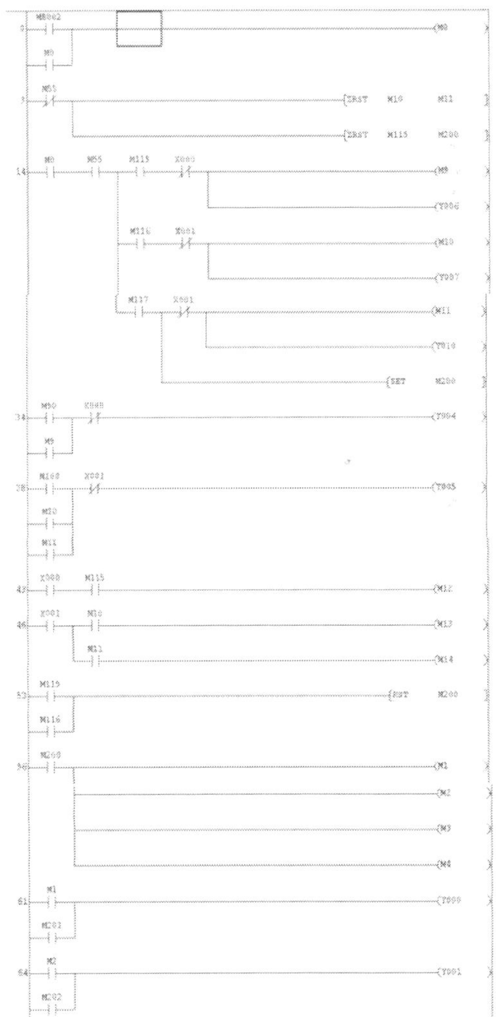

Fig. 3. The abstract of the related ladder program built into the PLC.

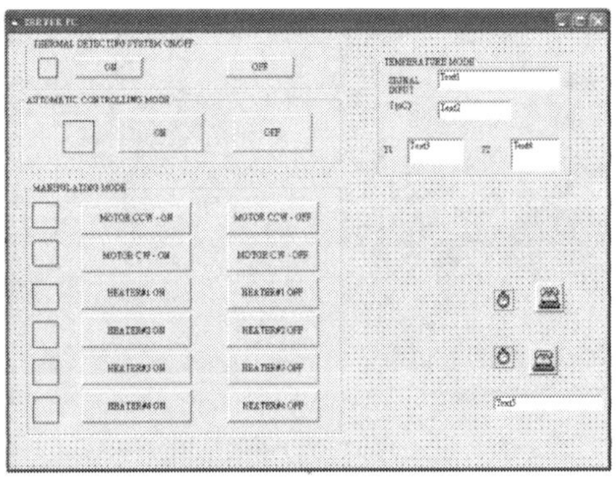

Fig. 4. The interface in the server PC.

3 System established and communication formated

As indicated in Fig. 1 and Table 1, a PLC is used as a controller to actuate the actuators (one wind-proofing motor and four heaters) of the energy-saving, wind-proofing controlling mechanism. Also, a thermal detecting device installed inside a pond and connected to the server PC will feedback the thermal data to the server PC. The related wiring diagram of the system is shown in Fig.5. In addition, both the solar generator and the fan generator linked to the system are used to provide the necessary electrical power. In order to allocate the wind-proofing device, two limit switch sensors are installed at both ends of the pond. To increase the pond's water temperature, four sets of heaters are installed around the pond. Moreover, in order to remotely monitor/control the energy-saving, wind-proofing, and thermal monitoring/controlling system, wireless communication between the server pc and the PLC is established using the ZigBee module via the RS232/RS422 protocol. Thereafter, an interface at the server PC is programmed by the VB and is used to implement the system by sending a command to the PLC controller. The abstract of the related interface program built in the server pc is depicted in Fig. 6. To facilitate communication between the PLC and the PC via a RS232/RS422, the picket format for the MISTISUE FX2n-PLC's primary port communication is depicted in Table 2. Also, the definition of the picket format is shown in Table 3. The MISTISUE FX2n-PLC's extended port communication is shown in Table 4. In addition, the communication parameters used in the extended port are illustrated in Table 5. Consequently, the command description of the MISTISUE FX2n-PLC's extended port communication protocol type 1 is shown in Table 6.

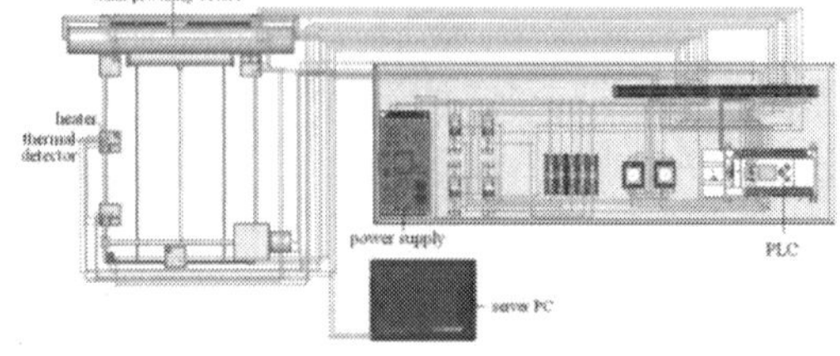

Fig. 5. The wiring diagram of the system.

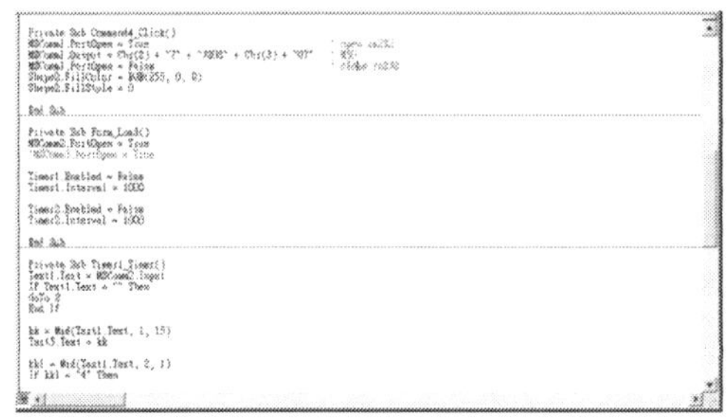

Fig. 6. The abstract of the related interface program built in the server PC.

Table 2. The communication packet format of the MISUBISHI FX2n-PLC's primary port.

Header code	Command code	Address and data	End code	Inspection code
STX (02)	CMD	DATA	ETX (03)	Checksum

Table 3. The definition of communication packet for MISUBISHI FX2n-PLC's primary port.

Command category	Command code	Object	Function description
Read	0	X,Y,M,S ,T,C,D	Read the object value
Write	1	Y,M,S,T, C,D	Write data to the object
Forcing on	7	Y,M,S,T, C	Forcing node to be on
Forcing off	8	Y,M,S,T, C	Forcing node to be off

Table 4. The communication packet format of the MISUBISHI FX2n-PLC's extended port.

Header code	Station number	Computer number	command	Object number/ data	End code	Insp-ection code

Table 5. The parameter of the extended port's communications protocol for the MISUBISHI FX2n-PLC.

D8120Bit	Name	Content
15	Protocol type	0: type 1, 1: type 4
14	protocol	0: No, 1: Yes
13	Inspection code	0: manipulate, 1: auto
12	Type of control line(II)	none
11	DTR inspection mode	none

10	Type of control line(I)	none
9	End code	none
8	Header code	none
7	Transmission speed	0011:300 0222:4800 0100:600 1000:9600 0101:1200 1001:19200 0110:2400
6		
5		
4		
3	Stop bit	0: 1 bit, 1: 2 bits
2	Parity check	00: none; 02: even; 01: odd
1		
0	Data bit	0: 7 bits ; 1: 8 bits

Table 6. The command of the extended port's communications protocol for the MISUBISHI FX2n-PLC.

Command	Description	Available object
BR	Read PLC's bit data	X, Y, M, S
WR	Read PLC's bit_set data	D, TN, CN
BW	Write data to PLC's bit	Y, M, S
WW	Write data to PLC's bit_set	D, TN, CN
BT	Hybrid type of the writing for bit	Y, M, S
WT	Hybrid type of the writing for bit_set	D, TN, CN
RR	Actuate PLC's running	
RS	Actuate PLC's stop	
PC	Read the PLC's category	

4 The program design for serial communication between the server PC and the client PC

The TCP/IP communication between the server pc and the client PC is performed using the WINSOCK object [5-9]. The host (server PC) can also be connected simultaneously by multiple client's PCs. The interface of the client pc is depicted in Fig. 7. Also, the related communication diagram is shown in Fig. 8. The communication process for the server port and the client port is described below.

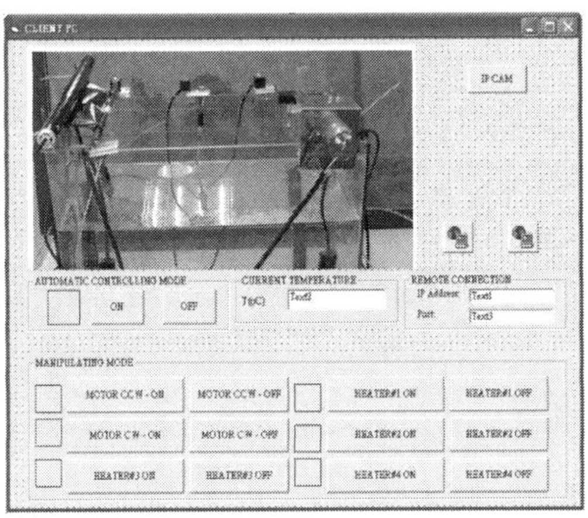

Fig. 7. The interface in the client PC.

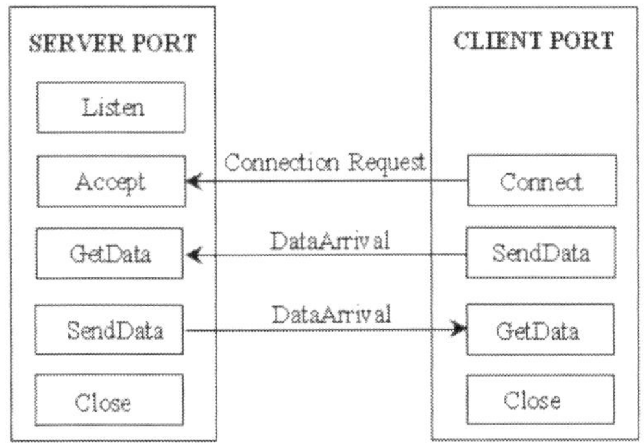

Fig. 8. The communication process between the client port and the server port.

4.1 The communication process (server PC viewpoint) [8]:

(1) Listen:
Because the server port will build a connection platform for the client port, a condition of "Listen" for the server pc is needed. The server port waits for the connection from the client port when "Listen" is ready. Moreover, because a WINSOCK object in the server port can be adopted for providing a connection for only one client, multiple WINSOCKs used in the program are requisite for a network of multiple clients connected to the server port. Hence, multiple WINSOCKs will be created using the "Load" command. One of the WINSOCKs (Winsock1(0)) is responsible for the "Listen" function. A "Local Port" will be stipulated as the port of "Listen."
(2) Accept:
The server port will receive an event ConnectionRequest from the client port when the client port submits a "Connect" for a further internet connection. The server port also accepts the client port's connection request using "Accept" via another WINSOCK.
(3) GetData:
The server port receives an event "DataArrival" from the client port when the client port sends data back to the server port. The server port will also receive data by using "GetData."
(4) SendData:
The server port sends data to the client port using "SendData."

4.2 The communication process (client PC viewpoint) [10]:

(1) Connect:
The WINSOCK in the client pc submits a connection request using "Connect." At this time, the WINSOCK will assign the server port's "Port Number" (at "Listen") to be the value of its "Remote Port."
(2) SendData:
The client port sends the data set to the server port using "SendData" when the internet connection between the server port and the client port is built.
(3) GetData:

The client port will generate an event "DataArrival" when the server port sends a data set to the client port. The client port can receive the data using "GetData."
(4) Close:
The end connection using "Close" is needed when the communication is finished.

5 Results and Discussion

5.1 Results

As indicated in Fig. 1, a prototype of a pc-base energy-saving, wind-proofing, and thermal monitoring/controlling system is established and shown in Fig. 9. A PLC is used as a controller to control the various actuators, including one motor and four heaters.
Here, the motor will be triggered by the PLC to forward or retrieve the wind proofing device when the current pond's water temperature detected by the thermal sensors reaches two thermal thresholds of pond water. A thermal sensor is adopted online to detect the temperature of the pond water. Two limit switches are installed at the two ends of the pond. A wireless communication between the server PC and the system is constructed using the ZigBee module connected to the PLC via the RS232/RS422 protocol. As indicated in Fig. 4, an interface at the server PC is established to implement the system by sending a command to the PLC controller. Moreover, to realize the current temperature of the pond water, a remote distributing, monitoring, and controlling network in conjunction with an IP CAM via the TCP/IP network is built and shown in Fig. 7. To assure the operation of energy-saving, wind-proofing, and the thermal monitoring/controlling system, a visual monitoring system using the IP CAM is also adopted.

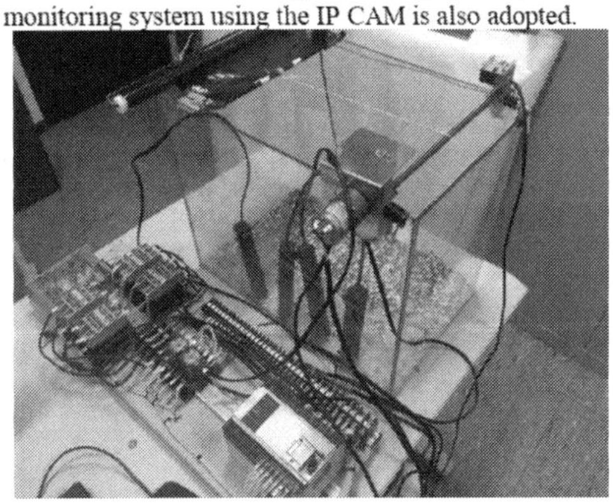

Fig. 9. A prototype of the automatic energy-saving, wind-proofing, and thermal monitoring/controlling system for a pond .

5.2 Discussion

The main issues concerning a pc-base energy-saving, wind-proofing, and thermal monitoring/controlling system follow :

(1) An electricity supplement extracted from solar energy and wind energy.
(2) Automation for keeping the pond water at an appropriate temperature and saving manpower.
(3) Establishing the client-server control system between the server PC and the client PC.
(4) Saving manpower on site by using an interface between the server PC and the system.
(5) Establishing the system's hardware and setting up a communication format between the PLC, ZigBee, and the server PC.
(6) Assuring the operation of energy-saving, wind-proofing, and the thermal monitoring/controlling system using an IP CAM visual monitoring system.

6 Conclusion

It has been shown that to reduce manpower and to maintain an appropriate temperature in the pond, a remote energy-saving, wind-proofing, and thermal monitoring/controlling system is necessary. This network is utilized using an inquiry between the server PC and the client PC via the TCP/IP network. To avoid interference from internet wiring with existing equipment, a wireless network for the system using the ZigBee module is adopted.

Consequently, to assure the operation of controlling temperature in pond water when cold currents arrive, an IP CAM visual monitoring system is established.

References

1. R. B. S. Sewell, *J. of the Asiatic Society of Bengal*, **22**, 177 (1927).
2. H. S. Pruthi, *Review of Hydrobiology Hydrographic*, **26**, 242 (1932)
3. Rossana Sallenave, New Mexico State University, Guide **W-105**, 1 (2013)
4. T. Zhu, C. Ringler, *International Food policy Research Institute*, 00961 (2010)
5. H. C. Cheng, M. C. Chiu, C. L. Huang, P. C. Chuo, *I. Tech. J.*, **11(6)**, 686 (2012)
6. H. C. Cheng, M. C. Chiu, *App. Mech. and Mat.*, **336-338**, 1211 (2013)
7. H. C. Cheng, M. C. Chiu, J. I. O. S., **34**(6), 373 (2013)
8. M. C. Chiu, H. C. Cheng, *J. Inter. Math.*, **18**(5), 493 (2015)
9. H. C. Cheng, M. C. Chiu, C. M. Chiu, C. Y. Yang, *Sylwan Journal*, 160(11), 147 (2016)
10. H. C. Cheng, M. C. Chiu, *I. J. on Smart Sens. and Intell. Syst.*, **7**(3), 1378 (2014)

How it outsourcing impacts on value creation of manufacturer?

Li-Shan Chen[1] and *Yen-Ming* Tseng[2,*]

[1]School of Management, Fujian University of Technology, China
[2]Key Laboratory of Big Data Mining and Application/ School of Information Science and Engineering, Fujian University of Technology, China

Abstract. A growing concern among the manufacturers who are actively involved in Information Technology outsourcing is post-contract management and the ensuing development of what many practitioners have coined the 'outsourcing partnership'. Over the past few decades, information technology outsourcing has become a widely used and researched means for manufacturers to enhance their performance. Increasing attention has been paid to building successful partnerships in information systems outsourcing. This study employs strategic social positioning, customer participation, customer orientation, and communication culture to develop research model. The quantitative method will be used to validate the research model and test hypotheses developed. This study uses survey method to collect data, and analysis of the reliability, validity and test the proposed hypotheses by Smart Partial Least Squares. The purpose of this study is to use above four factors to produce value co-creation between outsourcing vendors and manufacturers. The final aim is for the future collaboration. This combination will allow manufacturers and information technology outsourcing vendors to increase their value co-creations in the fields of these factors.

1 Introduction

Despite the development of IT outsourcing and its pledge to keep costs down, few studies have examined its impact on prices. In our global economy, successful e-commerce enterprises need to consider many challenges: a worldwide convergence of enterprises, voluntary enterprises, whacked data exercises, altering consumers' needs, saved production lifecycle, instant access to information, growing market divergence, heightened levels of in-depth business intelligence, online bargains, enhanced business harmonization and rapid variations in IT. [1-3]

Pagani commences by summarizing the results of a study on value networks and value-creation articles in the literature on information systems, strategic management and operations. In strategic management research, many scholars have researched the notion of value and sustainable competition. [4]

Mol et al. describe a path of composite value-increasing gradations that ultimately forms an entire value recommendation. In their model, each gradation contributes a special ratio to the entire value creation, while value seizing by contributors relies on participant's relative contracting authority. [5-9]

Woodard et al. define digital enterprise strategies as a planned competition by e-commerce enterprises through distributing digitally enabled products or services [10]. Bharadwaj et al. recognize four key areas that influence our reasoning on digital enterprise strategies and that help create a structure to describe the next generation of perceptions [11]. Oslo recognizes four principal groups of originality: product; process; organizational creation and marketing creation [12]. Nevertheless, most studies concentrate on the influence of personal creative actions in segregating or comprehending technological creation [13-14]. E-commerce enterprises seek a technology-based competing advantage, whereas society, academia and technological departments pursue an opportunity to comprehend the world in a unique manner [15].

To expand our knowledge of the link between outsourcing vendors and manufacturers, we have four main goals. First, we examine whether strategic social positioning positively influences value co-creation. Second, we examine whether customer participation positively influences value co-creation. Third, we examine whether customer orientation positively influences value co-creation. Fourth, we examine whether communication culture positively influences value co-creation.

2 Theory and hypotheses

We review the literatures related to the major concepts before developing hypotheses and suggested that strategic social positioning may be positively associated with value co-creation, customer participation may be positively associated with value co-creation, customer orientation may be positively associated with value co-creation, and communication culture may be positively associated with value co-creation. We present an

* Corresponding author: swk1200@qq.com

operational definition of the value co-creation construct, then turn to definitions found in the literature for strategic social positioning, customer participation, customer orientation, and communication culture. The research model of this research is shown in Fig.1.

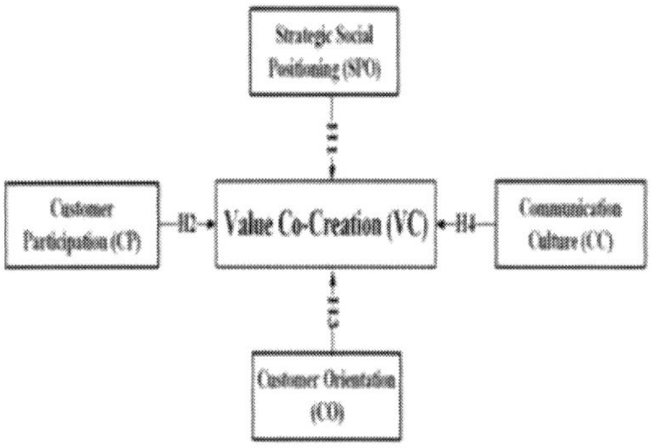

Fig.1. Research Model.

2.1 Value co-creation

The value co-creation construct has been introduced by Prahalad and Ramaswamy [16]. They suggest that service values are not created by manufacturers or suppliers alone but are co-created by them together with IT outsourcing vendors. Based on this construct, researchers have devised models to examine how values are co-created. For instance, Maglio and Spohrer use the value co-creation construct in clarifying the essentials of service science, whereas Wu and Wu apply it to construct a service process structure. Dong et al. investigate the contextual relation of service failures in a value co-creation situation [17-19].

2.2 Strategic social positioning

The positioning determination is frequently a critical strategic determination for e-commerce enterprises because the situation can be foremost in patrons' comprehension and choice determination [20]. Positioning is frequently regarded, erroneously, as a term interchangeable with the psychological concept of standing. While there are similarities between the concepts, there are also major differences [21]. Accordingly, we hypothesize as follows:

H1 : Strategic social positioning has a positive impact on value co-creation.

2.3 Customer participation

Customer participation may increase communication and relationship building between customers and employees [22-23]. According to health care literature, when patients work with doctors to incorporate their preferences and values, this improves the level of care delivered and encourages more empathetic, honest, and friendly interactions, which produces relational values. On the service provider side, employees may fulfill their social needs for approval when they cocreate services with customers, similar to the way their perceptions of being valued by the organization enable them to satisfy their social needs for approval, affiliation, and esteem [24-26]. Thus, every interaction between employees and customers represents an opportunity to co-create relational values for both parties [27]. Illustrative comments from service employees about camaraderie and social bonding with their customers support this assertion [28]. Based on the statements above, hypothesis 1 is proposed as follows.

H2: Customer participation has a positive impact on value co-creation.

2.4 Customer Orientation

Customer orientation is based on customer-driven value creation [29-31]. At the firm level, Slater and Narver emphasized that market-oriented firm's place highest priority on the profitable creation, and maintenance of superior customer value. In this study, value creation has been defined as customer's perspective of achieving desired benefits attributable to the salesperson and the company [32]. Saxe and Weitz suggest that customer orientation at an individually employee level is closely related to a 'concern for others' dimension [33]. Therefore, salespersons high on customer orientation are likely to show higher concern for self and for others, but salespeople with sales orientation or having low customer orientation would exhibit high concern for self, but low concern for others. Therefore, it is also likely that a customer oriented salesperson is more likely to create value for customers, since s/he is likely to be more concerned for others (customers). Therefore, the following hypothesis is proposed.

H3 : Customer orientation is positively associated with value co-creation.

2.5 Communication culture

Communication culture and structures - is the atmosphere within the organization that defines accepted communication behaviour which may facilitate or hinder the communication process [34-36]. The structures of communication and integration of knowledge are essential for enhancing the absorptive capacity [37]. Different authors regard the communication climate as an important factor impacting the communication process [38-39]. An open culture that views change as positive can facilitate communication and thus the learning process [38]. New knowledge creation requires routines that simplify and facilitate communication allows members of the firms to explore ways in which they can use existing knowledge or develop new knowledge [40]. They argue that structural, cognitive, behavioural and political barriers may stifle the effective sharing and integration of knowledge. Therefore, the following hypothesis is proposed.

H4: Communication culture has a positive impact on value co-creation.

3 Research methodology

This study is based on an online questionnaire and selected sample e-commerce enterprises from the Information Management Association of R.O.C., International Service Industry Association of R.O.C. and International Trade Executives Association of R.O.C. from July 2013 to October 2014. The sampling frame is set to all the enterprises in the three agencies including CIS 400 companies, IMA 115 companies, ITEA 653 companies. In accordance with the principle of simple random sampling, the 1168 enterprises in sampling box will be carried out in order, and based on the stochastic indicator compiled by the US RAND in 1955, it begins to sample. According to the experience Sample size, Gaythinks that the sample is at least accounted for 10% of overall in the description research; Tabachnick , Fidell, Velicer, and Fava think that in the SEM analysis ,the tested samples had better be in more than 200.Therefore ,the sample of study plan to determine the sample for 300 [41-42].

According to the sample plan, the sample company adopts method of questionnaire survey, and the sampling sample is 300.We design the questionnaire by considering related works and experts' views. After the first draft is completed, a pretest is undertaken with established leaders in outsourcing value co-creation in the IT industries to amend any part perceived to have a dubious presentation. Therefore, questionnaire recipients could understand the questions in the official survey, and a satisfactory completion of the questionnaire is ensured. All aspects are assessed using a five-point Likert scale from 1 = 'strongly disagree' to 5 = 'strongly agree'. We select a sample of 300 e-commerce enterprises; 293 e-commerce enterprises are removed from the initial sample because they have ceased trading, have a policy of not responding to surveys or requested their removal from the study. Sampling thus yields 207 useful questionnaires, a response valid rate of 97.67%

The method chosen to analyze the data is the structural equations method, using the Partial Least Squares (PLS) technique. This technique is designed to reflect the theoretical and empirical characteristics of social sciences and behavioral characteristics, in which we often find theories with insufficient support or little information available. More specifically, we use Smart PLS 2.0 build M3.

4 Research results

Structural equation modeling (SEM) technique is used to test the hypotheses. SmartPLS software is used to analyze the data.We first examine the instrument (the measurement model) to assess reliability and validity before testing the structural model using the level of significance of the path coefficients and the variance explained (R square measures), as shown in Table 1.

Table 1. Descriptive statistics of the constructs.

	AVE	CR	R square	CA	Communality	Redundancy
CC	0.93	0.97		0.96	0.93	
CO	0.94	0.98		0.97	0.94	
CP	0.93	0.98		0.97	0.93	
SPO	0.89	0.96		0.94	0.89	
VC	0.87	0.95	0.89	0.93	0.87	0.64

CC : Communication Culture
CO :Customer Orientation
CP :Customer Participation
SPO : Strategic Social Positioning
VC :Value Co-Creation
CR : Composite Reliability
CA : Cronbache Alpha

4.1 Measurement model

We assess Cronbach's alpha, the composite reliability of constructs, average variance extracted (AVE), R square, and communality to test convergent validity. As shown in Table 1, the Cronbach's alpha values of all constructs range from 0.93 to 0.97. Composite reliability values range from 0.95 to 0.98 and are above the benchmark value of 0.70, suggesting acceptable degrees of internal consistency of each construct. AVE values range from 0.87 to 0.94. Communality values range from 0.87 to 0.94 and are above the benchmark value of 0.50. This will indicate that the model explains most of the variation for those variables.

The second criterion requires that the square root of AVE shall be greater than the correlations between the constructs. As seen in Table2, the square roots of AVEs (diagonal elements) are higher than the correlation between constructs (off-diagonal elements). The results support the discriminant validity of the scale used.

Table 2. Square root of AVES.

	CC	CO	CP	SPO	VC
CC	0.96				
CO	0.92	0.97			
CP	0.88	0.92	0.96		

DOI:10.1051/matecconf/201712300022

SPO	0.83	0.82	0.80	0.94	
VC	0.92	0.90	0.88	0.85	0.93

4.2 Structural model

The structural model is assessed using the variance explained (R square measures) and the level of significance of the β-Values and T values, as shown in Table1 and Table3. The R square results for the dependent variable, value co-creation (R square =0.89), is high. This suggests that the model has high explanatory power, explaining 89% of the variance in the dependent variable respectively.

Table 3. β Value and T value.

	β Value	T Value
CP→VC	0.15	1.22
SPO→VC	0.24	2.50
CO→VC	0.19	1.24
CC→VC	0.40	3.76

The result for H1 linking strategic social positioning to value co-creation is supported (β-Value=0.24, T value=2.50, p<0.01). Consistent with this, we find that customer participation to value co-creation (H2) is supported (β-Value=0.15, T value=1.22). Customer orientation to value co-creation (H3) is supported (β-Value=0.19, T value=1.24). Communication culture to value co-creation (H4) is supported (β-Value=0.48, T value=3.76, p<0.001). The main effect model is shown as Fig. 2.

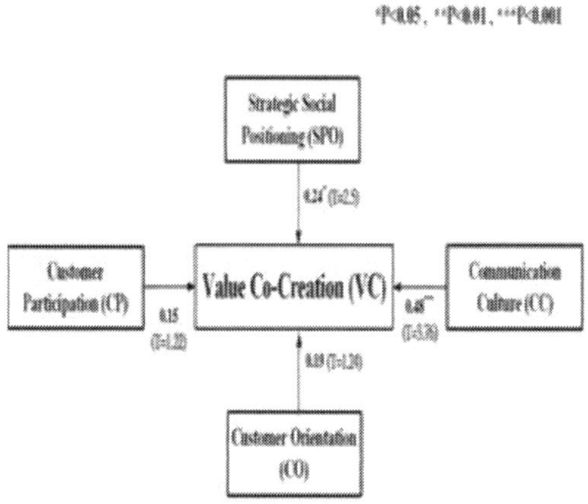

Fig.2. Research results.

5 Discussion and contribution

We formulate four hypotheses and conduct a field study to develop our understanding of how strategic social positioning, customer participation, customer orientation, and communication culture impact value co-creation in an organizational context. Our goal in this research is to document the association between IT outsourcing vendors and manufacturers. We find value co-creation has a strong impact on them.

The contributions of the paper are described as follows. First, this study contributes to manufacturers by integrating the perspective of economic and relational view in the study of the relational IT outsourcing vendors, which are not dealt with in previous studies. Second, this paper extends current research by highlighting the role of value-based relationships from the economic and relational view of manufacturers and IT outsourcing vendors.

In this study, we construct a new model to understand the factors influencing value co-creation between IT outsourcing vendors and manufacturers. Through studying Taiwan's IT outsourcing industries, we find that all factors modelled have an important effect on inter-organizational value creation and that there is a relation between vendors and manufacturers in these terms. We also find that these factors enhance the values of IT outsourcing vendors and manufacturers.

Future research will help identify the key factors that affect how consumers perceive and interpret website quality as a means for making product quality assessments when faced with high levels of information asymmetry. These findings provide a solid foundation for future investigations and practical insights for designing B2C e-commerce websites. This study also makes a contribution by addressing cognitive lock-in (Johnson, Bellman, and Lohse, 2003) and its effects on perceived value and purchase intentions. These insights not only help researchers better understand how PPRs influence consumers' shopping behavior in electronic markets, but also provide guidelines for online retailers to better adjust their IT strategies to improve customer retention.

A growing concern among the organisations who are actively involved in Information Technology outsourcing is post-contract management and the ensuing development of what many practitioners and scholars have coined the 'outsourcing partnership'. Numerous corporations today are outsourcing specific information systems (IS) functions. Outsourcing of information system functions is growing rapidly. The diversity of these outsourcing arrangements goes well beyond that associated with the more traditional facilities management. The effect of service quality of the provider and the ability of companies to build a partnership on these relationships are hypothesized and studied. Increasing attention has been paid to building successful partnerships in information systems outsourcing.

A classic question faced by technology suppliers and buyers are whether to compete in the product markets or to cooperate through licensing. Understanding what paying customers want and how to create products and services they value is the cornerstone of any business model. Customer value can be examined at different levels. At a low level, customer value can be viewed as the attributes of a product that a customer perceives to receive value from. Facilitated by a wide array of technological advancements, interaction between customers and companies is taking new forms and shapes that go beyond traditional exchange. Companies are no longer perceived as sole goods providers, nor customers regarded merely as sources of money. Thus, both companies and customers are initiating new ways to support each other's value creation, that is, new value co-creation mechanisms are developed.

References

1. C. Koo, N. Chung, K. Nam, *IJIM*, **35** (1), pp.64-79, (2015)
2. J. Y. Qi, Q. X. Qu, Y. P. Zhou, *ECRA*, **13** (5), pp.295-304, (2014)
3. A. Padilla-Meléndez, A. Rosa del Águila-Obra, *IJIM*, **33** (5), 892-898, (2013)
4. M. Pagani, *MISQ*, **37** (2), 617-632, (2013)
5. M. J. Mol, J. Birkinshaw, *OS*, **35** (9), 1287-1312, (2014)
6. M. J. Mol, R. J. M. van Tulder, P. R. Beije, *IBR*, **14** (5), 599-617, (2005)
7. L. Ogiela, *IJIM*, **35** (2), 154-159, (2015)
8. M. R. Ogiela, *IJIM*, **34** (3), 317-318, (2014).
9. D. A. Mirchandani, A. L. Lederer, *IJIM*, **34** (5), 622-633, (2014)
10. C. Woodard, R. N. Jason, F. T. Tschang, V. Sambamurthy, *MISQ*, **37**(2), 537-564, (2013)
11. A. E. I. Bharadwaj, O. A. Sawy, P. A. Pavlou, N. Venkatraman, *MISQ*, **37** (2), 471-482, (2013).
12. M. Oslo, The Measurement of Scientific and Technological Activities, *A joint publication of OECD and Eurostat*, (2005)
13. G. Battisti, P. Stoneman, *BJM*, **21** (1), 187-206, (2010)
14. F. Damanpour, R. M. Walker, C. N. Avellaneda, *JMS*, **46** (4), 650-675, (2009)
15. P. B. Goes, MISQ, **38 (1)**, piii-viii, (2014)
16. C. K. Prahalad, V. Ramaswamy, *S&L*, **36**(3), 4–9, (2004b)
17. P. P. Maglio, J. Spohrer, *JMAS*, **36**(11), 18-20, (2008)
18. L.C Wu, L.-H. Wu, TJCIS, **51**(2), 14-23, (2011)
19. B. Dong, K. R. Evans, S. Zou, *JAMS*, **36**(1), 123-137, (2008)
20. D. A. Aaker, G. G. Shansby, *BH*, **25** (3), 56-62, (1982)
21. L. J. Harrison-Walker, JIBR, **10** (2), 135-147, (2011)
22. C.C.A. Claycomb, H. Lengnick, L.W. Inks, *JBS*, **18** (1), 47–69, (2001)
23. H. Zhang, Y. Lu, B. Wang, S. Wu, *IM*, **52** (4), 468-482, (2015)
24. J. P. Foreyt, S. P. H. Walker, *JADA*, **98**(10), 27-30, (1998)
25. R. L. Street, K. Edward, A. Robert, R. L. Bell, H. Kravitz, & Paul, *JGIM*, **18** (8), 609-616,(2003)
26. R. Eisenberger, H. Robin, H. Steven, S. Debora, *JAP*, **71**(3), 500-507, (1986)
27. J. H. Fleming, C. Coffman, J. K. Harter, HBR, **83**(7-8), 106-114, (2005)
28. D. D. Gremler, K. P. Gwinner, *JSR*, **3**(1), 82-104, (2000)
29. M. A. Jolson, JPSSM, **17**(4), 75-88, (1997)
30. D. Korschun, C.B. Bhattacharya, S.D. Swain, *JM*, **78**(3), 20-37, (2014)
31. E. G. Harris, T.L. Brown, C. John, J.C. Mowen, A. Artis, P & M, **31**(3), 171-183, (2014)
32. S. F. Slater, J. C. Narver, *JM*, **59** (3), 63-74, (1985)
33. R. Saxe, B. A. Weitz, JMR, **19** (3), 343-351, (1982)

34. J. H. Adler, *Washington: Brookings Institution*, (1965)

35. S. L. Brown, K. M. Eisenhardt, *ASQ*, **42**(1), 1-34, (1997)

36. J. Littler, CCC, **6**(2), 227-243, (2013)

37. W. M. Cohen, D. A. Levinthal, *ASQ*, **35**(1), 152-162, (1990)

38. N. S. Levinson, M. Asahi, *OD*, **24**(2), 50-63, (1995)

39. W. Tsai, AMJ, **44** (5), 996-1005, (2001)

40. S.A. Zahra, G. George, AMR, **27** (2), 185-203, (2000)

41. B. G. Tabachnick, L. S. Fidell, *Boston, Massachusetts: Allyn and Bacon*, (2007)

42. W. F. Velicer, J. L. Fava, *PM*, **3** (2), 231-251, (1998)

MATEC Web of Conferences **123**, 00023 (2017) DOI:10.1051/matecconf/201712300023
ICPMMT 2017

Optimum design and research on novel vehicle hybrid excitation synchronous generator

Zhong-Shu Liu*

The Key Laboratory for Automotive Electronics and Electric Drive of Fujian Province /School of Information Science and Engineering, Fujian University of Technology, Minhou, Fuzhou City, 350118, Fujian, China

Abstract. Hybrid excitation is an organic combination of permanent magnet excitation and electric excitation. Hybrid excitation synchronous generator (HESG) both has the advantages of light quality, less losses and high efficiency like permanent magnet generator and the advantages of good magnetic field adjusting performance like electric excitation generator, so it is very suitable for the vehicle application. This paper presented a novel vehicle HESG which has skew stator core, permanent magnet rotor and both armature winding and field winding in the stator. Using ANSYS software, simulating the electric excitation field and the magnetic field, and finally the main parameters of HESG were designed. The simulation and the test results both show that the novel vehicle PMSG has the advantages of small cogging torque, high efficiency, small harmonic component output voltage and low waveform aberration, so as to meet the design requirements fully.

1 Introduction

With the rapid development of national economy, automobile industry has become one of the pillar industries of China economy, it plays an irreplaceable role in all aspects of transportation and human daily life. Generator is the key to the car electrical system components, the progressiveness and comfort of the car are closely related to the performance of the generator. At present, the silicon rectifier electrical excitation generator and permanent magnet generator are mostly used as vehicle generator in China, the losses of resistance in the former increased due to the existence of the excitation winding, thus the generator has low efficiency. Permanent magnet generator has the advantages of simple structure, low loss, but as its permanent magnet excitation magnetic field cannot be adjusted, it has the problem of the difficulty in voltage regulation. Particularly, when the output power of generator is greater than a certain value, to maintain the voltage stability would be even more difficult[1].

In this paper a new type of vehicle hybrid excitation synchronous generator was designed to solve the problem of electrical excitation generator and permanent magnet generator. It is a kind of promising vehicle generator, at same time the generator can make full use of the abundant rare earth resources in China, it is a novel vehicle generator which integrates motor, power electronics technology and new material technology, and it is able to adapt to the speed and load change in a wide range, while still maintaining a constant output voltage.

2 Generator structure and working principle

2.1 Structure

In this paper, a vehicle hybrid excitation synchronous generator which has rated power of 1kW, rated voltage of 400V and rated speed of 3000 r/min is designed and studied, and it is shown in Fig.1. The generator consists of two parts: permanent magnet part and electric excitation part, they share a common winding called three-phase stator winding [2]. Two kinds of poles are staggered distributed on the rotor surface, one is permanent magnet pole, and the other is core pole. The circular field winding located in the middle of the stator core. Two parts of the rotor share a shaft, with a certain gap between them used as magnetic separation slot. At the same time, around the stator core there has a magnetic sleeve, which provides axial ring pathways for excitation winding. The permanent magnets produce radial field, and the electric excitation winding produce axial field and radial field both. Thus the two excitation sources produce synthetic magnetic field. The permanent flux starts from N pole in rotor surface, passes through the air gap and stator tooth and then goes into the stator iron core, passes through the stator yoke and another stator tooth, again through the air gap going into the rotor S pole, finally returns to the N pole, thus the magnetic circuit forms. Electric excitation flux goes through the

*Corresponding author: lzs@fjut.edu.cn

shaft, air gap, the stator tooth, yoke, the magnetic sleeve, stator teeth, yoke and air-gap in turn, thus forms a closed path. The magnitude of flux of electric field in the air gap is controlled by the excitation current [3], when field current remains constant, electrical excitation magnetic field produces a one-way direction rectangular magnetic flux in the air gap.

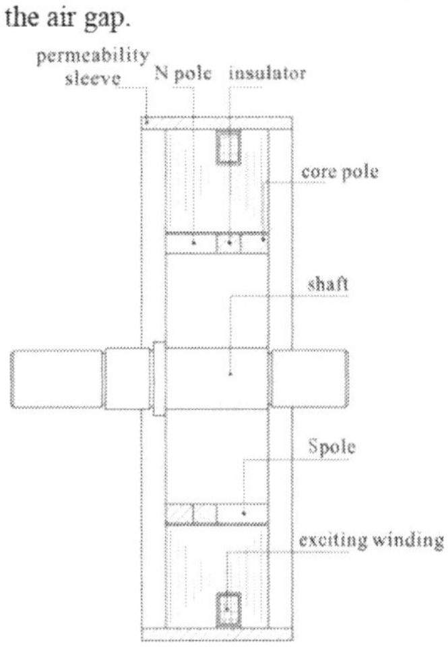

(a)Two-Dimension Structure

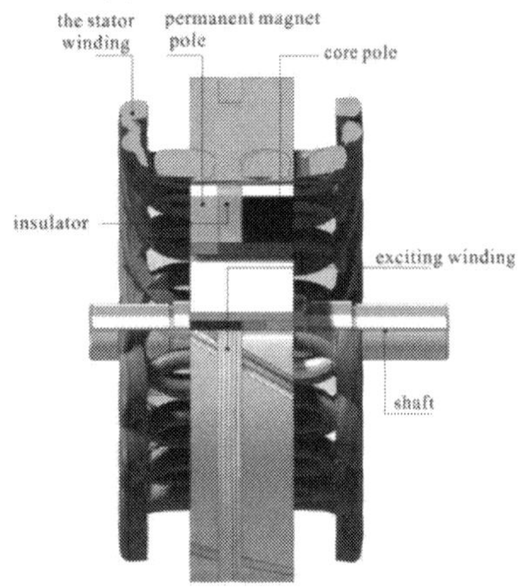

(b)Three-dimension structure

Fig.1. Hybrid excitation generator structure diagram.

2.2 Electromagnetic optimum design

The main technical indicators of vehicle HESG in this paper are as follows: rated power 1kW, rated voltage 240V, rated speed 3000r/min.

During design process, it should be noted that the load distribution between the permanent part and the electric excitation part determines the performance of the HESG. It requires that the HESG should have the voltage adjustment ability from no-load to rated load. And it can be seen from the operate principle of the HESG that we should ensure that the part of the magnetic circuit is in an unsaturated state when we design the HESG. While the HESG is with rated load, the electric excitation part plays a certain role in adding magnetism, and the permanent produces main magnetic field.

When the load is inductive load, the output voltage of the HESG is as follows:

$$U = \sqrt{E_d^2 + I_N^2 X_{aq}^2 \cos\psi_N - I_N^2 (R_1 \sin\varphi - X_1 \cos\varphi)^2} - I_N (R_1 \cos\varphi + X_1 \sin\varphi) \quad (1)$$

In above, U is the output voltage of HESG; E_d is the direct axis internal electric motive force. I_N is the output current of HESG; X_{aq} is the quadrature axis reactance of armature reaction; $\cos\psi_N$ is the internal power factor; R_1 is the armature resistance; X_1 is the armature leakage reactance and $\cos\varphi$ is the power factor.

It can be seen from formula (1) that we can increase the output voltage and reduce the voltage regulation by reducing the armature resistance R_1 and the armature leakage reactance X_1.

The optimal design measures used in this paper is to keep the total number of turns per phase of the armature unchanged, meanwhile, to decrease the number of turns per coil. And finally use optimal measure to design the groove.

Combined with the above optimum design rule, using the special software ANSYS and the method for designing permanent magnet part and electric excitation part separately, we can optimize the design of HESG, and the results were shown in Table 1.

Table 1. Structure parameters of the prototype.

Parameter	value	Parameter	value
Outer diameter of the stator	204/mm	Outer diameter of the rotor	118.4/mm
Inner diameter of the stator	120/mm	Inner diameter of the rotor	25/mm
The length of the core	50/mm	The air gap length	0.8/mm
The thickness of the magnet	8/mm	Slot number	36
The width of the magnet	121/mm	Pole	1

2.3 Chute stator core

The generator stator core is chute designed, by using ANSYS Maxwell; we analyze the influence of different chute width on the generator voltage waveform sine distortion rate. The results are shown in Table 2.

Table 2. Voltage waveform sine distortion rate of different chute width.

The chute width	0	0.25	0.5	0.75	1
Linear voltage waveform distortion rate (%)	1.762	1.248	1.126	0.893	0.674
The chute width	1.25	1.5	1.75	2	
Linear voltage waveform distortion rate (%)	0.316	0.221	0.353	0.42	

The table above shows that the stator chute can effectively reduce the magnitude of the sinusoidal output voltage waveform distortion rate, especially when the stator chutes width is 1.5 times the stator slot pitch, the generator voltage waveform sine distortion rate would be the minimum, and the sinusoidal of generator output voltage waveform would be the best. So, we chose the stator chute width as 1.5 times the stator slot pitch [4].

2.4 Operation principle

When the generator load reduces, the demagnetization effect of armature reaction would gradually weaken, it will lead to higher output voltage. The demagnetization current should be put into ring excitation winding current at this time, which makes the radial magnetic flux produced by the electric field at opposite direction of the radial air gap magnetic flux produced by permanent magnet, then the total air gap synthetic flux decreases, in this case the generator output voltage would reduce[4].

When the generator load increases, the demagnetization effect of armature reaction strengthen, results in the decrease of the output voltage. The enhancement magnetization current should be put into ring excitation winding current at this time, which makes the radial magnetic flux produced by the electric field at same direction of the radial air gap magnetic flux produced by permanent magnet, then the total air gap synthetic flux increases, in this case the generator output voltage would rise.

Above all, by adjusting the magnitude and direction of the hybrid excitation generator excitation current, we can change the syngas gap magnetic flux, so as to keep the output voltage constant when the load and speed change.

3 HESG electromagnetic field

3.1The field of permanent magnet

Without considering the magnetic steel bonding, the magnetic field of permanent magnet of generator will be two dimension fields, a solution of the grid part of its area and no-load magnetic field distribution are shown in Fig.2. (a) and Fig.2. (b) respectively.

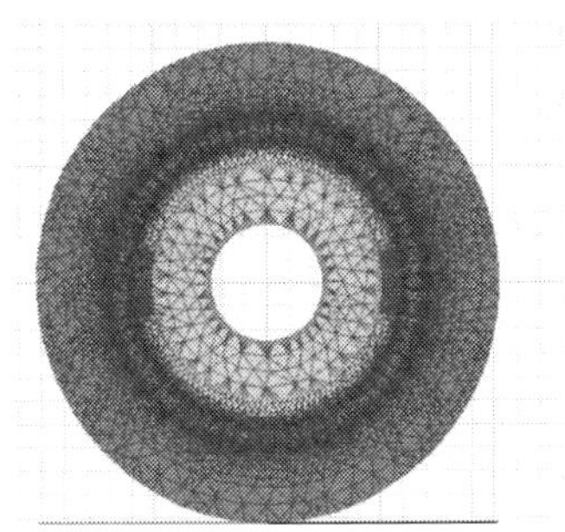

(a)the Grid Subdivision

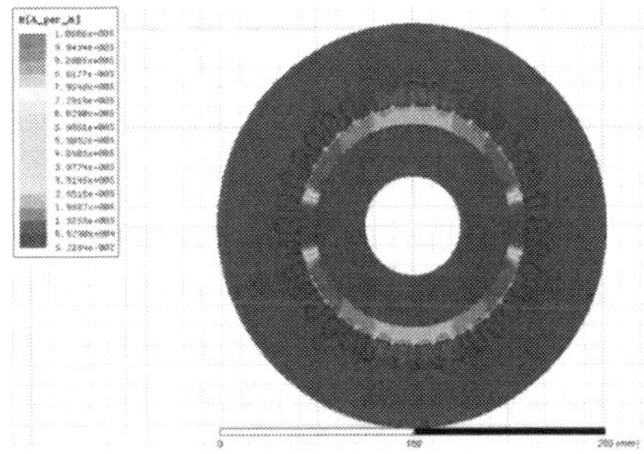

(b)Magnetic Field Distribution

Fig.2. Magnetic field distribution of permanent magnet.

3.2 The field of electric excitation

The grid subdivision and distribution of electromagnetic field of electric excitation are shown in Fig.3.(a) and Fig.3.(b) respectively. It can be seen from the diagram that the structure of the field of electric excitation is relatively complex, there exists both radial and axial magnetic flux in the solving region, and it is three dimension fields. The flux density in the shaft is higher, and prone to be saturation. At the same time, the field distribution of electric excitation along the axial is uneven; the field density on the boundary of the shaft surface between the slot and not slotted would be the maximum [5].

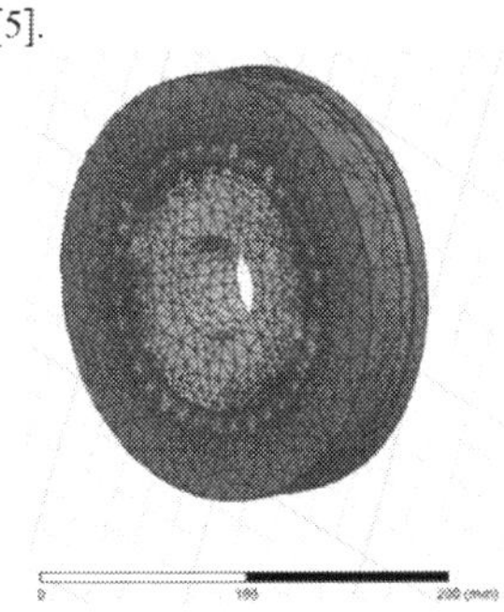

(a)The Grid Subdivision

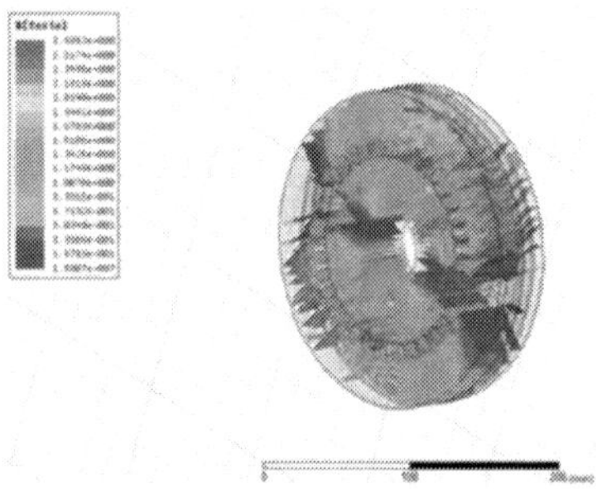

(b)Magnetic Field Distribution

Fig.3. Magnetic field distribution of electrical excitation.

4 Automatic voltage regulation circuit design

The permanent magnetic field and the electric field of the vehicle hybrid excitation generator synthesis are in the stator core. Because of its special structure, there are essential differences in .the principle of voltage control among hybrid excitation generator, traditional silicon rectifying generator and permanent magnet generator. In addition, due to the constantly changing of speed and load in the vehicle generators, to regulate the output voltage for maintaining voltage stability would be more difficult, the voltage regulating circuit is used to provide current to the hybrid excitation generator whose magnitude and direction can be adjusted[6]. When the voltage of the generator change with speed and load, the voltage regulating circuit can automatically adjust to keep the voltage of the generator at a given level and ensure the voltage regulation accuracy to a certain extent, by changing the magnitude and direction of excitation current, we can eliminate the influence of the armature reaction to make synthesis of magnetic field to be constant, and finally to meet the requirement of safe operation and normal generation. Its control principle is shown in Fig.4.

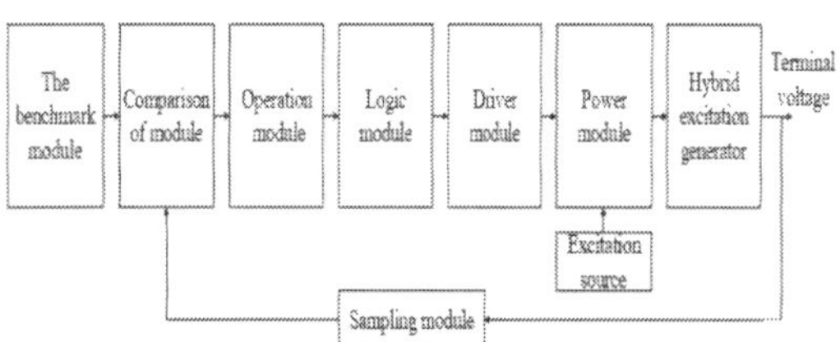

Fig.4. Hybrid excitation generator control principle diagram.

Voltage regulating control circuit designed in this paper consist of modules, such as the benchmark, comparison, operation, logical, drive, power and so on. when the output voltage of the generator is lower than the rating value of 400V, the control system through the control circuit and trigger circuit triggers the transistor so as to make the current flowing in the field winding is forward current[7], and then strengthen the excitation magnetic field, finally increase the output voltage of generator to ensure the output voltage of vehicle generator to reach to a specified value. When the output of the generator voltage is higher than the rating of 400 V, the control circuit to trigger circuit triggered another pair transistor so as to make the current flowing in the field winding is reversal, and then weaken the excitation magnetic field, finally decrease the output voltage of generator. The cycle then repeats itself to ensure that the output voltage of generator remain unchanged in condition of wide range speed, variable load.

According to the voltage regulating circuit designed above, and then to make the experiment of HESM on the motor testing stand, the results of generator no-load test and load test are obtained as follows.

4.1 No-load test

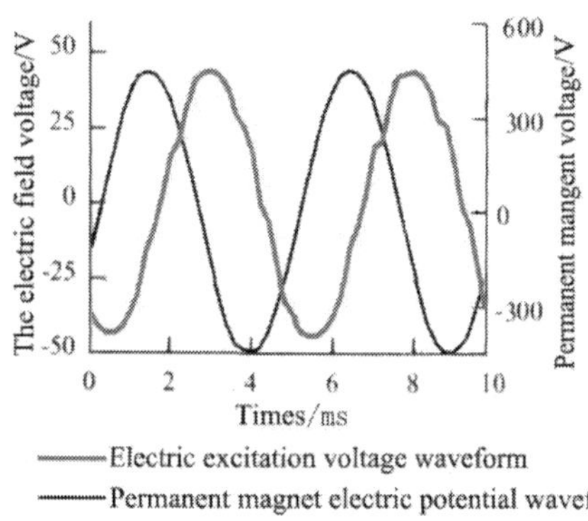

(a)Electric Field and Permanent Magnet Output Voltage Waveform

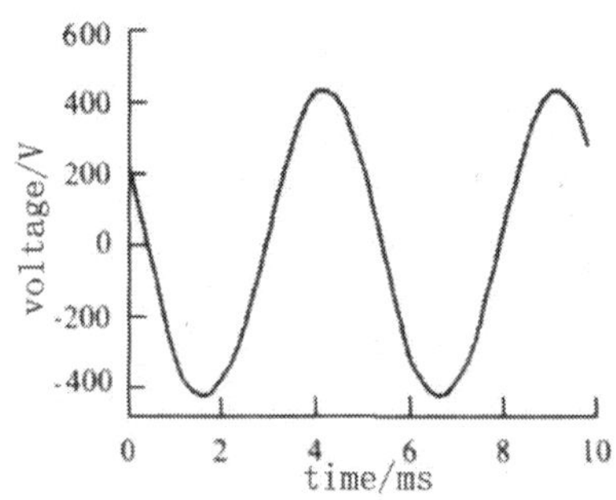

(b)Synthetic Voltage Waveform

Fig.5. Voltage waveform.

Generator operating in 3000 r/min, no-load running under the action of electric excitation alone output voltage waveform, permanent magnet excitation alone under the action of the output voltage waveform and the above two parts respectively synthesized under the joint action of magnetic potential output voltage waveform are shown in Fig.5 (a). As it can be seen from Fig.5 (b).that the output voltage of generator is equal to the difference between the permanent magnet voltage and electrical excitation output voltage [8].

4.2 The load test

4.2.1 External characteristic

To convert generator output from AC to DC, changing the generator speed and load at the same time [9], and keeping the output voltage maintaining in 400 V, we can get the relationship between output voltage and exciting current, and the curve is shown in Fig.6.

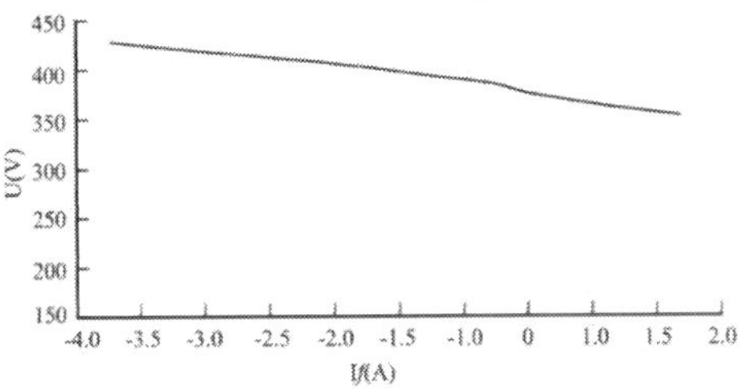

Fig.6. External characteristics.

Experimental result shows that: when the vehicle hybrid excitation generator load and speed change, the minimum output voltage is 352 V, the maximum value is 428 V, the voltage stabilizing circuit designed basically meets the design requirements, so we can obtain a stable voltage output in a wide range of load.

4.2.2 The effects of generator parameters on the external characteristics

Due to the changing of load and the stator current, the degree of saturation of the motor changes as while [10], thus affecting the generator armature reaction reactance, the inherent voltage regulation of the generator is:

$$\Delta U\% = \frac{E_0 - U}{U_N} \times 100\% \qquad (2)$$

The inherent voltage regulation of hybrid excitation generator is defined as, when the electric field current is zero, the ratio of the armature voltage to the rated voltage. Parameters that affect the load voltage regulation are the stator resistance R_a^*, power factor $\cos\varphi$.

1) The effect of the stator resistance on the load characteristic

Fig.7 shows the effect of stator resistance on external characteristic of generators, as can be seen in the Fig.7, the greater the stator resistance, the greater the inherent voltage regulation of the generator.

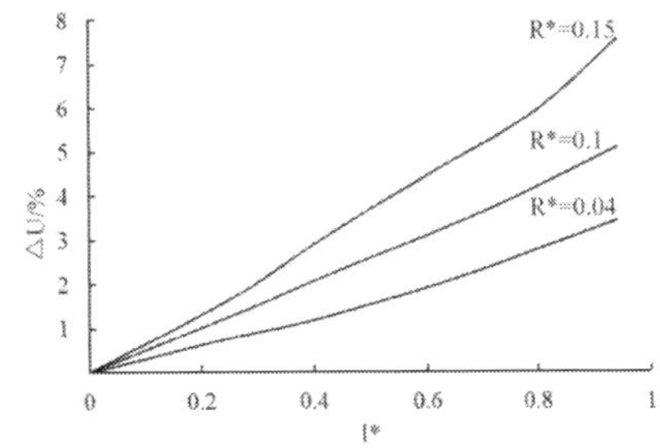

Fig.7. Effect of stator resistance on voltage regulation.

2) The effect of the power factor on external characteristic

From figure 8, it can be seen that the greater the power factor, the smaller the inherent voltage change rate [9].

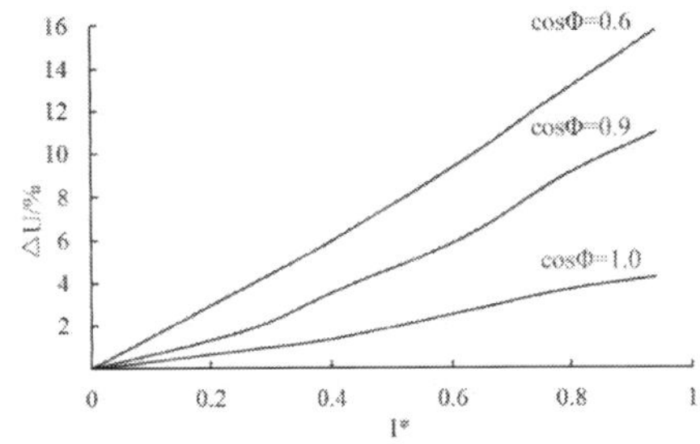

Fig.8. Effect of power factor on voltage regulation.

4.3 Short circuit characteristic

Short circuit characteristic refers to the relationship between short circuit current and excitation current when the three-phase synchronous generator is in short-circuits stability [11]. That the short circuit characteristics of hybrid excitation generator at different speeds result from the experiments are shown in Fig.9. Because the main flux does not equal zero when the excitation is zero, the short circuit characteristic curve does not pass through the origin [12]. It can be shown from the Fig.9 that the short circuit characteristic remains a straight line basically. It reflects that the magnetic circuit of the generator is in a state of unsaturation [13].

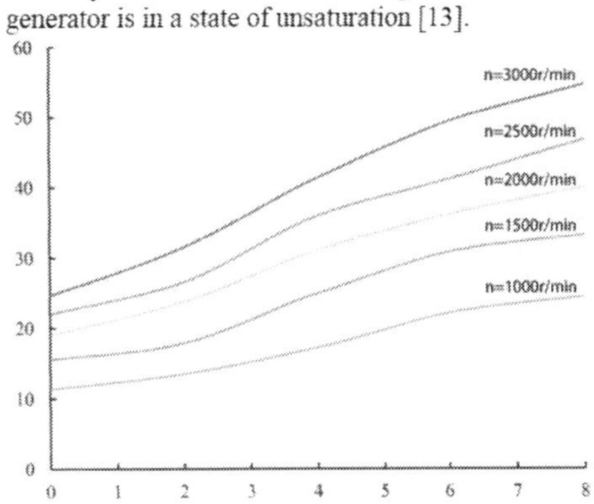

Fig.9. Short circuit characteristic under different speed.

5 Conclusion

This paper designed a new type of vehicle hybrid excitation synchronous generator according to the principle of operation, analyzed the magnetic field of permanent magnetic parts and electrical excitation part, and finally an automatic voltage regulation system was designed applicable to hybrid excitation generator. In general, the new vehicle hybrid excitation generator has characteristics as follows:

First, the field winding embedded in the stator core, realizing the brushless excitation, as to improve the reliability of the generator operation [14].

Second, due to the electric excitation flux mainly through the core, adjusting the excitation current does not cause irreversible short circuit of permanent magnet, and there is no additional air gap in the magnetic circuit, so the power of field winding can be very small [15];

Third, a chute hybrid excitation generator stator core brings lower cogging torque, small output of the generator voltage waveform harmonic component and low waveform sine distortion rate.

Finally, owing to the permanent magnet used in the HESM, it can be made with high efficiency and lower density. Through the reasonable design of permanent magnet part and electric excitation part, it can meet the design requirements of different application [16].

Acknowledgments my work was supported by the Natural Science Foundation of Fujian Province of China. (Grant No. 2017J01667)

References

1. Ma Jianwei, Research on permanent magnet and electro-magnetic hybrid excitation constant voltage generator, (2008)
2. Lin Nan, Transaction of China electrotechnical society, 31, 19 (2016)
3. Xiao Li, Ge Fahua, Liu Tong, High speed high power density of generator to improve the design and implementation, **48**, 18 (2015)
4. Jin Wanbing, Design and study on hybrid exciting permanent magnet generator and its control system, (2003)
5. Li Shenghu, An Rui, Xu Zhifeng, Dong Wangchao, Electric power automation, **35**, 21 (2015)
6. Gong Jianfang, Large Electric machine and hydraulic turbine, 15, 17 (2008)
7. Shi Mingming, Zhou Bo, Wei Jiadan, Mao Yiran, *Proceeding of the CSEE*, **32**, 1 (2012)
8. Wen Huihui, *Control engineering of China*, **21**, 142 (2014)
9. Shen Xijun, Zhao Qiaoe, Xu Pengtao, *Journal of Electric Power*, **29**, 28 (2014)
10. Zhang Zhuoran, Research on characteristics of novel hybrid excitation synchronous machines, (2008)
11. Feng Hao, Huo Zhihong, Hu Xuebin, *Renewable energy resources*, **33** , 1831 (2015)
12. Shen Jianxin, Miao Dongmin, *Transactions of China Electrotechnical Society*, **28**, 2 (2013)
13. An Quntao, Sun Li, Sun Lizhi, *Proceedings of the CSEE*, **35**, 5891 (2015)
14. Zhang Liguo, the Research and design on tangential structure hybrid excitation generator of magnetic multipled type for automobile, (2011)
15. Jian Zhijun, *Micromotors*, **46**, 16 (2013)
16. Fu Xinghe, Zou Jibin, Qi Wenjuan, *Journal of Southwest Jiao Tong Uniersity*, **45**, 99 (2010)

MATEC Web of Conferences **123**, 00024 (2017) DOI:10.1051/matecconf/201712300024
ICPMMT 2017

High current density on electroplating smooth alkaline zinc coating

Allen Bai[1], *Kai-Lin* Yang[2,*], *Hao-Long* Chen[2], *Yu-hong* Hong[2], and *Shyue-Bin* Chang[3]

[1]Department of Green energy Science and Technology, Kao-Yuan University, Kaohsiung, 821, Taiwan R.O. C.

[2]Department of Electronics Engineering, Kao Yuan University, Kaohsiung, 821, Taiwan R.O. C.

[3]Department of Mechanical & Automation Engineering, Kao-Yuan University, Kaohsiung, 821, Taiwan R.O. C.

Abstract. The behaviours of the Zinc coating would be subjected to the complex zinc anion, and also the additives, current densities and compositions of electroplating bath. In this study, the complex agents (anise aldehyde and Lugalvan P) were added into the electroplating bath to improve the anti-corrosion ability and inhibit dendritic structure on zinc coating. The Lugalvan P is added into the electroplating bath to inhibit dendritic structure during electroplating. The anise aldehyde and complex zinc ions plate a smooth zinc coating. Moreover, the crystalline structures of zinc coating analyzed by X-ray diffraction analyzer (XRD) and the surface morphology of zinc coating was observed by the scanning electron microscope (SEM). The results show that the Lugalvan P additive forms the relatively highest flatness, which inhibited the dendritic structure. High electroplating efficiency was achieved of current intensity of 3 A cm^{-2}, and a smoothness and gloss deposited surface could be obtained. With increasing the deposition current density, it can form a better zinc coating with low plating bath resistance to form a higher anti-corrosion zinc plate.

1 Introduction

Alkaline zinc plating zinc plating industry has always been one of the best options for the rust-preventive metallic iron [1-4], pure zinc and zinc alloy plating to replace the zinc plating has been published [5-7], but pure zinc plating is still the most simple, reliable and high-cost method of corrosion. Because corrosion of zinc coating can be isolated from the outside of the coated surface of the substrate, and also includes a cathodic protection [8].

Galvanized coating technology through research and evolution time, the current can be broadly divided into three types, respectively [9-12] acid zinc chloride, zinc cyanide and a basic alkaline non-cyanide zinc. Wherein cyanide alkaline zinc best effect (smooth, uniform thickness and high and low current region of the low-corrosion equipment), but it contains highly toxic cyanide is the [13-15], wastewater treatment cost is quite high, it is now have no acid zinc cyanide and technological development toward the basic, but its coverage (covering power) and deposition efficiency is still alkali cyanide and zinc [16,17].

Although acid zinc plating bath contains no toxic substances, but still faces two kinds of problems [18]:
(1) When plating system plating bath of an aqueous zinc cathode surface adsorption and hydrogen generation is easily formed, resulting in decreased current efficiency, but also results in the zinc coating unevenness.
(2) High current density zinc plating (High Current Density; HCD) accompanied by poor surface morphology and crystallinity.

Hull cell tank test at high current zinc-plated tip member, can easily form a dendritic zinc deposition [4], uneven coating thickness, since the adhesion of dendritic zinc layer deposited easy tear off in subsequent processing.

The HCD process is anticipated for the industry, because it can increase productivity with minimal cost. Conventional acidic galvanizing bath containing zinc salts (zinc sulfate or zinc chloride), buffers, or to improve the conductivity of the salt (chloride or sulfate) as well as other organic additives to improve coating coverage, covering power (throwing power), roughness, brightness, corrosion resistance, physical and chemical properties.

Although alkaline zinc cyanide can easily get intact zinc coating [19], however, its toxicity is too high. It could be obtained efficiency electroplating by the acid zinc electroplating process, but high corrosion equipment of plating bath as well as the poor plating coverage limited the applications [20,21]. Therefore, alkaline non-cyanide zinc plating efficiency although it is not high (only about 30 to 35% of zinc cyanide) [21], but with high covering power coating, and a low device resistance. Taking these various perspectives, the ability to improve the alkaline cyanide-free zinc, and industrial application can be extensively used.

* Corresponding author: t10056@cc.kyu.edu.tw

Alkaline zinc plating bath consists essentially of zinc ions, excess sodium hydroxide and water. However, in galvanized high alkalinity conditions, if no organic additives, the resulting zinc deposition is rather rough and cannot make any use of applications [19].

Generally, it contains a series of organic ammonium compound additives in the electrolytic solution in the industry to obtain bright coating with a smooth surface [21]. These additives have been identified as a leveling agent (Levelers) and brighteners (brighteners) in the plating bath. By these additives, it may inhibit the diffusion of zinc ions to the plating surface of the galvanized overvoltage increased to accelerate the nucleation rate of zinc [20].

Since the galvanized iron metal substrates have excellent anti-corrosion properties, as well as the low built-up cost, easy to operate, it is most widely used for the plating industry. Some problems limited the properties and the current efficiency on the zinc coating, hydrogen is the first problem, since a zinc electrode and the hydrogen generation will lead to the adsorption of the plating process will reduce the current efficiency. Second, the high current in the cathode region of the plating often cause a scorch generation or a dendritic structure, resulting a poor deposited quality. It would easy to fall off the galvanized layer, and the corrosion resistance deteriorates, so how to improve the plating production is motivated.

To overcome the high current electro-galvanized layer, the industry has been plagued by problems. Therefore, this experimentally pure zinc plating for the main process to do a variety of organic additives to improve plating of the cathode, and investigate the electrochemical behavior of these organic additives in an acidic or an alkaline bath with the case of the system performance. Via a linear scan voltammetry (the LSV) to understand the effect of complex organic additive effect on the plating at a high current density, and then additionally by a scanning electron microscope (SEM) and X-ray observation of organic additives on the galvanized surface morphology, growth affect the structure and crystalline characteristics.

Further zinc electrodes are some restrictions on the practical application of zinc on their own characteristics, such as it will generate a dendritic structure of the galvanized zinc alkaline solution.

In this study, the complexing agents anise aldehyde and Lugalvan P were added into the Zinc bath to inhibit the dendritic structure during electroplating. The electrochemical behaviour of the alkaline zinc coating layer would be analyzed and the materials properties would also be examined. The electrochemical analysis of the zinc coating with a linear voltammetry (linear sweep voltammetry, LSV) was to seek deposition potential, Chronopotentiometry discharge method (CP) was to obtain the coating of the discharge efficiency. The crystalline structures of zinc coating would be examined by X-ray diffraction analyzer (XRD) and the surface morphology of zinc coating would observed by the scanning electron microscope (SEM).

2 Experiment

In this experiment, the groove 267 mL Hull cell container was used for producing a cathode deposit that records the character of electroplate at variant current densities within the operating range. The cathode is electrically galvanized substrates to contact the test negative power.

This experiment is a test strip plating process prior to the same current Hull cell plating bath for one minute, then washed with water, and then subjected to surface pickling (pickling solution is a mixture of nitric acid and deionized water) to remove surface oxides, corrosion the product was then impurities and other contaminants, rinse with water, the same current level in the galvanizing bath 5 minutes out of Hull cell rinse with water, pickled surface (removal of oxide film), then try again with an electric hair drier sheet.

Pickling solution to 485 mL with deionized water and 15 mL of a mixture of nitric acid, nitric acid in the pickling solution are mainly bright surface, and the film dissolved corrosion products and the like.

The DC power supply was be carried out in this study and the anise aldehyde and Lugalvan P additives were be adapted as the complexing agent to inhibit the generation of hydrogen and dendritic structure formation during electroplating. In this experiment, the current densities are galvanized 1 A, 1.5 A, 2 A, 2.5 A, 3 A cm^{-2}, respectively, and the last recorded initial voltage and the voltage at different test strip galvanizing current electrical Harrington groove. These data used to analyze the plating rate in the test strip at different current Hull cell plating solution tank.

Electrochemical analysis linear voltammetry (linear sweep voltammetry, LSV) was used to seek deposition potential, and the Chronopotentiometry discharge method (CP) was used to obtain the coating of the discharge efficiency.

To study the crystalline deposits of plating, the X-ray diffraction analyzer (XRD) was used to analysis the crystalline characteristics and the scanning electron microscope (SEM) was used to observe the surface structure.

3 Experiments reaction principle

When the metal deposition or the like dissolved electrolytic reaction, the amount of electrolysis involving reactants or products with power through proportionally. Faraday's law quantitative description of this relationship:

1. When the amount of the electrolytic metal deposition or dissolution and is proportional to the power through

2. The quality of different metals deposited by the same amount, is proportional to its equivalent Moore. The number of charges 1mol proton called Faraday constant, is represented by F, F = NA e = 6.023 x 1023 / mol x 1.60218922 x 10-19C = 96500 C / mol.

Wherein, NA is a constant, e is the proton charge, C is the unit of coulomb electricity. Customary in the industry with "ampere - hours (A • h)" shows the power, 1F = 26.8A • h. Faraday's laws of electrolysis may be expressed as m = I t M / 96500z formula, M and z is the valence of plated metal mass moire metal ion, m is the current I through the elapsed time t after the weight of the metal deposited.

Flowing through the electrolytic cell during electrolysis power can be obtained from various methods, for the most accurate value, it is necessary to use coulomb meter. Coulomb meter is a 100% efficiency of the electrolysis cell, with the test series using the plating bath. Coulomb meter is a typically copper or silver iodide coulomb meter, electricity obtained by weighing the weight of metal electro-winning, iodine coulomb counter it is produced by electrolysis of iodometric titration to determine the amount. The copper coulomb counter is most widely used.

The working electrode substrate was used with industrial purity of 99.95% and a 1cm x 2cm of red copper substrate. Then drilled using a drill hole (orifice 1.2mm pore size) on a copper sheet, copper wire and then welding (at both ends of the enameled paint scraped off the table, while the welding part, the other side is in communication with the power supply, copper about 15cm). The electrode welded using Teflon (PTFE) is wound, showing only the total surface area 2cm X 2cm.

The copper surface treatment of the substrate is first copper electrode in making a goodlye, red copper by electrolytic polishing to make the surface smooth. In this step, the red copper anode is placed in position. Electrolytic solution of 1 M NaOH, through which current is 0.12 amperes, electropolished 5 minutes. After electrolytic polishing, the surface of the electrode with deionized water clean. Wash the electrodes, and then subjected to electrolytic degreasing grease removing the electrode surface. When the copper is subjected to electrolytic degreasing cathode electrode in position. Electrolytic solution of 1 M NaOH solution, the current through the electrodes is 0.3 amps. Electrolytic degreasing for 10 seconds, using the generated hydrogen bubbles away from the electrode surface of the grease.

After completion of electrolytic degreasing, washing with deionized water to a clean electrode. And placed in ultrasonic oscillator shaking cleaning is carried out for 5 minutes, and then pickled to remove the oxide electrode surface, so as to facilitate the activity of the plating is performed. Pickling solution is 1: 1 concentrated sulfuric acid mixed with ultra-pure water dilution. Pickling time is 1 minute. And then subjected to electrolytic degreasing 10 seconds. After completion of pickling electrode surface washed with deionized water, and then shaking using ultrasonic oscillator for 5 minutes.

The electrochemical plating bath analysis linear sweep voltammetry experiments with constant current chronopotentiometry assay. Characteristics of linear sweep voltammetry of the electrode potential over time as a forward scanning or reverse scanning, while the measurement results will be related to the rate or direction of change in the potential of this method is typically performed when the solid electrode used for study [15]. Through the action of the linear sweep voltammetry, the scanning speed is controlled at 5 mV / s. The current detection voltage related to the bath at various conditions. Galvanized zinc plating studied under different conditions, the crystal growth of zinc precipitation state using a current efficiency change chronopotentiometry. Structure identified object material composition of the coating itself understanding, microstructure, and other properties, crystalline form, X-ray diffraction experiments (XRD) a crystalline coating is responsible for detection, growth situations plane. A scanning electron microscope (FE-SEM), the surface patterns can be observed, while usually accompanied by EDS, WDS and elemental mapping functions, may be analyzed and the individual elements of the multi-component, trying to understand the composition, surface morphology and distribution of pores.

4 Experimental Results and discussion

Acid zinc is generally used as the primary brightener aromatic aldehydes and ketones, but because of its poor water solubility, it is necessary a surfactant or wetting agent to help dissolved and dispersed in the plating bath. Additives acting between these quite complex, these viewpoints, this study will mainly affect the analysis of the current density, the primary brightener and surfactant traits and zinc coating the crystalline structure. Characters crystalline structure of zinc-plated surface, respectively will be X-ray diffraction analyzer for detection by a scanning electron microscope (SEM). Further, linear voltammetry method to understand these complex organic additives influence of electro-galvanized baths.

Ammonium chloride bath main components 22wt% (4.1M) and 6 wt% (0.44M) of zinc chloride, using 1M HCl pH was adjusted to 5.0. And the flow efficiency of the electro-galvanized η (CE) by the equation (1) is obtained:

$$\eta\,(\%) = (2\,w / ZnAW) \times 100\% / [(I \times t) / 96{,}485] \quad (1)$$

Wherein, w is a zinc deposit weight, ZnAW zinc molecular weight, I is the current through, t is the elapsed time.

Experimental conditions in the preparation of zinc plating temperature controlled 25∘ C, pH = 2, during the deposition of zinc ions generally accompanied by generation of hydrogen gas, at this time will be a competitive reaction shown in the following two:

$$Zn + 2 + 2e^- \rightarrow Zn \quad (2)$$

$$2H + + 2e^- \rightarrow H_2 \quad (3)$$

Due to the occurrence of hydrogen can cause the overall current efficiency decreases and the surface of dendritic structure.

The technical development of a weakly acidic galvanizing system, many of the organic additives are added to the plating bath is necessary to obtain a broader current density range and bright region richer zinc coating ductility. These foreign additives are grouped into three categories: primary additive, and auxiliary minor additives additive. The general main additive concentration are much higher than the galvanizing bath additives or auxiliary minor additives, which main purpose is to enhance the flatness and dispersibility of the zinc coating bath. Further, the main feature is the additive aqueous chemotactic effect, helping the minor additions (or brightener) was dissolved in the plating bath. While the secondary brightener mainly zinc coating so that light, under certain conditions additionally also help improve the dispersibility of the plating bath. Auxiliary additives are added to the main purpose is to increase the range of light, so that the bath easier to operate, further additives can also help some auxiliary brightener dissolved in the plating bath. Maybe it has a part in the plating bath additives in the role not only one of the above, maybe two or more, especially when excessive dosage, but in any case, in order to control the galvanizing bath, three listed above additives are indispensable.

This study is plating bath composition of 15wt% (3.75M) of sodium hydroxide and 1.5% (0.18M) of zinc oxide further added polyamines Poly quaternary ammonium salt-Lugalvan P 0.2wt% to completion of the final plating.

The polyethyleneimine has been widely used in the galvanizing bath alkali cyanide, which can effectively improve the dispersion capacity and coverage of the plating bath at low current density, but also part of the glittering effect. Further, ethyleneimine in alkaline zinc-nickel alloy plating bath is also very good complexing agent.

Lugalvan P alkaline cyanide zinc bath at a high current density effective additive. Galvanizing bath of substantially 15wt% 1.5wt% sodium hydroxide and zinc oxide are consisting of high current density (100mA cm^{-2}). In the absence of any additives added condition that the deposition of zinc can be seen clearly only a very rough surface.

Figure 1 shows the alkaline zinc coating in the 1M sulfuric acid by the Line Scan voltammetry (LSV) detection, scanning potential range from -2V scan to 3V, scanning rate of 1mV s-1, 1 M sulfuric acid in this experiment is mainly used to test alkaline zinc coating. Accordingly, the open circuit potential of the alkaline zinc coating is -0.8V, and the passivation region is between 0.2 and 0.8V with an etching current density of about 0.0001 A cm^{-2} when the corrosion potential is adjusted from 1V to 3V. Moreover, the corrosion current density increases by ten times to 0.001A cm^{-2}. Hence, it could be judged that the corrosion potential is more than 1V, the surface structure of the alkaline zinc coating began to loosen. When the corrosion potential increased up to 3 V, the structure is damaged by ten times the speed. According to the Fig.1 result, the alkaline zinc coating is strong enough to protect sulfuric acid.

The anti-corrosion ability of a zinc coating is affected by a combination of characteristics, including current density, and additives. In this study, the zinc coatings were prepared in the different plating period with (a) 5 min, (b) 10 min and (c) 20 min during a plating process with different plating period. The anti-corrosion abilities of those oxide coatings were examined in a 5 wt.% NaCl solution for the potentiodynamic polarization test. In addition, the polarization curves of the oxide coatings are shown in Fig. 2.

The zinc coating in the electrolyte with additive contained a passive potential range from -1800 to -1200 mV (see curve c with 20 min period). Moreover, the corrosion current density of the coated metal in the electrolyte with additive was 1.2 × 10^{-6} A/cm^2. Accordingly, the right additive can structure a dense, uniform, thick, and high crystalline zinc coating on substrate.

Figure 3 presented the impedance of the bath resistance of an alkaline zinc plating at different current densities. With the increase in current from 1A cm^{-2} to 3A cm^{-2}, the resistance is reduced from 3.3 ohms to 1.6 ohms. The overall electroplating solution becomes low impedance, which is due to Lugalvan P complexing agent at a low density of zinc ions active malocclusion before alkali zinc plating solution, and when the working

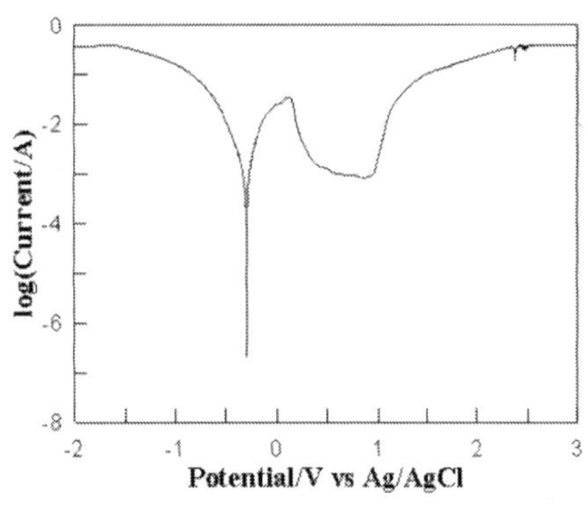

Fig. 1. LSV examination for zinc plating in 1M H_2SO_4 solution.

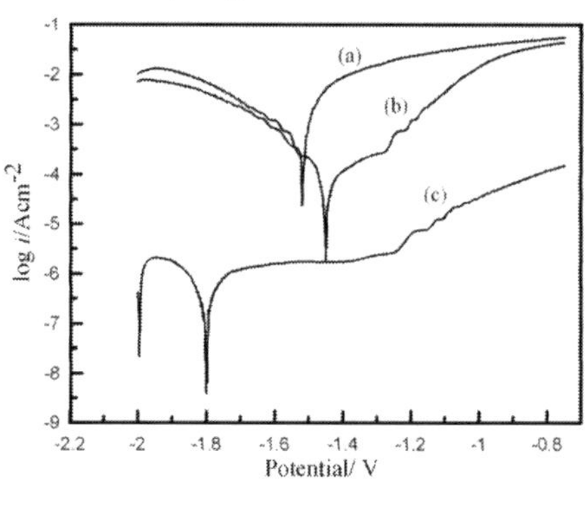

Fig. 2. A zinc coting is prepared in the different plating period with (a) 5 min, (b) 10 min and (c) 20 min.

DOI:10.1051/matecconf/201712300024

electrodes as the current density increases, the adsorption of wrong engagement Lugalvan P zinc ions and the like are also fold increase in the rate of release, can be seen Lugalvan P suitable for alkaline zinc plating bath known as leveling agent.

Figure 4 shows the impedance of the zinc plating thickness at different current densities. The data presented in the study were deposited in an alkaline bath with different current densities of 1, 1.5, 2, 2.5, and 3A cm^{-2} in a Hull cell. The data found by the Hatschek test, the current position at a high current density areas tend to be stable uniform thickness of 0.3 ~ 0.4 μm. Thus, the Lugalvan P complexing agent system acts as an effective smooth zinc coating having a consistent performance.

The SEM images (Fig. 5) show the surface morphology of the high-conductivity zone of the Hull Cell substrate at different current densities of 1 A, 1.5 A,

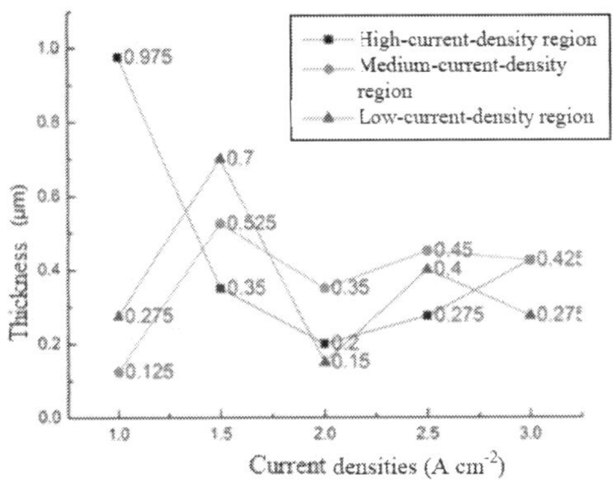

Fig. 3. Relative solution resistance of the Zinc coatings in accordance with current densites during 5 mins.

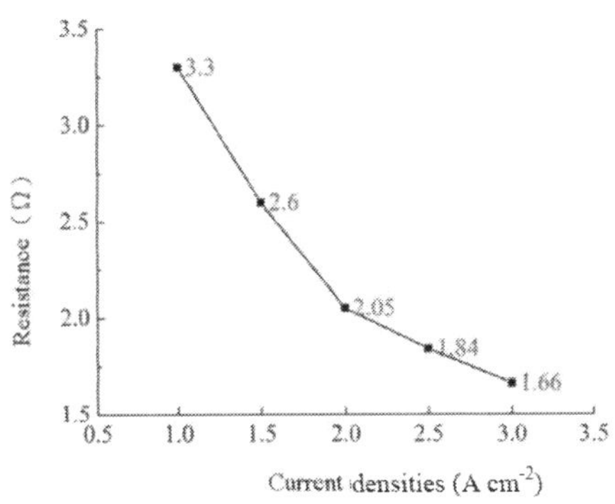

Fig. 4. Relative thickness of the Zinc coatings in accordance with current densites during 5 mins.

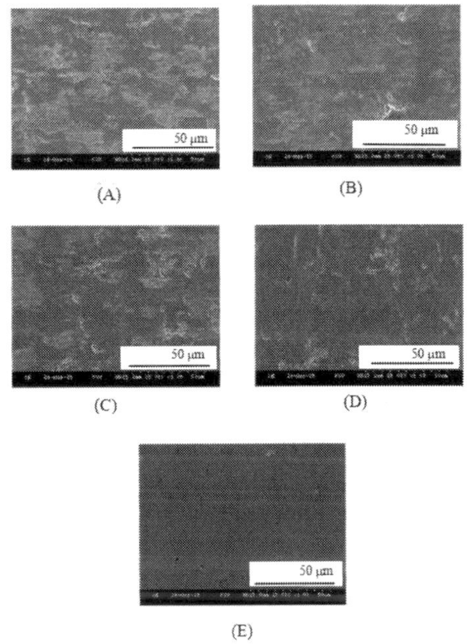

Fig. 5. The SEM morphologies of Zinc coatings are prepared at different current densities of (A) 1, (B) 1.5, (C) 2, (D) 2.5 and (E) 3 A cm^{-2} at high-density region.

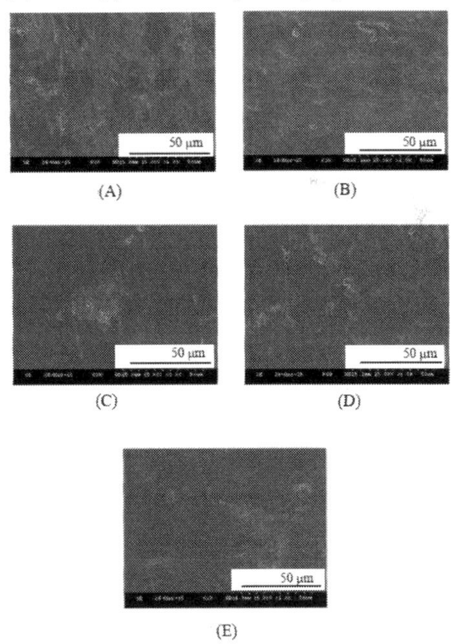

Fig. 6. The SEM morphologies of Zinc coatings are prepared at different current densities of (A) 1, (B) 1.5, (C) 2, (D) 2.5 and (E) 3 A cm^{-2} at high-density region.

2 A, 2.5 A and 3 A. The brightness corresponds to the different current density formed on the Hull Cell sheet. With Lugalvan P addition, high-conductivity zone of the zinc coating cathodes exhibit excellent brightness at high current density of 3A. The uneven pores are relatively smaller and less.

The surface morphology images of the alkaline zinc plating of the medium-conductivity zone of the Hull Cell substrate at different current densities are shown in Fig. 6. It is found that the flatness of the plating surface is the worst, and it is clear that there are a large number of uneven pores formed. As the current densities increases to an relatively high of 2.5A and 3 A, better flatness of zinc coating surfaces could be obtained, as sown in Fig. 6(D) and 6(E), respectively. However, compared with the surface flatness at the high-conductivity zone, the pores are unfortunately larger in the medium-conductivity zone at the same current densities.

Figure 7 shows basic zinc coated surface morphology at low-conductive region of the Hatschek sheets at different current densities. It was found that Lugalvan P plating solution in alkaline systems, flatness at the low density region behave of poor conductivity. The existing large pores were highly appeared, resulted the uneven surface.

A smoothness and gloss deposited surface could be obtained at a high electroplating efficiency of current intensity of 3A (in Fig. 5(E)) of the high-conductivity zone of the Hull Cell substrate. The higher the charge density of the additive, the physical stability adsorbing electrode, is formed. The Lugalvan P additive forms the relatively highest flatness, which inhibited the dendritic structure.

Figure 8 shows the X-ray diffraction results at high-current zone, medium-current zone and low-current zone of the zinc coating of current intensity of 3A during 5 mins. The mainly Fe (110), Zn (101), and Fe (200) peaks are appeared, and no other crystal phases were found. The anisaldehyde and the zinc ions complexes effectively, inhibiting the other crystal phases formatted during electroplating.

The results show that the type of complexing agent will be affected with the plating pattern to the coating surface morphology and crystal lattice arrangement, the anodic dissolution behaviour of the zinc coating by the addition of zinc salt anion and the effect of additives, but also by the current density and the plating pattern influences. Thus, an additive to Lugivam P having a higher flatness, and suppression of hydrogen generation with a dendritic structure of the plating.

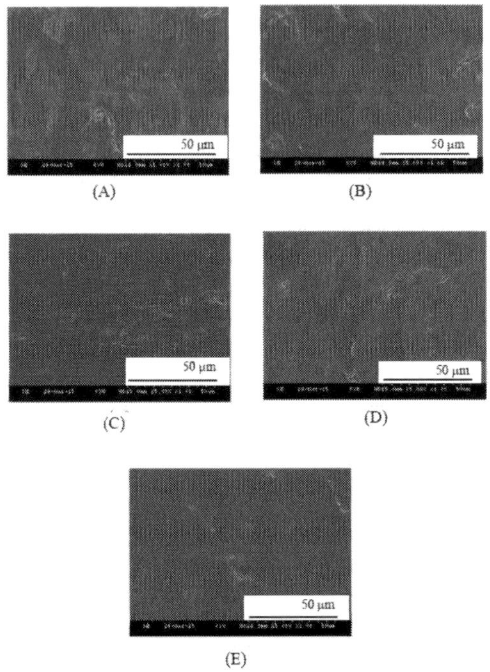

Fig. 7. The SEM morphologies of Zinc coatings are prepared at different current densities of (A) 1, (B) 1.5, (C) 2, (D) 2.5 and (E) 3 A cm^{-2} at high-density region.

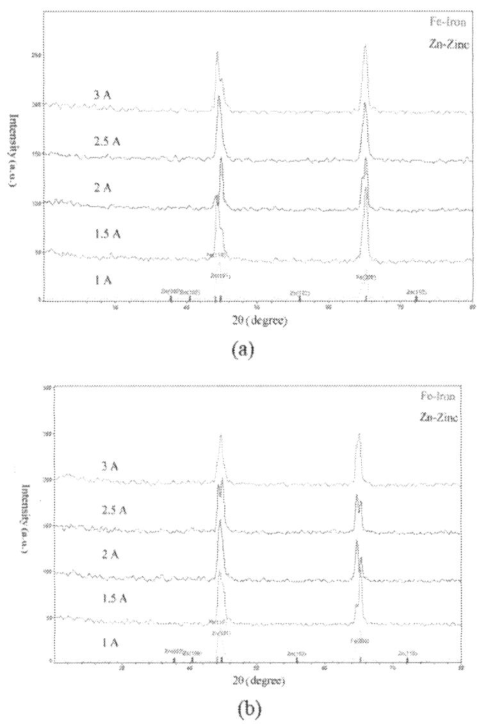

DOI:10.1051/matecconf/201712300024

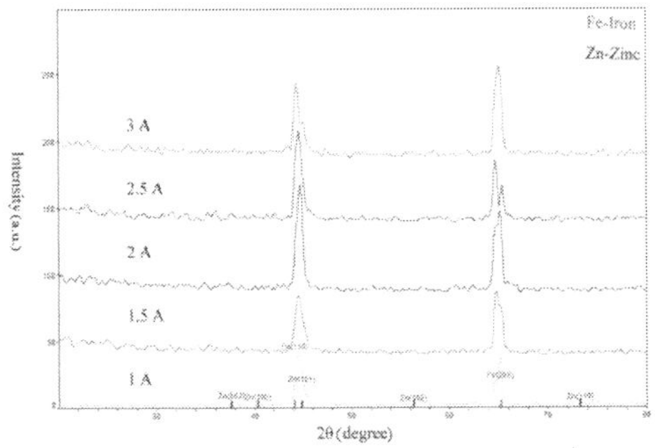

Fig. 8. The XRD pattens of Zinc coatings at different current densities (1, 1.5, 2, 2.5, and 3A cm^{-2}) during 5 mins deposited at (a) high-current zone at (b) medium-current zone and (c) low-current zone.

(c)

5 Conclusions

Polymeric additives are strongly affected the galvanized zinc coating quality. The higher the charge density of the additive, the physical stability adsorbing electrode is formed. Such that steric hindrance phenomenon in suppressing zinc coating unevenness is generated. After testing, the optimal current density was found of 3A cm^{-2}, which could perform a relatively high plating efficiency to obtain a dense and smooth plating surface.

When quaternized Pyridine carboxylic acid inner salt is added to the system containing Lugalvan P, the original granular form zinc aggregates may be planarized. Thereby that between the quaternized polyamines has a strong additive effect of the addition of zinc alkaline cyanide bath, can effectively improve the flatness and grain boundary structure of the zinc coating.

References

1. R. Ramanauskas, *Appl. Surf. Sci.*, **153**, 53 (1999)
2. H. Kim, B.N. Popov, K.S. Chen, *Corros. Sci.*, **45**, 1505 (2003)
3. B. Veeraraghavan, B. Haran, S. Prabhu, B. Popov, *J. Electrochem. Soc.*, **150**, B131 (2003).
4. N. M. Martyak, J. E. McCaskie, U.S. Pat. 6585812 (2003)
5. R. Albalat, E. Gomez, C. Muller, J. Pregonas, M. Sarret, E. Valles, *J. Appl. Electrochem.*, **20**, 635 (1990)
6. R. Fratesi, G. Roventi, *J. Appl. Electrochem.*, **22**, 657 (1992)
7. G. Barcelo, J. Garcia, M. Sarret, C. Müller, J. Pregonas, *J. Appl. Electrochem.*, **26**, 1249 (1994)
8. N. M. Martyak, J. E. McCaskie, U.S. Pat. 5718818 (1998)
9. F. Passal, U.S. Pat. 4014761 (1977)
10. W. E. Rosenberg, U.S. Pat. 4251331 (1981)
11. C. W. Skimin, U.S. Pat. 4405413 (1983)
12. J. D. Fellman, H. L. Kaufman, U.S. Pat. 4218292 (1980)
13. C. W. McFarland, U.S. Pat. 4146442 (1979)
14. K. Glaser, W. Streit, R. Fikentscher, G. Gotsmann, U.S. Pat. 4178217 (1979)
15. C. V. Bishop, U.S. Pat. 4210500 (1980).
16. R. K. Lowery, T. W. Starinshak, U.S. Pat. 4229268(1980)
17. R. J. Lash, R. W. Herr, U.S. Pat. 4366036(1982)
18. C. Steinecker, U.S. Pat. 4444630(1984)
19. W. J. Willis, N. Royalton, U.S Pat. 4730022(1988)
20. J. D. Fellman, H.L. Kaufman, U.S. Pat. 4218292(1980)
21. C. Muiller, M. Sarret, M. Benballa, J. of Electroanal. Chem., 519, 85 (2002)

A robust and fast control technology of AC power conditioning for high-speed micromachining

En-Chih Chang*

Department of Electrical Engineering, I-Shou University, Kaohsiung City, Taiwan, R.O.C.

Abstract. In this paper, a robust and fast control technology is used to AC power conditioning, thus increasing the performance of the high-speed micromachining. The robust and fast control technology is made up of a robust sliding function (RSF) and a computationally fast grey forecasting model (GFM). The RSF without singularity problem admits system state converged to zero within finite time so that the output-voltage with low harmonic distortion in AC power conditioning is obtained. Nevertheless, while a severe non-linear loading is applied to high-speed micromachining, the needless chattering around robust sliding function occurs. The chattering results in thermal breakdown and serious voltage distortion in AC power conditioning output, and the reliability and stability of the high-speed micromachining will be worsened. Therefore, the GFM with Fourier series is introduced as a computationally fast and algorithmically easy means of removing the chattering existing in RSF when the system uncertainty bound is overestimated. By using this presented control technology, the AC power conditioning provides a high-quality AC output-voltage with accurate steady state and fast transience under various loading conditions, thus obtaining the excellent reliability and stability of the high-speed micromachining. Experimental results are performed in support of the proposed control technology.

1 Introduction

The AC power conditioning has widely been applied in high-speed micromachining [1-4]. High performance AC power conditioning must supply the output AC voltage to the reference sinusoidal with low total harmonics distortion (THD) and fast transient response. To minimize THD, some control schemes have been proposed for AC power conditioning. PI control is frequently used in industry due to simple control structure and ease of design. But, PID control cannot give good control performance as the controlled plant is severe non-linear and uncertain [5], [6]. Deadbeat control can improve the shortcoming of the PI control, but it is highly dependent on the accuracy of the parameters [7]. The repetitive control and mu-synthesis approach can overcome system uncertainties. However, they have implementation difficult and algorithm complexity [8], [9]. Sliding mode control (SMC) is being given attention due to its robustness characteristics [10-13]. A number of SMC associated with AC power conditioning have been reported [14-18]. A fixed switching frequency sliding mode controlled AC power conditioning is presented in [14]. However, the control design uses a typical SMC and cumbersome analog implementation, and thus incurs distorted output voltage during steady-state operation with a non-linear load. In [15], the multiple-sliding-surface is suggested to improve the incomplete system dynamics of classic sliding surface. Though the system performance is improved, the proposed methodology has time-consuming operation in algorithms. The control scheme based on fixed-frequency SMC has also been applied to the design of grid-connected AC power conditioning. In this case, the resulting output voltage makes a concession between steady-state and transience [16]. Reference [17] employs an integral SMC law to achieve AC power conditioning. But, the system trajectory could not hit the desired sliding surface fast and accurately. The noticeable distortion exists in the output voltage waveform. A modified SMC with the elimination of the disturbances for AC power conditioning is developed by [18]; this technology has complicated hardware design and a chattering problem. As mentioned by [14-18], linear sliding surface is adopted. Its characteristic is that the system tracking error converges to zero asymptotically. A robust sliding function (RSF), which employs nonlinear sliding surface is developed instead of linear sliding surface. Compared with linear sliding-surface-based control, the RSF can drive the system tracking error to converge to zero in finite time and there is no singular problem [19-23]. From the point of view in practical high-speed micromachining application, if the load disturbance is a severe non-linear condition, the chattering around RSF occurs. The chattering leads to thermal breakdown and serious voltage distortion in AC power conditioning output, thus deteriorating the reliability and stability of the high-speed micromachining. A computationally fast grey forecasting model (GFM) with Fourier series is employed to describe and analyze the future trend of sequence numbers according to the past and nowadays data for

* Corresponding author: enchihchang@isu.edu.tw

dynamic system. The GFM improves the accuracy of the basic grey forecasting model through the Fourier series, and has been successfully applied in many areas of engineering [24-28]. Thus, a mathematically simple and accurately forecasted GFM is employed to eliminate the chattering while the system uncertainty bounds are overestimated. Combining RSF with GFM, the proposed control technology yields a closed-loop AC power conditioning with low total harmonic distortion and fast transience under different types of loading, thus increasing the performance of the high-speed micromachining. Experimental are shown to certify the performance of the proposed control technology.

2 System description

Fig. 1 shows a commonly used AC power conditioning. The $L_f C_f$ filter and resistive load R can be regarded as a plant. Choosing the state variables $x_1 = v_o$ and $x_2 = \dot{v}_o$, where v_o is the output voltage and $\dot{v}_o$ is its derivative, the state equation yields

$$\begin{cases} \dot{x}_1 = x_2 \\ \dot{x}_2 = -a_1 x_1 - a_2 x_2 + b_1 u \end{cases} \tag{1}$$

where $a_1 = 1/L_f C_f$, $a_2 = 1/RC_f$, $b_1 = K_{pm}/L_f C_f$ and u is the control signal. Owing to the unpredictability of the load condition, the parameters a_2 is almost impossible to know exactly, however to have reasonable approximations, the following assumptions are made as $a_2 = \bar{a}_2 + \Delta a_2$, where the $\bar{a}_2$ represents nominal parameters of the system, and Δa_2 is parameter uncertainties. The bound of the a_2 can be given by $a_{2\min} \le a_2 \le a_{2\max}$. If the switching frequency is much higher than the fundamental frequency of the AC output, K_{pm} is regarded as a proportional gain of a pulse-width modulation (PWM) full-bridge converter and equals $V_s/\hat{v}_{tr}$; $\hat{v}_{tr}$ is the amplitude of the triangular wave v_{tr} in the PWM. Once u is determined, by comparing u with v_{tr}, the PWM gating signals is produced and controls the power switches. Based on the state-space averaging and linearization technique, the product of u and K_{pm} is equivalent to output voltage v_i of the full-bridge converter. In equation (1), the output voltage v_o is desired to be maintained as close as possible to a sinusoidal reference voltage v_{re}. Generally, $v_{re}(t) = \sqrt{2} \cdot V_{rms} \cdot \sin(2\pi f_r \cdot t)$ in which V_{rms} and f_r are the root-mean-square and frequency values of the desired sinusoidal reference voltage, respectively. Therefore, the tracking requirement $v_o(t) \to v_{re}(t)$, $t \to \infty$ must be maintained, the design problem of the AC power conditioning can be regarded as a path-following control problem.

Define the tracking errors as

$$\begin{cases} e_1 = x_1 - v_{re} \\ e_2 = x_2 - \dot{v}_{re} \end{cases} \tag{2}$$

From (1) and (2), the error state equations of the AC power conditioning can be derived as

$$\begin{cases} \dot{e}_1 = e_2 \\ \dot{e}_2 = -a_1 e_1 - a_2 e_2 + b_1 u - f \end{cases} \tag{3}$$

where $f = a_1 v_{re} + a_2 \dot{v}_{re} + \ddot{v}_{re}$ is the disturbance.

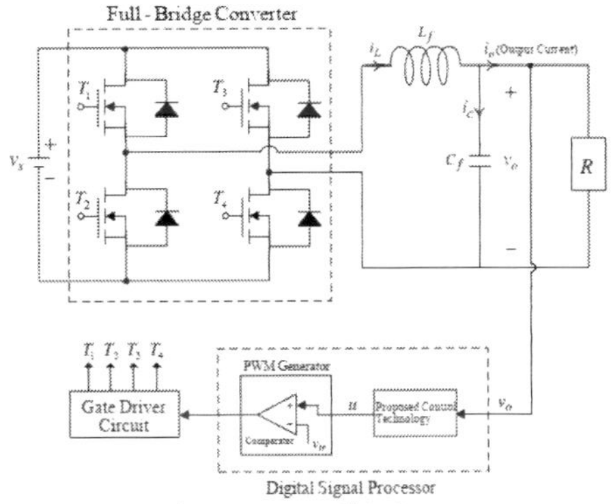

Fig. 1. Structure of AC power conditioning.

As can be seen from (3), the control signal u must be designed so that e_1 and e_2 can be converged to zero. This paper proposes a RSF by analyzing Lyapunov stability criterion and designing RSF parameters carefully. The RSF will drive the system tracking error to converge to zero within finite time and is without singular problem, thus the closed-loop stability of the RSF can be guaranteed. However the loading is not fixed, the RSF system easily has the chattering problem and may present a high THD, especially when the loading is a large step change or an uncertainty or even a severe non-linear condition. Therefore, the GFM is employed to eliminate the chattering for producing higher-performance AC output voltage and the control design is represented in the following. The design concept of this proposed control technology is to modify the classic SMC by introducing nonsingularity criterion and GFM, so as to resolve infinite-time convergence and chattering.

3 Control technology design

For the error dynamics (3), the robust sliding function is defined as

$$\sigma(t) = e_1(t) + \frac{1}{\varepsilon} e_2^{\frac{p}{q}}(t) \tag{4}$$

where $\varepsilon > 0$, $p > q$, p and q are positive odd numbers ($1 < p/q < 2$), and a sliding-mode reaching equation $\dot{\sigma} = -k_1|\sigma|^{\alpha} sign(\sigma) - k_2|\sigma|^{\beta} sign(\sigma)$ is constructed.

Then, the control law u can be expressed as

$$u(t) = u_e(t) + u_s(t) \tag{5}$$

with

$$u_e(t) = ^{-1}[a_1e_1 + a_2e_2 - \varepsilon\frac{q}{p}e_2^{2-\frac{p}{q}}] \quad (6)$$

$$u_s(t) = -b^{-1}\left[k_1|\sigma|^{\alpha} sign(\sigma) + k_2|\sigma|^{\beta} sign(\sigma)\right] \quad (7)$$

where $k_1, k_2 > 0$, $1 > \alpha$ and $\beta > 0$, u_e denotes the equivalent control with non-singularity, and u_s displays the sliding control for compensating the perturbation influences. Thus, the system will be driven to the sliding mode $\sigma = 0$ and converged within finite time, and the perturbation $f(t)$ is bounded as $|f(t)| < \psi$, $\forall t \geq 0$.

Proof: Let us use the following Lyapunov candidate function:

$$V = 0.5 \cdot \sigma^2 \quad (8)$$

Along the trajectory of the dynamic system (3) with the control law (5), and using (4), the time derivative of V is given by

$$\begin{aligned}\dot{V} &= \sigma\dot{\sigma} \\ &= \sigma\left(\dot{e}_1 + \frac{1}{\varepsilon}\frac{p}{q}e_2^{\frac{p}{q}-1}\dot{e}_2\right) \\ &\leq -\sigma\left(\frac{1}{\varepsilon}\frac{p}{q}e_2^{\frac{p}{q}-1} + k_1|\sigma|^{\alpha} sign(\sigma) + k_2|\sigma|^{\beta} sign(\sigma)\right)\end{aligned} \quad (9)$$

Since $e_2^{\frac{p}{q}-1} > 0$ holds, $\dot{V} \leq 0$, which infers that the FCSMC manifold σ in (4) converges to zero within finite time. On the other hand, when $\sigma = e_1 + \frac{1}{\varepsilon}e_2^{\frac{p}{q}}$ is reached, then the states of the system (3) will converge to zero within finite time. However, FCSMC has chattering in AC power conditioning system design. This is because of the changeable load, so once the loading is a severe uncertain condition, the system (3) will not provide accurate tracking performance. Thus, the control signal $u(t)$ (5) is modified by the addition of the Fourier modified grey control (u_{gfm}), which eliminates chattering in AC power conditioning system. The modeling steps of the GFM are described below.

Step 1: Input the original sample data sequence

Letting the original data sequence be denoted as

$$x^{(0)} = \{x^{(0)}(k),\ \ k = 1, 2, \cdots, n\} \quad (10)$$

where $x^{(0)}$ stands for the set of n original sample data.

Step 2: Accumulated generating operation (AGO)

By taking the AGO on $x^{(0)}$, the following first-order AGO sequence is expressed as

$$x^{(1)} = \left\{\sum_{i=1}^{1} x^{(0)}(i), \sum_{i=1}^{2} x^{(0)}(i), \ldots, \sum_{i=1}^{k} x^{(0)}(i)\right\}, k = 1, 2, \cdots, n \quad (11)$$

Step 3: Grey model

Based on the accumulated data sequence, $x^{(1)}$, a first-order ordinary differential grey model, GM(1,1) is formed as

$$\frac{dx^{(1)}}{dt} + ax^{(1)} = b \quad (12)$$

where a and b stand for model coefficients, and need to be decided.

By employing MEAN generating operation to $x^{(1)}$, namely $\mathrm{MEAN}(x^{(1)}) = (x^1(k+1) + x^1(k))/2$, the (12) is restated as

$$x^{(0)}(k+1) = \begin{bmatrix} -0.5(x^{(1)}(k+1) + x^{(1)}(k)) & 1 \end{bmatrix}\begin{bmatrix} a \\ b \end{bmatrix} \quad (13)$$

where $x^{(0)}(k+1) = x^{(1)}(k+1) - x^{(1)}(k)$.

Letting $k = 1, 2, \cdots, n-1$, and (13) is represented as

$$\begin{bmatrix} x^{(0)}(2) \\ x^{(0)}(3) \\ \vdots \\ x^{(0)}(n) \end{bmatrix} = \begin{bmatrix} -0.5(x^{(1)}(2) + x^{(1)}(1)) & 1 \\ -0.5(x^{(1)}(3) + x^{(1)}(2)) & 1 \\ \vdots & \vdots \\ -0.5(x^{(1)}(n) + x^{(1)}(n-1)) & 1 \end{bmatrix}\begin{bmatrix} a \\ b \end{bmatrix} \quad (14)$$

The following matrices can be presumed as

$$M = \begin{bmatrix} x^{(0)}(2) \\ x^{(0)}(3) \\ \vdots \\ x^{(0)}(n) \end{bmatrix}, \quad N = \begin{bmatrix} -0.5(x^{(1)}(2) + x^{(1)}(1)) & 1 \\ -0.5(x^{(1)}(3) + x^{(1)}(2)) & 1 \\ \vdots & \vdots \\ -0.5(x^{(1)}(n) + x^{(1)}(n-1)) & 1 \end{bmatrix},$$

$$\Psi = \begin{bmatrix} a \\ b \end{bmatrix}.$$

We can solve the estimated parameters Ψ via means of least square method below.

$$\Psi = \begin{bmatrix} a \\ b \end{bmatrix} = (N^T N)^{-1} N^T M \quad (15)$$

To solve the grey differential equation, substituting (15) into (12), and the forecast output is computed as

$$\hat{x}^{(1)}(k+1) = \left(x^{(0)}(1) - \frac{b}{a}\right)e^{-ak} + \frac{b}{a} \quad (16)$$

where '^' denotes forecasted value, and $\hat{x}^{(1)}(k+1)$ symbols the approximation solution of differential equation in (16).

Step 4: Inverse accumulated generating operation (IAGO)

By the use of the IAGO, the data sequence $\hat{x}^{(0)}(k)$ can be estimated as

$$\begin{aligned}\hat{x}^{(0)}(k) &= \hat{x}^{(1)}(k) - \hat{x}^{(1)}(k-1) \\ &= (1 - e^{a})\left(x^{(0)}(1) - \frac{b}{a}\right)e^{-a(k-1)}\end{aligned} \quad (17)$$

Let $k = 1, 2, \ldots, n$, the forecasted value yields

$$\hat{x}^{(0)} = \{\hat{x}^{(0)}(1), \hat{x}^{(0)}(2), \ldots, \hat{x}^{(0)}(n)\} \quad (18)$$

DOI:10.1051/matecconf/201712300025

To improve the accuracy of forecasting models, the Fourier series is used in modifying the residuals in grey forecasting model GM(1,1) so that a Fourier modified grey model (FGM(1,1)) can be obtained.

Step 5: Get the residual series from GM(1,1)

Based on the forecasted series, a residual series is defined as

$$\varepsilon_r = \{\varepsilon_r(2), \varepsilon_r(3), \ldots, \varepsilon_r(n)\}^T \tag{19}$$

where $\varepsilon_r(k) = x(k) - \hat{x}(k)$.

Step 6: Defining Fourier modified grey model (FGM(1,1))

The Fourier series can approximate the residual series as

$$\varepsilon_r = \frac{1}{2}a_0 + \sum_{i=1}^{k_f}[a_i \cos(\frac{i \cdot 2\pi}{T}k) + b_i \sin(\frac{i \cdot 2\pi}{T}k)] \tag{20}$$

where $k = 2, 3, \ldots, n$, $T = n-1$ and $k_f = (\frac{n-1}{2}) - 1$.

Therefore, the residual series is restated as

$$\varepsilon_r = PC \tag{21}$$

where $P = \begin{bmatrix} \frac{1}{2} & \cos(\frac{2\pi \cdot 1}{T} \cdot 2) & \sin(\frac{2\pi \cdot 1}{T} \cdot 2) & \cdots & \cos(\frac{2\pi \cdot k_f}{T} \cdot 2) & \sin(\frac{2\pi \cdot k_f}{T} \cdot n) \\ \frac{1}{2} & \cos(\frac{2\pi \cdot 1}{T} \cdot 3) & \sin(\frac{2\pi \cdot 1}{T} \cdot 2) & \cdots & \cos(\frac{2\pi \cdot k_f}{T} \cdot 3) & \sin(\frac{2\pi \cdot k_f}{T} \cdot n) \\ \vdots & \vdots & \vdots & \vdots & \vdots & \vdots \\ \frac{1}{2} & \cos(\frac{2\pi \cdot 1}{T} \cdot n) & \sin(\frac{2\pi \cdot 1}{T} \cdot 2) & \cdots & \cos(\frac{2\pi \cdot k_f}{T} \cdot n) & \sin(\frac{2\pi \cdot k_f}{T} \cdot n) \end{bmatrix}$

and $C = [a_0, a_1, b_1, a_2, b_2, \ldots, a_{k_f}, b_{k_f}]^T$.

The parameters $a_0, a_1, b_1, a_2, b_2, \ldots, a_{k_f}, b_{k_f}$ are obtained through the least squares method as follows:

$$C = (P^T P)^{-1} P^T \varepsilon_r \tag{22}$$

While the parameters are computed, the modified residual series is fulfilled as

$$\hat{\varepsilon}_r = \frac{1}{2}a_0 + \sum_{i=1}^{k_f}[a_i \cos(\frac{i \cdot 2\pi}{T}k) + b_i \sin(\frac{i \cdot 2\pi}{T}k)] \tag{23}$$

Step 7: Correct original forecast series

The original forecast series of FGM (x_f) can be corrected as

$$\hat{x}_f^{(0)}(k) = \hat{x}^{(0)}(k) + \hat{\varepsilon}_r^{(0)}(k), \quad k = 2, 3, \ldots, n \tag{24}$$

Therefore, the control law of (5) is rewritten as

$$u(k) = u_e(k) + u_s(k) + u_{gfm}(k) \tag{25}$$

where the added compensation component is Fourier modified grey control, u_{gfm} that can eliminate the chattering.

$$u_{gfm}(k) = \begin{cases} 0 & , \quad |\hat{\sigma}(k)| < \delta \\ K\hat{\sigma}(k) sign(\sigma(k)\hat{\sigma}(k)) & , \quad |\hat{\sigma}(k)| \geq \delta \end{cases} \tag{26}$$

where $\hat{\sigma}(k)$ represents for the forecasted value of $\sigma(k)$. K is a constant, and δ symbols the system boundary.

4 Experimental results

To evaluate the performance of the proposed control technology, the results of the proposed control technology are compared with the results of the classic SMC. The system parameters are listed in Table 1.

Table 1. Parameters of the AC power conditioning.

Filter inductor	L=0.2 mH
Filter capacitor	C=5 μF
DC supply voltage	V_s=200 V
Full resistive load	R=12 Ω
Output voltage and frequency	v_o=110 V_{rms}, f_o=60 Hz
Switching frequency	f_{sw} =15 kHz

Figure 2 and Fig. 3 show the output voltage and the load current of the AC power conditioning with the proposed control technology and the classic SMC, respectively, under full resistive load. Their output-voltages are close to sinusoidal waveforms. In order to verify the control technology under transient circumstances, step load change with linear resistive load is explored. Figure 4 shows the waveform obtained using the proposed control technology under step load change from no load to full load. Note that the transient behaviour is satisfactory, i.e., the output voltage dip is small and the recovery time is very speedy. On the contrary, the waveform obtained using the classic SMC, displayed in Fig. 5 has a significant voltage dip and a slow recovery time at the firing angle. Under rectifier load shown in Fig. 6, the output-voltage waveform with the proposed control technology is almost sinusoidal (%THD=1.72%), but that with the classic SMC shown in Fig. 7 has a high %THD of 10.51%. Also, Fig. 8 plots the error convergence time and it clearly demonstrates that the proposed control technology does give for reaching $\sigma(k) = 0$ in finite time and therefore reduces the distortion of the waveform.

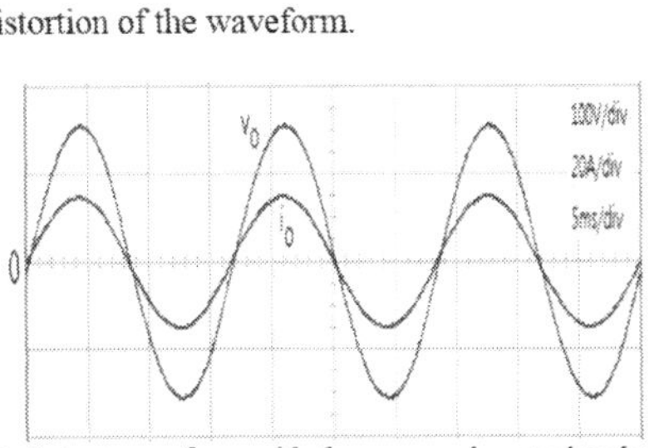

Fig. 2. Output waveform with the proposed control technology under full resistive load.

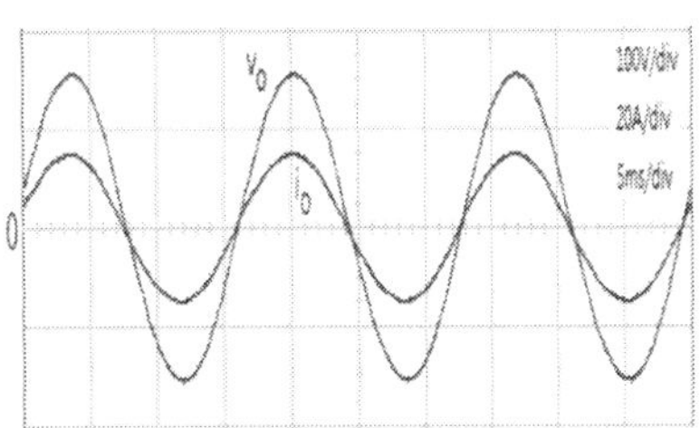

Fig. 3. Output waveform with the classic SMC under full resistive load.

DOI:10.1051/matecconf/201712300025

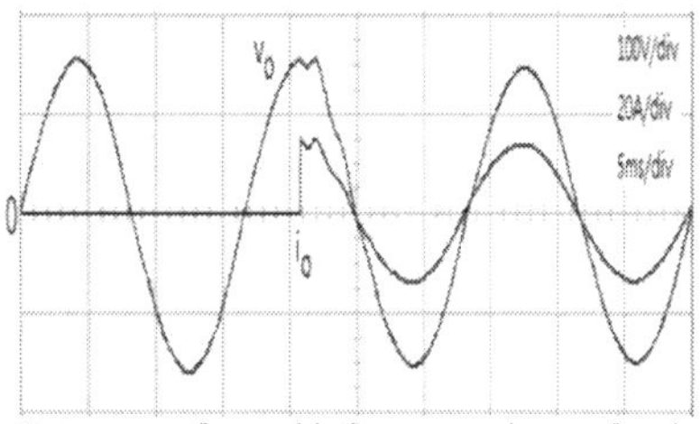

Fig. 4. Output waveform with the proposed control technology under step load change.

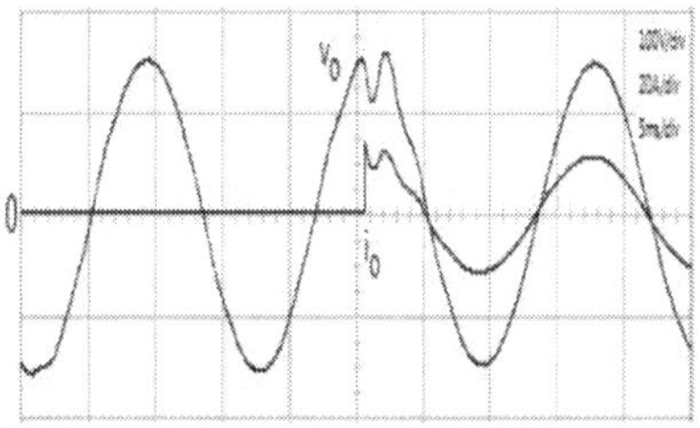

Fig. 5. Output waveform with the classic SMC under step load change.

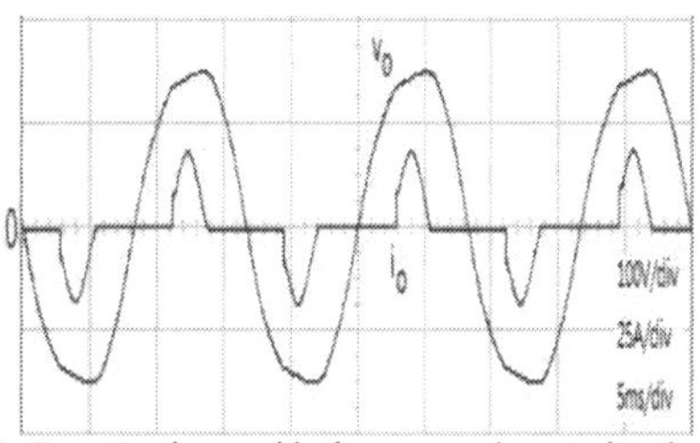

Fig. 6. Output voltage with the proposed control technology under rectifier load.

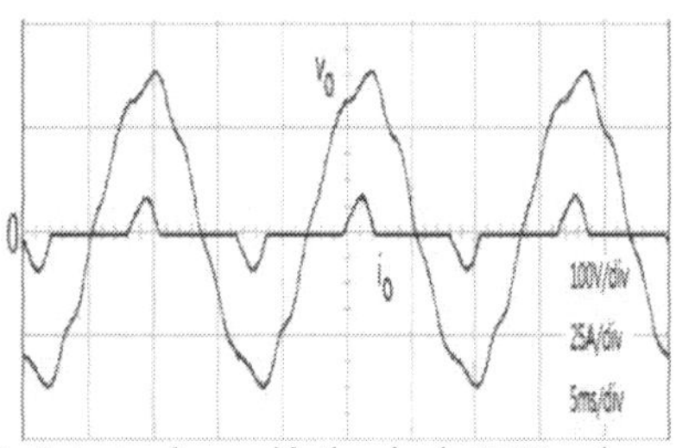

Fig. 7. Output voltage with the classic SMC under rectifier load.

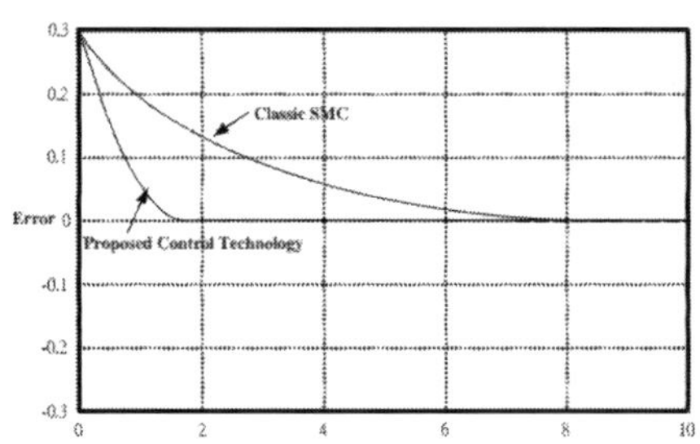

Fig. 8. Error convergence time.

5 Conclusions

This paper describes a high-performance AC power conditioning controlled high-speed micromachining by associating RSF with GFM. Classic SMC is intrinsically robust against internal parameter variations and external disturbances, but it will undergo infinite system-state convergence time. The RSF guarantees finite system-state convergence time and is singularity-free. But, while the system uncertainty bounds are overestimated, the chattering may occur. A computationally fast and algorithmically easy GFM is employed to resolve the chattering problem. Experimental results display that THD, transient response and chattering elimination results from an AC power conditioning under the presented system exceed the results achieved under the SMC system with various loads.

Acknowledgment

The authors gratefully acknowledge the financial support of the 2017 ISU Research Project, under contract number ISU-106-01-01A.

References

1. F.L. Luo, H. Ye, *Power Electronics: Advanced Conversion Technologies* (CRC Press, Boca Raton, FL, USA, 2010)
2. N. Mohan, T.M. Undeland, W.P. Robbins, *Power Electronics: Converters, Applications, and Design* (Wiley, New York, USA, 2003)
3. B.M. Wilamowski, J.D. Irwin, *Power Electronics and Motor Drives* (CRC Press, Boca Raton, FL, USA, 2011)
4. D.W. Hart, *Power Electronics* (McGraw-Hill, New York, NY, USA, 2010)
5. R.S. Rebeiro, M.N. Uddin, Performance Analysis of an FLC-Based Online Adaptation of Both Hysteresis and PI Controllers for IPMSM Drive, *IEEE Trans. on Industry Applications* **48**, 1, pp. 12-19 (2012)
6. I. Rasoanarivo, S. Brechet, A. Battiston, B. Nahid-Mobarakeh, Behavioral Analysis of a Boost Converter with High Performance Source Filter and a Fractional-Order PID Controller, *Proc. IEEE Int. Conf. Industry Applications Society Annual Meeting*, pp. 1-6 (2012).
7. Y. A.-R. I. Mohamed, M. A-Rahman, R. Seethapathy, Robust Line-Voltage Sensorless Control and Synchronization of LCL-Filtered Distributed Generation Inverters for High Power Quality Grid Connection, *IEEE Trans. on Power Electronics* **27**, 1, pp. 87-98 (2012)
8. S. Dasgupta, S.K. Sahoo, S.K. Panda, G. Amaratunga, Single-Phase Inverter-Control Techniques for Interfacing Renewable Energy Sources With Microgrid—Part II: Series-Connected Inverter Topology to Mitigate Voltage-Related Problems Along With Active Power Flow

Control, *IEEE Trans. on Power Electronics* **26**, 3, pp. 732-746 (2011)
9. T.S. Lee, K.S. Tzeng, M.S. Chong, Robust Controller Design for a Single-Phase UPS Inverter Using μ-Synthesis, *IEE Proc. Electr. Power Appl.* **151**, 3, pp. 334-340 (2004)
10. H. Sira-Ramirez, Sliding Regimes in General Non-Linear Systems: A Relative Degree Approach, *Int. J. Contr.* **50**, 4, pp. 1487-1506 (1989)
11. Y. Shtessel, C. Edwards, L. Fridman, A. Levant, *Sliding Mode Control and Observation* (Springer, New York, 2014)
12. J.K. Liu, X.H. Wang, *Advanced Sliding Mode Control for Mechanical Systems Design, Analysis and MATLAB Simulation* (Springer-Verlag, Heidelberg, Berlin, 2012)
13. L.G. Wu, P. Shi, X.J. Su, *Sliding mode control of uncertain parameter-switching hybrid systems* (Wiley, United Kingdom, 2014)
14. S.C. Tan, Y.M. Lai, C.K. Tse, Sliding Mode Control of Switching Power Converters: Techniques and Implementation (Taylor & Francis, Boca Raton, FL, 2012)
15. A. Abrishamifar, A.A. Ahmad, M. Mohamadian, Fixed Switching Frequency Sliding Mode Control for Single-Phase Unipolar Inverters, *IEEE Trans. on Power Electronics* **27**, 5, pp. 2507-2514 (2012)
16. X. Hao, X. Yang, T. Liu, L. Huang, W.J. Chen, A Sliding-Mode Controller with Multiresonant Sliding Surface for Single-Phase Grid-Connected VSI with an LCL Filter," *IEEE Trans. on Power Electronics* **28**, 5, pp. 2259-2268 (2013)
17. S.W. Kang, and K.H. Kim, Sliding Mode Harmonic Compensation Strategy for Power Quality Improvement of a Grid-Connected Inverter under Distorted Grid Condition, *IET Power Electronics* **8**, 8, pp. 1461-1472 (2015)
18. R. Aghatehrani, R. Kavasseri, Sensitivity-Analysis-Based Siding Mode Control for Voltage Regulation in Microgrids," *IEEE Trans. on Sustainable Energy*, **4**, 1, pp. 50-57 (2013)
19. S.D. Xu, C.C. Chen, and Z.L. Wu, Study of Nonsingular Fast Terminal Sliding-Mode Fault-Tolerant Control, *IEEE Trans. on Industrial Electronics* **62**, 6, pp. 3906-3913 (2015)
20. Z. Galias, X.H. Yu, Dynamical Behaviors of Discretized Second-Order Terminal Sliding-Mode Control Systems, *IEEE Trans. on Circuits and Systems II: Express Briefs* **59**, 9, pp. 597-601 (2012)
21. A. Mohammadpour, A. Ajami, S.G. Zadeh, M.P. Aghababa, B.M. Mahboubi, Robust Terminal Sliding Mode Power Flow Controller Using Unified Power Flow Controller with Adaptive Observer and Local Measurement, *IET Generation, Transmission & Distribution* **8**, 10, pp. 1712- 1723 (2014)
22. X.H. Yu, M. Efe, *Recent Advances in Sliding Modes: From Control to Intelligent Mechatronics* (Springer, 2015)
23. X.H. Yu, J.X. Xu, *Variable Structure Systems: Towards the 21st Century* (Springer, 2002)
24. S.F. Liu, Y. Lin, *Advances in Grey Systems Research* (Springer-Verlag, Heidelberg, Berlin, 2010)
25. Y.T. Hsu, M.C. Liu, J. Yeh, H.F. Hung, Forecasting the Turning Time of Stock Market Based on Markov-Fourier Grey Model, *Expert Systems with Applications* **36**, 8597-8603 (2009)
26. C.N. Wang, V.T. Phan, An Improved Nonlinear Grey Bernoulli Model Combined with Fourier Series, *Mathematical Problems in Engineering* **2015**, 1-7 (2015)
27. S.F. Liu, Y. Lin, *Grey Information: Theory and Practical Applications* (Springer, 2006)
28. S.F. Liu, Y.J. Yang, J. Forrest, *Grey Data Analysis: Methods, Models and Applications* (Springer, 2017)

The measuring technique developed to evaluate the thermal diffusivity of the multi-layered thin film specimens

Tse-Chang Li[1,*], *Yang-Ru* Li[1], and *Jen-Fin* Lin[1]

[1]Department of Mechanical Engineering, National Cheng Kung University, Tainan 701, Taiwan

Abstract. In the present study, the thermal diffusivities of the Al, Si and ITO films deposited on the SUS304 steel substrate are evaluated via the present technique. Before applying this technique, the temperature for the thin film of the multi-layered specimen is developed theoretically for the one-dimensional steady heat conduction in response to amplitude and frequency of the periodically oscillating temperature imposed by a peltier placed beneath the specimen's substrate. By the thermal-electrical data processing system excluding the lock-in amplifier, the temperature frequency a_3 has been proved first to be independent of the electrical voltage applied to the peltier and the contact position of the thermocouples. The experimental data of phase difference for three kinds of specimen are regressed well by a straight line with a slope. Then, the thermal diffusivity of the thin film is thus determined if the slope value and the film-thickness are available. In the present arrangements for the thermocouples, two thermal diffusivity values are quite close each other and valid for every kind of specimen. This technique can provide an efficient, low-cost method for the thermal diffusivity measurements of thin films.

1 Introduction

Measurement of thermal diffusivity in diversity materials had been studied for various sizes of samples. Recently, much attention had been paid to the technique for the samples with multi-layer thin films. Apart from the intrinsic importance of thermal diffusivity, its determination gives the value of the thermal conductivity (k), if the material density ρ and the thermal capacity at constant pressure (C_p) are available.

A simple apparatus was described in the studies of Asams and Kirkbright [1, 2] that permitted heat-flow characteristics of a sample to be determined, and its application of the thermal diffusivity or thickness of polymer films on a copper substrate was reported. The amplitude and phase of observed optoacoustic signal depend on the thermal diffusivity and thickness of the material irradiated. The use of time-resolved temperature measurement techniques for the evaluation of the thermal diffusivity of material had been documented, and most methods were based on the original work undertaken by Ångström [3] and later modification by King [4] and Abeles et al. [5]. Development of the phase-lag methods for use with thin slabs or discs of material had been reviewed by Touloukian et al. [6].

There had been a few experiments using an ac calorimetric method on thermal diffusivity so far [7, 8]. A method to measure thermal diffusivity of thin sample no matter how thin a sample may be. It was useful in the measurement for thin materials with a variety of thermal diffusivities. A variation of the thermal conductance measurements was described using the dynamic 3ω method [9]. By using the 3ω method, the measurements of temperature oscillation needed to be obtained and calculated. Thus, the experimental data would taken into account the thermal resistance between the sample and the thermocouples and the heat capacity of the contact point of the thermocouples. In order to use this method, a fixed sample geometry should be provided. Simple temperature oscillation techniques were adopted in the study of Czarnetzki and Roetzel [10] to describe for the fast measurement of thermal diffusivity and thermal conductivity of liquids. The thermal diffusivity and conductivity of the specimen were deduced by measuring and evaluating the amplitude attenuation and/or the phase shift between the fundamental temperature oscillations at the surface of the liquid specimen and a well-defined position inside the specimen. The results agreed well with the data obtained from the literature.

An experimental technique, namely the amplitude method, was proposed in the study of Zhang and Grigoropoulos [11], and applied to measurement of the thin film thermal diffusivity. The thermal diffusivity was determined by three independent experimental methods. Good agreement among the measured thermal diffusivities obtained by the three methods confirmed the validity of the amplitude method. The work of Yu et al. [12] investigated and compared three laser-based ac heating methods for the thermal diffusivity measurement of free-standing thin film structures. Both the phase and the amplitude signals of the ac temperature rise of the sample were ultilized to deduce the thermal diffusivity of the thin film structures. The results showed that the

* Corresponding author: TCL92@mail.ncku.edu.tw

uniform illumination method was more susceptible to the error than the other two configurations due to two-dimensional effects. In the study of Jansen and Obermeier [13], an accurate technique to measure the thermal conductivity of the thin films in a wide temperature range was presented. Measurements on polycrystalline diamond films due to the variations of thermal conductivity with temperature had been explained by the effect of the small grain size of the diamond film. Experimental and theoretical studies of heat transport by lattice vibrations in thin film materials were reviewed [14]. Selected examples of experimental methods are discussed to illustrate a variety of methods that have been developed for thin film measurements. Dielectric thin films have significant atomic-scale disorder, and the heat transport is dependent upon this disordered material. The study of Bhusari et al. [15] presented the design of a simple setup for measuring the thermal conductivity of thin films. This method was based on the principle of phase lag of a traveling thermal wave. The phase lag ($\Delta\theta$) between the excitation wave and resulting thermal wave, at a variable distance d from the edge of the sample, was determined by measuring the deflection of another "probe-laser". This method offered an accuracy of better than $\pm 0.5\%$. The thermal diffusivities of thin films of evaporated metal less than 1×10^{-6} m thick had been measured in the study of Yamane et al. [16] using the light-irradiation ac calorimetric method. The thermal diffusivities of metallic thin films become smaller than those of the bulk material as the thickness of the metallic thin films decreased. In the study the Chung et al. [17], the photothermal deflection technique is used to measure the thermal diffusivity of a metal-ceramic decomposite. The Kapitza radius is used as the effective value of the thermal boundary resistance of heterointerfaces. Possible reasons for the data from metal-ceramic composite materials were discussed in terms of the acoustic mismatch model for thermal boundary resistance. A picosecond thermoreflectance measurement system had been developed in the study of Taketoshi et al. [18] in order to measure thermal diffusivities of metal thin films. In-plane thermal diffusivities of the thin films are close to those of the bulk materials although out-of-plane electrical resistivities measured by the four-probe method were larger than that of the bulk materials. The development of an approach to the Angstrom's method for the measurement of the thermal diffusivity was reported in the study of Muscio et al. [19]. In this method, the diffusivity is determined from the damping and the phase shift of a periodic thermal signal during its propagation along the specimen. A direct-contact source was employed, with a temperature-oscillation signal supplied on a portion of one of the two main surfaces of the specimen. The main difficulty about its use is to obtain a perfectly harmonic and well-balanced thermal signal. Photothermal beam deflection technique had been employed for the measurement of thermal diffusivity of different compositions of amorphous Ge_xSe_{100-x} thin films [20]. Slope from the tangetial component of the deflection signal with the pump-probe offset was used to evaluate the numerical value of thermal diffusivity. The thermal diffusivity and conductivity of the EB-PVD 7YSZ thermal barrier coating were evaluated on the basis of its photothermal infrared emission [22, 23]. Photopyroelectric spectroscopy (PPE) was used to study the thermal and optical properties of melanins. The photopyroelectric intensity signal and its phase were independently measured as a function of wavelength and chopping frequency for a given wavelength in the saturation part of the PPE spectrum. The thermal diffusivity coefficient, thermal conductivity, the specific heat of the samples, and a value for the condensed phase optical gap could be obtained [24]. Thermal properties, such as thermal conductivity, thermal diffusivity, thermal effusivity and specific heat capacity, of geometrically frustrated multiferroic $ErMnO_3$, having range of grain sizes from 32 to 220 nm has been measured by the photopyroelectric technique [25]. The coatings at high-temperature exposure with thermal cycling brought in substantial increases in thermal diffusivity and conductivity. The thermal conductivity of the films was measured using the dynamic 3ω method [9]. This method was employed to measure thermal conductivity of Ni, Co, and Fe-doped $La_5Ca_9Cu_{24}O_{41}$ thin films.

The thermal conductivity of multi-layer composites can differ markedly from that of the corresponding bulk materials. Many methods for the measurement of the thermal conductivity of multi-layer composites have been reported [26-28]. A simple arrangement of double-layer composites has the component materials arranged as parallel slabs. In such composites, the thermal conductivities have been reported for the two extreme cases of heat flow parallel and perpendicular to the plane of the slabs. Multilayers consisting of transition-metal silicides have been achieved by the size of growing silicides strongly depends on the annealing conditions [29].

The advantages of the present measurements are: (1) the thermal diffusivity (α) is determined dependent upon the film thickness and the oscillating temperature frequency (ω) only, stable and accurate measures can be obtained from the measurements so long as the ω parameter is proved to be independent of the thermocouples contact position and the electrical voltage applied to the peltier; (2) the specimens can be prepared in a simple geometry with multilayer films; (3) the components in the thermal-electrical detecting system are low cost and thermal diffusivities of all layers can be achieved in just one specimen and simple way compared to the other techniques. In the present study, the phase lag method is adopted to evaluate the thermal diffusivity of each layer of two multi-layered specimens. A container of multi-layered specimen with a property of good heat insulation at the specimen's lateral surfaces is appropriately designed with a ventilation system placed beneath the peltier such that the oscillating temperature at everywhere of the specimen is presented to be uniform and operating at the steady condition. The data processing system applied to the temperature signals of the three thermocouples is composed of a data acquisition system and a lock-in amplifier. The theoretical temperature solution in the multi-layered specimens is developed for one-dimensional, oscillating heat conduction in terms of the amplitude and frequency of the oscillating temperature imposed by a peltier placed beneath the specimen's substrate. The effects of electrical voltage applied to the peltier and different thermocouple contact position on the frequency of the periodically oscillating temperature at various layers are first evaluated using the thermal-electrical data processing system excluding the lock-in amplifier. Theoretically, different phase lag in temperature exists at every layer of the specimen w.r.t the oscillating temperature of the peltier. The difference in the phase lag between the top and bottom surfaces of every layer can be expressed as a function of layer thickness (d), oscillating temperature frequency (ω), and thermal diffusivity (α) of this layer. Two of three thermocouples are placed to have their tip in contact with the top surface of the thin film at arbitrary positions in order to achieve the phase lags occurring at two different positions of this layer. The third thermocouple is placed on the top surface of the lower layer adjacent to this measured film in order to evaluate the temperature phase difference across this thin film with a thickness of d. The thermal-electrical data processing system with a lock-in amplifier was then used to obtain the temperature solution as a function of time. The phase difference between any two of these three thermocouples can be measured as a function of wave frequency (ω) in a form of $a + b\sqrt{\omega}$. The differentiation of this phase difference w.r.t. $\sqrt{\omega}$ is obtained equal to $d/\sqrt{2a}$ (= b). The thermal diffusivity (α) of the film layer is thus determined if the b value (slope of the straight line) and the measured film thickness d are available. The thermal diffusivities of three thin film materials determined by the present technique are then compared with those reported in the literatures. The effect of the surface roughness on the measures of thermal diffusivity is investigated.

2 Theoretical models for thermal diffusivity

In order to measure the thermal diffusivity of each layer in the multi-layered specimen, the container of the specimen should be properly designed to have good thermal insulation such that the heat transfer in the specimen can be to the utmost the one-dimensional heat conduction. The temperature at every layer of specimen is regarded to be uniformly distributing but oscillating with time in a thermally stable form if one-dimensional heat transfer is assumed. If the one-dimensional heat transfer is satisfied, it can be obtained as

$$\frac{\partial^2 T}{\partial x^2} = \frac{1}{\alpha}\frac{\partial T}{\partial t} \quad (1)$$

The x-axis is defined to be normal to the specimen and in the upward direction, with its origin point set at the interface of the measured film and its adjacent lower layer. In the present study, a peltier is placed beneath the substrate of the multi-layered specimen. The heat produced by the peltier is released in an oscillating form with a frequency ω (= $2\pi/t_p$; t_p: period). Fins and fan are placed in sequence beneath the peltier. The composition of the specimen + peltier + heat ventilation system is then held firm by the container made of Acrylic material. It should be mentioned that this material is a good insulator with a thermal conductivity of 0.2 $W \cdot m^{-1} \cdot K^{-1}$ only. Air was sucked into the chamber (see Fig. 1) by the fan to cool the fins protruding out of the bottom. Appropriate design for the air flow rate provides the temperature variations in the peltier to be oscillating but in a stable condition; that is, the temperature oscillations are operating in a stable band almost invariant with time. If so, the heat transfer in the multi-layered specimen can be regarded to be one-dimensional at every layer. Define $\xi \equiv x\left(\frac{\omega}{2\alpha}\right)^{1/2}$ and $\tau \equiv \omega t$. ξ is then specified as the dimensionless phase lag of the oscillating temperature arising at specimen's layer. Define d as the thickness of thin film (or substrate). $\xi d \leq 0.1$ is the additional prerequisite of one-dimensional heat transfer in the multi-layered specimens. The dimensionless form of Eq. (1) is then rewritten as:

$$\frac{\partial^2 T}{\partial \xi^2} = 2\frac{\partial T}{\partial \tau} \quad (2)$$

(a)

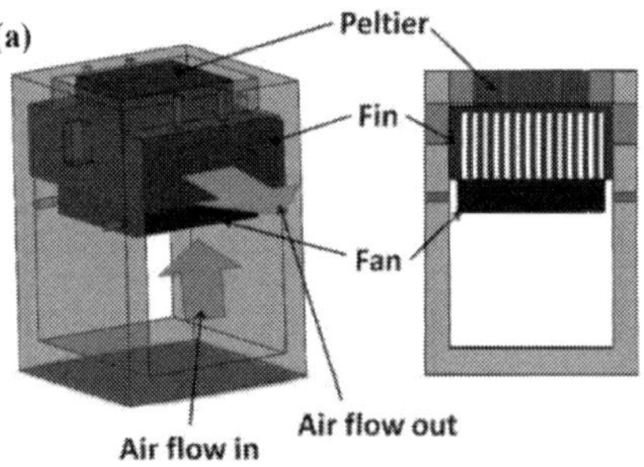

(b)

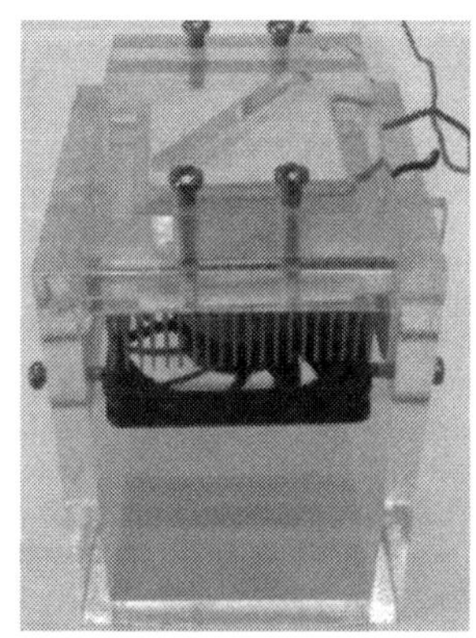

Fig. 1. (a) Ventilation and cooling system for peltier; (b) the photograph of the specimen heating system.

The temperature solution of Eq. (2) is obviously obtained as a function of time. Define the x-axis in the direction normal to the thin film top surface, and x = 0 is positioned at the bottom surface of the measured film. If the heat transfer is controlled to be one-dimensional in the x-direction, the temperature solution in an oscillating form with a fixed frequency ω is almost uniformly distributing over every cross section of the multi-layered specimen. With this condition, Eq. (2) is conveniently considered as the quasi-steady state, and the solution of the film layer with the factor of time is obtained simply from the boundary condition imposed on the bottom surface of the layer in an oscillating form. Due to the complicate geometries of the multi-layered specimen, it is indeed difficult to attain the temperature solution in an analytical form for the entire specimen. Instead of this, the analytical solution of temperature is now established for every individual layer by imposing appropriate boundary conditions for the top and bottom surfaces of this layer. For example, Figure 2 shows film A deposited on the substrate with d as the thickness. The coordinate set for film A is shown in Fig. 2. The boundary conditions given for the bottom and top surfaces of film A are [10]:

$$\xi = 0, \quad T(x = 0, \tau) = T_M + u_0 \cos \tau \tag{3.a}$$

$$\xi = \infty, \quad \frac{\partial T}{\partial \xi} = 0 \tag{3.b}$$

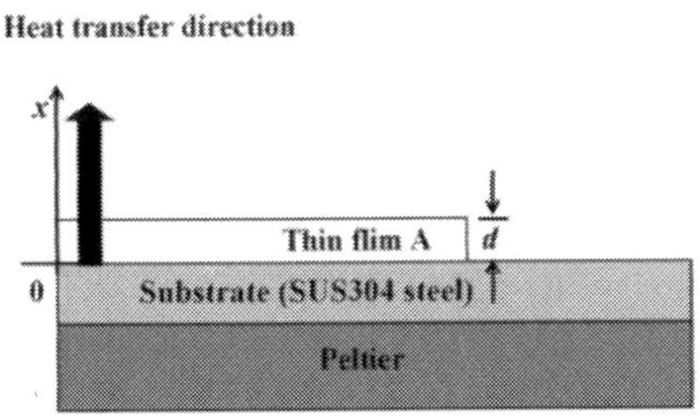

Fig. 2. The schematic diagram of the multilayered specimen with its origin point located at the interface of film A and substrate.

where T_M denotes the mean value of the oscillating temperature occurring at the interface of the substrate and film A ; u_0 denotes the amplitude of the oscillating temperature. T_M and u_0 are actually two unknowns in the current situation. Solution of Eq. (2) satisfying these two boundary conditions is thus expressed as [10]

$$T(\xi, \tau) = T_M + u_0 e^{-\xi} \cos(\tau - \xi) \tag{4}$$

ξ denotes the phase difference in temperature w.r.t. the oscillating heat transfer at ξ = 0. It should be mentioned that this heat transfer is actually unknown. In the present study, a reference signal corresponding to a fixed electrical voltage (10 V in this study) but a variable frequency was given. Then, the phase difference w.r.t. the reference signal can be obtained for all three thermocouples. Then, the phase difference ϕ of temperature wave across the film with a thickness of d, according to the ξ definition, is written as

$$\phi = d\sqrt{\frac{\omega}{2\alpha}} \tag{5}$$

It should be mentioned that Eq. (5) is valid for the case that the top surfaces of all layers are prepared to be perfectly smooth such that the temperature measurement is irrelevant to the measuring position of the same surface.

3 Experimental details

The system for thermal diffusivity (α) measurement is composed of one-dimensional steady heat transfer design for multi-layered specimen and peltier's fan cooling system, the circuit control system for oscillating temperature amplitude and phase, thermocouples, and the lock-in amplifier for the grasps of the temperature signals with specified frequency and phase via reference signal. The schematic diagram of the system for the measurement of thermal diffusivity (α) is shown in Fig. 3. A lock-in amplifier (NI-4431, National Instruments, USA) is used to detect and measure very small AC signals all the way down to a few nanovolts. Accurate measurements may be made even when the small signal is obscured by noise sources many thousands of times larger. The lock-in amplifier uses a technique known as phase-sensitive detection to single out the component of the signal at a specific reference frequency and phase. Noise signals, at frequencies other than the reference frequency, are rejected and do not affect the measurement. A NE555 chip collaborating with the peltier was used for the controls of the oscillating temperature amplitude and frequency of peltier in an approximately sinusoidal form. Its circuits are adjustable for electrical capacitance and resistance to create various oscillating signals in a square form. An approximately sinusoidal temperature oscillation is thus created in the peltier. Beside the NE555 chip, a capacitor and a variable resistor are provided such that the product value of the electrical resistance (R) and capacitance (C) in a wide range is ensured to be linearly proportional to the oscillation period/frequency.

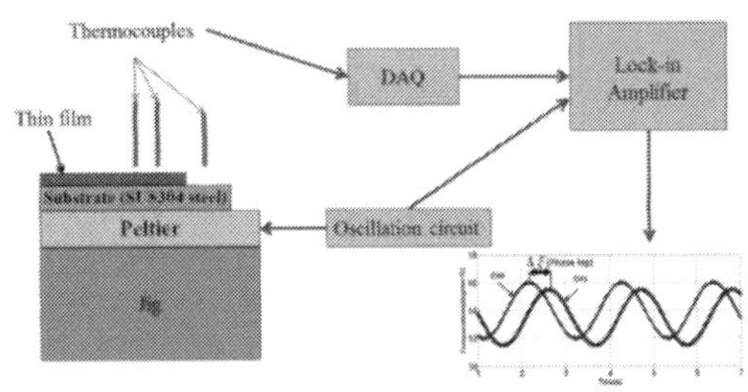

Fig. 3. Schematic diagram of the entire thermal-electrical system for the measurements of thin film thermal diffusivity.

As Fig. 3 shows, the multi-layered specimen is placed on the top surface of peltier in order to accept the oscillating heat waves uniformly distributing over the substrate-peltier interface. Uniform distribution of heat waves in the substrate is important in order to sustain the heat transfer in the multi-layered specimen to be considerably one dimensional. In order to achieve this goal, the jig for the peltier placement, as Fig. 1 shows, is designed to have a fence surrounding the lateral surfaces of the square peltier. The jig is made of the Acrylic material such that the temperature distribution on the peltier top surface is nearly uniform after heating for a while. On another hot surface of the peltier, the heat was dissipated by a row of fins and a fan installed for air ventilations. The peltier was provided as a cooling heat source to create an oscillating temperature wave. The choice of this kind of peltier is made for the purpose of achieving the specimen temperature a rapid thermal equilibrium with the environment. Fig. 1(b) shows the photograph of the peltier heating equipment, and a specimen has been placed on the top surface of the peltier. In the present study, each of all specimens is prepared to be a ladder form in a small area near the specimen edge. That is, there is a small area of the substrate surface without coating thin film. Three K-type thermocouples (AWG 40 ANSI, USA) with a temperature resolution of ±1.1 °C were used as the temperature measurements at the top surfaces of the measured film and the substrate. Two thermocouples were placed to have contacts with the top surface of the measured film at arbitrary positions in order to examine the surface roughness effect on the phase of temperature waves. The third thermocouple was in contact with the blank area of the lower layer adjacent to the measured film. The temperature wave signals from these three thermocouples were acquired by a NI-4431 data acquisition system with a high-speed USB carrier (NI, USA); the synchronous sampling rate is up to 102 kS/s. A "LabView" software was used to have a collocation with the data acquisition system to form the lock-in amplifier for the original temperature signals in a random form from these thermocouples.

4 Results and discussions

The frequency of oscillating temperatures is controlled by the peltier. Therefore, the temperature period $\overline{T}$ is governed by the product value of electrical resistance (R) and electrical capicity (C) applied in the peltier controller. The relationship between the $\overline{T}$ and $R \cdot C$ parameters is provided in Fig. 4(a). A linear behavior is demonstrated in the period range used in this study. It implies that the temperature period, ant thus the period can be determined from $\overline{T} = 1.635\ R \cdot C$ if the $R \cdot C$ value applied to the peltier is available. Then, the frequency of oscillating ω is defined as $1/\overline{T}$ and the period can be determined from $\omega = 0.6581/R \cdot C$ in Fig. 4(b).

Due to the steady oscillating temperature output from the peltier, the temperature (T) data measured by the thermocouples with their tip placed on the coating film and its substrate surface are able to express in a general form of $T(t) = a_0 + a_2\cos(a_3t + a_4)$, where t represents the time of heat conduction. The general T(t) solution is obtained as the sum of the constant and Cosine terms. The linear term (a_1t) is neglected here if the peltier is controlled without the linear temperature rise arising at the thermocouples shown in Fig. 5(a). Figure 5(b) shows the temperature data acquired from these two thermocouples respectively. Since these data are presented without a linear increase with time, T(t) solution is simplified as $T = a_0 + a_2\cos(a_3t + a_4)$. The data obtained from the CH0 for the substrate can be regressed well as $T = 15.29 - 1.773\cos(0.7328t + 0.0424)$; whereas the data form the CH1 for the Si film can be regressed as $T = 15.97 - 1.651\cos(0.7328t + 0.0572)$.

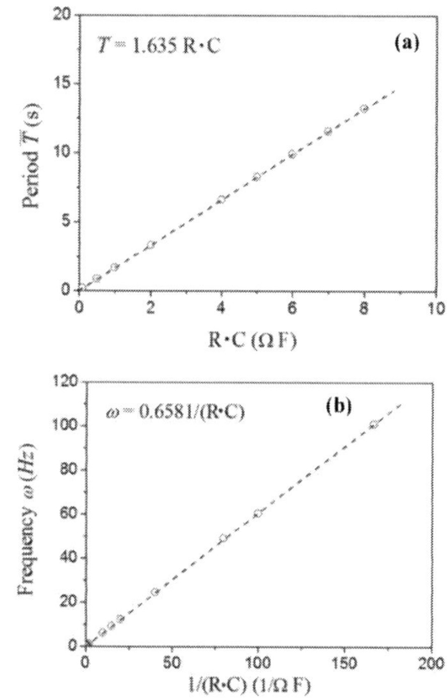

Fig. 4. Results of the time period (a) and frequency (b) of oscillating temperature expressed as a function of the product value of electrical resistance and capacity.

The a_3 parameter in the T(t) expression represents the frequency of oscillating temperature, and a_4 denotes the phase lag arising at the measured layer with a distance from the peltier. In order to investigate the variations of the a_3 parameter due to the changes in the electrical voltage and the measured point at the same surface, the thermal-electrical system was applied intently without the lock-in amplifier. The SUS304 steel and Si (200 nm)/SUS304 steel specimens are prepared for the temperature measurements. Two thermocouples are arranged to contact with either the SUS304 steel substrate or the peltier simultaneously at different contact positions. The T(t) expression is thus applicable to evaluate the temperature arising at different position. Then, the a_3 values associated with the above two thermocouple arrangements can be thus obtained from these temperature solutions.

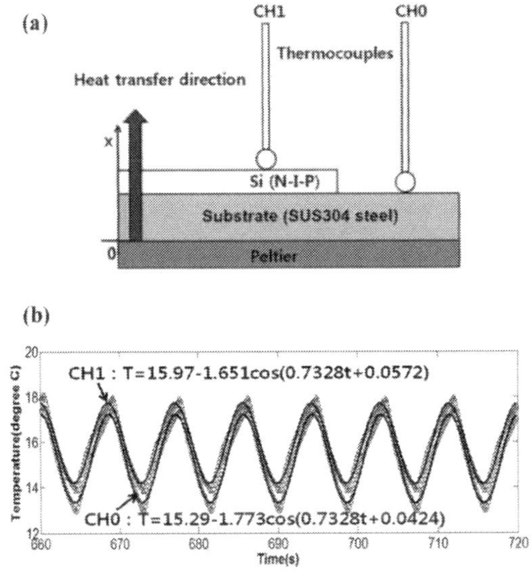

Fig. 5. (a) Schematic diagram of the thin film specimen heated by the peltier and the temperature at different layer detected by the CH1 and CH0 thermocouples; (b) the experimental data of the temperatures obtained from these two thermocouples and their regression.

Figure 6(a) shows the a_3 values obtained from the arrangements of these two thermocouples noted by CH0 and CH1 at two peltier oscillating frequencies and measuring at two arbitrary positions of the SUS 304 steel substrate. $a_3 \cong 0.586$ Hz is obtained from the measurements with peltier oscillating frequency of 0.586 Hz; whereas $a_3 \cong 100.552$ Hz is obtained from the measurements of the two thermocouples by applying peltier's oscillating frequency at 100.552 Hz. It is interestingly found that the a_3 data obtained from the CH0 and CH1 are quite close each other and nearly a value invariant with the measuring position and the peltier's oscillating frequency. The Si film was then deposited on the SUS304 steel substrate as the coating layer. The thermocouple labelled as CH0 was placed in contact with the top surface of the SUS304 steel substrate; whereas the thermocouple noted by CH1 was placed on the top surface of the Si film. The a_3 values obtained from these two thermocouples at different positions are shown in Fig. 6(b). Theses data can be regressed well with $a_3 \cong 0.586$ Hz in response to the peltier's oscillating frequency of 0.586 Hz, and $a_3 \cong 100.552$ Hz is obtained as the applied oscillating frequency is 100.552 Hz. The provisions of Fig. 6(a) and Fig. 6(b) for the a_3 results can prove that: (1) a_3 is a constant value nearly invariant with the contact positions of two thermocouples placed at the same surface; (2) a_3 is a constant value invariant with the contact positions at two different contact surfaces; (3) a_3 is obtained very close to the peltier's oscillating frequency. It is thus concluded that the frequency of oscillating temperature in the peltier is sustained to be almost unchanged in the heat conduction for any layer of the specimen. Therefore, the a_3 parameter in the T(t) expression is regarded as a constant in the solution of thermal diffusivity for any coating film, irrespective of the contact position of thermocouples.

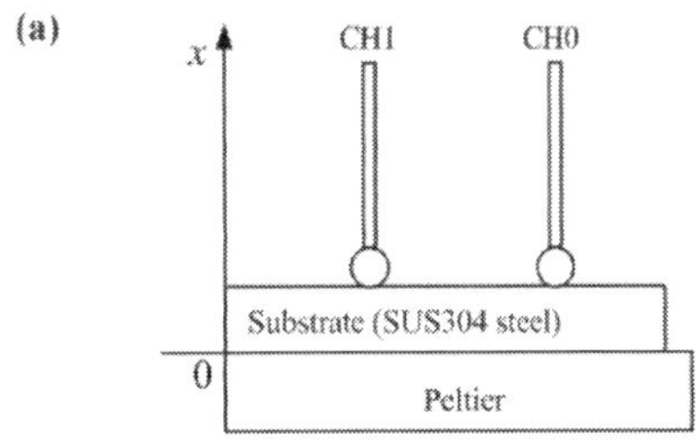

Applied ω in peltier (Hz) \ Different positions		1	2	3	4	5
0.586	CH0	0.58625	0.58528	0.58675	0.58588	0.58674
	CH1	0.58703	0.58635	0.58612	0.58647	0.58549
100.552	CH0	100.552	100.552	100.554	100.555	100.556
	CH1	100.553	100.555	100.553	100.555	100.554
Applied ω in peltier (Hz) \ Different positions		**6**	**7**	**8**	**9**	**10**
0.586	CH0	0.58519	0.58688	0.58703	0.58635	0.58612
	CH1	0.58675	0.58588	0.58674	0.58519	0.58635
100.552	CH0	100.555	100.553	100.555	100.552	100.551
	CH1	100.555	100.552	100.554	100.555	100.556

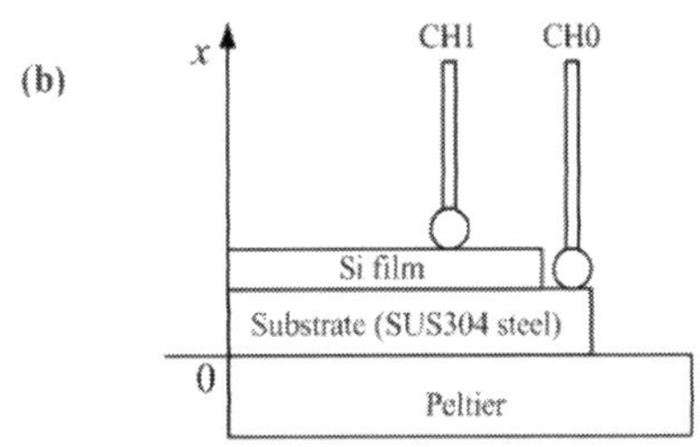

Applied ω in peltier (Hz) \ Different positions		1	2	3	4	5
0.586	CH0	0.58625	0.58528	0.58712	0.58621	0.58642
	CH1	0.58668	0.5871	0.58523	0.58647	0.58549
100.552	CH0	100.552	100.554	100.555	100.552	100.556
	CH1	100.553	100.554	100.552	100.552	100.554
Applied ω in peltier (Hz) \ Different positions		**6**	**7**	**8**	**9**	**10**
0.586	CH0	0.58568	0.58688	0.58703	0.58635	0.58612
	CH1	0.58675	0.58588	0.58674	0.58519	0.58635
100.552	CH0	100.555	100.553	100.555	100.552	100.551
	CH1	100.555	100.552	100.554	100.555	100.556

Fig. 6. (a) Schematic diagram of the CH1 and CH0 thermocouples arranged to measure the temperatures of the SUS 304 steel substrate at arbitrary contact positions and the a_3 values evaluated for these positions; (b) Schematic diagram of the CH1 and CH0 thermocouples arranged to measure the Si film and the steel substrate respectively at various contact positions and the a_3 values corresponding these positions.

Instead of contact position, a_3 is also obtained from the change in the electrical voltage applied to the peltier. Fig. 7(a) shows the a_3 values varying with electrical voltage (unit: V) for both the CH0 and CH1 in contact with the SUS304 steel substrate. The oscillating amplitude of temperature is varied with electrical voltage but frequency is kept as same as those shown in Fig. 6. The a_3 values due to the change in electrical voltage have their magnitude almost as same as the peltier frequency, and they are extremely close to those (0.586 Hz and 100.522 Hz) shown in Fig. 6(a) respectively. This implies that the a_3 value remained to be invariant with the electrical voltage so long as the peltier oscillating temperature frequency is unchanged. Figure 7(b) shows the a_3 values varying with electrical voltage for the contact positions of CH0 and CH1 are as same as those shown in Fig. 6(b). These data can be averaged with a_3 = 0.586 Hz and 100.552 Hz respectively too. The provisions of Fig. 6 and Fig. 7 make the conclusion that the oscillating temperature frequency a_3 created in the specimen, so long as the frequency of oscillating temperature applied to the peltier, remains to be the same

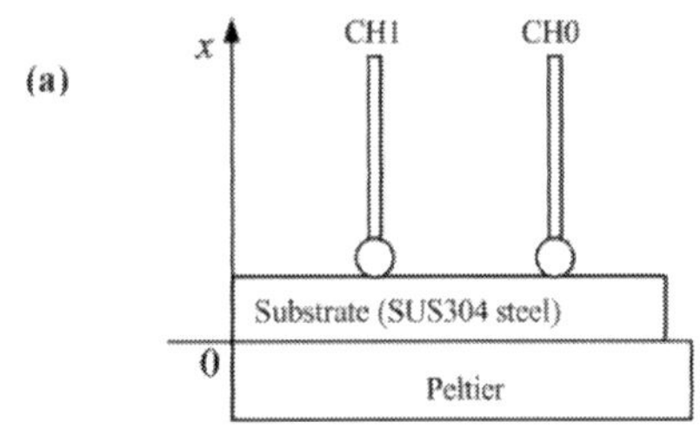

Applied ω in peltier (Hz) \ Electrical voltage of peltier		2	4	6	8	10
0.586	CH0	0.58625	0.58528	0.58675	0.58588	0.58674
	CH1	0.58703	0.58635	0.58612	0.58647	0.58549
100.552	CH0	100.552	100.552	100.554	100.555	100.556
	CH1	100.553	100.555	100.553	100.555	100.554

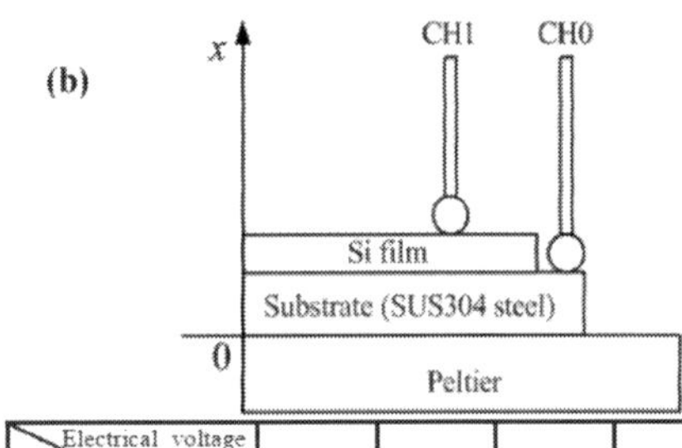

Applied ω in peltier (Hz) \ Electrical voltage of peltier		2	4	6	8	10
0.586	CH0	0.58621	0.58642	0.58568	0.58688	0.58703
	CH1	0.58647	0.58549	0.58675	0.58588	0.58674
100.552	CH0	100.554	100.555	100.552	100.556	100.555
	CH1	100.554	100.552	100.552	100.554	100.555

Fig. 7. (a) Schematic diagram of the CH1 and CH0 thermocouples arranged to measure the temperatures of the SUS 304 steel substrate at various electrical voltages and the a_3 values evaluated for these voltages; (b) Schematic diagram of the CH1 and CH0 thermocouples arranged to measure the Si film and the steel substrate respectively at various electrical voltage and the a_3 values corresponding these voltages.

constant, irrespective of the applied electrical voltage and the contact positions of thermocouples. This characteristic is certainly valid for the temperature measurements using the lock-in amplifier.

The wave pattern of the reference signal can be set by the thermal-electrical data processing system with a lock-in amplifier in a flexible way; and the reference frequency in the present study is taken as a variable in order to achieve phase differences varying with the frequency. After processing by the lock-in amplifier, an oscillating temperature profile with its frequency (ω) as same as that of the reference signal can be achieved. The amplitudes of these two temperature curves in Fig. 3 may be different; however, the phase change due to the temperature waves passing through the film with a thickness of d is determined equal to the difference in the time of any two adjacent peaks on these two curves. With the change in frequency at various values, the experimental data for the phase difference between the two thermocouples can be expressed as a function of $\sqrt{\omega}$. A regression of these data have been made to achieve the slope value (b). This b value will be employed to determine the thermal diffusivity (α) of this film. In the present study, temperature profile was processed by the lock-in amplifier with a reference signal in a rectangular pattern. This temperature profile is presented to have the same frequency as that of the reference signal. It should be stressed that the noises in the original temperature signals partly attributable to the unavoidable thermal drift in the measurements can be removed from the use of the lock-in amplifier. Before the measurements, specimen's temperature in the equilibrium state (steady state), as Fig. 8 shows, was demanded.

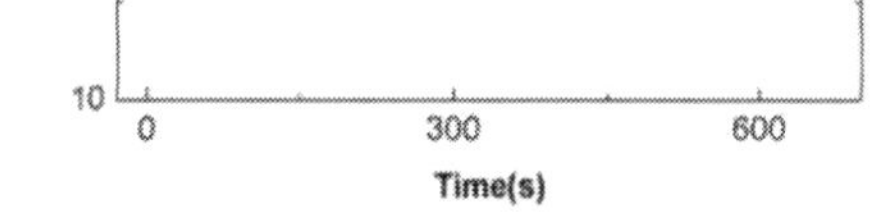

Fig. 8. Oscillating temperature profile obtained at thermally equilibrium state.

In the present study, three kinds of specimen with different coating film(s) were prepared. The SS304 stainless steel with a thickness of 5 mm was commonly used as the substrate of these three specimens. Al and Si films with a thickness of 200 nm were deposited on the steel substrate respectively by the technique of electron beam evaporation (EBE, VT1-10CE, Japan) with a deposition rate of 0.2 nm/s. In order to remain a small blank area in the steel substrate without the film coating, this area was stuck by an adhesive tape before conducting the thin film coating. The third type of specimen was provided by a research institution. The Si-film with a thickness of 200 nm was deposited as the N-I-P buffer layer by the technique of magnetron sputtering deposition first, then an indium tin oxide (ITO) film with a thickness of 200 nm was deposited successively as the top layer of the multi-layered specimen. This layer is deposited intermittently such that a periodic blank area of the buffer layer (Si(N-I-P) layer) was formed between any two ITO circular plateaus. The thermal diffusivity (α) of the top layer was able to determine if two thermocouples were placed to contact with the ITO film and one thermocouple was placed in the blank area of the buffer layer.

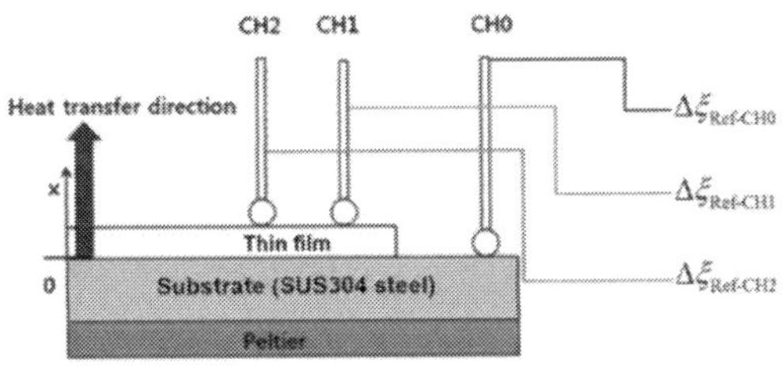

Fig. 9. Schematic diagram presented to show the placements of the three thermocouples adopted in the present study.

In the present study, the CH0 thermocouple was arranged to contact with the steel substrate or the Si buffer layer (200 nm in thickness), and the other two thermocouples, CH1 and CH2, were arbitrary positioned to contact with the top surface of the measured film. As Fig. 9 is shown as an example, the temperature signals of CH0 were processed by the lock-in amplifier to obtain the phase difference $\Delta\xi_{Ref-Ch0}$ from the reference signal. By the similar way, $\Delta\xi_{Ref-Ch1}$ and $\Delta\xi_{Ref-Ch2}$ are able to achieve via the channels of CH1 and CH2 respectively. Define $\phi_{CH0-CH1}$ as

$$\phi_{CH0-CH1} \equiv \phi_{Ref-CH0} - \phi_{Ref-CH1} \tag{6}$$

then, $\phi_{CH0-CH1}$ represents the phase change (lag) of temperature waves due to the thin film with a thickness of d, it is created between the CH0 and CH1 channels. Similarly, $\phi_{CH0-CH2}$ can be achieved following the same method.

By changing the frequency (ω) of the reference signal, the phase differences were measured as a function of $\sqrt{\omega}$. Figure 10(a) and Fig. 10(b) show the experimental data of $\phi_{CH0-CH1}$ and $\phi_{CH0-CH2}$ varying with $\sqrt{\omega}$ respectively for the specimen with an Al film as the top layer. The ω is varying in a range of 0.586 Hz to 100.522 Hz. These data can be regressed well by a straight line with an adj. R^2 value higher than 0.96. These two straight lines are not passing through the origin point, but rather intercepting the ordinate with a value of a. The slope of the straight line is denoted as b. Define y as the temperature phase change ($\phi_{CH0-CH1}$ or $\phi_{CH0-CH2}$) and x as the $\sqrt{\omega}$ parameter. Then, the data of phase difference can be expressed well in a form of y = a + bx = a + b $\sqrt{\omega}$. Taking the differentiations of $\phi = d\sqrt{\omega/2\alpha}$ and $\phi = a + b\sqrt{\omega}$ with respect to $\sqrt{\omega}$ obtains

$$\frac{\partial\phi}{\partial\omega} = \frac{d}{\sqrt{2\alpha}} = \frac{\partial(a + b\sqrt{\omega})}{\partial\sqrt{\omega}} = b \tag{7}$$

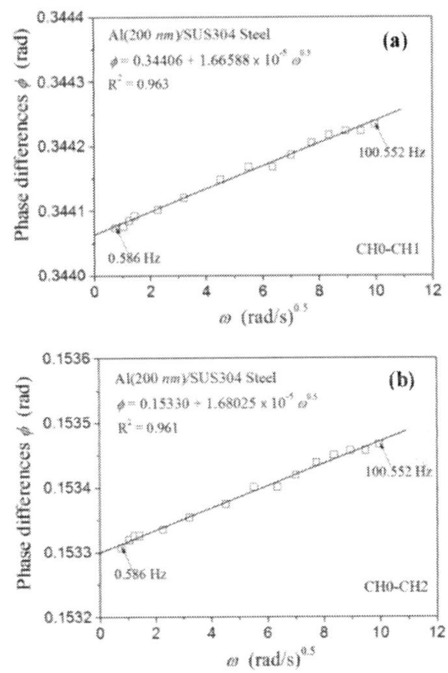

Fig. 10. Variations of (a) $\phi_{CH0-CH1}$; (b) $\phi_{CH0-CH2}$ with $\sqrt{\omega}$ for the Al/steel specimen.

Eq. (7) can be rewritten to obtain the thin-film thermal diffusivity (α) as

$$\alpha = \frac{d^2}{2b^2} \tag{8}$$

if the slope (b) of the straight line is available. By the method mentioned above, $\alpha_{CH0-CH1}$ and $\alpha_{CH0-CH2}$ values are obtained as $7.21 \times 10^{-5}\ m^2/s$ and $7.08 \times 10^{-5}\ m^2/s$, respectively. These two α values are seen within the reported range, $(6.8\sim9) \times 10^{-5}\ m^2/s$ [8].

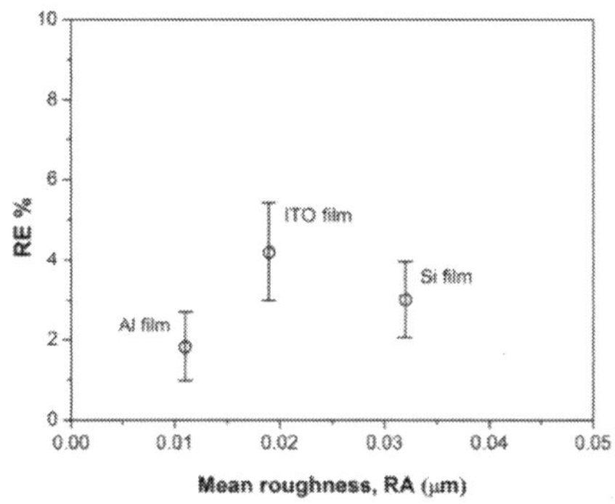

Fig. 11. The relative errors of the thermal diffusivity varying with the surface roughness of the three films.

The $\phi_{CH0-CH1}$ and $\phi_{CH0-CH2}$ v.s. $\sqrt{\omega}$ regressions for the Si/SUS304 Steel specimen have been obtained following the same way. Two slopes, $b = 1.67\times10^{-5}$ $(s\cdot rad)^{0.5}$ and $b = 1.68\times10^{-5}$ $(s\cdot rad)^{0.5}$, are obtained respectively. The thermal diffusivities corresponding to these two b values are $6.07\ \times10^{-5}$ m^2/s and $5.89\ \times10^{-5}$ m^2/s, respectively. Quite small difference between the two α values is believed in relation to the surface roughness difference of the Si film, and it will be discussed in later section.

Table 1. The thermal diffusivities of the three films including the predictions by the current method and the reported data.

Specimen	Current α results (m^2/s)	Reported data (m^2/s)
ITO (200 nm)/Si/Steel (1)	3.41×10^{-6}	$(1.5\sim2.3)\times10^{-6}$ [17]
ITO (200 nm)/Si/Steel (2)	3.27×10^{-6}	
Si (200 nm)/Steel (1)	6.07×10^{-5}	$(0.95\sim8.2)\times10^{-5}$ [8]
Si (200 nm)/Steel (2)	5.89×10^{-5}	
Al (200 nm)/Steel (1)	7.21×10^{-5}	$(6.8\sim9)\times10^{-5}$ [8]
Al (200 nm)/Steel (2)	7.08×10^{-5}	

The $\phi_{CH0-CH1}$ and $\phi_{CH0-CH2}$ values of the ITO film as a function of $\sqrt{\omega}$ for the ITO/Si/Steel specimen have been obtained respectively too. The b values for these two sets of ϕ values are evaluated to be $7.65\ \times10^{-5}$ $(s\cdot rad)^{0.5}$ and $7.81\ \times10^{-5}$ $(s\cdot rad)^{0.5}$. The thermal diffusivities, 3.41×10^{-6} m^2/s and 3.27×10^{-6} m^2/s, are thus determined respectively. These two α values are quite close each other. The reported data of α are varying in a range of $(1.5\sim2.3)\times10^{-6}$ m^2/s [21]. Therefore, the two α values are somewhat outside of the reported range. This difference can be attributable to several factors including the different chemical compositions and microstructures of the ITO film prepared by different deposition techniques. The α values are achieved from the measures at two different positions of the top surface in the same film. Small difference in these two α values are thus due to different surface roughness at different position of the same film. Table 2 shows the mean surface roughness (Ra) of the three kinds of thin film and the SUS304 steel substrate. Difference in the thermal diffusivity between $\alpha_{CH0-CH1}$ and $\alpha_{CH0-CH2}$ is found to be related to the surface roughness of the thin film as the top layer. However, it is the relative error (RE) of thermal diffusivity defined as $RE \equiv |\alpha_{CH0-CH1}-\alpha_{CH0-CH2}|/\frac{1}{2}(\alpha_{CH0-CH1}+\alpha_{CH0-CH2})$, rather than the value of $|\alpha_{CH0-CH1}-\alpha_{CH0-CH2}|$, becomes more meaningful to establish its relationship with the thin film roughness (Ra). Figure 11 shows the RE values as a function of Ra for these three thin films. The three RE values show the sequence that $(RE = 1.8 \pm 0.86\ \%)_{Al\ film}$ < $(RE = 2.97 \pm 0.91\ \%)_{Si\ film}$ < $(RE = 4.25 \pm 1.10\ \%)_{ITO\ film}$. It can be noticed that small RE occurred in these three film materials with their value lower than 6%.

Table 2. Surface roughnesses of the three films and the SUS304 steel substrate.

Thin film/substrate material	Roughness, Ra (μm)
ITO	0.019
Si	0.032
Al	0.011
SUS304 steel	0.073

5 Conclusions

The thermal-electrical data acquisition and processing system without lock-in amplifier is efficiently employed to examine the effects of the electrical voltage applied to the peltier and the contact position of thermocouples on the frequency of the oscillating temperature solution created at thin coating films. As long as a constant frequency is fixed in the oscillating temperature of the peltier, the oscillating frequency of oscillating temperature at various layers remains to be the same constant, irrespective of the changes in the electrical voltage of the peltier and the contact position of thermocouples. The above characteristic is also valid for the thermal-electrical system including the lock-in amplifier, which is applied to evaluate the thermal diffusivity of coating films. The present method provides an efficient way to determine the film thermal diffusivity via the measures of the phase difference of one-dimensional oscillating temperature wave across the film with a thickness d. With the proper design of the peltier heating system, the phase difference (ϕ) results are proved to be linearly proportional to $\sqrt{\omega}$ (ω : temperature wave frequency) from the regression of the experimental data of different thin film material. The thin-film thermal diffusivity can be determined accurately if the slope of the regressed line and the film thickness are available. The relative errors of thermal diffusivity resulting from different thermocouple contact position for these three coating films with different surface roughness are quite small (<6%). The experimental results of thin film thermal diffusivity obtained from the current technique are consistent with those reported in the literatures.

This work has been granted by Frontier Materials and Micro/Nano Science and Technology Center, National Cheng Kung University, Taiwan, R.O.C.

References

1. M.J. Adams, C.F. Kirkbright, *Analyst*, **102**, 281 (1977)
2. M.J. Adams, C.F. Kirkbright, *Analyst*, **102**, 678 (1977)
3. A.J. Ångström, *Phil. Mag.*, **26**, 161 (1863)
4. R.W. King, Phys. *Rev.*, **6**, 437 (1915)
5. B. Abeles, G.D. Cody, D.S. Beers, *J. Appl. Phys.*, **31**, 1585 (1960)
6. Y.S. Touloukian, R.W. Powell, C.Y. Ho, M.C. Nicolasu, *Thermal Diffusivity* (IFI/Plenum, New York, 1973)
7. M.B. Salamon, P.R. Garnier, B. Golding, E. Buehler, *J. Phys. Chem. Solids*, **35**, 851 (1974)
8. I. Hatta, Y. Sasuga, R. Kato, A. Maesono, *Rev. Sci Instrum.*, **56**, 1643 (1985)
9. D.G. Cahill, M. Katiyar, J.R. Abelson, Phys. Rev. B **50**, 6077 (1994)
10. W. Czarnetzki, W. Roetzel, *Internat. J. of Thermophys.*, **16**, 413 (1995)
11. X. Zhang, C.P. Grigoropoulos, *Rev. Sci. Instrum.*, **66**, 1115 (1995)
12. X.Y. Yu, L. Zhang, G. Chen, *Rev. Sci. Instrum.*, **67**, 2312 (1996)
13. E. Jansen, E. Obermeier, *J. Micromech. Microeng.*, **6**, 118 (1996)
14. D.G. Cahill, Review Article, *Microscale Thermophys. Egin.*, **1**, 85 (1997)
15. D.M. Bhusari, C.W. Teng, K.H. Chen, S.L. Wei, L.C. Chen, *Rev. Sci. Instrum.*, **68**, 4180 (1997)
16. T. Yamane, Y. Mori, S.I. Katayama, M. Todoki, *J. Appl. Phys.*, **82**, 1153 (1997)
17. Y.D. Chung, A.P. Chojnacka, C.T. Avedisian, R. Raj, *Acta Mater.*, **45**, 2983 (1997)
18. N. Taketoshi, T. Baba, A. Ono, *Measur. Sci. and Techn.*, **12**, 2064 (2001)
19. A. Muscio, P.G. Bison, S. Marinetti, E. Grinzato, *Internat. J. of Thermal Sci.*, **43**, 453 (2004)
20. A. Mathew, J. Ravi, K.N. Madhusoodanan, K.P.R. Nair, T.M.A. Rasheed, *Appl. Surf. Sci.*, **277**, 410 (2004)
21. T. Ashida, A. Miyamura, N. Oka, Y. Sato, T. Yagi, *J. Appl. Phys.*, **105**, 73709 (2009)
22. T.R. Kakuda, A.M. Limarga, T.D. Bennett, D.R. Clarke, *Acta Mater.*, **57**, 2583 (2009)
23. G.I. Athanasopoulos, E. Svoukis, M. Pervolaraki, R. Saint-Martin, A. Revcolevschi, J. Giapintzakis, *Thin Solid Films*, **518**, 4684 (2010)
24. J.E.D. Albuquerque, C. Giacomantonio, A.G. White, P. Meredith, *Appl. Phys. Lett.*, **87**, 061920 (2005)
25. B. Raneesh, H. Soumya, J. Philip, N. Kalarikkal, *J. Alloys Comp.*, **579**, 243 (2013)
26. P.H.M. Böttger, A.V. Gusarov, V. Shklover, J. Patscheider, M. Sobiech, *Int. J. Therm. Sci.*, **77**, 75 (2014)
27. H. Wang, Q.H. Qin, *Appl. Math. Compt.*, **268**, 311 (2015)
28. B. Weidenfeller, S. Kirchberg, *Compos.*, Part B **92**, 133 (2016)
29. R. Verma, C. Lal, I.P. Jain, *J. Mater. Res. Technol.*, **3**, 257 (2014)

Preparation and properties of the (Sr,Ba)Nb_2O_6 thin films by using the sputtering method

Chien-Chen Diao[1], *Cheng-Fu* Yang[2], and *Chia-Ching* Wu[1,*]

[1] Department of Electronic Engineering, Kao Yuan University, Kaohsiung, Taiwan
[2] Department of Chemical and Materials Engineering, National University of Kaohsiung, Kaohsiung, Taiwan

Abstract. Strontium barium niobate ($Sr_{0.3}Ba_{0.7}Nb_2O_6$, SBN) thin films were deposited on silicon substrate by using the radio frequency magnetron sputtering and under different deposition power and time at room temperature. Surface morphology and thicknesses of the SBN thin films were characterized by field emission scanning electron microscopy. The crystallization films at different deposition power and time were analyzed by X-ray diffraction (XRD) using CuKα radiation from a Rigaku rotating anode with an incident angle of 2°. The remnant polarization (Pr), saturation polarization (Ps), and minimum coercive field (Ec) properties of the metal-ferroelectric-metal (MFM) structure were measured using ferroelectric material test instrument. The SBN thin films deposited at 90 min and 125 W had the maximum Pr, Ps, and minimum Ec of 1.26 μC/cm^2, 2.41 μC/cm^2, and 201.6 kV/cm, respectively. From above results, it knows that the SBN thin films suit for application on ferroelectric random access memory (FeRAM).

1 Introduction

Ferroelectric materials are an important class of materials whose main characteristic is the presence of a spontaneous polarization that can be changed with an external electric field. The tungsten bronze family is one of several ferroelectric materials that includes niobates such as (Sr,Ba)Nb_2O_6 (SBN), (Pb,Ba)Nb_2O_6 (PBN) and (Pb,K)Nb_2O_6 (PKN). In February 2003, the European Union adopted the Restriction of Hazardous Substances (RoHS) directive, limiting the use of certain hazardous substances in electrical and electronic equipment. In accordance with environmental concerns and to avoid problems related to use of lead (Pb), researchers have moved toward using lead-free materials. The SBN is a lead-free ferroelectric material, widely considered very useful in diverse device applications, including memory devices [1], waveguide devices [2], and others. An important representing of ferroelectric material is the SBN, known by excellent pyroelectric coefficient [3, 4], piezoelectric [5] and electro optic properties [6, 7]. SBN presents a tetragonal or orthorhombic phase, depending of composition and temperature.

The SBN solid-solution series crystallize in the tetragonal tungsten bronze (TTB) structure. At room temperature it has the symmetry of the space group P4bm for $0.25 < x < 0.75$. It belongs to the 4mm point group in the ferroelectric phase and changes to the 4/mmm when in the paraelectric phase. Investigation of $Sr_xBa_{1-x}Nb_2O_6$ ceramics using X-ray diffraction suggests that a morphotropic phase boundary (MPB) exist at round the composites of $Sr_{0.25}Ba_{0.75}Nb_2O_6$, which is characterized by the coexistence of the tetragonal and orthorhombic phases [8]. $Sr_{0.2}Ba_{0.8}Nb_2O_6$ ceramic presents a tetragonal ferroelectric phase between 120 and 293°C, an orthorhombic phase below 120°C and a tetragonal paraelectric phase above 293°C.

SBN thin films have been prepared by several techniques like sol-gel processing [9], pulsed laser deposition [10] and metalorganic chemical vapor deposition (MOCVD) [11]. Recently, it has been proposed to prepare SBN thin films by a chemical method based on a polymeric resin containing metallic ions [12]. However, the radio frequency (RF) magnetron sputtering is a simple method to deposit the needed thin films [13]. In the research, $Sr_{0.7}Ba_{0.3}Nb_2O_6$ (SBN) thin films were obtained by a radio frequency (RF) magnetron sputtering method at room temperature. The influence of deposition power and time on the crystallinity and ferroelectricity of SBN thin films was studied using the analysis of scanning electron microscopy and X-ray diffraction, and the measured of leakage current and polarization, respectively.

2 Experimental procedures

The conventional solid-state reaction method was used to prepare the $Sr_xBa_{1-x}Nb_2O_6$, x=0.3 (SBN) ceramic. Reagent-grade oxide or carbonate powders of $SrCO_3$, $BaCO_3$, and Nb_2O_5 with greater than 99.5% purity were used as starting raw materials. The oxides were mixed with deionized water by ball-milling for 1 h. After drying and grinding, the powder was calcined at 1100 °C for 2 h, and the formation of the tungsten bronze structure was confirmed by the XRD pattern. Then, ground again and mixed with polyvinyl alcohol (PVA)

* Corresponding author: t10068@cc.kyu.edu.tw

as binder. the calcined powder was uniaxially pressed in a steel die into a 2-inch disk, and the ceramic was sintered at 1340 °C in air for 2 h. The SBN thin films were deposited under various deposition power and time on a silicon wafer by an radio frequency (RF) sputtering method. The deposition pressure was 5×10^{-3} torr, and pure argon (Ar) was used as the reaction gas. Finally, the aluminum (Al) top electrodes were deposited by an electron beam evaporation method on the SBN thin films, with a 1 mm diameter shadow mask, to form the metal-ferroelectric-metal (MFM) structures. The structure of the SBN ceramic and thin films were analyzed by X-ray diffraction (XRD) using CuKα radiation from a Rigaku rotating anode with an incident angle of 2°. Field emission scanning electron microscopy (FE-SEM) was used to study the surface morphologies of the SBN ceramic and thin films. The leakage current density versus applied electric field (J-E) was measured using an HP 4156 semiconductor parameter analyzer. The saturation polarization, remanent polarization, and coercive field properties of the MFM structure were measured using ferroelectric material test instrument.

3 Results and discussion

The FE-SEM morphology of 1340 °C sintered SBN ceramic as shown in Fig. 1. In Fig. 1, as the SBN ceramic sintered at 1340°C, the several bar-type grains developed and no any pores are observed. Figure 2 shows the X-ray diffraction (XRD) patterns of 1340 °C sintered SBN ceramic in the 2θ range of 20 to 80° as shown in Fig. 2. The XRD results revealed at 2θ values of 22.4°, 25.6°, 27.6°, 28.3°, 29.3°, 30.3°, 31.9°, 34.2°, 36.6°, 38.9°, 42.1°, 45.8°, 48.7°, 49.4°, 51.5°, 52.3°, 53.5°, 54.3°, 55.4° and 57.1° were corresponded to the (001), (320), (221), (440), (410), (221), (331), (321), (401), (520), (530), (260), (630), (601), (550), (541), (332), (631), (412) and (711) planes of SBN ceramic, respectively. No second phase was present in this ceramic.

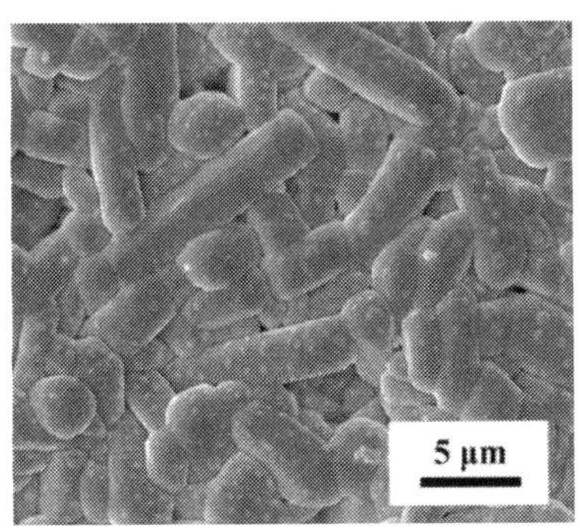

Fig. 1. SEM images of the SBN ceramic sintered at 1340 °C.

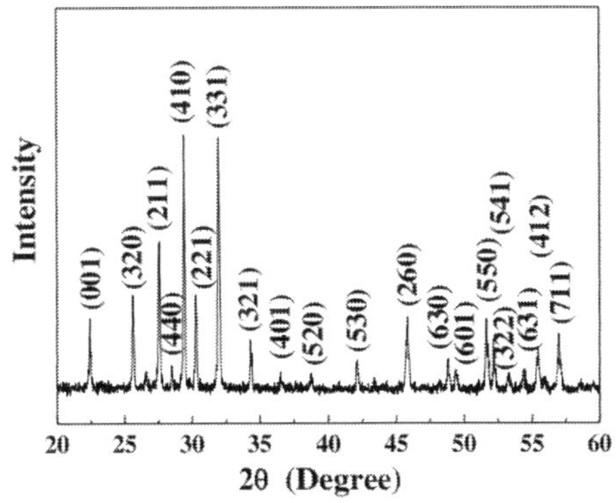

Fig. 2. XRD patterns of the SBN ceramic sintered at 1340 °C.

Figure 3 shows the temperature dependence on the dielectric constants (ε_r-T) of the SBN ceramic sintered at 1340 °C. It fount that the dielectric constant increased as the measured temperature increased at 135 °C, than, the dielectric constant decreased as the measured temperature increases from 140 to 200 °C. The SBN ceramic have a dielectric peak (K_{max}) at 135 °C, which is the Curie temperature (T_c) of SBN ceramic, and the maximum dielectric constant is ε_r=4322.

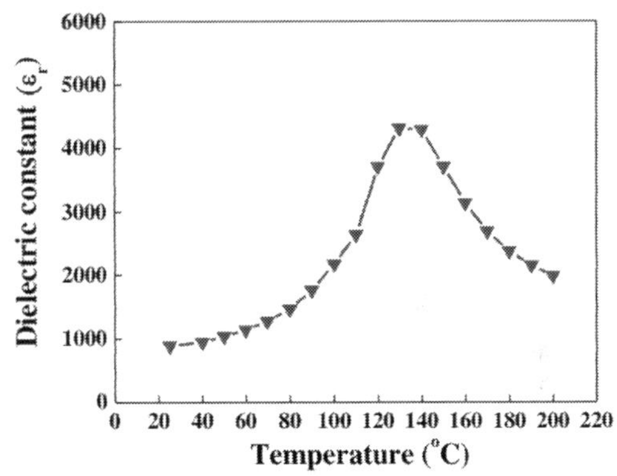

Fig. 3. Temperature dependence on the dielectric constants of the SBN ceramic sintered at 1340 °C.

Figure 4 shows surface FE-SEM images of the SBN thin films with different deposition power. The SBN surface morphology varies greatly with the different sputtering power, as shown in Fig. 4(a) to Fig. 4(d). In the case of the SBN thin film deposited at 75 W (Fig. 2(a)), the surface is smooth and no grain growth is observable. Because the kinetic energy of the sputtered particles arriving at the substrate surface is weak. As the deposition power increases from 75 W to 125 W, the surface roughness of the SBN thin films slightly increased, and no grain growth is observable. However, the grain in the SBN thin film was observed as the deposited power at 150 W.

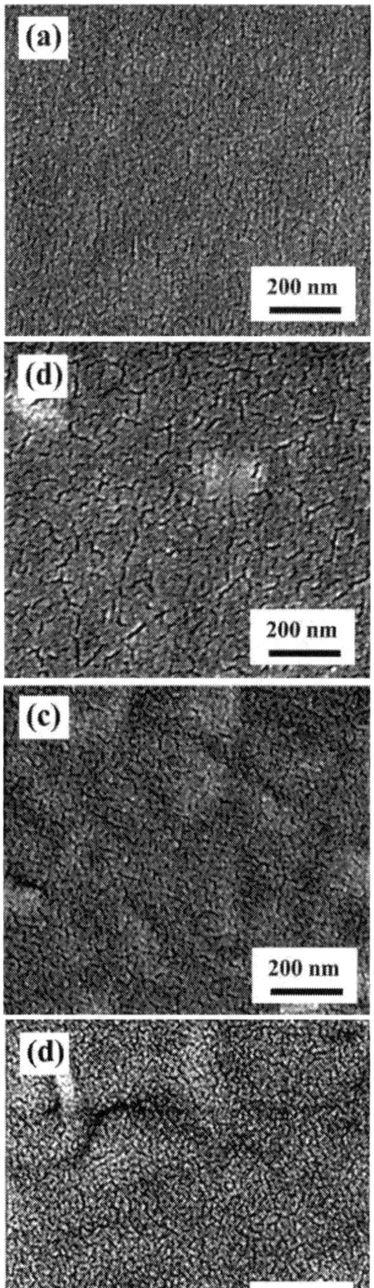

Fig. 4. SEM images of the SBN thin films deposited with different power. (a)75 W, (b)100 W, (c)125 W and (d)150 W.

Figure 5 shows the influence of deposition power on the thickness of the SBN thin films. At deposition power were 75, 100, 125 and 150 W, the thickness of the SBN thin films were 277.19, 333.33, 350.87, and 282.46 nm, respectively. The thinckness of the SBN thin films increased as the depoaition power increases from 75 W to 125 W. This results cause by the number of sputtered particles arriving at the substrate increased. As the deposition power increasing from 125 W to 150 W, the thinckness of the SBN thin film decreased. This results cause by the higher energy of sputtered particles arriving at the substrate and etching the film surface as the deposition power at 150 W.

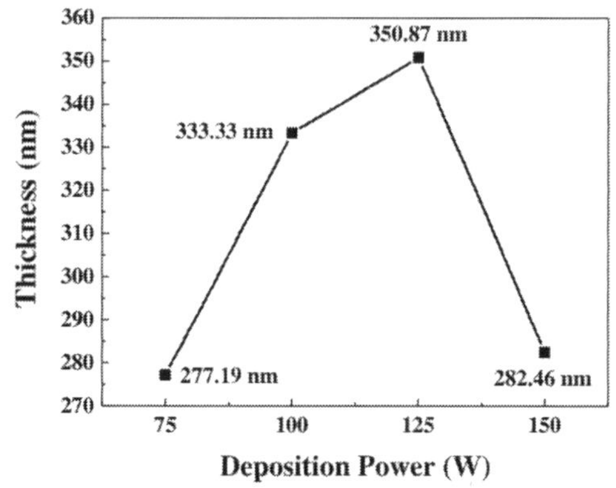

Fig. 5. Thickness of the SBN thin films deposited with different power.

The XRD analysis of the SBN thin films deposited at different power are shown in Fig. 6. All the SBN thin films showed an amorphous structure because they were deposited at room temperature. Figure 6(a) is the standard peak of the aluminum (Al). A crystalline peak of 38.5° is revealed in the Al electrode, and the diffraction peaks' intensity decreases as the deposition power increases from 75 to 150 W, as shown in Figs. 6(b) to 6(e). This result confirms that the thickness of the SBN thin films increases as the deposition power increases from 75 W to 150 W.

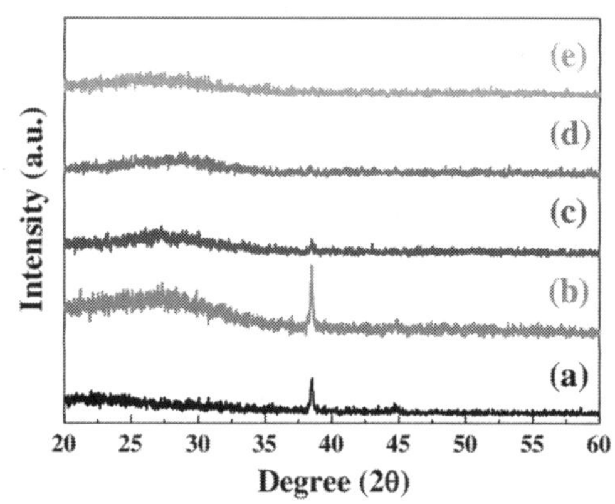

Fig. 6. XRD analysis of the SBN thin films deposited with different power. (a) standard peak, (b)75 W, (c)100 W, (d)125 W and (e)150 W.

Figure 7 shows the leakage current density versus electrical field (J-E) curves of the SBN thin films

DOI:10.1051/matecconf/201712300027

deposited with different power. The leakage current density of the SBN thin films slightly decreased as the deposition power increased from 75 W to 125 W. There are many reasons will influence the leakage current densify, including the thickness and defects of the deposited this films [14, 15].

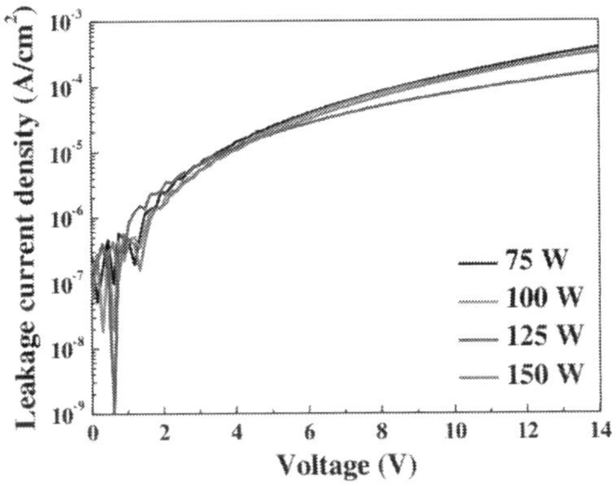

Fig. 7. Leakage current density of the SBN thin films deposited with different power.

Figure 8 shows surface FE-SEM images of the SBN thin films with different deposition time. In the case of the SBN thin film deposited at 30 min, the surface is smooth and no grain growth is observable as shown in Fig. 8. As the deposition time increases from 60 to 120 min, the surface roughness of the SBN thin films increased, and grain growth is observable at the 120 min deposition time.

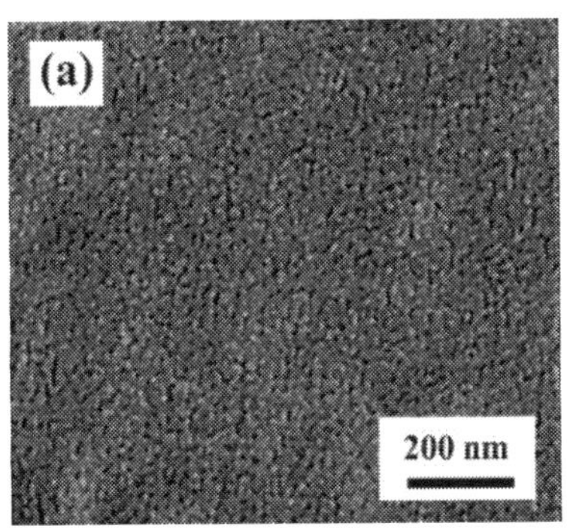

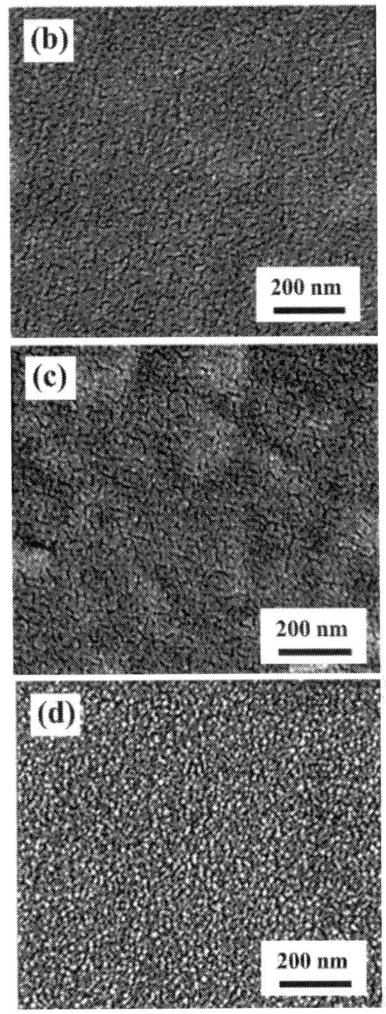

Fig. 8. SEM images of the SBN thin films deposited with different time. (a)30 min, (b) 60 min, (c) 90 min and (d) 120 min.

Figure 9 shows the influence of deposition time on the thickness of the SBN thin films. At deposition times were 30, 60, 90 and 120 min, the thickness of the SBN thin films were 175.43, 189.47, 292.98, and 350.87 nm, respectively. The thinckness of the SBN thin films increased as the depoaition times increases from 30 min to 120 min. This results cause by the number of sputtered particles arriving at the substrate increased as the deposition time increased.

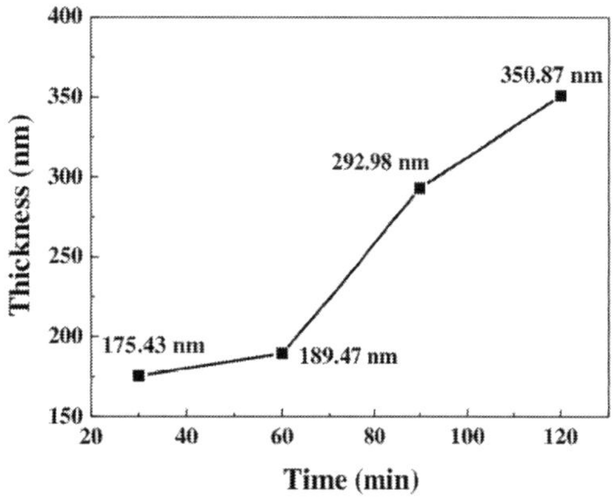

Fig. 9. Thickness of the SBN thin films deposited with different time.

The XRD analysis of the SBN thin films deposited at different time are shown in Fig. 10. All the SBN thin films showed an amorphous structure because they were deposited at room temperature. A crystalline peak of 38.5° is revealed in the Al electrode, and the diffraction peaks' intensity decreases as the deposition time increases from 30 min to120 min, as shown in Figs. 10(a) to 10(d). This result confirms that the thickness of the SBN thin films increases as the deposition time increases from 30 min to 120 min.

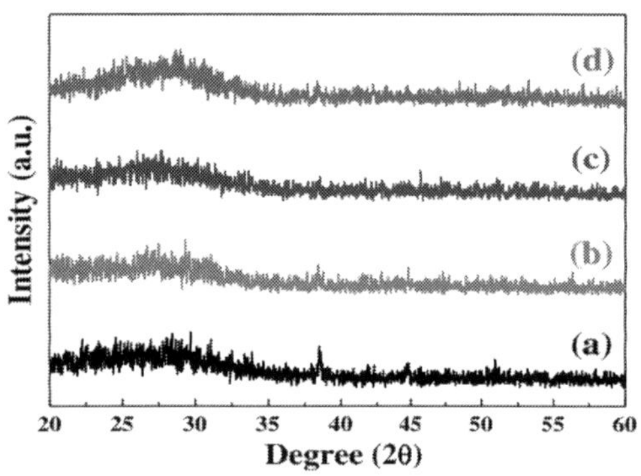

Fig. 10. XRD analysis of the SBN thin films deposited with different time. (a)30 min, (b) 60 min, (c) 90 min and (d) 120 min.

Figure 11 shows the leakage current density versus electrical field (J-E) curves of the SBN thin films deposited with different time. The leakage current density of the SBN thin films decreased as the deposition time increased from 30 min to 90 min. As the deposition at 120 min, the leakage current density of the SBN thin films slightly increased. The larger leakage current density led to less residual charge in the ferroelectric layer.

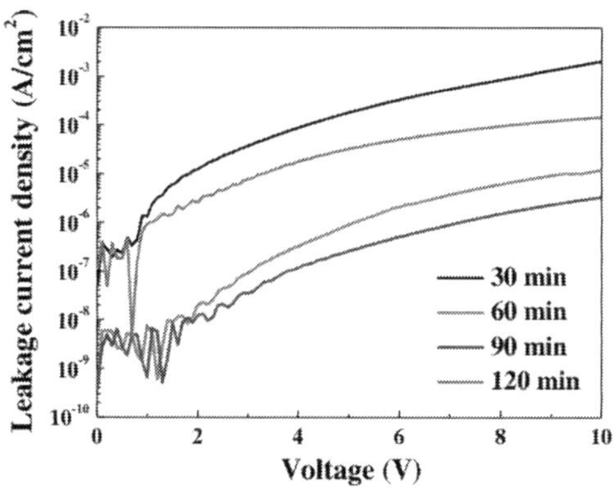

Fig. 11. Leakage current density of the SBN thin films deposited with different time.

Figure 12 shows the polarization versus applied electrical field (P-E) curves of the SBN thin film. The development of hysteresis loops in the SBN thin film confirms the ferroelectric nature. The SBN thin films deposited at 90 min and 125 W had the maximum remnant polarization (Pr), saturation polarization (Ps), and minimum coercive field (Ec) of 1.26 μC/cm², 2.41 μC/cm², and 201.6 kV/cm, respectively. The Pr, Ps and Ec values of the SBN thin film deposited at 90 min, 125 W and room temperature are slightly larger than those in other reports [22-24].

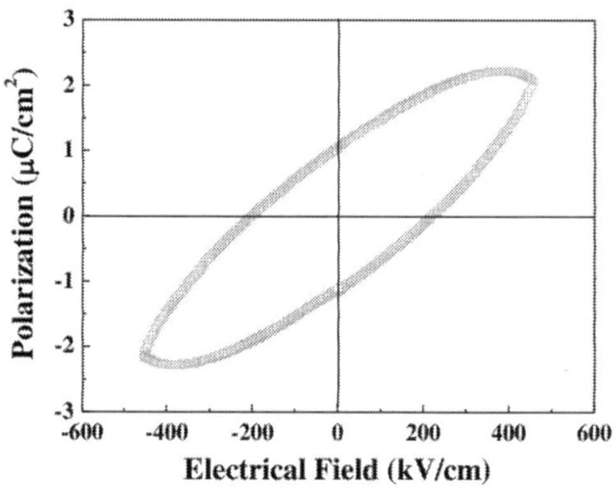

Fig. 12. Polarization versus electrical field curves of the SBN thin film deposited at 90 min and 125 W.

4 Conclusion

In this study, high-quality $Sr_{0.7}Ba_{0.3}Nb_2O_6$ (SBN) thin films were deposited on silicon wafer using an radio frequency (RF) magnetron sputtering. When 90 min and 125 W were used as deposition time and power, the smaller leakage current density of the SBN thin film was obtained. In addition, the remnant polarization (Pr), saturation polarization (Ps), and minimum coercive field (Ec) of 1.26 μC/cm², 2.41 μC/cm², and 201.6 kV/cm, respectively. We demonstrated that the SBN thin film will play an important role in applications for nonvolatile memory devices.

Acknowledgements

The authors acknowledge the financial support of the Ministry of Science and Technology, R.O.C. with the contract Nos. 105-2221-E-244-013- and 105-2221-E-244-014-.

References

1. M. C. Kuan, K. H. Chen, W. C. Tzou, C. M. Cheng, Y. J. Lin, *Advanced Materials Research*, **239-242**, 2628 (2011)
2. J. Koo, J. H. Jang, B. S. Bae, *J. Mater. Res.*, **16**, 430 (2001)
3. P. B. Jamieson, S. C. Abrahams, J. L. Bernstein, *J. Chem. Phys.*, **48**, 5048 (1968)
4. A.M. Glass, *J. Appl. Phys.*, **40**, 4699 (1969)
5. J. D. Zook, S. T. Liu, *J. Appl. Phys.*, **49**, 4604 (1978)
6. S. Sakamoto, T. Yazaki, *Appl. Phys. Lett.*, **22**, 429-431 (1973)
7. M. Horowitz, A. Bekker, B. Fischer, *Appl. Phys. Lett.*, **62**, 2619 (1993)
8. S. Nishiwaki, J. Takahashi, K. Kodaira, M. Kishi, *Jap. J. Appl. Phys.*, **35**, 5137 (1996)
9. Y. Xu, C. J. Chen, R. Xu, J. D. Mackenzie, *Phys. Review B*, **44**, 35 (1991)
10. S. S. Thöny, K. E. Youden, J. S. Harris Jr., L. Hesselink, *Appl. Phys. Lett.*, **65**, 2018-2020 (1994)
11. M. Lee, R. S. Feigelson, *J. Crystal Growth*, **180**, 220 (1997)
12. R. G. Mendes, E. B. Araújo, H. Klein, J. A. Eiras, *J. Materials Sc. Letters*, **8**, 1941 (1999)
13. F. H. Wanga, C. F. Yang, M. C. Liu, *J. Sci. Innovation*, **6**, 7 (2016)
14. Y. K. Wang, T. Y. Tseng, and L. Pang, *Appl. Phys. Lett.*, **80**, 3790 (2002)
15. R. Balachandran, B. H. Ong, H. Y. Wong, K. B. Tan, M. Muhamad Rasat, *Inter. Jour. Electrochem. Sci.*, **7**, 11895 (2012)

MATEC Web of Conferences **123**, 00028 (2017) DOI:10.1051/matecconf/201712300028
ICPMMT 2017

Design of books inventory with RFID antenna in library management system

Ching-Chien Cheng[1,*], *Sung-Tsun* Shih[2], *Kong-Xin* Cheng[1], *Huang-Kuang* Kung[1], and *Chin-Yu* Wang[1]

[1]Institute of Mechatronics Engineering, Cheng Shiu University, Kaohsiung City, Taiwan
[2]Department of Electronic Engineering, Cheng Shiu University, Kaohsiung City, Taiwan

Abstract. A 915 MHz printed radio-frequency-identification (RFID) antennas with the characteristics of good gain and omnidirectional beam wave is constructed and evaluated in this study. The objective is to find out their best reading rates for providing effective wireless communications among RFID antenna during the library book inventory process. And an optimal library inventory system which is based on electromagnetic identification (EMID) technology is proposed, which is constructed to find the optimal tag location for a book, test the tag readability for bookshelves, and connect a couple of multi-layer bookshelves with multiplexers and updat the tag reading status in the database of the computer terminal. The fabricated antenna reader and the proposed system are embedded into different locations of bookshelves and tested at the library of Cheng Shiu University in Taiwan. According to the experimental results, the designed prototype of the antenna reader has the characteristics of the directional radiation pattern, good gain, simple shape, low cost and is easy to be integrated into the bookshelf. And the designed library inventory system can authenticate the location of a book automatically . They can benefit administrating librarians with the capabilities of decreasing the library inventory processing time and reducing the possibility of the books being misplaced.

1 Introduction

In the past, barcode labels and electromagnetic (EM) strips are two common technologies used to support the identification and anti-theft capabilities in the library inventory system. However, the barcode labels could be only stuck on the book once, manually scanned, easily damaged and the EM strips don't support resource identification capability [1]. In recent years, in order to overcome the weaknesses mentioned above, RFID wireless communication technology has replaced the barcode label and magnetic strip with the identification and anti-theft detection capabilities in the library inventory management. Compared with using barcodes in the library inventory management, a RFID tag could be read more than ten-thousand times and could support more convenient and efficient advantages on check-out and return processes [2-6].

Several literatures have been proposed for the smart shelf application in library, including utilizing mobile robot to improve inventory management [7], applying micro-strip lines to improving tag detecting capability [8]. However, these methods are not good for the needs of real time and low cost. As for the use of RFID technology in the bookshelf applications, Kim and Choi [9] proposed a wideband rectangular loop tag antenna for ultra-high frequency RFID library management systems. Another paper proposed a small wideband circularly polarized patch antenna printed on the low-cost FR-4 material for RFID smart bookshelves in libraries [10]. The proposed antenna was operated at a center frequency of 915 MHz and the wave-beam width is 115 °.

Reviewing some references applying UHF RFID Antenna in the library inventory managements, Kim et al. [11] designed a loop antenna for UHF band RFID tag used for a library management system. To achieve the wide-band characteristic, both vertical sides of a loop antenna are widened. Lukas et al. [12] and Kaiwen et al. [13] proposed UHF omnidirectional RFID antenna researches to increase the wireless identification and tracking capabilities for the item-level RFID application. Both literatures presented the designs of UHF RFID tag antennas which supported omnidirectional reading patterns and provided great potential for item-level RFID applications in libraries for tracking the books.

Golding and Tennant [14] investigated the factors that may affect the reading rate of an inventory reader in a library. The investigated factors were reading distance, tag location, number of sweeps and sweep direction. Wang [15] studied the RFID-based methodology and approaches that support library services and management, including sensor gate control, circulation, inventory, searches and utilization statistics. The study also discussed barriers, challenges and future work about RFID applications in libraries and concluded that RFID-based technology would improve digital archives and

* Corresponding author: chching@kmvs.km.edu.tw

digital humanities in library systems. Zhang and Shi [16] and Bin Abdullah, etc. [17] proposed two literatures related to self services of library management systems. Zhang and Shi strongly pointed out the necessity of the new RFID technology of self-service book borrowing and returning system in library, which would replace the barcode technology and accelerate the library's selfservice process. Bin Abdullah, etc. developed an integrated RFID system with graphical user interface (GUI) at the host PC. The study aims to develop an automatic library shelf-management system to assist the librarians for more efficient shelf-management to find any misplaced books on the shelf.

In addition, when reviewing some other RFID related literatures [18-23], two comments were made: (1) such an intelligent plan may not meet the library demand because library internal disturbances, which could affect the RFID operation, such as metal conditions, electric motion-communication wave, and electric wireless network wave should be taken into consideration; (2) the proposed inventory system should be convenient to all readers. As mentioned above, driven by the advanced ubiquitous computing technology, the authors propose a library inventory system with printed RFID antenna in which the characteristics of good gain and omnidirectional beam wave are constructed and evaluated. The objective is to find out their best reading rates for providing effective wireless communications among RFID antenna during the inventory process. What follows is the application of the RFID antenna design as shown in section 2, and section 3 which describe the proposed library inventory system based on EMID technology. Then the experimental results are shown in section 4. Finally, the conclusions and discussions are summarized in section 5.

2 RFID antenna design (omnidirectional beam antenna)

Efficacy of reading antenna identification is one of the key factors in the accuracy of the overall RFID inventory management system. So the antenna design is the most important part in whether library inventory management system can read correctly or not. So in this section our proposed RFID antenna design and feature evaluations will be introduced.

In the library inventory system, because the distance among bookshelves is very close and the individual bookshelf has a wider surface with multi-layer factors, the antenna should be designed with the characteristics of a relatively wide beam to cover the detction on all books of the same bookshelves, and should has a smaller radiation power to avoid the detection of shelf books nearby. In order to achieve good detection accuracy, we designed a flatbed printed RFID reader antenna used in the library books inventory. This antenna has a wide beamwidth with omnidirectional radiation characteristics and has good gain features. With reference to the designed patch mentioned in the literature, we first design patch antenna with the center frequency at 915MHz, and then do the size fine-tuning to obtain the RFID antenna mentioned above.

2.1 RFID antenna design procedure

The geometry plane and the prototype of the antennas are shown in Fig. 1. Two antennas consist of an upside down patch and rectangle ground substrates. Fig. 1. (a) shows the traditional patch antenna, and Fig. 1. (b) shows our proposed RFID antenna. Here we use dielectric constant 4.4 and FR4 substrate with substrate thickness h = 1.6 for designing, and the steps of these two antennas are as follows.

Step 1: Calculation of Width (W) of top Patch[13] The width of the MSTPA is given by

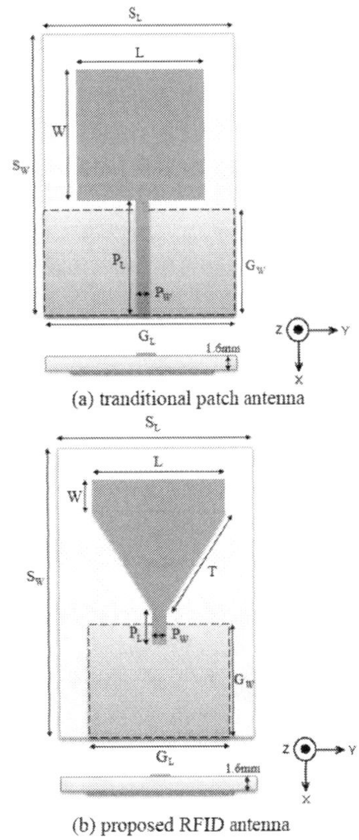

Fig. 1. Structure of proposed RFID antenna filter (a) tranditional patch antenna (b) proposed RFID antenna.

$$W = \frac{c}{2f_r}\sqrt{\frac{2}{\varepsilon_r + 1}}$$

By substituting C = 3x108 m/s, ε r = 4.4 and fr = 915 MHz, W = 100 mm can be easily determined.

Step 2: Define the effective dielectric constant (ε reff) [24] The equation below illustrates the effective dielectric constant,

$$\varepsilon_{reff} = \frac{\varepsilon_r + 1}{2} + \frac{\varepsilon_r - 1}{2}\sqrt{1 + 2\frac{h}{W}}$$

By substituting $\varepsilon_r = 2.2$, W = 100 mm and h = 1.6 mm, $\varepsilon_{reff} = 4.556$ can be determined.

Step 3: Define the effective length (Leff) and the effective length is illustrated as [24] :

$$L_{eff} = \frac{c}{2f_r\sqrt{\varepsilon_{reff}}}$$

By substituting $\varepsilon_{reff} = 4.556$, c = 3x108 m/s and fr = 915 MHz, Leff = 76.8 mm can be determined.

Step 4: The length extension is determined.(ΔL) The length of extension can be illustrated as [24]:

$$\Delta L = 0.412h\frac{(\varepsilon_{reff} + 0.3)(\frac{W}{h} + 0.264)}{(\varepsilon_{reff} - 0.258)(\frac{W}{h} + 0.8)}$$

By substituting $\varepsilon_{reff} = 4.446$ and other corresponding values, ΔL = 0.7385 mm can be determined.

Step 5: Actual length of top patch (L): is determined.
The actual length of top patch is obtained as [24]:

$$L = L_{eff} - 2\Delta L$$

By substituting Leff = 76.8 mm and length extension ΔL = 0.7385 mm, the actual length can defined as L = 76 mm.

Step 6 : Following the top patch chamfering and size fine-tuning, circuit style by RFID antenna is shown in Fig. 1 (b).

2.2 RFID antenna fabricated

With the above design of the antenna step, the center frequency of the antenna can be estimated, and the traditional patch antenna parameters based on the dielectric constant 4.4 and the FR4 substrate with substrate thickness h = 1.6 can be calculated. Similarly, the RFID omni-directional radiation antenna size can be obtained after fine-tuning and chamfering. Table 1 shows the parameter table of the calculated tranditional patch antenna size and the fine-tuned Proposed RFID antenna size.

By starting from the top view of Fig. 1 (b), the lengths of each side of upside down triangle patch are L= 41 mm, W= 14mm, G_L=55 mm, G_W=64.5 mm, P_L=3 mm, P_W=14.5 mm and T=60 mm, respectively, with a two-layer FR4 board there. The photograph of the proposed omnidirectional beam RFID antenna is shown in Fig. 2. The antenna consists of an upside down triangle patch and rectangle ground substrates. The antenna is printed on a 65 mm × 150 mm dielectric substrate. The detecting point is located at the center area of the upside down triangle printed patch.

Table 1. Comparison of patch parameter for antennas.

Parameter	Tranditional Patch Antenna	Proposed RFID Antenna
L	76	41
W	100	14
S_L	150	65
S_W	220	150
G_L	150	55
G_W	75	64.5
P_L	4	3
P_W	80	14.5
T		60

(unit: mm)

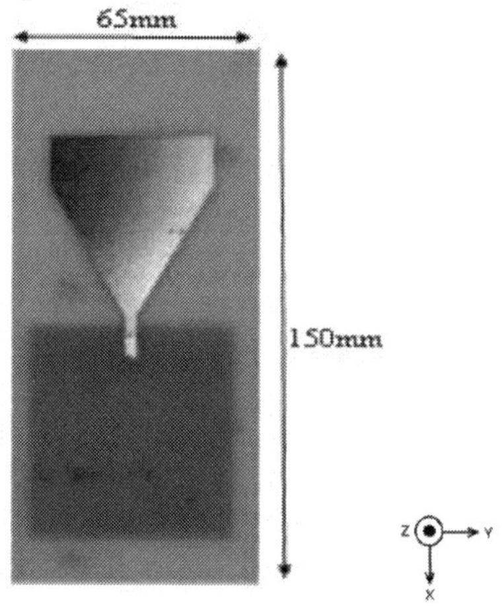

Fig. 2. photograph of the proposed RFID antenna.

2.3 RFID antenna measurement results

RFID antenna readers have been fabricated on the printed circuit board (PCB) with the operating frequency at 915 MHz. In this paper, three performances, including the radiation, return losses, and peak gain, are tested by using vector network analysis device named Agilent PNA N5230A and simulated by using antenna design kit named ANSOFT HFSS. Fig. 3 shows the types of proposed antenna radiation patterns which are displayed in YZ and XZ planes with the operating frequency at 915 MHz. The wavebeam width of antenna is omnidirectional and the wave-beam for the power is greater than 0dB is 360 degree. Fig. 4 shows the measured -10 dB return losses from 795MHz up to 1 GHz. This indicates that three antenna are all suitable for the ultra high frequency (UHF) RFID application. Fig. 5 shows the measured peak gain of antenna. For our proposed antenna, the peak gain for the proposed antennas is 3.01dB at 915MHz operating frequency. Fig. 6 and Fig. 7 are the radiation patterns of RFID antenna from +X and +Y.

DOI: 10.1051/matecconf/201712300028

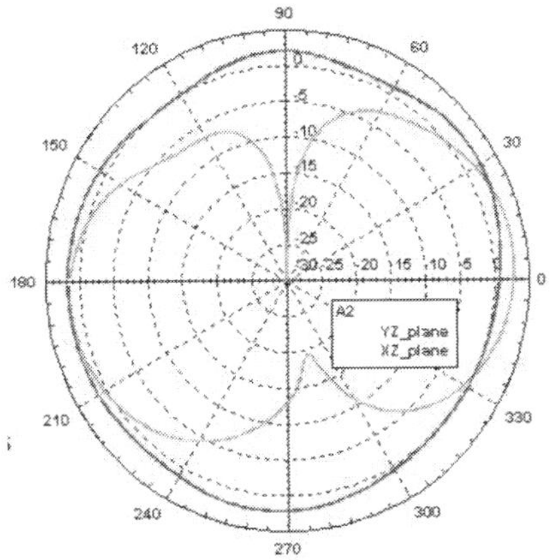

Fig. 3. Measured radiation patterns for XZ and YZ plane at 915MHz.

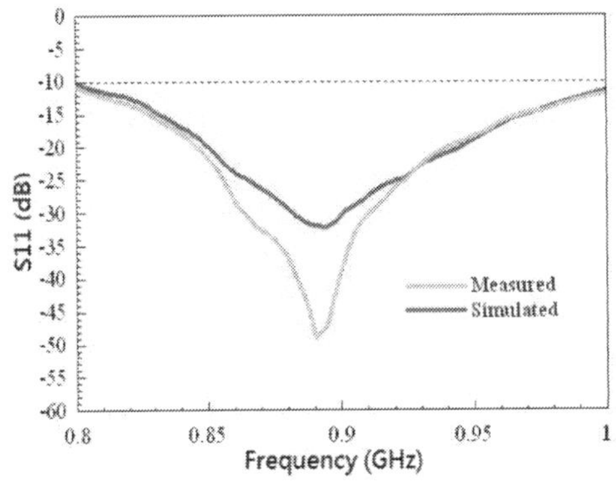

Fig. 4. Measured return losse of RFID antenna.

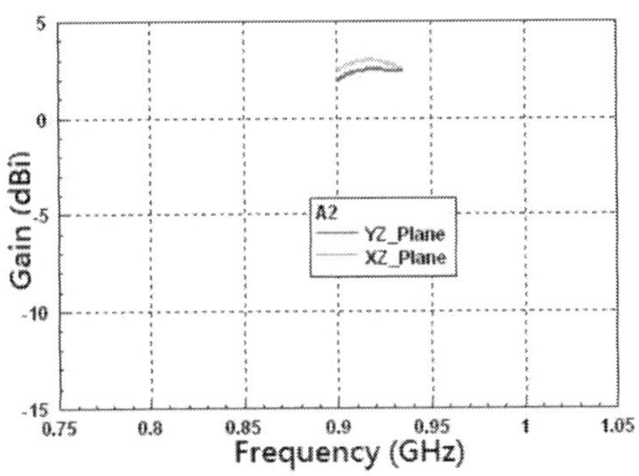

Fig. 5. Measured total field of RFID maximum gain.

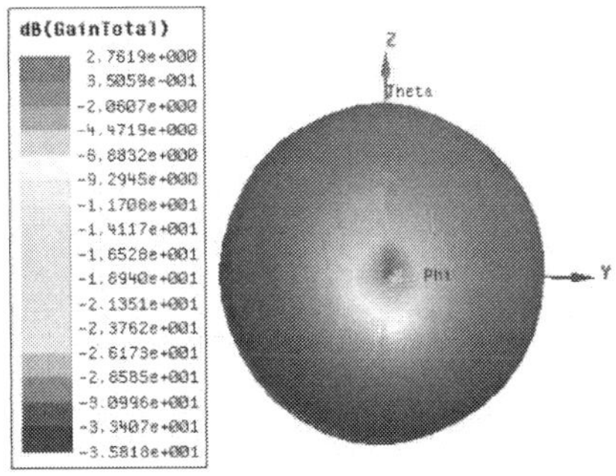

Fig. 6. Radiation pattern of RFID antenna (from +X).

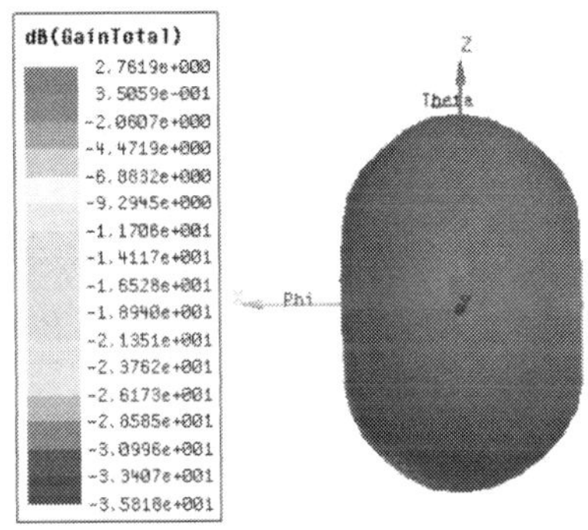

Fig. 7. Radiation pattern of antenna (from +Y).

3 Library inventory system design

Fig. 8 illustrates the flow chart of the optimal library inventory technology with four steps: deciding the tag type, finding the optimal tag location for a book, testing the tag readability for multi-layer bookshelves, and connecting couples of multi-layer bookshelves with multiplexers and u updating the tag reading status in the computer terminal database.

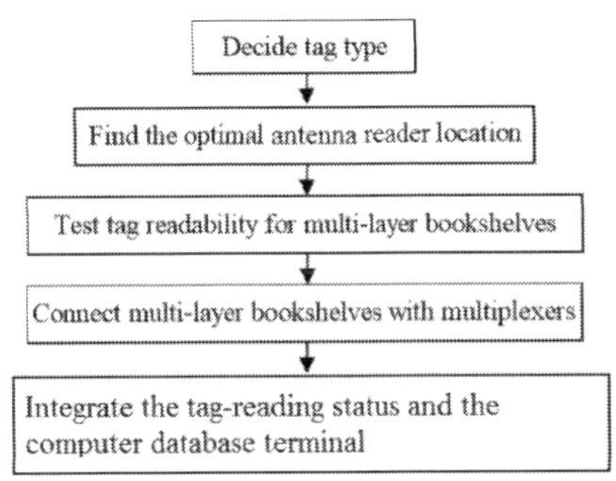

Fig. 8. Flow chart of optimal library inventory technology.

3.1 Deciding tag type and finding optimal antenna reader location

In this study, the authors propose an optimal library inventory system based on electromagnetic identification

(EMID) technology with the advantages of using electromagnetic (EM) tag and radio frequency identification (RFID) tag. Besides, embedded electromagnetic identification (EMID) security strips, as shown in Fig. 9, are used for securing and managing the stocks on the library bookshelves. With security advantage and radio frequency identification (RFID) tags with advantages of managing distribution and miniature, this kind of tag combines the technology of electromagnetic (EM) strips.

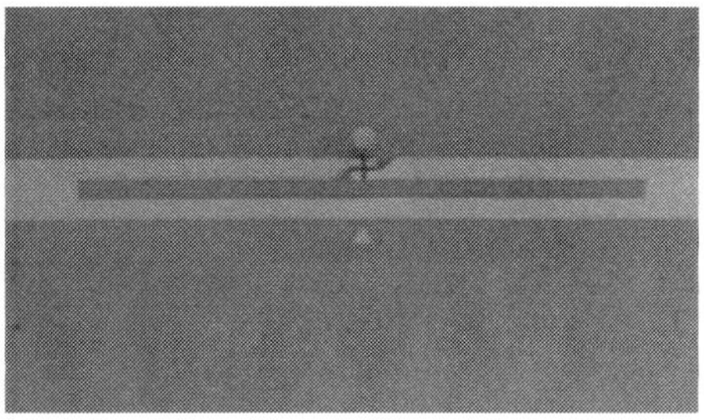

Fig. 9. EMID security strips.

Fig. 10 is the architecture of a bookshelf with the width and height of 90 cm and 19 cm, respectively. In this study, three places labeled as A, B and C in Fig. 2, are tested to find the optimal antenna reader location. The place A is located on the left hand side of the bookshelf; the place B is located on the middle of the bookshelf; the place C is located either on the 30 cm from the distance of left or right hand sides of the bookshelf.

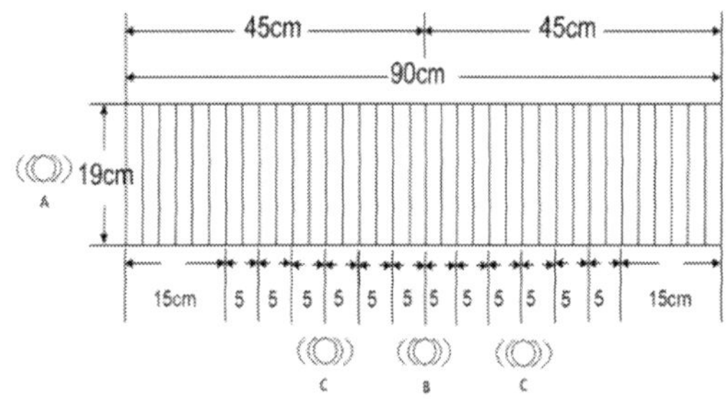

Fig. 10. Antenna locations on the one layer bookshelf.

3.2 Connecting multi-layer bookshelves and testing tag readability

Fig. 11 illustrates the structure of connecting multi-layer bookshelves in a library by using multiplexers. As shown in Fig. 11, by rolling every EMID tag, the library computer terminal through multiplexers and readers obtains antenna's signals. The database of the computer will authenticate the location of the book after the computer obtained the EMID tag signal. This indicates that whether the book is on the bookshelf or not the library inventory work can be automatically done. Fig. 12 shows antenna's locations in a multi-layer bookshelf to test the tag readability.

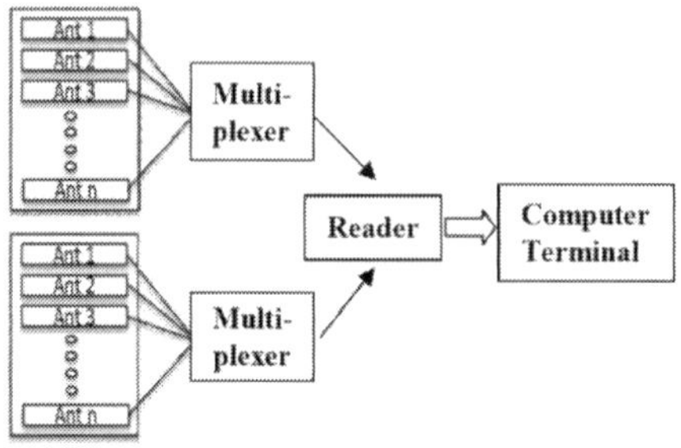

Fig. 11. Connection of multi-layer bookshelves.

Fig. 12. Antenna locations in multi-layer bookshelves.

3.3 Integrate tag-reading status and computer database terminal on software

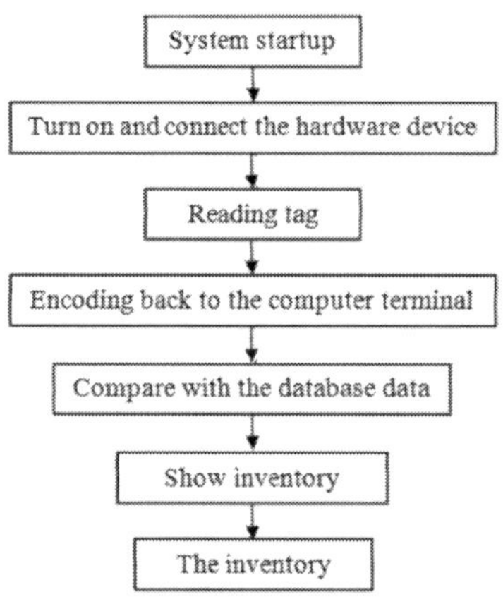

Fig.13. Flow chart of inventory system implementation.

Fig. 13 illustrates the flow chart of the inventory system technology implementation which includes seven steps: system startup, turning on and connecting the hardware device, reading tag code, encoding back to the computer terminal, comparing with the database data, showing inventory results, and the inventory ends. Fig. 14 shows the user interfaces of reading EMID tag strips and administrator's database management in Chinese on the part of the computer system in which the user interface and database are built by C++ codes. There are six main functions shown on the screen of Fig. 14(a), including IP address setting of the main-control-unit Multiplex Master, the subordinate-unit-multiplexer (Slave) of the main-control-unit Multiplex Master, determining the antenna port the RFID Reader will transmit and receive, and displaying the antenna connection status (the antenna number hole will display a green light when the antenna is connected). As for Fig. 14(b), there are five main functions shown on the screen, including EMID tag number, book title, author's name, publisher, and reading status. According to the experimental results, the proposed system with easy operations, by checking the screen of Fig. 14(b), allows the library administrator to identify whether a book is on the bookshelf or not and to accomplish library inventory work efficiently and effectively. The relevant test results will be described in the next section.

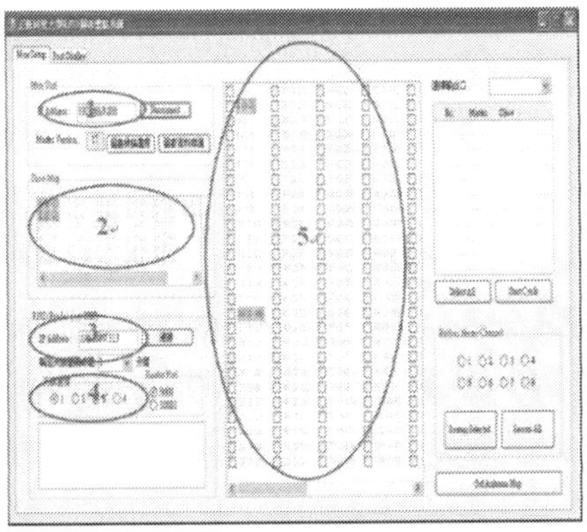

(a) reading EMID tag

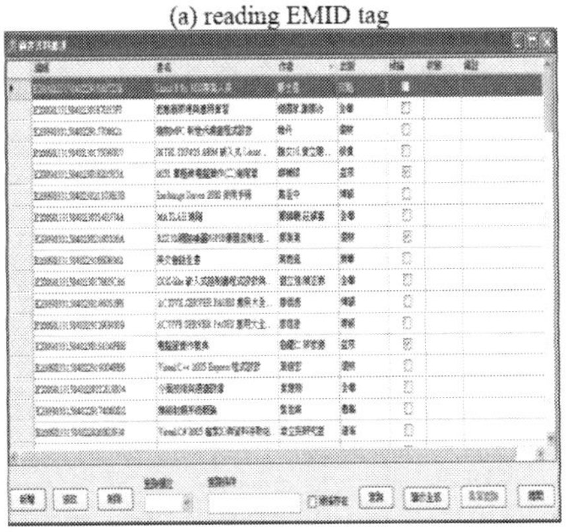

(b) book database management

Fig. 14. User interfaces for managing stocks in library (a) reading EMID tag (b) book database management.

4 Experimental results and discussion

In order to minimize the disturbance during testing, the proposed system is experimented at night during the library closing time interval. Based on the polling inquiry methodology, the system would automatically read every EMID tag individually by switching different multiplexers gradually. On the other side, the user interface of the computer terminal would show the reading status of each EMID tag and the reading data is then stored in the computer terminal database.

Table 2. Reading rate for different RFID tag locations.

EMID Tag Quantity	Location	Reading Quantity	Reading Rate (%)
30	A	15	50%
	B	30	100%
	C	30	100%

Table 2 shows the reading rate of EMID tags located at A, B, and C places as shown in Fig. 3. From Table 1, the reading rate of 30 EMID tag strips at the location A, B, and C are 50%, 100%, and 100%, respectively. This implies that either location B or C gets better reading rate. Table 2 gives the reading rate of 25 EMID tag strips with different measuring time for one-sided bookshelf.

Table 3. Reading rate for one side bookshelf.

EMID Tag Quantity	Time Interval(Min.)	Reading Quantity	Reading Rate (%)
25	1	15	60%
	3	23	88%
	5	25	100%

From Table 3, the reading rate of 25 EMID tag strips with the time interval of 1, 3, and 5 minutes are 60%, 88%, and 100%, respectively. This implies that measuring time with 5 minute interval gets the best reading rate.

Table 4. Reading rate for two side bookshelves.

EMID Tag Quantity	Time Interval(Min.)	Reading Quantity	Reading Rate (%)
30	1	17	56.6%
	3	24	80%
	5	28	93.3%

Table 4 illustrates the reading rate of 30 EMID tag strips with different measuring time for two-sided bookshelves. From Table 4, the reading rate of 30 EMID tag strips with the time interval of 1, 3, and 5 minutes are 56.6%, 80%, and 93.3%, respectively. This implies that measuring time with 5 minute interval gets the best reading rate.

As for experimental results, when the bookcase with width and depth being 90 cm and 19 cm is chosen, the receiving angle must exceed 136.4 degree. And when the choice of the bookshelf length is 90 cm, the receiving angle can be reduced to 112.6 degree. In general, the width of the library shelves is mostly between 90-150 cm, so the omni-directional radiation antenna has a good performance of wide beam and omni-directional radiation advantage, more suitable for library inventory management.

5 Conclusion

This paper presents the RFID system used in the library inventory management, RFID antenna design and the actual test results. We propose a RFID antenna design method and focus on the diffusion range of omnidirectional field radiation. The center frequency of the designed RFID antenna works in the UHF 915 MHz, and is made on the printed circuit board as well as the plane structure, making it easy to be applied in the book inventory system. The features of the real-made antenna include a good gain and omni-directional radiation pattern, a simple architecture and low manufacturing cost. The results show that the omni-directional radiation antenna has good performance of omnidirectional radiation advantage and good detection accuracy, suitable for use in library shelves for large-scale book inventory operations and it can save a lot of manpower and reduce the risk of manual errors. In addition, an optimal library inventory system based on EMID technology has been proposed in this study. The main contribution of the proposed system is to improve traditional human labor-based library inventory methodology by decreasing the inventory processing time and reducing the possibility of the books being misplaced.

References

1. C.W. Chou, The application of radio frequency identification technology to the library real-time storage, Department of Engineering Science, National Cheng-Kung University, Kaohsiung, Taiwan, 2006
2. M.S. Selamat, B.Y. Majlis, Challenges in implementing RFID tag in a conventional library, *IEEE Conf. on Semiconductor Electronics*, 258-262 (2006)
3. S.F. Chiou, S.K. Lin, S.L. Chen, Dual-radiation pattern RFID tag design for application in library management, *IEEE 4th Asia-Pacific Conf. on Antennas and Propagation*, 176-178 (2015)
4. R.J. Miao, J. Xu, J.S. Wu, Y.R. Zheng, H.H. Gao, Design and implement of libaray intelligent management system based on RFID, *International 4th Conf. on Intelligent Systems Design and Engineering Applications*, 288-291 (2015)
5. C. Feng, Research for application of RFID in library, *IEEE Conf. on Computer and Communication Technologies in Agriculture Engineering*, 262-264 (2010)
6. A. Boursianis,T. Samaras, A.C. Polycarpou and J.N. Sahalos, A UHF RFID reader antenna for searching tagged items, *IEEE Conf. on RFID Technology and Applications*, 193-198 (2014)
7. I. Ehrenberg, C. Floerkemeier, S. Sarma, Inventory management with an RFID-equipped mobile robot. *IEEE Conf. on Automation Science and Engineering*, 1020-1026 (2007)
8. C. R. Medeiros, J. R. Costa, C. A. Fernandes, RFID smart bookshelf with confined detection range at UHF. *European Conf. on Antenna and Propagation*, 2779- 2783 (2009)
9. U. Kim, J. Choi, Design of a wideband rectangular loop tag antenna for UHF RFID library management, *Microwave and Optical Technology Letters*, 1439-1445, (2010)
10. P.Y. Lau, K.K. Yung, E.K. Yung, A low-cost printed CP patch antenna for RFID smart bookshelf in library, *IEEE Transactions on Industrial Electronics*, **57**,1583-1589 (2010)
11. U. Kim, J. Koo, J. Choi, Design of a UHF band RFID tag antenna for library management system application, *Asia-Pacific Microwave Conference*, 1-4 (2008)
12. I.J. Lukas, X. Qing, Z.N. Chen, A. Alphone, UHF omnidirectional bent dipole antenna for item-level RFID applications, *IEEE International Symposium on Antennas and Propagation Society*, 1-4 (2008)
13. K.T. Kaiwen, X. Qing, C.K. Goh, L. Zhu, *IEEE International Symposium on Antennas and Propagation Society*, 1-4 (2010)
14. P. Golding and V. Tennant, Performance characterization for two radio frequency identification inventory readers within a university library environment, International Journal of Radio Frequency Identification Technology and Applications, *Special Issue on RFID-Enhanced Technology Intelligence and Management*, 107-123 (2011)
15. L. Wang, RFID-based technology intelligence in libraries, *International Journal of Technology Intelligence and Planning*, 32-41 (2010)
16. D. Zhang, X. Shi, Self-service management platform design for library based on RFID. *2nd International Conference on Artificial Intelligence, Management Science and Electronic Commerce*, 7237-7240 (2011)
17. A.T. Bin Abdullah, I.B. Ismail, A.B. Ibrahim, M.Z. Hakim Bin Noor, Library shelf management system using RFID technology, *IEEE International Conf. on System Engineering and Technology*, 215-218 (2011)
18. D.Y. Li, S.D. Xie, R.J. Chen, H.Z. Tan, Design of Internet of Things System for Library Materials Management using UHF RFID, *IEEE International Conf. on RFID Technology and Applications*, 44-48 (2016)
19. X.Z. Huang, Application practice and analysis research of RFID technology in digital library, *International Symposium on Information Technologies in Medicine and Education*, 473-475 (2012)
20. P. Golding, V. Tennant, Work in Progress: Performance and reliability of radio frequency identification (RFID) library system, *International Conference on Multimedia and Ubiquitous Engineering*, (1143-1146) 2007
21. I. Ehrenberg, C. Floerkemeier, and S. Sarma, Inventory management with an RFID-equipped mobile robot, *IEEE Conf. on Automation Science and Engineering*, 1020-1026 (2007)

22. Z.A. Mohideen, S. Muhamad, M.P. Ghadzali and M.R.M. Arshad, A practical approach to radio frequency identification library management system, *IEEE Conf. on RFID-Technologies and Applications*, 323-329 (2012)
23. H.M. Cheng, L. Huang, H. Xu, Y.F. Hu, X.A. Wang, Design and implementation of library books search and management system using RFID technology, *International Conference on Intelligent Networking and Collaborative*, 329-327 (2016)
24. Osama M. Haraz,Abdel-RazikSebak, Quadband planar PCB antenna for WLAN and mobile WiMAX applications, *IEEE International Symposium on Antennas and Propagation*, 416-418 (2011)

MATEC Web of Conferences **123**, 00029 (2017) DOI:10.1051/matecconf/201712300029
ICPMMT 2017

Quality prediction modeling for multistage manufacturing based on classification and association rule mining

Hung-An Kao[1,2,*], *Yan-Shou* Hsieh[1], *Cheng-Hui* Chen[1], and *Jay* Lee[2]

[1] Central Industry Research & Service Division (CID), Institute for Information Industry, Nantou, 540, Taiwan
[2] NSF I/UCRC for Intelligent Maintenance Systems (IMS), University of Cincinnati, Cincinnati, OH 45221, USA

Abstract. For manufacturing enterprises, product quality is a key factor to assess production capability and increase their core competence. To reduce external failure cost, many research and methodology have been introduced in order to improve process yield rate, such as TQC/TQM, Shewhart Cycle、Deming's 14 Points, etc. Nowadays, impressive progress has been made in process monitoring and industrial data analysis because of the Industry 4.0 trend. Industries start to utilize quality control (QC) methodology to lower inspection overhead and internal failure cost. Currently, the focus of QC is mostly in the inspection of single workstation and final product, however, for multistage manufacturing, many factors (like equipment, operators, parameters, etc.) can have cumulative and interactive effects to the final quality. When failure occurs, it is difficult to resume the original settings for cause analysis. To address these problems, this research proposes a combination of principal components analysis (PCA) with classification and association rule mining algorithms to extract features representing relationship of multiple workstations, predict final product quality, and analyze the root-cause of product defect. The method is demonstrated on a semiconductor data set.

1 Introduction

For manufacturing enterprises, product quality is a key factor to assess production capability and increase their core competence. To reduce external failure cost, many research and methodology have been introduced in order to improve process yield rate, such as TQC/TQM, Shewhart Cycle、Deming's 14 Points, etc. Nowadays, impressive progress has been made in process monitoring and industrial data analysis because of the Industry 4.0 trend. Industries start to utilize quality control (QC) methodology to lower inspection overhead and internal failure cost. Currently, the focus of QC is mostly in the inspection of single workstation and final product, however, for multistage manufacturing system (MMS), many factors (like equipment, operators, parameters, etc.) can have cumulative and interactive effects to the final quality. When failure occurs, it is difficult to resume the original settings for cause analysis.

Currently, most state-of-art manufacturing lines are actually quite 'smart' in themselves, which means more and more build-in sophisticated sensors and computerized components while they are designed or increasingly various add-on sensors to monitor tools in real-time. The data delivered via both build-in and add-on sensors are highly related with process's status and performance. However, it is difficult for field engineers and management staff to get the information of product quality just through checking the big amount of mixed data, not mention being able to track the degradation trend, which will eventually lead to a catastrophic failure. Therefore, it is necessary to develop a data-to-information conversion tools that are able to convert process data into final quality and performance-related information. The output of these analytics is the real-time quality index after each workstation, which shows the current performance, for decision makers to effectively aware of the situation and make compensation decisions before potential failures occur. It can largely prevent waste in terms of time, spare parts and personnel.

To address these problems, statistical methods and machine learning technique are two main research target. Multivariate statistics was used to build quality prediction model for MMS, however, the correlation cannot be easily revealed for further verification and improvement [1, 2]. Some research treated variables from different stages as one single point and applied machine learning technique such as Naïve-Bayes, PCA, SVM, ANN, KNN, etc [3-5]. Naïve Bayes is a classifier it based on statistical knowledge. Naïve Bayes assumes independency between attributes and it attempts to maximize the posterior probability in determining the class. The advantage of using naive bayes classifiers is it have works well in many complex real-world situations. The k-nearest neighbor (k-NN) algorithm is a classifier based on closest data in feature space. It is amongst the simplest of all machine learning algorithms. But the

* Corresponding author: hakao@iii.org.tw

accuracy of classification will be effected by noisy or irrelevant features. It is one of the methods that are receiving increasing attention with remarkable results. SVM algorithm constructing an optimal separating hyper plane in the hidden feature space using quadratic programming to find a unique solution. The Artificial neural networks (ANNs) is based on human brain by making the right connections. It is a flexible mathematical structure. The structure is capable of identifying complex nonlinear relationships between input and output data sets. Like other machine methods, ANN have been used to solve a wide variety of tasks [5]. On the other side, some research consider the characteristics of MMS and conduct learning after extract related features [6-8].

In this research, we propose a combination of principal components analysis (PCA) with classification and association rule mining algorithms to extract features representing relationship of multiple workstations, predict final product quality, and analyze the root-cause of product defect. The method is demonstrated on a semiconductor data set.

2 Methodology

In this research, classification and rule mining algorithms are implemented and compared for quality prediction. To consider the characteristics of MMS, single point approach and multi-point approach are both discussed. Firstly, the original manufacturing dataset will be pre-processed to remove noise and segmented into training and testing dataset. Secondly, we build the quality prediction model through classification and rule mining process. For classifier learning, the relationship among workstations and the impact to the final product are included in feature selection so the feature set can better represent MMS property. For rule mining, a priori algorithm is applied on the training dataset after minimum support and confidence threshold are defined. The output classifiers and association rules are evaluated by testing dataset. Finally, the model can be used online for quality prediction and defect cause analysis in multiple workstation scenarios. Figure 1 shows the flowchart of the proposed methodology.

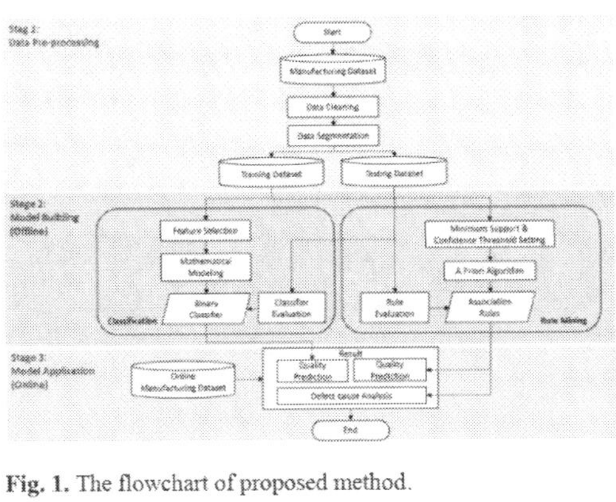

Fig. 1. The flowchart of proposed method.

2.1 Data pre-processing

Since manufacturing environment may have lot of anomaly events and situations, data pre-processing including cleaning and normalization is important. To deal with "Missing Value" issue, in this research we replace missing values with the mean value that calculated from all instances. Besides, if values from all instances of a certain variable are the same, the variable will be removed since it does not carry any discriminative information.

Another issue for quality prediction research is the insufficient instances for failed products. In recent manufacturing lines, typical shop floor is quite stable that the yield rate is usually higher than 85%. Hence, to balance the size of positive and negative datasets, over-sampling [9] and boosting technique [4] are introduced. In this research, we adopt the boosting technique by simply duplicate the fail dataset to the same amount of pass dataset.

2.2 Model building based on classification

To consider the characteristic of MMS, Cascade Quality Prediction Method (CQPM) [7, 10] is implemented to mine the hidden relationship. Figure 2 shows the scenario of MMS and idea of CQPM. In CQPM, three types of relationship are existed in MMS:

- R1: Relationship among manufacturing operation in a workstation.
- R2: Relationship among workstations.
- R3: Relationship between manufacturing operation variables and final product quality.

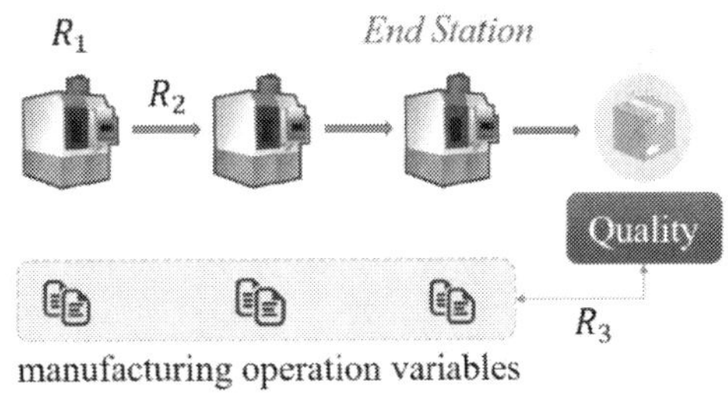

Fig. 2. Scenario of MMS and relationships identified with CQPM.

Based on CQPM definition, it is necessary to find the relationships of inter-correlated variables. PCA can then be used to transform latent variables into a new set of dimensions while they carry MMS characteristics [7]. After the feature set is selected, several classification algorithm, including decision tree, support vector machine (SVM), and Naive Bayes, will then be applied in order to train the binary classifier. For decision tree learning, iterative dichotomiser 3 (ID3), recursive partitioning and regression trees (RPART), and C50 are potential algorithms for decision tree learning.

2.3 Model building based on rule mining

Dependency modeling is to mine the association rules between the target variable and the independent variables. This area of research is very active in financial data mining and is usually referred to as "association rule" mining. The most classical algorithm is the a priori algorithm proposed by Agrawal and Srikant in 1994 [11]. This approach finds the frequent items and builds relations aggressively. Then the algorithm will prune the rules by removing the less frequent itemsets. Huang et al. [12] proposed a method based on the physical facts. Dependency modeling was to build physics-based model to describe the relationship between important IC chip power parameters and temperature. Other approaches include principal component analysis contribution plot [13], or partial least squares analysis [14].

In this research, a priori algorithm is applied to the manufacturing dataset in order to mine the relationship model between operation parameters and quality outcomes. Firstly, assume there is a set T formed from a manufacturing dataset, a priori algorithm will find several association rules X ($X \in T$) that can impact final quality Y and meanwhile satisfy minimum property of support. The sets are called frequent item sets. Secondly, a support threshold M is defined and each frequent item set consists at least M items. Then, breadth-first search (BFS) is used to prune the items that do not meet the requirement. The equation of a priori algorithm to calculate support property of item sets is as follows:

$$Support(X) = |t \in T; X \in T| / |T| \tag{1}$$

Support(X) indicates the support degree of the association rule X, and t stands for a single manufacturing rule. Therefore, the support degree can show the proportion of the frequency of each rule appears in the dataset. The confidence value can then be determined by the probability of Y result once X rule is true. The equation is as follows:

$$Confidence(X,Y) = Support(X \cup Y) / Support(X) \tag{2}$$

To describe the correlation degree between X and Y, *Lift* is also defined as:

$$Lift(X,Y) = \frac{Support(X \cup Y)}{Support(X) * Support(Y)} \tag{3}$$

When *Lift(X,Y)* is 1, X and Y are independent and no association rules can be extracted. The correlation between X and Y is higher if *Lift* value is larger. Generally, a priori algorithm is a filtering process based on minimum support and minimum confidence, which are user-defined thresholds. Hence, the involvement of expert knowledge is encouraged in order to have a better filtering process.

3 Case study

3.1 Dataset description

To verify the proposed methodology, a semiconductor dataset SECOM (SEmiCOnductor Manufacturing) [15] is utilized. This dataset includes 1567 samples, and each consists of 590 manufacturing operation variables and 1 quality variable. Among the dataset, only 104 samples represent failure case. Each operation variable data is collected from a certain process control sensor from semiconductor manufacturing line and is named based on sensor ID. To balance positive and negative dataset, boosting is applied on the dataset so the number of failed instance is 1456. The training and testing dataset are segmented from the original dataset on a 3:1 basis.

For data pre-processing, firstly, we remove 115 redundant features during data pre-processing stage. Secondly, after feature extraction stage, 40 most important features are kept for the following classifier learning [15]. To simulate the scenario of MMS, the 40 features are separated into five groups (each representing a workstation) following the property of semiconductor monitoring process described in [16]. Before classification, PCA is applied on the dataset to acquire features with correlations. For each workstation, the features will be combined with previous workstations and through PCA we can get principle components (PC). The final data set includes 14 correlation features and one quality result label.

3.2 Results

Results of the six models learned from SECOM dataset are given in Table 1. To compare our methodology with state-of-the-art work, two references that have done research on SECOM dataset are listed. The evaluation metrics include: accuracy, true positive rate (TP rate), geometric mean (g-mean), balanced error rate (BER), F1 score (also F-score or F-measure), and false alarm rate (FAR). To definition of the confusion matrix for SECOM dataset and the equations for the above metrics are shown in Table 2 and equations (4) to (9).

Table 1. Results of the proposed method and related work.

	ID3	CART	C5.0	SVM	Naïve Bayes	A Priori	Rough Set	Ref. [7]	Ref. [4]
Consider MMS?	Y	Y	Y	Y	Y	N	N	Y	N
Accuracy	0.682	0.818	0.954	0.683	0.681	0.773	**0.984**	0.900	0.859
TP rate	0.698	0.790	0.915	0.712	0.697	0.668	**1.000**	0.209	**1.000**
G-mean	0.682	0.820	0.957	0.687	0.682	0.765	**0.992**	0.445	0.916
BER	0.683	0.821	0.958	0.687	0.682	0.765	**0.992**	0.578	0.919
F1 score	0.666	0.826	0.956	0.659	0.665	0.746	**0.992**	0.209	0.641
FAR	0.333	0.148	0.000	0.338	0.333	0.124	**0.0164**	0.053	0.161

Table 2. Confusion matrix for quality prediction dataset.

		Predicted Condition	
		Class = 1 (Fail)	Class = -1 (Pass)
True Condition	Class = 1 (Fail)	TP (True Positive)	FN (False Negative)
	Class = -1 (Pass)	FP (False Positive)	TN (True Negative)

$$Accuracy = \frac{TP + TN}{TP + TN + FP + FN} \quad (4)$$

$$TP\ rate = \frac{TP}{TP + FN} \quad (5)$$

$$G - mean = \sqrt{\frac{TP}{TP + FN} \times \frac{TN}{TN + FP}} \quad (6)$$

$$BER = 1 - 0.5 \times \left(\left(\frac{FN}{TP + FN} \right) + \left(\frac{FP}{TN + FP} \right) \right) \quad (7)$$

$$F1\ score = \frac{2 \times TP}{(2 \times TP + FP + FN)} \quad (8)$$

$$FAR = \frac{FP}{FP + TN} \quad (9)$$

From the results shown in Table 1, Rough Set learning performs best among our seven models and two previous work. With a 100% TP rate, all failed product can be detected during manufacturing stage. Moreover, the false alarm rate is acceptable (about 1.64%). For quality prediction, the motivation is to relief the overhead of quality inspection and detect product failure at early stage. Therefore, high TP rate with low false alarm rate is the most important thing to consider. Compared with the work of Kerdprasop, K. et at. [4], who used decision tree with boosting technique on the same dataset, the proposed method gain better performance that indicate the introduction of MMS characteristics can benefit the model building process. The result from the work of Arif, F. et al. [7] has the good FAR, however, the very low TP rate is not applicable for quality prediction purpose. The confusion matrix of Rough Set result is shown in Table 3.

Table 3. Confusion matrix of Rough Set learning result.

		Predicted Condition	
		Class = 1 (Fail)	Class = -1 (Pass)
True Condition	Class = 1 (Fail)	1456	0
	Class = -1 (Pass)	24	1439

3.3 Discussion on Defect Cause Analysis

Beyond detect product quality, it is also a key to understand the root cause of product failure. In this case, association rule mining is a useful tool for mining the rules representing the relationship between manufacturing operation variables and final quality. The experiment set minimum confidence to 0.2 and minimum support to 2. The repetitive rules are eliminated, and there are 978,664 rules extracted. To avoid rules for exceptions and less frequently occurs, In training dataset, the minimum support is set to be 0.012 so only 483 valid rules remain. Table 4 shows the top 16 rules extracted.

Table 4. Top 16 association rules that affect product quality.

LHS	Support	Confidence	Lift
{V65=very-high,V81=low}	0.0283	1	2.005
{V29=normal-high,V65=very-high}	0.0247	1	2.005
{V65=very-high,V320=normal-high}	0.0238	1	2.005
{V386=high,V547=low}	0.0224	1	2.005
{V354=low,V576=high}	0.0210	1	2.005
{V67=very-low,V512=high}	0.0206	1	2.005
{V67=very-low,V344=normal-high}	0.0206	1	2.005
{V274=very-high,V384=normal-low}	0.0206	1	2.005
{V246=normal-low,V274=very-high}	0.0206	1	2.005
{V274=very-high,V518=normal-low}	0.0206	1	2.005
{V274=very-high,V383=normal-low}	0.0206	1	2.005
{V245=normal-low,V274=very-high}	0.0206	1	2.005
{V274=very-	0.0206	1	2.005

high,V517=normal-low}			
{V63=very-low,V78=low}	0.0201	1	2.005
{V67=very-low,V78=low}	0.0201	1	2.005
{V71=very-low,V78=low}	0.0201	1	2.005

Right hand side (RHS) represents instances of failed product and left hand side (LHS) is the association rule set that can affect RHS. Among 483 association rules, {V65=very-high,V81=low} is the most frequent rule. All the rules extracted have confidence value 1. To visualize the relationship of top 16 rules, Figure 3 shows the balloon plot. Each balloon indicates a valid rule. The bigger size is related to the higher support degree, and the darker color means higher lift and confidence value. The arrows pointing to the balloon are the features that form the rule. If one feature connect to many balloons, the feature can be commonly seen in many different rules. From Figure 3, "V65=very-high" and "V274=very-high" are two important rules.

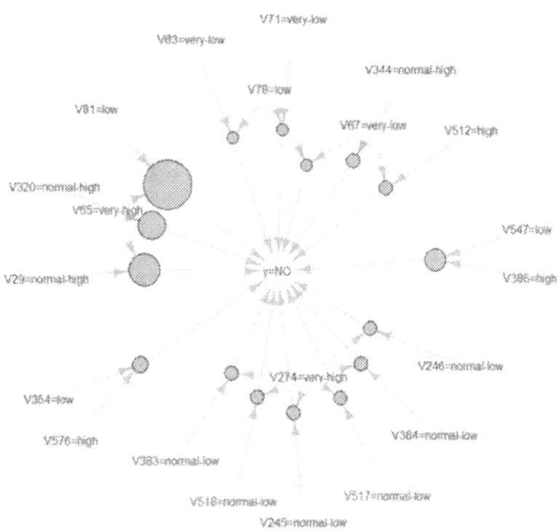

Fig. 3. Balloon plot for visualizing extracted association rules.

The quality prediction model is then build with the rules that pass the confidence and minimum support threshold. After applying rules on the testing dataset, 290 product failures are captured, while 74 failed instances are misclassified. Five rules with accurate prediction are listed in Table 5. From the results, many instances meet more than one rules. If the physical meaning of variables is known, the rules can be improved by expertise knowledge involvement and have more accurate prediction performance.

Table 5. Results of defect cause analysis.

Instance ID	Cause
367	{V441=high,V520=normal-high},{V170=high,V520=normal-high},{V80=very-high,V170=high},{V7=high,V432=high},{V60=normal-high,V432=high},{V80=very-high,V497=high},{V424=high,V426=very-high},{V295=normal-high,V426=very-high},{V160=normal-high,V426=very-high},{V426=very-high,V490=normal-high},{V426=very-high,V471=normal-high},
368	{V157=high,V456=low},{V157=high,V184=low},{V157=high,V320=low},{V356=normal-high,V386=high},{V2=low,V430=high},
370	{V355=low,V586=high},{V317=low,V568=low},{V217=low,V586=high},{V355=low,V584=high},
384	{V438=high,V563=very-low},{V167=high,V563=very-low},{V81=high,V438=high},{V13=low,V201=low},{V60=normal-high,V201=low},{V201=low,V286=normal-high},
390	{V124=low,V561=very-high},

4 Conclusion

While information communications technologies become mature and easily adopted, factories tend to have more data from add-on sensors, machine controllers, metrology, maintenance history, and all kinds of e-manufacturing systems. How to process the data for a deep understanding and insight on product quality is definitely a key for manufacturing industries to be competitive in this Industry 4.0 revolution.

In this paper, a framework for quality prediction modeling in MMS environment is introduced. The proposed methodology with the capability on predictive analysis can benefits manufacturers to acquire the real-time quality information of their manufacturing lines earlier and improve current process immediately. Eventually, it can optimize the manufacturing process and also can extend to the collaboration among manufacturers, suppliers and customers.

This study is conducted under the "III Innovative and Prospective Technologies Project (1/1)" of the Institute for Information Industry which is subsidized by the Ministry of Economic Affairs of the Republic of China.

References

1. Y. Qi, P. Wang, and X. Gao. Enhanced batch process monitoring and quality prediction using multi-phase dynamic PLS, in *Control Conference (CCC), 2011 30th Chinese*. 2011. IEEE
2. C. Zhao et al., Stage-based soft-transition multiple PCA modeling and on-line monitoring strategy for batch processes. *Journal of Process Control*, 2007. **17**(9): pp. 728-741
3. M. McCann et al., Causality Challenge: Benchmarking relevant signal components for effective monitoring and process control, in *NIPS Causality: Objectives and Assessment*. 2010
4. K. Kerdprasop, N. Kerdprasop, Feature selection and boosting techniques to improve fault detection accuracy in the semiconductor manufacturing process, in *Proceedings of the International MultiConference of Engineers and Computer Scientist*. 2011
5. S. Munirathinam, B. Ramadoss, Predictive Models for Equipment Fault Detection in the Semiconductor Manufacturing Process, *International Journal of Engineering and Technology*, 2016. **8**(4): p. 273
6. D. Djurdjanovic, J. Ni, Stream of variation based analysis and synthesis of measurement schemes in multi-station machining systems, Ann Arbor, 2001. **1001**: pp. 2109-2125
7. F.Arif, N. Suryana, and B. Hussin, A data mining approach for developing quality prediction model in multi-stage manufacturing, *International Journal of Computer Applications*, 2013. **69**(22)
8. P. Jiang et al., Real-time quality monitoring and predicting model based on error propagation networks for multistage machining processes, *Journal of Intelligent Manufacturing*, 2014. **25**(3): pp. 521-538
9. K. Chomboon, K. Kerdprasop, and N. Kerdprasop, Rare class discovery techniques for highly imbalance data, in *Proc. International multi conference of engineers and computer scientists*. 2013
10. K. Salahshoor, H.K. Alaei, and H.K. Alaei. A new on-line predictive monitoring using an integrated approach adaptive filter and PCA, in *Soft Computing Applications (SOFA), 2010 4th International Workshop on*. 2010. IEEE
11. R. Agrawal, R. Srikant. Fast algorithms for mining association rules, in *Proc. 20th int. conf. very large data bases, VLDB*. 1994.
12. H. Huang, G. Quan, and J. Fan, Leakage temperature dependency modeling in system level analysis, in *Quality Electronic Design (ISQED), 2010 11th International Symposium on*. 2010. IEEE
13. K. Friston et al., Functional connectivity: the principal-component analysis of large (PET) data sets, Journal of Cerebral Blood Flow & Metabolism, 1993. **13**(1): pp. 5-14
14. H. Wold, *Partial least squares*. Encyclopedia of statistical sciences, 1985
15. P. Murphy, D. Aha, UCIML repository secom dataset
16. G.S. May, C.J. Spanos, *Fundamentals of semiconductor manufacturing and process control.*, 2006: John Wiley & Sons

Influence of thin film thickness of working electrodes on photovoltaic characteristics of dye-sensitized solar cells

Yeong-Lin Lai[1,*], *Hung-Ru* Hsu[2], *Yeong-Kang* Lai[3], *Chun-Yi* Zheng[1], *Yung-Hua* Chou[1], *Nai-Kun* Hsu[1], and *Guan-Yun* Lung[1]

[1]Department of Mechatronics Engineering, National Changhua University of Education, 50007 Changhua, Taiwan
[2]Green Energy and Environment Research Laboratories, Industrial Technology Research Institute, 31040 Hsinchu, Taiwan
[3]Department of Electrical Engineering, National Chung Hsing University, 40227 Taichung, Taiwan

Abstract. This paper presents the study of the influence of thin film thickness of working electrodes on the photovoltaic characteristics of dye-sensitized solar cells. Titanium dioxide (TiO_2) thin films, with the thickness from 7.67 to 24.3 μm, were used to fabricate the working electrodes of dye-sensitized solar cells (DSSCs). A TiO_2 film was coated on a fluorine-doped tin oxide (FTO) conductive glass substrate and then sintered in a high-temperature furnace. On the other hand, platinum (Pt) solution was coated onto an FTO substrate for the fabrication of the counter electrode of a DSSC. The working electrode immersed in a dye, the counter electrode, and the electrolyte were assembled to complete a sandwich-structure DSSC. The material analysis of the TiO_2 films of DSSCs was carried out by scanning electron microscopy (SEM) and ultraviolet-visible (UV-Vis) spectroscopy, while the photovoltaic characteristics of DSSCs were measured by an AM-1.5 sunlight simulator. The light transmittance characteristics of the TiO_2 working electrode depend on the TiO_2 film thickness. The thin film thickness of the working electrode also affects the light absorption of a dye and results in the photovoltaic characteristics of the DSSC, including open-circuited voltage (V_{OC}), short-circuited current density (J_{SC}), fill factor, and photovoltaic conversion efficiency.

1 Introduction

The rapid economic development and rapid population expansion have caused a deteriorating energy problem. The energy problem has a significant impact on industrial development. The BP Statistical Review of World Energy in 2017 reported that by the end of year 2016, the reserves/production ratio (R/P ratio) in the world for coal was 153 years, for natural gas 52.5 years, and for oil 50.6 years [1]. On the other hand, burning too much coal, oil, and natural gas will release a large amount of carbon dioxide, cause air pollution, and deteriorate global climate warming. These have a serious impact on the environment and life. Satisfying energy requirements and avoiding fuel crises are the current global focus. There is an urgent need to develop sustainable energy to solve these imperative problems. Solar power is a popular sustainable energy source to help solve the energy problem [2, 3].

The dye-sensitized solar cell (DSSC) is one of the most promising potential solar power sources [4-19]. In 1996, Grätzel's research group proposed low-cost novel photovoltaic modules based on monolithically series-connected dye sensitized photoelectrochemical cells. The photovoltaic modules were produced in a continuous non-vacuum process by simple printing techniques [20].

In 2008, Kim et al. reported on novel thixotropic gel electrolytes [21]. The novel thixotropic gel electrolytes were prepared by oligomeric poly(ethylene oxide) (PEO)-based bis-imidazolium diiodide salts and hydrophilic silica nanoparticles for application in quasi-solid-state DSSCs. A quasi-solid-state DSSC with the proposed electrolyte exhibited highly improved properties, such as easy penetration and long-term stability.

In 2010, Grätzel's group discussed the effect of heat and light on the performance of DSSCs based on organic sensitizers and nanostructured TiO_2 [22]. The main reasons for the loss of performance were a decrease in open circuit photovoltage under light soaking conditions and a drop of photocurrent under heat stresses.

The fabrication cost of DSSCs is lower than that of other solar cells, so that the development of DSSCs can greatly reduce the cost of solar power generation [23]. In addition, DSSCs have a low light effect that can also generate electricity in an environment where it is cloudy or sunlight is insufficient. Furthermore, DSSCs are suitable for building materials, such as windows, shades, and glass curtains of buildings. DSSCs can be used to improve the energy efficiency of buildings. Owing to the numerous advantages of DSSCs compared to other types of solar cells, DSSCs are promising in relation to energy generation.

This paper discusses the studies on the advanced manufacturing technology of DSSCs. The key process technologies for DSSCs, including working electrodes,

* Corresponding author: yllai@cc.ncue.edu.tw

DOI:10.1051/matecconf/201712300030

dyes, counter electrodes, electrolytes, were investigated. This paper probes into the influence of TiO_2 film thickness on photoelectric conversion efficiency of DSSCs.

2 Experimental

The structure of the DSSC is shown in Fig. 1. A DSSC consisted of a fluorine-doped tin oxide (FTO) conductive glass substrate, a TiO_2 film as a working electrode, a dye, a counter electrode and an electrolytic solution. The structure components of the DSSC were stacked like a sandwich.

2.1 Preparation of working electrodes

The working electrode of the DSSC was fabricated by the doctor blade method. A TiO_2 thin film was used for the fabrication of the working electrode of the DSSC.

At first, an invisible tape was pasted on the FTO substrate. The circle area of the tape with an inner radius of 0.5 cm was cut out. Then, the TiO_2 film was coated on the FTO substrate by the doctor blade method. The TiO_2 working electrode was placed on a 70-°C hot plate for 10 min. The working electrode was then placed in a 450-°C high temperature furnace for 20 min for high temperature sintering. A fabricated TiO_2 working electrode is shown in Fig. 2. There were three thickness conditions of TiO_2 films, indicated by A, B, and C, corresponding to one layer, two layers, and three layers of the invisible tape, respectively.

2.2 Preparation of counter electrodes

The platinum (Pt) material was used for the fabrication of the counter electrode. The Pt counter electrode was fabricated by the spin coating method.

Firstly, the Pt solution was dropped onto a drilled FTO substrate. Secondly, the Pt was uniformly coated on the FTO substrate with a spin speed of 500 rpm for 90 s utilizing a spin coater. Finally, the Pt counter electrode was placed in a high temperature furnace at 450 °C for 30 min for high-temperature annealing. A fabricated Pt counter electrode is shown in Fig. 3.

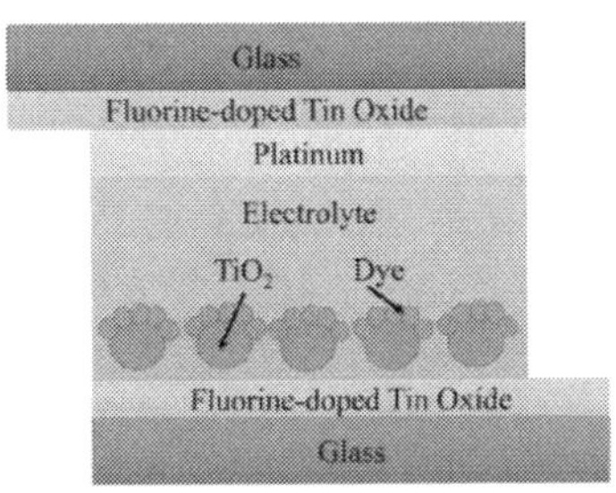

Fig. 1. The structure of DSSCs.

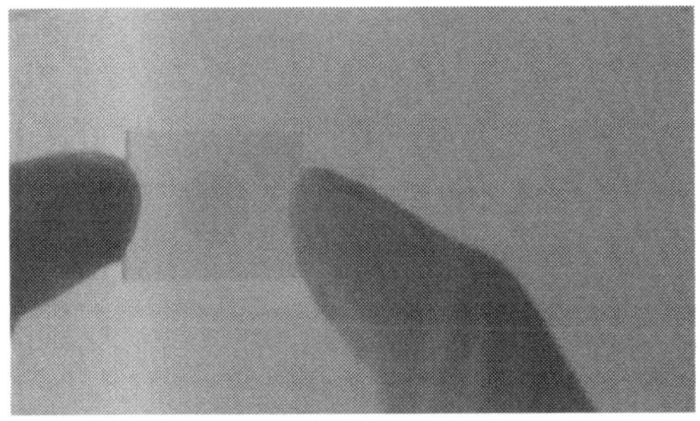

Fig. 2. Fabricated TiO_2 film working electrode.

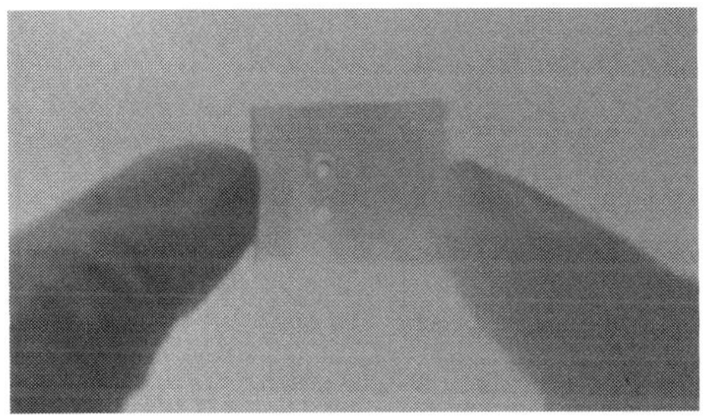

Fig. 3. Fabricated Pt counter electrode.

2.3 Preparation of dyes

The chemicals utilized to prepare the dye included N719 dye powders, *tert*-Butyl alcohol (TBA), and acetonitrile. TBA and acetonitrile were mixed at a ratio of 1:1. The dye concentration was 5×10^{-4} mol/L. The prepared TiO_2 working electrodes were then immersed in the prepared dye solution for 24 h.

2.4 Preparation of electrolyte

The chemicals utilized to prepare the electrolyte included 0.1 M lithium iodide (LiI), 0.6 M 1,2-dimethyl-3-propylimidazolium iodide, 0.005 M iodine (I_2), and 0.5 M tert-butylpyridine. These chemicals were added to acetonitrile.

2.5 Cell assembly

DSSC components were assembled utilizing a membrane (Surlyn) with an area of 1.5 cm × 1.5 cm and a thickness of 60 μm. Firstly, a circle of the membrane with a radius of 0.5 cm was cut out. Secondly, the membrane was placed on the TiO_2 working electrode without covering the TiO_2 layer. Thirdly, the counter electrodes were then aligned and stacked on top of the membrane. The stacked components were placed in the heat press sealing machine with the settings of 145-°C temperature and 1.5-kg/mm^2 pressure. Fourthly, after the stacked components of DSSCs were heat press sealed, the electrolyte was then injected through the drilled opening hole of the Pt counter electrode. Finally, the drilled hole of the Pt counter electrode was sealed with a 3M invisible tape. A fabricated DSSC is shown in Fig. 4.

3 Results and discussion

The surface morphology of the thin film under various coating conditions was studied by scanning electron microscopy (SEM) analysis. The thickness, uniformity, porosity, and grain size, of the thin film were observed. The SEM photographs of the cross-sectional views of the TiO_2 working electrodes fabricated under different conditions are shown in Fig. 5. The thickness values of the TiO_2 working electrodes fabricated under the conditions A, B, and C are 7.67, 18.5, and 24.3 μm, respectively.

The light transmittance characteristics of TiO_2 working electrodes, fabricated under the conditions A, B, and C, were measured utilizing ultraviolet-visible (UV-Vis) spectroscopy. The measured light transmittance characteristics are shown in Fig. 6. The light transmittance characteristics of the TiO_2 working electrodes fabricated under the conditions A, B, and C are 73%, 69%, and 66%, respectively, at a wavelength of 850 nm. It was found that the increasing thickness of the TiO_2 film reduced the light transmittance of the TiO_2 working electrode.

The photovoltaic characteristics of the DSSCs were measured by an AM-1.5 sunlight simulator with a light intensity of 100 mW/cm^2. The photovoltaic characteristics of the DSSCs, including open-circuited voltage (V_{OC}), short-circuited current density (J_{SC}), fill factor, and photovoltaic conversion efficiency were obtained. Table 1 lists the photovoltaic characteristics of the DSSCs. Fig. 7 shows the photocurrent density versus voltage (*J-V*) characteristics of the DSSCs. The short-circuited current density of the DSSCs with the TiO_2 working electrodes fabricated under the conditions A, B, and C were 16.28, 14.27, and 14.07 mA/cm^2, respectively. The photoelectric conversion efficiency of the DSSCs with the TiO_2 working electrodes fabricated under the conditions A, B, and C were 6.01%, 5.29%, and 4.98%, respectively. It was seen that the increasing thickness of the TiO_2 film from 7.67 to 18.5 and 24.3 μm decreased the photocurrent density and photovoltaic conversion efficiency of the DSSC.

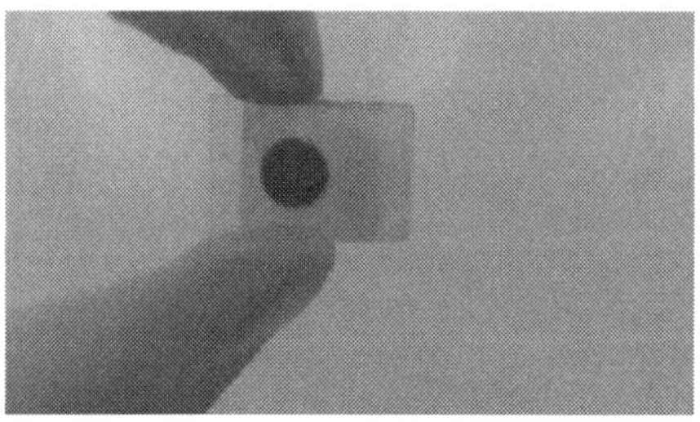

Fig.4. Fabricated DSSC.

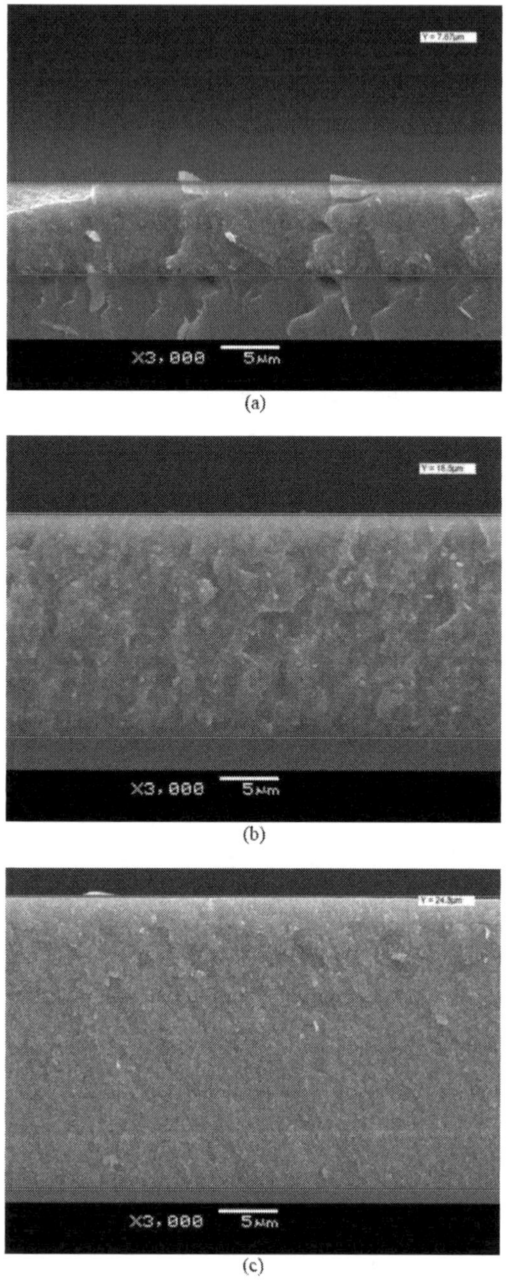

Fig. 5. Cross-sectional views of the TiO_2 working electrodes fabricated under different conditions. (a) condition A, (b) condition B, and (c) condition C.

The thicker TiO_2 film resulted in the lower light transmittance. The lower light transmittance reduced the amounts of dye molecules excited by the photons. The fewer dye molecules excited by the photons led to the lower photocurrent density and photoelectric conversion efficiency of the DSSC.

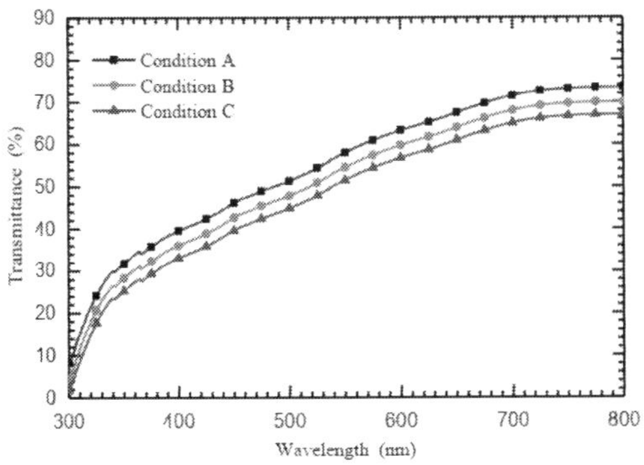

Fig. 6. Light transmittance characteristics of TiO_2 working electrodes.

Table 1. Photovoltaic characteristics of DSSCs.

Condition	*Voc* (V)	*Jsc* (mA/cm^2)	FF (%)	Efficiency (%)
Condition A	0.76	16.28	48.00	6.01
Condition B	0.75	14.27	48.82	5.29
Condition C	0.73	14.07	48.32	4.98

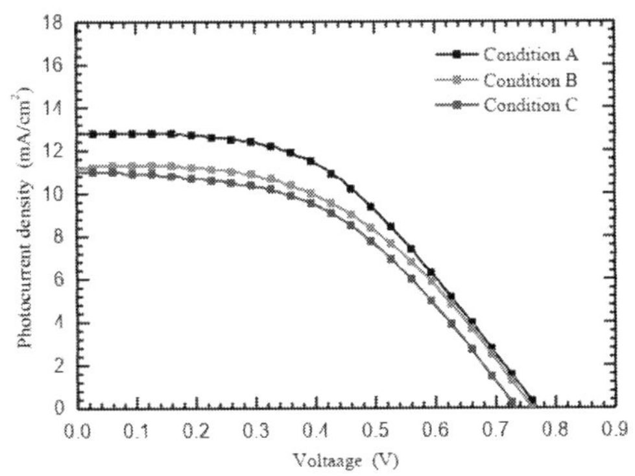

Fig. 7. Photocurrent density versus voltage (*J-V*) characteristics of the DSSCs.

4 Conclusion

The influence of the TiO_2 film thickness (from 7.67 to 24.3 μm) of the working electrodes on the DSSC performance, including V_{OC}, J_{SC}, fill factor, and photovoltaic conversion efficiency, was studied herein. Both the SEM and UV-Vis were used for the material analysis of the TiO_2 working electrodes. The dye of the DSSC used the chemicals of N719 dye powders, TBA, and acetonitrile. The photovoltaic characteristics of the DSSCs were measured by an AM-1.5 sunlight simulator. The relations among the TiO_2 film thickness, light transmittance characteristics, and photovoltaic characteristics were investigated. The TiO_2 film thickness of the working electrodes affected the light transmittance and the light absorption of the dye, and gave rise to the final photovoltaic conversion properties.

Acknowledgment

This work was supported in part by the Ministry of Science and Technology of Taiwan, R.O.C. under Contracts MOST 104-2221-E-018-016, MOST 105-2218-E-018-001, MOST 106-2218-E-018-002, and MOST 106-2218-E-018-003.

References

1. BP Statistical Review of World Energy June 2017.
2. S. Mozaffari, M. R. Nateghi, M. B. Zarandi, *Renew. Sustainable Energy Rev.*, **71**, 675 (2017)
3. S. Sharma, B. Siwach, S.K. Ghoshal, D. Mohan, *Renew. Sustainable Energy Rev.*, **70**, 529 (2017)
4. D. M. Chapin, C. S. Fuller, and G. L. Pearson, *J. Appl. Phys.*, **25**, 676 (1954)
5. J. M. Nunzi, *C. R Phys.*, **3**, 523 (2002)
6. Y.L. Lai, L.Y. Tang, *J. Precis. Mach. Manuf. Technol.*, **6**, 1 (2016)
7. K. Hara, K. Sayama, Y. Ohga, A. Shinpo, S. Suga, H. Arakawa, *Chem. Comm.*, 569 (2001)
8. Y.L. Lai, L.W. Chen, C.-H. Chen, *Appl. Mech. Mater.*, **865**, 54 (2017)
9. H. Lindström, A. Holmberg, E. Magnusson, S. E. Lindquist, L. Malmqvist, A. Hagfeldt, *Nano Lett.*, **1**, 97 (2001)
10. W. Kubo, T. Kitamura, K. Hanabusa, Y. Wada, S. Yanagida, *Chem. Comm.*, 374 (2002)
11. M. Grätzel, J. Photochem. & Photobio. *A: Chem.*, **164**, 3 (2004)
12. Y.L. Lai, Y.H. Huang, *Lect. Notes Electr. Eng.*, **234**, 987 (2013)
13. Y. Li, J. Hagen, W. Schaffrath, P. Otschik, D. Haarer, *Sol. Energy Mater. Sol. Cells*, **56**, 167 (1999)
14. K. Srikanth, Md. M. Rahman, H. Tanaka, K.M. Krishna, T. Soga, M. K. Mishra, T. Jimbo, M.Umeno, *Sol. Energy Mater. Sol. Cells*, **65**, 171 (2001)
15. K. Kalyanasundaram, M. Grätzel, Coord. *Chem. Rev.*, **177**, 347 (1998)
16. Y.L. Lai, S.H. Chen, J.H. Lu, J.-S. Ting, T.Y. Tsai, *Lect. Notes Electr. Eng.*, **234**, 975 (2013)
17. M. S. Suait, M. Y. A. Rahman, A. Ahmad, *Sol. Energy*, **115**, 452 (2015)
18. A. Goetzberger, C. Hebling, H. W. Schock, *Mater. Sci. Eng.: R: Rep.*, **40**, 1 (2003)
19. A. I. Kontos, A. G. Kontos, D. S. Tsoukleris, M. C. Bernard, N. Spyrellis, P. Falaras, *J. Mater. Process. Technol.*, **196**, 243 (2008)

20. A. Kay, M. Grätzel, *Sol. Energy Mater. Sol. Cells*, **44**, 99 (1996)
21. J. Y. Kim, T. H. Kim, D. Y. Kim, N.-G. Park, K.-D. Ahn, *J. Power Sources*, **175**, 692 (2008)
22. J.H. Yum, R. Humphry-Baker, S. M. Zakeeruddin, M. K. Nazeeruddin, M. Grätzel, *Nano Today*, **5**, 91 (2010)
23. V. Kislyuk, O. Dimitriev, *J. Nanosci. Nanotechnol.*, **8**, 1 (2008)

Fabrication of working and counter electrodes on plastic substrates for flexible dye-sensitized solar cells

Yeong-Lin Lai[1,*], *Yeong-Kang* Lai[2], *Chun-Yi* Zheng[1], *Guo-Hui* Xu[1], *Yi-Ming* Wang[1], and *Shan-Ching* Chen[1]

[1]Department of Mechatronics Engineering, National Changhua University of Education, Changhua 50007, Taiwan
[2]Department of Electrical Engineering, National Chung Hsing University, Taichung 40227, Taiwan

Abstract. This paper presents the fabrication technologies related to flexible titanium-dioxide (TiO_2) working electrodes and graphite counter electrodes on plastic substrates for dye-sensitized solar cells (DSSCs). The DSSC on a plastic substrate had a sandwich structure consisting of a TiO_2 working electrode, an electrolyte, and a graphite counter electrode. Nano-crystalline TiO_2 powder, titanium (IV) bis (ammonium lactato) dihydroxide (TALH), de-ionized water (DI water), and Triton X-100 were added to the mixture of PDMS base and curing agents to form TiO_2 colloids for the TiO_2 working electrode. The material analysis of the TiO_2 working electrode and the graphite counter electrode employed field-emission scanning electron microscopy (FE-SEM), atomic force microscopy (AFM), and X-ray diffraction (XRD). Nano-crystalline TiO_2 powder was used to form the TiO_2 working electrode. The surface characteristics of the TiO_2 and graphite films were investigated herein. The low-cost fabrication technologies of the TiO_2 working electrode and graphite counter electrode on the plastic substrate were clearly demonstrated.

1 Introduction

According to the BP Statistical Review of World Energy in 2017, the oil reserves/production ratios in North America, Europe/Eurasia, and the Middle East were 32.3, 24.9, and 69.9 years, respectively. As energy reserves decline while energy consumption increases globally, the energy situation continues to deteriorate [1]. In addition, the use of energy derived from the fossil fuels has caused serious climate problems, such as global warming and abnormal weather, as well as pollution and respiratory diseases.

In response to the energy and climate problems, research on renewable energy has become an urgent global topic. The renewable energy power sources include solar power, hydroelectric power, tidal power, wind power, etc. Among the renewable energy sources, solar power has great development potential and application in renewable energy.

Dye-sensitized solar cells (DSSCs) [2-28] offer numerous advantages compared to other types of solar cells. Firstly, the manufacturing processes of DSSCs are simple and do not need expensive vacuum equipment; sencondly, DSSCs have low material cost; thirdly, DSSCs can be flexible; and fourthly, DSSCs technology is environmentally friendly. Owing to their excellent advantages, DSSCs are promising for a wide range of applications; for instance, DSSCs can be utilized as the windows and glass curtains of buildings.

In this paper, the fabrication technologies of flexible TiO_2 working and graphite counter electrodes for DSSCs were studied. The DSSCs adopted the flexible indium-tin-oxide PET (ITO-PET) plastic materials. The material analysis of TiO_2 working and graphite counter electrodes were analyzed by field emission electron scanning microscopy (FE-SEM), atomic force microscopy (AFM), and X-ray diffraction (XRD). The low-cost manufacturing technology of TiO_2 working and graphite counter electrodes of DSSCs was demonstrated.

2 Structure and mechanism

2.1 Structure

The structure and mechanism of a DSSC are shown in Fig. 1. A DSSC was composed of a working electrode, a counter electrode and an electrolyte between the working and counter electrodes, which were stacked into a sandwich structure.

The working electrode contained TiO_2 film grown on a flexible PET plastic substrate with an indium tin oxide (ITO) conductive film. The energy absorption range of TiO_2 is in the ultraviolet spectrum. When ultraviolet light is irradiated on the surface of TiO_2, the electrons of the valence band are excited to the conduction band and produce electron-hole pairs. The working electrode mainly provided a dye adsorption area and a path of electron conduction. The working electrode needed to have a large surface area, high roughness, high porosity, high stability, and good adhesion to the substrate.

An electrolyte consisted of negatively charged iodide ions (I^-) and triiodide ions (I_3^-) in organic solvents as a reduction-oxidation couple.

* Corresponding author: yllai@cc.ncue.edu.tw

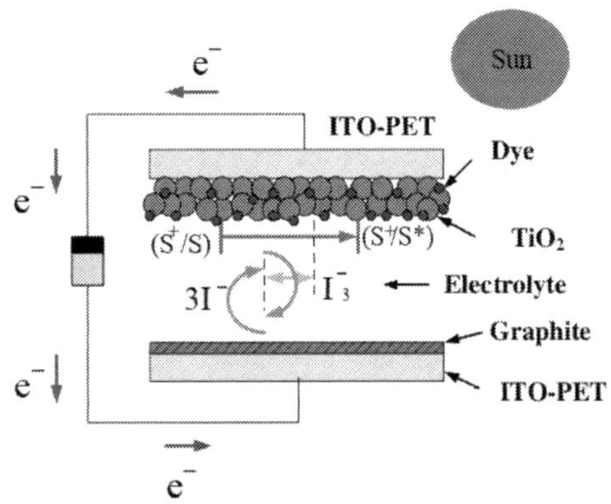

Fig. 1. Structure of dye-sensitized solar cell.

The graphite counter electrode had a graphite film grown on a flexible ITO-PET plastic substrate. Graphite was a useful counter electrode material due to its good conductivity. The counter electrode with regard to the reduction-oxidation reaction needed to have good catalytic properties, corrosion resistance, and low equivalent series resistance.

2.2 Mechanism

The mechanism regarding the DSSC is based on the processes of photoelectrochemistry, which include photoexcitation, electron injection, oxidation, and reduction, as follows [28]:

•Photoexcitation

The light irradiates the working electrode, and the dye molecules absorb photons and change from the ground state (S) to the excited state (S^*):

$$S+hv \rightarrow S^* \qquad (1)$$

•Electron injection and oxidation

The electrons on the excited state dye molecules are injected into the conduction band of the TiO_2 semiconductor. At this moment, the dye molecules are present in an oxidized state (S^+):

$$S^* \rightarrow e^-(TiO_2)+S^+ \qquad (2)$$

•Reduction

The electrons on the conduction band of the TiO_2 semiconductor are conducted into the transparent conductive film. The electrons on the transparent conductive film are then conducted to the counter electrode via an external circuit. The triiodide ions in the electrolyte obtain electrons from the counter electrode and convert into iodide ions:

$$I_3^- + 2\,e^- \rightarrow 3\,I^- \qquad (3)$$

The oxidized state dye molecules react with the iodide ions in the electrolyte and convert to ground state dye molecules. The iodide ions lose electrons and then convert it into triiodide ions:

$$2\,S^+ + 3\,I^- \rightarrow 2\,S + I_3^- \qquad (4)$$

The photoelectrochemistry through the photoexcitation, electron injection, oxidation, and reduction achieves the required complete loop for electron conduction.

3 Experimental

The fabrication of DSSCs, including the PDMS, working electrode, counter electrode, and electrolyte processes, was conducted.

3.1 PDMS

The PDMS used was the Sylgard 184 silicone from Dow Corning Corp. The PDMS was made up of a base agent and a curing agent. The 10:1 weight ratio of the base agent to the curing agent was adopted. The physical properties of the PDMS were determined by the ratio.

3.2 Working electrode

The TiO_2 colloid for the working electrode was formed by nano-crystalline TiO_2 powder, titanium (IV) bis (ammonium lactato) dihydroxide (TALH), de-ionized water (DI water), Triton X-100, a PDMS base agent, and a PDMS curing agent, indicated by A, B, C, D, E, and F, respectively. In this work, the weight ratio of the nano-crystalline TiO_2 powder, TALH, DI water, and Triton X-100 was kept at 2:10:5:1. The weight ratio of the PDMS base and curing agents, E:F, was kept at 10:1. The weight ratios of the PDMS base and curing agents were varied so that the weight ratios of A:B:C:D:E were changed from 2:10:5:1:1 to 2:10:5:1:5.

In this work, the conditions of the TiO_2-PDMS mixtures studied are listed in Table 1. There were three conditions. In the first condition, the nano-crystalline TiO_2-PDMS mixture was mixed with 0.2-g TiO_2 powder, 1-g TALH, 0.5-g DI water, 0.1-g Triton X-100, a 0.1-g PDMS base agent, and a 0.01-g curing agent. In the second condition, the nano-crystalline TiO_2-PDMS mixture was mixed with 0.2-g TiO_2 powder, 1-g TALH, 0.5-g DI water, 0.1-g Triton X-100, a 0.3-g PDMS base agent, and a 0.03-g curing agent. In the third condition, the nano-crystalline TiO_2-PDMS mixture was mixed with 0.2-g TiO_2 powder, 1-g TALH, 0.5-g DI water, 0.1-g Triton X-100, a 0.5-g PDMS base agent, and a 0.05-g curing agent.

Table 1. Conditions of TiO2-PDMS mixtures.

	A	B	C	D	E	F
	TiO_2 (g)	TALH (g)	DI water (g)	Triton X-100 (g)	PDMS base agent (g)	PDMS curing agent (g)
I	0.2	1	0.5	0.1	0.1	0.01
II	0.2	1	0.5	0.1	0.3	0.03
III	0.2	1	0.5	0.1	0.5	0.05

The uniform TiO_2-PDMS mixture without air bubbles was obtained by gently stirring with a glass rod for 30 min. The active region of DSSC was surrounded by Scotch-tape (3M). The TiO_2-PDMS mixture was coated by the doctor blade technique. The tape was then removed and the electrode was heated at 70 °C for 30 min for curing. The TiO_2-PDMS electrode was dipped in the N719 dye for 24 h.

3.3 Counter electrode

The graphite colloid for the counter electrode was formed by 0.6-g graphite powder, a 3-g PDMS base agent, and a 0.3-g PDMS curing agent. The weight ratio of the graphite powder, PDMS base agent, and PDMS curing agent was kept at 2:10:1.

3.4 Electrolyte

An electrolyte was made by iodine (I_2), potassium iodide (KI), and propylene carbonate.

At first, a beaker contained 20 ml (about 23.4 g by weight) of propylene carbonate. Then, 0.254-g iodine and 1.66-g potassium iodide were added. The beaker containing the mixer was vibrated by an ultrasonic vibrator for 30 min twice. Finally, the electrolyte was then formed.

4 Results and discussion

In this work, the SEM analysis was conducted by a JEOL JSM-6700F FE-SEM. The FE-SEM analysis of the nano-crystalline TiO_2 powder and graphite powder were studied. Fig. 2 shows the surface morphology of the nano-crystalline TiO_2 powder. The surface morphology of the graphite powder is shown in Fig. 3. The TiO_2 powder was also analyzed using the XRD. The nano-crystalline TiO_2 powder was observed. Fig. 4. shows the XRD characteristics of the TiO_2 powder. Two mineral forms of TiO_2: anatase and rutile titanium dioxide, were obtained.

Fig. 2. Surface morphology of nano-crystalline TiO_2 powder.

Fig. 3. Surface morphology of graphite powder.

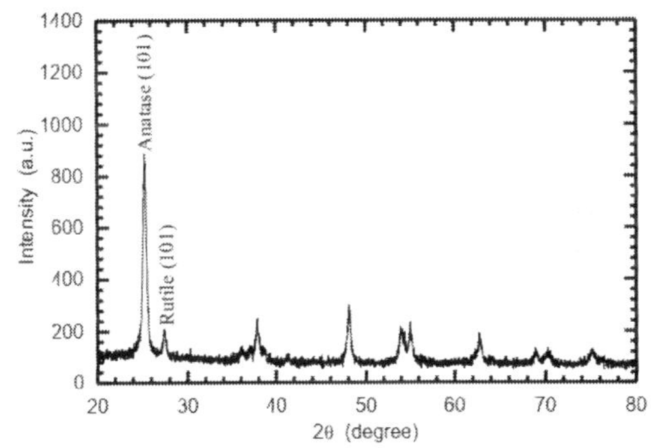

Fig. 4. XRD characteristics of TiO_2 powder.

The transmission property of the flexible ITO-PET substrate was analyzed by a spectrum analyzer. The transmission property of the flexible ITO-PET substrate coated with the PDMS was also studied. Fig. 5 shows the transmission characteristics of the flexible ITO-PET substrate and the ITO-PET substrate coated with the PDMS. The ITO-PET substrate had better transmission characteristics than the ITO-PET/PDMS substrate.

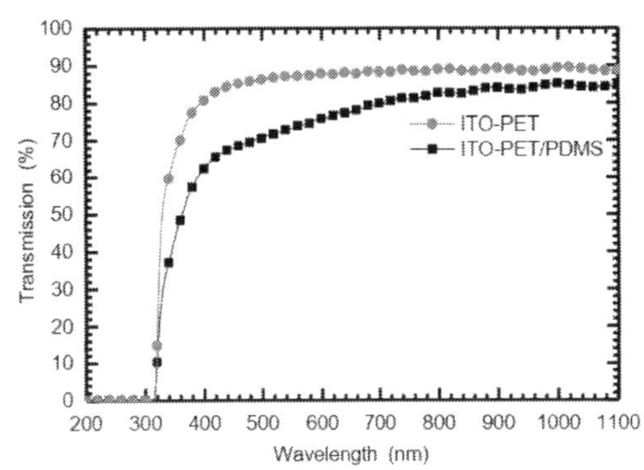

Fig. 5. Transmission characteristics of ITO-PET and ITO-PET/PDMS substrates.

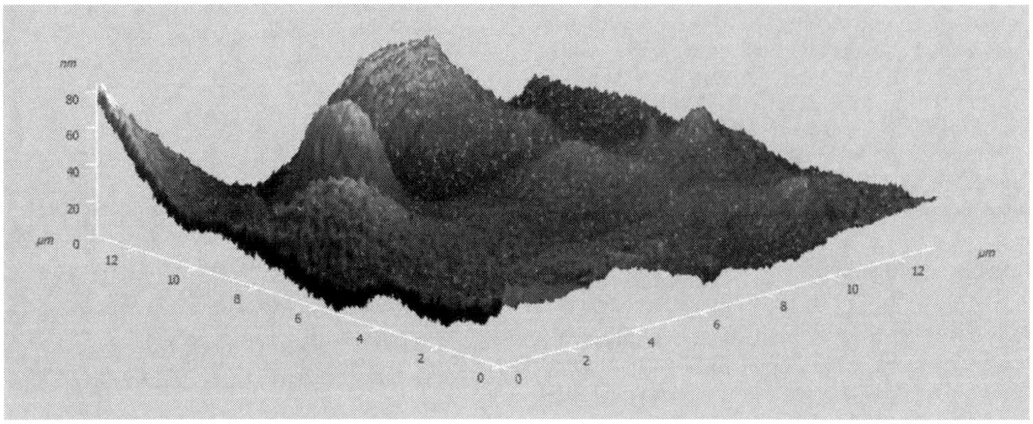

Fig. 6. AFM analysis result of graphite colloid.

The surface analysis of the flexible ITO-PET substrate was conducted by an NT-MDT Solver P47 AFM. The AFM analysis of the graphite counter electrode was also investigated. Fig. 6 shows the AFM analysis result of the graphite colloid.

The surface analysis of the TiO_2 working electrod was conducted by the FE-SEM. The surface observation of the TiO_2 colloid with the A:B:C:D:E ratio of 2:10:5:1:1 is shown in Fig. 7. The TiO_2-PDMS colloid had a porous surface.

Fig. 8 shows the fabricated flexible DSSC consisting of a flexible IOT-PET substrate, a TiO_2-PDMS working electrode, a graphite-PDMS electrode, and an electrolyte.

The investigation of the electrical characteristics of the TiO_2 colloids was carried out. Fig. 9 shows the electrical characteristics of the DSSCs with different A:B:C:D:E ratios.

Fig. 8. Fabricated flexible DSSC.

Fig. 7. Surface morphology of TiO_2-PDMS colloid.

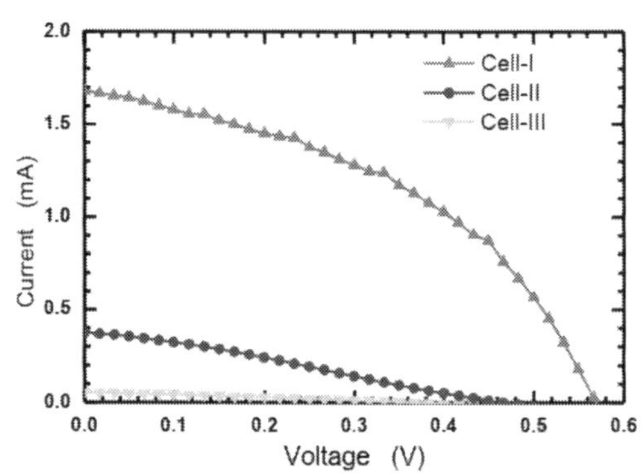

Fig. 9. Electrical characteristics of DSSCs with different A:B:C:D:E ratios.

The DSSCs corresponding to the A:B:C:D:E ratios of 2:10:5:1:1, 2:10:5:1:3, and 2:10:5:1:5 were indicated by Cell-I, Cell-II, and Cell-III, respectively. It was found that the increase in the TiO_2 percentage of the colloid enhanced the current and voltage characteristics. The DSSC with the A:B:C:D:E ratio of 2:10:5:1:1 exhibited the best current and voltage characteristics.

5 Conclusion

The flexible DSSCs on the ITO-PET substrates were fabricated. The TiO_2 and graphite colloids were fabricated for the working and counter electrodes of the DSSCs, respectively. The TiO_2 colloid was made from the nano-crystalline TiO_2 powder, TALH, DI water, Triton X-100, PDMS base agent, and PDMS curing agent. The graphite colloid was made from the graphite powder, PDMS base agent, and PDMS curing agent. The transmission characteristics of the ITO-PET and ITO-PET/PDMS substrates were studied. The fabrication techniques of the flexible TiO_2 working and graphite counter electrodes on the plastic substrate demonstrated the potential for flexible green energy applications.

Acknowledgment

This work was supported in part by the Ministry of Science and Technology of Taiwan, R.O.C. under Contracts MOST 104-2221-E-018-016, MOST 105-2218-E-018-001, MOST 106-2218-E-018-002, and MOST 106-2218-E-018-003.

References

1. BP Statistical Review of World Energy June 2017.
2. Y.L. Lai, L.W. Chen, C.-H. Chen, *Appl. Mechanics Mater.*, **865**, 54 (2017)
3. L. W. Veldhuizen, Y. Kuang, R. E. I. Schropp, *Sol. Energy Mater. Sol. Cells*, **158**, 209 (2016)
4. M. Grätzel, J. Photochem. & Photobio. *A: Chem.*, **164**, 3 (2004)
5. K. Kalyanasundaram, M. Grätzel, *Coord Chem. Rev.*, **177**, 347 (1998)
6. Y.L. Lai, L.Y. Tang, *J. Precis. Machinery Manuf. Technol.*, **6**, 1 (2016)
7. H. C. Weerasinghe, F. Huang, Y.B. Cheng, *Nano Energy*, **2**, 2 (2013)
8. J. Huang, S. Varlamov, J. Dore, J. S. Yun, M. A. Green, *Sol. Energy Mater. Sol. Cells*, **132**, 282 (2015)
9. W. A. Badawy, *J. Adv. Res.*, **6**, 123 (2015)
10. Y.L. Lai, S.H. Chen, J.-H. Lu, J.-S. Ting, T.Y. Tsai, *Lect. Notes Electr. Eng.*, **234**, 975 (2013)
11. J. N. Clifford, E. Martínez-Ferrero, A. Viterisi, E. Palomares, *Chem. Soc. Rev.*, **40**, 3 (2011)
12. S. Kim, D.M. Geum, M.S. Park, C. Z. Kim, W. J. Choi, *Sol. Energy Mater. Sol. Cells*, **141**, 372 (2015)
13. M. S. Suait, M. Y. A. Rahman, A. Ahmad, *Solar Energy*, **115**, 452 (2015)
14. Goetzberger, C. Hebling, H. W. Schock, Mater. Sci. Eng.: R: Reports **40**, 1 (2003)
15. D. M. Chapin, C. S. Fuller, G. L. Pearson, *J. Appl. Phys.*, **25**, 676 (1954)
16. S. Y. Chae, S. J. Park, O.S. Joo, B. K. Min, Y. J. Hwang, *Sol. Energy*, **135**, 821 (2016)
17. K. Hara, K. Sayama, Y. Ohga, A. Shinpo, S. Suga, H. Arakawa, *Chem. Comm.*, 569 (2001)
18. K. Srikanth, Md. M. Rahman, H. Tanaka, K.M. Krishna, T. Soga, M. K. Mishra, T. Jimbo, M.Umeno, *Sol. Energy Mater. & Sol. Cells*, **65**, 171 (2001)
19. J. M. Nunzi, *C. R Phys.*, **3**, 523 (2002)
20. S. Mozaffari, M. R. Nateghi, M. B. Zarandi, *Renew. Sustainable Energy Rev.*, **71**, 675 (2017)
21. S. Sharma, B. Siwach, S.K. Ghoshal, D. Mohan, *Renew. Sustainable Energy Rev.*, **70**, 529 (2017)
22. I. Kontos, A. G. Kontos, D. S. Tsoukleris, M. C. Bernard, N. Spyrellis, P. Falaras, *J. Mater. Proce Tech.*, **196**, 243 (2008)
23. V. Kislyuk, O. Dimitriev, J. Nanosci, *Nanotechnol.*, **8**, 1 (2008)
24. H. Lindström, A. Holmberg, E. Magnusson, S. E. Lindquist, L. Malmqvist, A. Hagfeldt, *Nano Letters*, **1**, 97 (2001)
25. Y.L. Lai, Y.H. Huang, *Lect. Notes Electr. Eng.*, **234**, 987 (2013)
26. W. Kubo, T. Kitamura, K. Hanabusa, Y. Wada, S. Yanagida, *Chem. Comm.*, 374 (2002)
27. H. Tsubomura, M. Matsumura, Y. Nomura, T. Amamiya, *Nat.*, **261**, 402 (1976)
28. B. O'Regan, M. Grätzel, *Nat.*, **353**, 737 (1991)

Managing self congruity to influence behavioral intention in organic food contexts in Fujian province, China

Tsai-Fa Yen*
Anxi College of Tea Science, Fujian Agriculture and Forestry University, Fujian, China; No.15 Shangxiadian Rd., Cangshan District, Fuzhou City, Fujian, 350002, China (P.R. China)

Abstract. Regarding China, this country is considered as one of the most potential markets in terms of economic growth and population expansion in organic food. Due to the improvement in living standards such as increase in income and healthy consciousness, Chinese consumers start to focus more on food quality, and hence create the chance for the development of organic food market. How to attract consumers' attention has become an emergency issue for both academia and practioners. This study was aims at investigating the effects of gender on self congruity- behavioral intention relationships. Questionnaire survey was employed for collecting data and 200 valid samples were received from those consumers in Fujian province, China. Findings show that self congruity has significantly positive impacts on behavioral intention. However the effects of gender on self congruity- behavioral intention relationships were not significant. Firms should hold more family reunions of organic food in order to enhance its image and maintain the interaction and linkage among consumers.

1. Introduction

Organic food has developed impressively, from a neglected niche market to the food market mainstream [1]. More and more researchers and practitioners are attracted and involved. As Marian, Chrysochou, Krystallis, and Thogersen [2] ascertained from several studies, the attitude of consumers towards organic food is in general positive with typically associated benefits being superior taste, more environmental-friendliness, improved health, safer food, and more animal welfare. The demand towards organic food is · growing dynamically in China. China organic food market is estimated to cross RMB 2.48 billion in 2015 and its production will have a growth rate of an average of 20% to 20% in the future ten years [3]. Understanding the attitude and intentions of consumers towards organic food in China will benefit the academia and practitioners.

The foci of prior research into organic foods consumption included identifying motivations behind their purchase to understand the "whys" of consumers' choices. The key drivers previously identified include health benefits and environmental concerns (e.g., [4,5,6]), assafety concerns regarding conventional foods (e.g., [4]), as well psychological motives [7].

However, despite the valuable insights provided by previous studies, at least two research gaps exist, which must be addressed. That is, previous studies have neglected psychological motives (i.e., self-congruity) and an important role of organic foods, gender. In particular, initial psychological motives, such as self-congruity, deserve attention given their impact on consumers' product choices. Consumers purchase goods/services because they are driven by a motivation to express their own self-images [8,9,10]. Thus, a product or brand that defines, sustains, and enhances the self-images of consumers is more likely to be selected and purchased [11]. This possibility that psychological motives can play an important role in organic product consumption has not been researched in this context.

Second, gender difference is an important variable in the marketing literature [12,13,14]. However, the studies on gender differences in the context of organic foods, as well as its influence on formation of intention, have rarely been carried out. Thus, examination of the role of gender in intention formation contributes to the existing organic foods literature. More importantly, gender could be an easy recognized variable for marketers. Rather than other demographic characteristics (e.g., family income, or level of education), gender is can be quickly judged by consumers' appearance in most situations [15]. Therefore, accurate marketing strategies could be done to provide a better service, which may further increase the level of consumers' intention and purchase behaviors.

Therefore,this study examines the role of a psychological motive(i.e.,self-congruity) and gender that havebeenidentified inpreviousorganicfoodstudies or marketing literature. The current research demonstrates the relative importance of this variable in stimulating consumers' consumption of organic foods. To our knowledge, this is the first study to clarify the importance of self-congruity in predicting consumers' intention to purchase organic foods. Therefore, the results enrich our understanding of key drivers of consumers' organic food purchase.

A core proposition is that the effects of self-congruity on purchase intention depend on the magnitude of gender in the organicfood context. Self-congruity should play a lesser role when genderismale and a greater role when genderisfemale. This proposition, if supported would (1) augment existing psychological motive(i.e.,self-congruity)- purchase intention models

* Corresponding author: 1722997311@qq.com

which focus mostly on self-congruity; (2) help to explain variability in the self-congruity- purchase intention relationshipevidenced in prior research; and (3) provide guidance to organicfoodindustry in developing self-congruity- purchase intention programs.

2. Methodology

2.1 Population

Having considered the data collection requirements of this study such as a need of large sample of organic food consumers in Fujian province, it would be appropriate to employ the field survey with a self- administered questionnaire as the primary data collection technique. The population was set on those organic food consumers who consume organic food during the past six month. Organic firms were conducted for collecting data. After giving them phone call/interview by the researcher, finally four organic firms agreed to give us help in collecting data. They were found more than 5 years, stood on a standard of operation, and could be the representation of organic food firms in Fujian province. Their main products of those firms were fresh organic vegetables, organic tea, and others.

2.2 Instrument

The field study method was chosen in order to gain information directly from organic food consumers who consume organic food during the past six month. As such, their perceptions about the setting with respect to self-congruity (SC) and behavioral intention (BI) are likely to be clearly in mind.

The survey instruments were borrowed from previous studies and adapted for this study. Self-congruity was defined as how a consumer hopes others see her/him. A consumer's self-image congruity comprises their actual self-congruity, ideal self-congruity, social self-congruity, and ideal social self-congruity [16]. Four items for self-image congruity are distinguished, based on the studies of Ryu & Lee (2013) [17] and Yen (2013) [18]. Behavioral intention was defined as a consumer's evaluation of the intents to word-of-mouth, recommendation, and repurchase [19].Three items for behavioral intentions are drawn from Lee & Yun (2015) [19] and Yen (2013) [13].

2.3 Survey

Quota sampling was adopted for collecting data based upon firms scale. Regarding to the feasibility and accessibility, four firms were adopted for collecting data after conducting to legal organic firms in Fujian province. They have offered the member list for sending questionnaires or space for surveyors to interview the organic food consumers. Survey was down during May to June in 2016. Finally a total of 218 samples were received. After non-qualified and incomplete surveys were excluded, the remaining 200 respondents were valid reporting a responds rate of valid samples was 91.7%.

Of 200 questionnaires obtained, about 45.5% were male and 55.5%were from female respondents. At about 16.5% of respondents were below 22 years of age,20.5% of respondents were 22-32 years of age, 20% of respondents were 32-42 years of age, 28% of respondents were 42-52 years of age, and 9% exceed52 years of age. Regarding to married status, 38.5% of respondents was single while 56% of respondents were married with children. Approximately 8% of respondents were graduated from junior high school and 19% of respondents were graduated from senior high school. At about 72.5% respondents were graduated from college or above. With regard to the occupation, 16% of respondents work at government related section, 7.5% of respondents belong to industrial section, 14% of respondents were businessman, 10.5% of respondents were farmers, and 34.5% of respondents were retired. Approximately 20% of respondent's monthly income was below 3000 RMB, 43.5% of respondent's monthly income was 3000-6000 RMB, 22.5% of respondent's monthly income was 6000-8000 RMB, and 8.5% of respondent's monthly income exceeds 8000 RMB. With regard to the frequency consumed within six month, 27% of respondents were less than three times, 31% of respondents were 4-6 times, and 39.5% of respondents were more than seven times.

3. Result

3.1 Descriptive statistics

Normality was tested by means of SPSS 17.0, based on the skewness and kurtosis of the observed variables [20]. Both samples revealed acceptable kurtosis (-0.838 ~ -0.384) and skewness (-0.223~ -0.028) for most observed variables (see Table 1).

3.2 Analysis for measurement model

Data analysis was conducted using structural equation modeling with AMOS 19.0. According to the guidelines offered by Anderson and Gerbing (1988) [21], a confirmatory factor analysis (CFA) was performed and then a structural model was estimated. During the process of CFA, several items were excluded from the model due to low standardized factor loadings, low squared multiple correlations, and high standardized residuals. The final CFA results (see Table 2) indicated that the fit indices showed a satisfactory fit to the data: χ^2=20.95, df=13(p=.074), χ^2/df=1.61, GFI=.972, AGFI=.939, CFI=.987, RMSEA=.055. Means, standard deviations, and Cronbach's α among the constructs are presented in Table 1.

3.3 Convergent validity

As shown in Table 2, all estimated factor loadings in the measurement model showed high factor loading coefficients and significant t-values (all loadings ranging from 0.66 to 0.919, $p < 0.001$). The results of composite reliability and the average variance extracted (AVE) value of each multi-item scale showed that high reliabilities were achieved for all scales with values ranging from .791 to .864; the AVE was between .486 and .683. Therefore, convergent validity was met.

Table 1. Descriptive statistics.

Measure items (reference)	M	SD	Sk	Ku	α
Self congruity (Ryu & Lee. 2013; Yen, 2013)					
SC1:Eating organic food fits well with my image.	3.70	1.018	-.223	-.834	.79
SC2: Eating organic food is highly consistent with how I see myself.	3.54	1.041	-.094	-.838	
SC3: Eating organic food helps reflect who I am.	3.32	1.049	-.054	-.526	
SC4: People similar to me eat organic food.	3.17	1.023	-.091	-.384	
Behavioral intention (Lee & Yun, 2015; Yen, 2013)					
BI1: I try to buy organic food because it is the best choice for me	3.56	1.016	-.093	-.737	.86
BI2:I will recommend organic food to others	3.43	1.049	-.077	-.610	
BI3: If I had to do it again, I would buy organic food	3.60	.956	-.028	-.653	

M: mean; SD: standard deviation; Sk: skewness; Ku: kurtosis; α: Cronbach's α

Table 2. Reliability and validity.

Conpt.	Items	λ	t-values	SMC	CR	AVE
SC	SC1	0.660	9.562	0.435	0.791	0.486
	SC2	0.725	10.763	0.525		
	SC3	0.684	10.011	0.469		
	SC4	0.718	10.636	0.516		
BI	BI1	0.863	14.523	0.745	0.864	0.683
	BI2	0.919	15.922	0.845		
	BI3	0.678	10.506	0.459		

SC: Self congruity; BI: Behavioral intention; λ: Standardized factor loading; CR: Composite reliability; AVE: Average variance extracted; χ^2=20.95, df=13(p=0.074), χ^2/df=1.61, GFI=0.972, AGFI=0.939, CFI=0.987, RMSEA=0.055

3.4 Discriminant validity

To examine the discriminant validity (i.e., the distinctiveness of constructs) for all constructs, the AVEs related to each latent construct were compared to the squared correlations between the corresponding constructs. The evidence of discriminant validity exists when AVE from each construct exceeds the squared correlations [22]. The AVEs (0.486 for SC and 0.683 for BI) by the underlying construct should be larger than the shared variance (i.e., the squared correlation coefficient) between two constructs (0.396). Based on the tests, therefore, all constructs were considered to be distinctively different, confirming discriminate validity.

3.5 Analysis for structural model

The next step was the structural model test of the causal relationships between constructs in the proposed model. The results revealed a good level of fit for the structural model: χ^2=20.95, df=13(p=0.074), χ^2/df=1.61, GFI=0.972, AGFI=0.939, CFI=0.987, RMSEA=0.055. Based on these fit measures, it can be concluded that the model is acceptable. The standardized parameter estimates, t-values, and fit indices are reported in Table 3. Next, the individual parameters/paths in the structural path model were evaluated. Specifically, the hypotheses were tested by evaluating the relationships between the exogenous and endogenous variables. Table 3 reports the findings of the main effects, and it was found that self congruity was significantly and positively related to behavioral intention (β= 0.628), supporting H1.

3.6 The moderating effects of gender on SC-BI relationship

Next, the moderating effects of gender were assessed using the multi-group analysis in the SEM framework. The full sample was divided into male (n=65) and

DOI:10.1051/matecconf/201712300032

female (n=135) sub-groups. First, the theoretical model, which contains both groups, was examined without constraining any parameters (i.e., the unconstrained model or base model). The next step was to estimate a "fully constrained" structural path model in which all regression path coefficients were constrained to be equal across both groups. Comparing the goodness-of-fit statistics for the constrained model and unconstrained model using a χ^2 difference test, in which a $\Delta\chi^2$ value can be used to examine whether a particular path coefficient estimation is statistically different across the male and female groups. The $\Delta\chi^2$ values for constrained model is not significantly increased ($\Delta\chi^2$=2.65, Δdf=10, $p > 0.05$), which means the two-group analysis is not theoretically supported. Therefore, the H2 is not supported.

Moreover, this study compares the path coefficient estimation between genders. Table 4 reports that the effects of self congruity on behavioral intention within the relationship between genders, the path coefficient estimation was larger in male than in female. The largest indicator of self congruity for male respondents was "People similar to me eat organic food", however it's "Eating organic food is highly consistent with how I see myself" for female respondents. Within the four types of self-concept as well as self congruity, the preference between genders could be met. Male respondents preferred ideal social self-concept but female respondents preferred social self-concept. In summary, ideal self-concept was fond to be the most important one in influencing behavioral intention and followed by ideal social self-concept, social self-concept and actual self-concept when purchasing organic food. This finding was different from previous studies.

Table 3. Results of SEM.

Path and items	β	t	SMC
SC-BI	0.628	6.696	0.395
SC1	0.660	9.562	0.435
SC2	0.725	10.763	0.525
SC3	0.684	10.011	0.469
SC4	0.718	10.636	0.516
BI1	0.863	12.758	0.745
BI2	0.919	13.292	0.845
BI3	0.678	9.758	0.459

χ^2=20.95, df=13(p=0.074), χ^2/df=1.61, GFI=0.972, AGFI=0.939, CFI=0.987, RMSEA=0.055

Table 4. the comparison among gender.

Paths and items	All(n=200)	Male(n=65)	Female(n=135)
	β (t-values)	β (t-values)	β (t-values)
SC-BI	0.628 (6.736)	0.837(3.975)	0.543(6.736)
SC1:Eating organic food fits well with my image.	0.660(9.562)	0.733(6.445)	0.622(9.562)
SC2: Eating organic food is highly consistent with how I see myself.	0.725(10.763)	0.735(6.467)	0.738(10.763)
SC3: Eating organic food helps reflect who I am.	0.684(10.011)	0.743(6.560)	0.664(10.011)
SC4: People similar to me eat organic food.	0.718(10.636)	0.771(6.907)	0.677(10.636)
BI1: I try to buy organic food because it is the best choice for me	0.863(12.758)	0.814(5.130)	0.878(12.758)
BI2:I will recommend organic food to others	0.919(13.292)	0.889(5.265)	0.915(13.292)
BI3: If I had to do it again, I would buy organic food	0.678(9.758)	0.736(4.788)	0.646(9.758)
R^2	0.395	0.700	0.295
χ^2	20.95	62.403	15.680
df	13(p=0.074)	13(p=0.000)	13(p=0.267)
χ^2/df	1.61	4.8	1.206
GFI	0.972	0.809	0.968
AGFI	0.939	0.588	0.930
CFI	0.987	0.815	0.993
RMSEA	0.055	0.244	0.039

4. Conclusion

The main purpose of this study was to propose and examine a theoretical model of behavioral intention formation and to address the differences between males and females in terms of how they form behavioral intention toward organic food, using data obtained from consumers in Fujian province, China. In summary, the findings indicate that self congruity have a positive impact on behavioral intention. This finding is consistent with previous studies regarding behavioral intention formation[10, 17, 18]. However, the moderating effect of gender on self congruity - behavioral intention relationship is not supported indicating that gender failed to moderate the relationship between self congruityand behavioral intention; and the influence of self congruity on behavioral intention was stronger for males.

In terms of theoretical implications, this study contributed to the body of knowledge on self congruityand behavioral intention in organic food context in two major ways. First, a theoretical model of behavioral intention formation is presented and empirically examined with a relatively large sample of real organic food consumers in Fujian province, China.This study empirically examined howself congruity influence behavioral intention. That is, within the four type of self image, the order of its importance/components can be found. According to Sirgy (1982) [16], self-concept is a multidimensional construct having four major components, namely, actual self-concept, ideal self-concept, social self-concept and ideal social self-concept. Ideal self-concept was fond to be the most important one in influencing behavioral intention and followed by ideal social self-concept, social self-concept and actual self-concept when purchasing organic food.

Moreover, the study demonstrated that the influence of self congruityonbehavioral intention was stronger for males than for females but it's not significant indicating that male and female consumer form different behavioral intentiontoward the organic food depending on different self congruity components. For males, ideal social self-concept was the key component, but ideal self-concept would be important for females when predicting behavioral intentiontoward the organic food.

Finally, the marketers of organic food need to pay more emphasize on ideal conceptand ideal social concept of organic food. This would enhance consumers'ideal self-concept and ideal social self-concept and thus bridge the gap between organic food image and consumers' self image. Consequently they will have positive image/words of mouth on organic food, will recommendorganic food to others, and most importantly they will buy it again and again. With regard to future research, they can extend the model to other countries and collect more samples to improve the litmiation of the sample size used in the study.

References

1. J. Van Doorn, P.C.Verhoef, Drivers and barriers to organic purchase behavior, *J. Retailing*, **91**(3), 436-450 (2015)

2. L. Marian, P. Chrysochou, A. Krystallis, J. Thogersen, The role of price as a product attribute in the organic food context: An exploration based on actual purchase data, *Food Quality and Preference*, **37**, 52-60 (2014)

3. MOA, 2015, Ministry of Agriculture, People's of Republic of China, http://www.moa.gov.cn/

4. N. Michaelidou, L.M. Hassan, The role of health consciousness, food safety concern and ethical identity on attitudes and intentions towards organic food, *Int. J. Consumer Study*, **32**(2), 163-170 (2008)

5. E. Tsakiridou, C. Boutsouki, Y. Zotos, K. Mattas, Attitudes and behaviour towards organic products: an exploratory study, *Int. J. Retailing Distribution Management*, **36** (2), 158-175 (2008)

6. M. Wandel, A. Bugge, Environmental concern in consumer evaluation of food quality, *Food Quality and Preference*, **8**(1), 19-26 (1997)

7. C. C. Teng, C. H. Lu, Organic food consumption in Taiwan: Motives, involvement, and purchase intention under the moderating role of uncertainty, *Appetite*, **105**, 95-105 (2016)

8. D. Aaker, Building Strong Brands. Free Press, New York. (1996)

9. F. Kressmann, M. Sirgy, A. Herrmann, F. Huber, S. Huber, D. Lee, Direct and indirect effects of self-image congruence on brand loyalty, *J. Business Research*, **59**, 955-964(2006)

10. P.H. Chiou, T.F. Yen, K.K. Chu, Tourists' behavioral intentions in rural area: An integrated aspect of SCT and TPB, *J. Food, Agriculture & Environment*, **13**(3&4), 168-174 (2015)

11. T. Graeff, Using promotional messages to manage the effects of brand and self-image on brand evaluations, *J. Consumer Marketing*, **13** (3), 4-18 (1996)

12. C. Wang, H. Qu, M. K. Hsu, Toward an integrated model of tourist expectation formation and gender difference, *Tourism Management*, **54**, 58-71(2017)

13. J. Hwang, H. Han, W. Kim, Organic food as self-presentation: The role of psychological motivation in older consumers' purchase intention of organic food, *J. Retailing and Consumer Services*, **28**, 281-287 (2016)

14. Y. King, P. Wan, Minding the gap in the gaming industry, *Int.J. Contemporary Hospitality Management*, **26** (1), 130-153 (2014)

15. H. Han, B. Meng, W. Kim, Bike-traveling as a growing phenomenon: Role of attributes, value, satisfaction, desire, and gender in developing loyalty, *Tourism Management*, **59**, 91-103 (2017)

16. M. J. Sirgy, Self-concept in consumer behavior: a critical review, *J. Consumer Research*, **9**, 287-300 (1982)

17. K. Ryu, J. S. Lee, Understanding convention attendee behavior from the perspective of self-congruity: The case of academic association convention, *Int. J. Hospitality Management*, **33**, 29-40 (2013)

18. T.F. Yen, Understanding Tourists' Intention from self-image congruity: An empirical study from B&B in Kenting, *Leisure Industry Research*, **11**(3), 1-14 (2013)

19. H.J. Lee, Z.S. Yun, Consumers' perceptions of organic food attributes and cognitive and affective attitudes as determinants of their purchase intentions toward organic food, *Food Quality and Preference*, **39**, 259-267(2015)

20. K. A. Bollen, Structural equations with latent variables, New York: John Wiley & Sons. (1989)

21. J.C. Anderson, D.W. Gerbing, Structural equation modeling in practice: a review and recommended two-step approach, *Psychological Bulletin*, **103**(3), 411-423 (1988)

22. C. Fornell, D. Larcker, Evaluating structural equation models with unobservable variables and measurement error, *J. Marketing Research*, **18**(1), 39-50 (1981)

FEM simulation analysis of wire rod drawing process using the rotating die under Coulomb friction

Gow-Yi Tzou[1], *Un-Chin* Chai[1], *Chao-Ming* Hsu[2,*], and *Hung-Ying* Hsu[2]

[1]Department of Mechanical and Automation Engineering, Chung-Chou University of Science and Technology, Yuanlin, Changhua, 510, Taiwan
[2]Department of Mechanical Engineering, National Kaohsiung University of Applied Sciences, Kaohsiung, 807, Taiwan

Abstract. This study proposes a cold drawing technology of wire rod with rotating die; it carries out an FEM simulation on rotating drawing using DEFORM-3D commercial software. Frictions among the die and the wire material are assumed as Coulomb friction. The effective stress, the effective strain, the velocity field, various stress distributions, the die stress, the drawing force, and the rotating torque can be determined from the FEM simulation. In this study, effects of various drawing conditions such as the rotating angular velocity, the half die angle, the frictional coefficient, the die filet on the drawing forming characteristics are explored effectively. From this FEM simulation, it is noted that the effect of rotation die and die filet can reduce the drawing force and increase the material flow.

1 Introduction

Wire rod drawing process is a high-efficiency and high-precision processing technology. Wire rod is become to the required geometry and dimension after the single pass or multi-passes. Wire rod drawing process has been widely used in various industrial fields, and it is an important indispensable product in industrial countries. Therefore, how to obtaining energy-efficient and rapid production of high-precision wire process technology is actively pursuing at present industries. Kim et al [1] and Yoshida et al. [2] used 3D rigid-plastic finite element method (FEM) to improve the unsteady drawing process of straight-pipe. Lee [3] et al. used the finite element simulation to investigate the manufacturing process of precision linear guide rail through shape rolling and shape Drawing. Haddi et al. [4] studied the temperature increase and tensile stress under different drawing process using a thermocouple and stress sensing system. Zhang et al. [5] explored the temperature increase of wire rod and dies due to the friction at high drawing velocity to calculate the heat of material and friction in the plastic deformation under various lubrications. Due to the high drawing velocity, the material considered the strain rate in flow stress. Tiernan et al. [6, 7] and Huh et al [8] proposed a new load control system of dieless drawing to improve the die wear and lubrication requirements in traditional drawing process. Norasethasopon et al. [9-11] used the finite element simulation to verify the dimensional accuracy and formability of impurities on the drawn copper-containing impurities wire rod. Tiernan et al [12] carried out the experiment of dieless drawing to achieve the ability of reducing wire diameter. Feldera et al [13] proposed the stress analysis due to the friction under temperature of 20 ℃ -150 ℃ in a variety of metal soaps to realize effect of various lubrications to the drawing process. Chen et al [14] used DEFORM-2D with the Taguchi method to design the drawing process of CuZn37 brass alloy considering four control factors such as area reduction, lubrication, drawing velocity and die angle; the optimization of drawing process could be obtained. Based on previous references, traditional drawing process did not consider the die rotation. Tzou et al [15] proposed a cold drawing technology of wire rod with rotating die; it carried out an FEM simulation on rotating drawing using DEFORM-3D commercial software under constant shear friction. The effective stress, the effective strain, the velocity field, and the drawing force have been determined from the FEM simulation. This study assumes the friction to be Coulomb friction different from constant shear friction to explore a cold drawing of wire rod with a rotating die using FEM simulation.

2 FEM simulation

Fig. 1 is a schematic diagram of the drawing process with rotation die. The die is rotated with angular velocity ω (rad/s), the front end of the wire is pulled by the cross head with an initial velocity V_0 (mm/s).

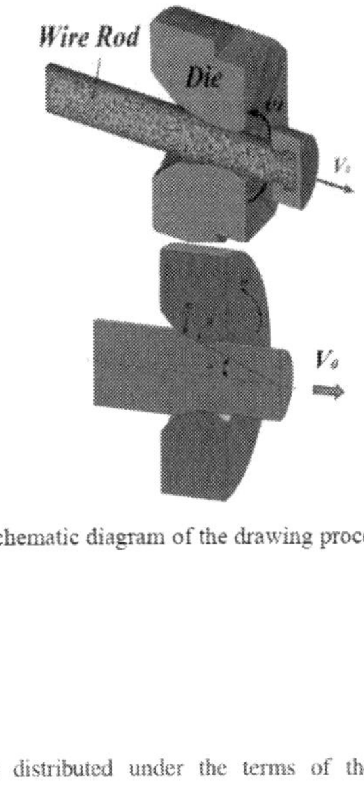

Fig. 1. Schematic diagram of the drawing process with rotation die.

* Corresponding author: jammy@kuas.edu.tw

Table 1 shows FEM simulation parameters of wire rod material in the drawing process with rotating die. The wire rod material is SWRCH22AB (where SWRCH22AB is a carbon steel for Cold Heading and Cold Forging, the chemical composition: C=0.18-0.23%, Si=0.1%, Mn=0.7%-1.0%, P=0.03%, S=0.035%; the Young's modulus 210GPa and Poisson's ratio 0.3. The flow stress shown in Fig. 2 can be obtained by the compress test, σ = 832.233 ε 0.382. The initial diameter (Do) is 6.4mm; the number of mesh element is 60,000; the drawing velocity is 33.50mm/s. Table 2 shows FEM simulation parameters of dies in the drawing process with rotation die. The die diameter (Dd) is 5.6mm; the rotating angular velocity (ω) is set to 0~4.5 rad/s; the half die angle (α) is set to 10 ° ~ 18 °; the Coulomb frictional coefficient (μ) is assumes as 0.05, 0.1, and 0.15, respectively, and the die filet (R) is 3 ~ 7mm.

Table 1. FEM simulation parameters of wire rod material in the drawing process with rotation die.

Wire Rod Conditions	
Material	SWRCH22AB
Wire Diameter D_o (mm)	6.4
Wire Velocity V_0 (mm/s)	33.50
Elements	60000

Table 2. FEM simulation parameters of dies in the drawing process with rotation die.

Die Conditions	
Die Diameter D_d (mm)	5.6
Half Die Angle α (°)	10°,14°,18°
Frictional Coefficient μ	0.05, 0.1, 0.15
Die Filet R (mm)	3, 5, 7
Rotating Angular Velocity ω (rad/s)	0.5, 2.5, 6.5

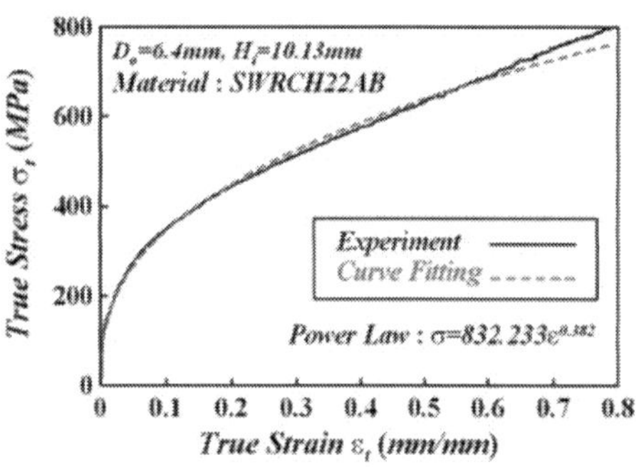

Fig. 2. Flow stress for SWRCH22AB wire rod in compression test.

Because considering the rotation effect, the Deform 2D software cannot be used, therefore in the study the Deform 3D software should be used in the drawing process with rotation die.

3 Results and Discussions

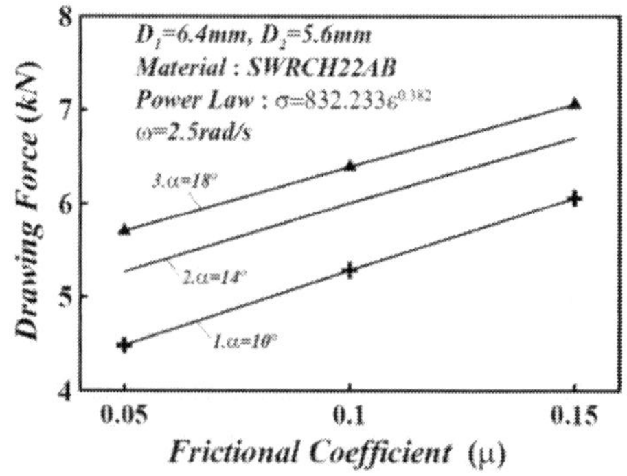

Fig. 3. Variations of drawing force with frictional coefficient for various half die angles.

Fig. 3 shows variations of drawing force with frictional coefficient for various half die angles. The drawing force is increased with an increase of the frictional coefficient under the fixed half die angle. Under the fixed frictional coefficient, the drawing force is increased with an increase of half die angle.

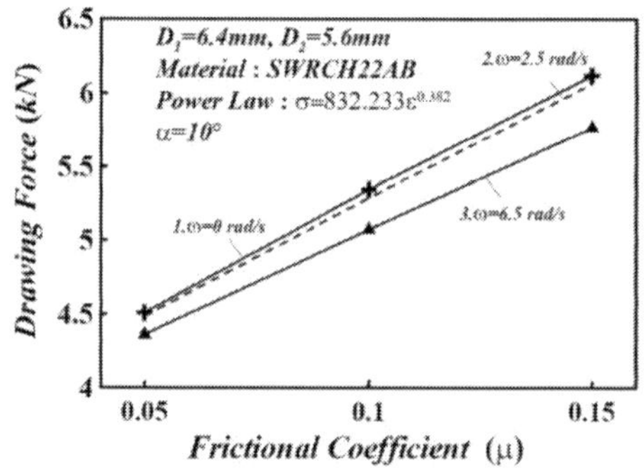

Fig. 4. Variations of drawing force with frictional coefficient for various angular velocities.

Fig. 4 shows variations of drawing force with frictional coefficient for various angular velocities. The drawing force is reduced with an increase of rotating angular velocities. As the rotating angular velocity increases to make the hoop frictional stress increased.

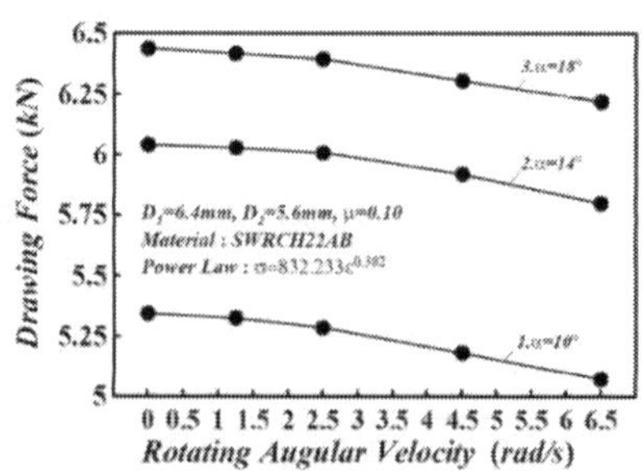

Fig. 5. Variations of drawing force with rotating angular velocity for various half die angles.

Fig. 5 shows variations of drawing force with rotating angular velocity for various half die angles. It reveals that the drawing force is reduced as the rotating angular velocity increases.

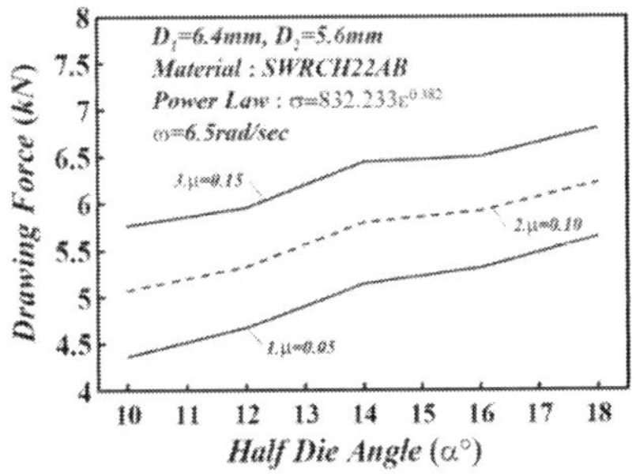

Fig. 6. Variations of drawing force with half die angles for various frictional coefficients.

Fig. 6 shows variations of drawing force with half die angles for various frictional coefficients. As the half die angle increases the drawing force increases under the fixed frictional coefficient.

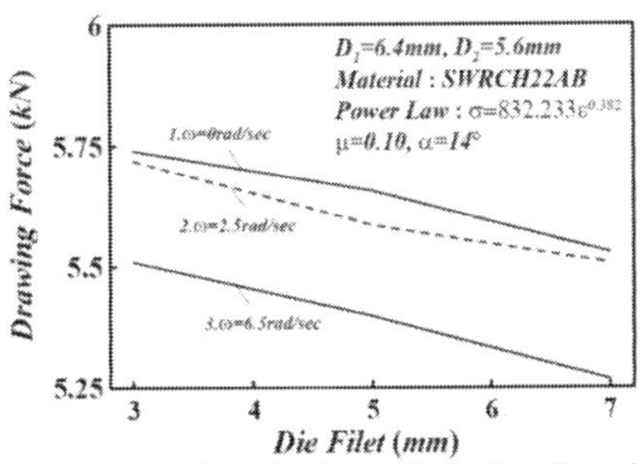

Fig. 7. Variations of drawing force with die filets for various angular velocities.

Fig. 7 shows variations of drawing force with die filets for various angular velocities. The drawing force is reduced with an increase of die filet. As the die filet increases the material is easy to deform.

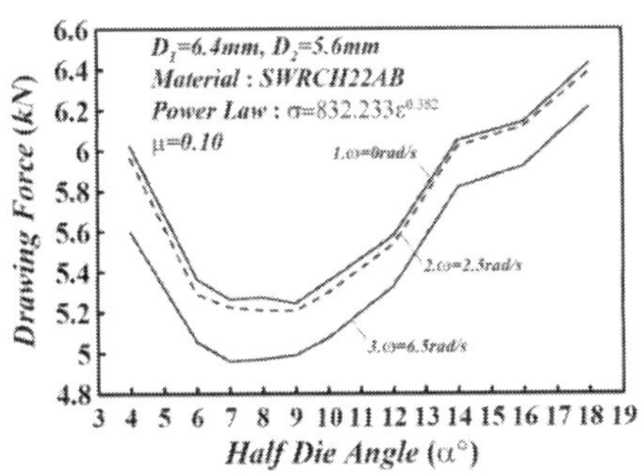

Fig. 8. Variations of drawing force with half die angle for various angular velocities.

With a view to realizing what the optimum half die angle under different rotating angular velocities is. Fig. 8 shows the optimum half die angle decreases with an increase of rotating angular velocity. As the half die angle does not exceed the optimum half die angle, the drawing force is reduced due to a decrease of frictional stress along the interface between die and material. However, as the half die angle exceeds the optimum half die angle, the material deformation effect is larger than the frictional stress effect so that the drawing force is increased.

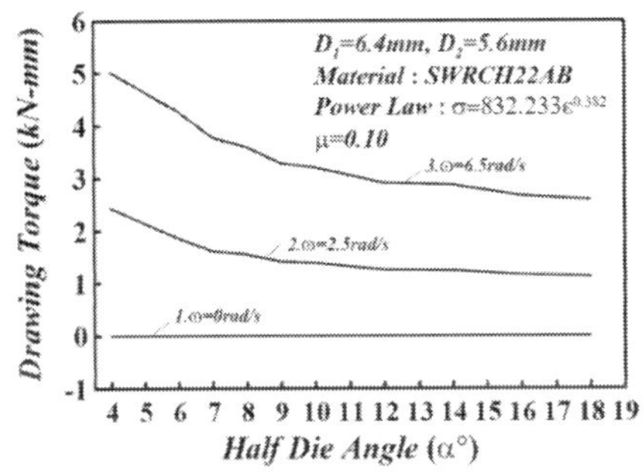

Fig. 9. Variations of drawing torque with half die angle for various angular velocities.

Fig. 9 shows variations of drawing torque with half die angle for various angular velocities. Without the die rotation, no drawing torque is occurred. The rotating angular velocity increases to make the drawing torque increased. Under the fixed angular velocity, the drawing torque is reduced with the increase of half die angle.

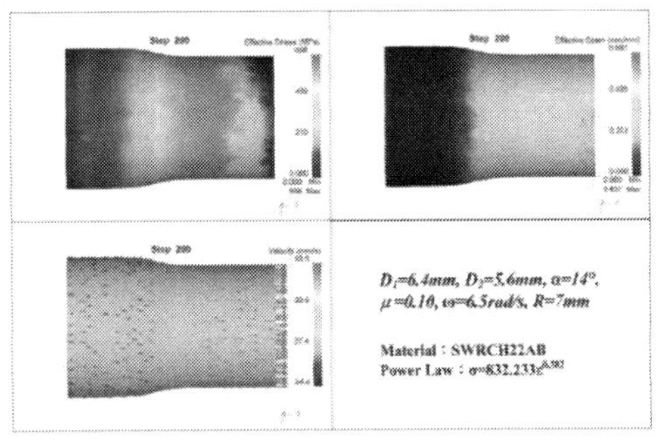

Fig. 10. Effective stress, effective strain, and velocity field with the rotating die.

Fig. 10 shows effective stress, effective strain, and velocity field with the rotating die. Under these conditions, the maximum effective stress is 658MPa; the maximum strain is 0.637 mm/mm; the maximum flow velocity is 33.5 mm/s; the drawing force is 5.799 kN.

4 Conclusions

This study uses FEM simulations to obtain the effective stress, the effective strain, the velocity field, drawing force, and drawing torque. The major results can be summarized as follows:

1. Optimum half die angle can be obtained from FEM simulation, and it is decreased with an increase of rotating angular velocity.

2. Drawing torque is changed with the frictional stress along the interface between die and material.

3. After the optimum half die angle, as the half die angle (α) increases, the internal plastic deformation is more intense so that the drawing force increases.

4. As the frictional coefficient (μ) increases, the frictional resistance is large so that the drawing force increases.

5. As the die filet (R) increases, the material flow is smoother so that the drawing force decreases.

6. As the rotating angular velocity (ω) increases, the material flow is smoother so that the drawing force decreases.

Acknowledgments

Thanks for MOST project's financial supports (MOST 104-2221-E-235-002) to complete this research.

References

1. H. C. Kim, Y. Choi, B. M. Kim, Three-Dimensional Rigid-Plastic Finite Element Analysis of Non-Steady-State Shaped Drawing Process, *International Journal of Machine Tool and Manufacture*, **39** (1999) 1135-1155

2. K. Yoshida, H. Furuya, Mandrel Drawing and Plug Drawing of Shape-Memory-Alloy Fine Tubes Used in Catheters and Stents, *Journal of Materials Processing Technology*,**153-154** (2004) 145-150

3. S. H. Lee, S. K. Lee, B. M. Kim, Study on The Manufacturing Process of Precision Linear Guide Rail Through Shape Rolling and Shape Drawing, *Journal of Mechanical Science and Technology*, **24** (2010) 1647-1653

4. A. Haddi, A. Imad, G. Vega, Analysis of Temperature and Speed Effects on the Drawing Stress for Improving the Wire Drawing Process, *Materials and Design*, **32** (2011) 4310-4315

5. J. Zhang, J. Sun, J. Long, S. Tang, Computer Simulation on Temperature Rise in Process of Wire Drawing, *Journal of Shanghai Jiaotong University*,**E-5** (2000) 242-247

6. P. Tiernan, R. Carolan, E. Twohig, S. A. M. Tofail, Design and Development of A Novel Load-Control Dieless Rod Drawing System, *CIRP Journal of Manufacturing Science and Technology*, **4** (2011) 110-117

7. P. Tiernan, M. T. Hillery, Dieless Wire Drawing-An Experimental And Numerical Analysis, *Journal of Materials Processing Technology*,**155-156** (2004) 1178-1183

8. Y. Huh, B. K. Ha, J. S. Kim, Dieless Drawing Steel Wires Using A Dielectric Heating Method and Modeling the Process Dynamics, *Journal of Materials Processing Technology*, **210** (2010) 1702-1708

9. S. Norasethasopon, H. Yoshida, Finite Element Simulation of Inclusion Size Effects on Copper Shape-Wire Drawing, *Materials Science and Engineering*, **422** (2006) 252-258

10. S. Norasethasopon, H. Yoshida, Influences of Inclusion Shape and Size in Drawing of Copper Shaped-Wire, *Journal of Materials Processing Technology*, **172** (2006) 400-406

11. S. Norasethasopon, K. Yashida, Prediction of Chevron Crack Initiation in Inclusion Copper Shaped-Wire Drawing, *Engineering Failure Analysis*,**15** (2008) 378-393

12. P. Tiernan, M. T. Hillery, An Analysis of Wire Manufacture Using the Dieless Drawing Method, *Journal of Manufacturing Processes*,**10** (2008) 12-20

13. E. Feldera, C. Levraua, M. Mantel, N. G. T. Dinh, Identification of the Work of Plastic Deformation and the Friction Shear Stress in Wire Drawing, *Wear*, **286-287** (2012) 27-34

14. D. C. Chen, J. Y. Huang, Design of Brass Alloy Drawing Process Using Taguchi Method, *Materials Science and Engineering*,**464** (2007) 135-140

15. G. Y. Tzou, D. C. Chen, S. H. Lin, Study on Wire Rod Drawing Process using the Rotating Die, *Key Engineering Materials*,**76** (2016) 63-67

The implement of a remote control system for automated warehouse by ZigBee networking

Ho-Chih Cheng[1,*] and *Min-Chie* Chiu[1]

[1] Department of Mechanical and Automation Engineering, Chung Chou University of Science and Technology, 6, Lane 2, Sec.3, Shanchiao Rd., Yuanlin, Changhua 51003, Taiwan, R.O.C.

Abstract. Excessive raw materials and products in stock unduly increase an enterprise's overhead. However, insufficient raw materials and products will also influence the efficiency and output of the manufacturing process. Therefore, to largely reduce stock cost and facilitate adequate production for plants with multiple manufacturing processes and warehouses, the adjustment of raw material and products between warehouses is necessary. In addition, in order to keep track of current raw materials and products, remote online monitoring and control of the storage/retrieval system is compulsory. To simplify the search in an AS/BS process, a graph-aided strategy that manually selects the location within the stock map is presented. Moreover, to avoid interference from internet wiring allocation with existing equipment, a wireless storage/retrieval system using the ZigBee module is adopted. Consequently, a prototype of the remote wireless monitoring and control storage/retrieval system for sixteen storage spaces is exemplified.

1 Introduction

It is important to have an automated warehouse in the modern manufacturing field [1, 2]. From a mechanical perspective, a 3-D automated warehouse will save space; therefore, an assortment of automated warehouses has been developed [3, 4]. The automated Storage/Retrieval System (AS/RS) used in manufacturing, warehousing, and distribution applications is an efficient system. Order selection is fundamental for the retrieval function in automated warehouses. However, it is costly and time-consuming [5]. Many researchers have assessed the order selection problems in an Automated Storage/Retrieval System (AS/RS) and in automated warehousing systems. Ratliff and Rosenthal [6] proposed a graph-based algorithm to find the shortest path for selecting location. Linn and Wysk [7] proposed an expert-system-based AS/RS controller to search for storage and retrieval. Egbelu [8] minimized the service response time in the AS/RS by selecting the dwell point of the storage/retrieval machine. Mansuri [9] proposed a computerized algorithm to find the dedicated storage allocation.

However, for a plant with multiple manufacturing processes and warehouses, research for a more flexible and efficient distributing, monitoring, and controlling network has rarely been addressed. So, to reduce stock cost and to improve the manufacture process, a remotely distributing, monitoring, and controlling network to analyze online the status of the AS/RS and remotely adjust the stock is proposed. In addition, to simplify the search for an AS/BS process, a graph-aided method that manually selects the location within the stock map is presented. Moreover, to avoid interference from internet wiring allocation with existing equipment, a wireless storage/retrieval system using the ZigBee module is adopted. A prototype of the remote wireless monitoring and control storage/retrieval system with sixteen storage spaces (used for storage/retrieval of one kind of raw material) will be exemplified in this study. The related architecture for a distributing, monitoring, and controlling network for multiple AS/RS systems is shown in Fig. 1.

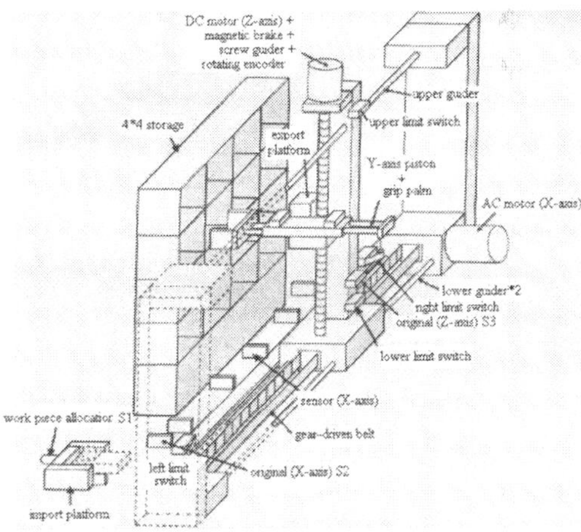

Fig. 1. The structure of the AS/RS system.

2 Graph-aided location selection for the storage/retrieval process

* Corresponding author: hccheng@gm.ccut.edu.tw

The structure of the AS/RS system is depicted in Figs. 1 and 2. In addition, the hardware for the remotely distributing, monitoring, and controlling network for multiple AS/RS systems with sixteen storage spaces (used for storage/retrieval for one kind of raw material) is shown in Fig. 3. The structure of the server pc for an individual AS/RS system is depicted in Fig. 4.

Fig. 2. The prototype of the remote distributing, monitoring, and controlling network for multiple AS/RS systems with sixteen storage spaces (used for storage/retrieval for one kind of raw material).

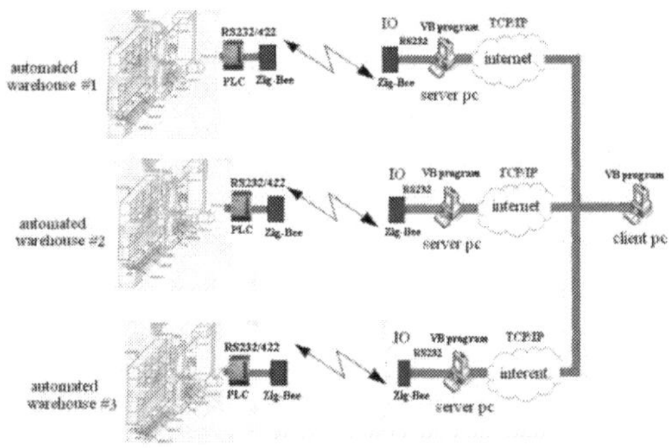

Fig. 3. The related architecture of the remote distributing, monitoring, and controlling network for multiple AS/RS systems.

Fig. 4. The structure of the server pc for an individual AS/RS system.

The related wiring diagram of the PLC connection is shown in Fig. 5. In order to remotely manipulate the AS/BS system via the server pc, the auxiliary relays defined in Table 1 have been added into the PLC program. As indicated in Fig. 6, the auxiliary relays remarked in red are in place of the button functions. The server pc can then submit the command to actuate the related function of the AS/BS system.

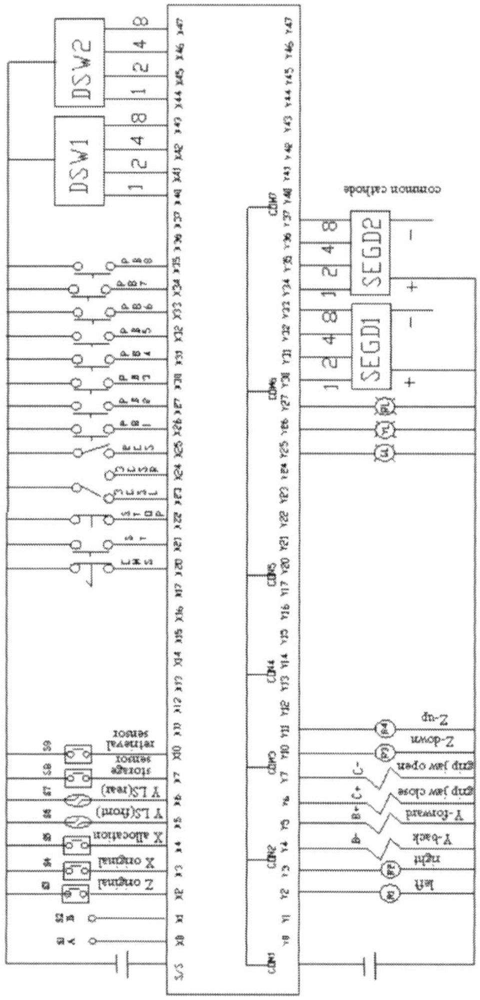

Fig. 5. The related wiring diagram of the PLC connection.

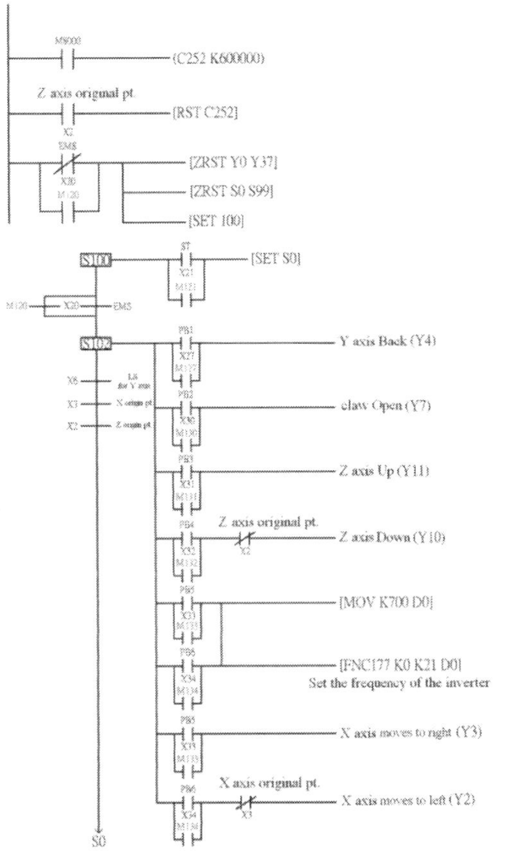

Fig. 6. An abstract of the ladder program built in the PLC used to actuate the AS/RS system by the auxiliary relays.

Table 1. The auxiliary relays used to remotely control the action of the PLC program.

Auxiliary Relay Number	Action	PLC Program
M120	Emergent open/close	SET · RST M120
M121	Start the AS/BS system	SET · RST M121
M123	Object retrieval	SET · RST M123
M124	Object storage	SET · RST M124
M127	Y axis Back	SET · RST M127
M130	Claw Open	SET · RST M130
M131	Z axis Up	SET · RST M131
M132	Z axis Down	SET · RST M132
M133	X axis move to right	SET · RST M133
M134	X axis move to left	SET · RST M134

Moreover, the interface of the server pc programmed in VB is also illustrated in Fig. 7. In order to simplify the search for the AS/RS process, a graph-aided method that manually selects the location within the stock map is presented. As indicated in Fig. 7, there is a 4x4 stock map. Each location reveals the current status of the stock in the AS/RS system. A location with a red light ON represents an occupied storage space; therefore, only the retrieval operation can be executed. With the red light OFF, the storage location is empty and will be available for storage. Each location has coordinates and is built into the system program.

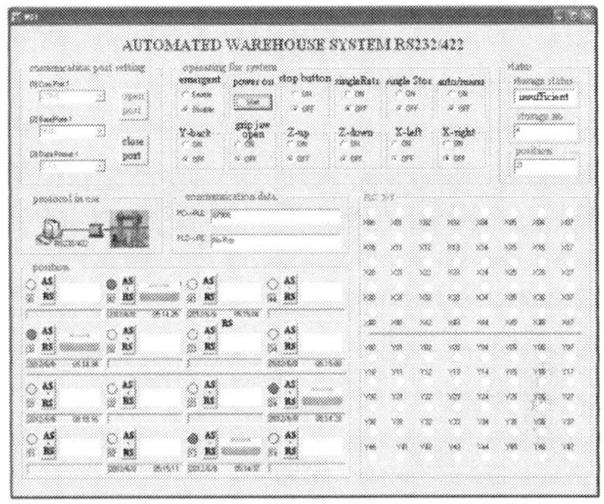

Fig. 7. The interface of the server pc programmed in VB.

3 System established and communication format

As indicated in Figs. 1 and 2, a PLC is used as a controller to actuate the actuators (an X-axis, Y-axis, and Z-axis air-driven piston) of the storage/ retrieval mechanism. In addition, a grip palm is used to catch the object. There are two sensors used to detect the object at both the feeding and the output platforms. In order to monitor online the storage status and remotely control the storage/retrieval with a server pc at the control room of the individual AS/RS system, a wireless communication between the server pc and the AS/RS system is established using the ZigBee module connected to the PLC via RS232/RS422 protocol. Thereafter, an interface at the server pc is programmed in VB and is used to implement the AS/RS system by sending a command to the PLC controller. The abstract of the related interface program built in the server pc is depicted in Fig. 8. To facilitate communication between the PLC and the pc via a RS232/RS422, the picket format for the MISTISUE FX2n-PLC primary port's communication is ruled and shown in Table 2. The definition of the picket format is depicted in Table 3. In addition, the MISTISUE FX2n-PLC extended port's communication is ruled and shown in Table 4. The communication parameter used in the extended port is shown in Table 5. Finally, the command description of the MISTISUE FX2n-PLC extended port's communication protocol type 1 is illustrated in Table 6.

```
Private Sub Timer1_Timer()
  Dim tmpStr As String
  Dim StatusStr As String
  Dim StatusInt As Integer
  Dim Wait As Double
  Dim i As Integer
  If Not (MSComm1.PortOpen) Then
     Exit Sub
  End If
  tmpStr = MSComm1.Input
  CmdStr = "0008005" + ETX
  MSComm1.Output = STX + CmdStr + CheckSum(CmdStr)
```

Fig. 8. The abstract of the related interface program built in the server pc.

Table 2. The communication packet format of MISUBISHI FX2n-PLC's primary port.

Header code	Command code	Address and data	End code	Inspection code
STX(02)	CMD	DATA	ETX(03)	Checksum

Table 3. The definition of communication packet for MISUBISHI FX2n-PLC's primary port.

Command category	Command code	object	Function description
Read	0	X,Y,M,S,T,C,D	Read the object value
Write	1	Y,M,S,T,C,D	Write data to the object
Forcing on	7	Y,M,S,T,C	Forcing node to be on
Forcing off	8	Y,M,S,T,C	Forcing node to be off

Table 4. The communication packet format of MISUBISHI FX2n-PLC's extended port.

Header code	Station number	Computer number	command	Object number/data	End code	Inspection code

Table 5. The parameter of the extended port's communications protocol for the MISUBISHI FX2n-PLC.

D8120Bit	Name	Content
15	Protocol type	0: type 1 1: type 4
14	protocol	0: No 1: Yes
13	Inspection code	0: manipulate 1: auto
12	Type of control line(II)	none
11	DTR inspection mode	none
10	Type of control line(I)	none
9	End code	none
8	Header code	none
7	Transmission speed	0011:300 0222:4800 0100:600 1000:9600 0101:1200 1001:19200 0110:2400
6		
5		
4		
3	Stop bit	0: 1 bit 1: 2 bits
2	Parity check	00: none 02: even 01: odd
1		
0	Data bit	0: 7 bits 1: 8 bits

Table 6. The command of the extended port's communications protocol for the MISUBISHI FX2n-PLC.

Command	Description	Available object
BR	Read PLC's bit data	X, Y, M, S
WR	Read PLC's bit_set data	D, TN, CN
BW	Write data to PLC's bit	Y, M, S
WW	Write data to PLC's bit_set	D, TN, CN
BT	Hybrid type of the writing for bit	Y, M, S
	Hybrid type of the writing for bit_set	D, TN, CN
	Actuate PLC's running	
	Actuate PLC's stop	
	Read the PLC's category	

4 ZigBee wireless network

ZigBee is a short-distance communication technique. It is based on the Physical Layer (PHY), the Media Access Control Layer (MAC), and the Data Link, which are specified by the IEEE 802.15.4. Moreover, the logic network, data transmission, applied interface specification, and the communication specification between the products are specified by the ZigBee alliance.

The effective distance for the ZigBee transmission is 150~200 meters. The data transmission rate is 20K~250K bps. Because it has a little lower effective transmission distance and data transmission rate than the Wi-Fi and the Blue-tooth, the ZigBee is confined to the lower data transmission. However, low electricity exhaust, large network nodes, and low device cost are the advantages of ZigBee. ZigBee has been widely adopted in the field of environmental monitoring, automatic control, plant monitoring process, building automation, and medical sensoring.

The ZigBee wireless module used in this paper is the IP-Link series (Helicomm brand). It provides a RS-232/RS-485 interface. In addition, it adopts a full mesh in the network structure. It also possesses Master/Slave characteristics and two-way communication. The frequency of 900MHz and 2.4GH (free of charge) are also used.

IP-Link provides three kinds of modes for the client: the binary protocol mode, the transparent-broadcast mode, and the transparent – P2P mode.

4.1 Binary protocol mode

Based on the command format of the ZigBee, a binary command is written by the user. The data packet submitted from the pc (server pc) will be transmitted to the IP-Link Master via the RS-232/RS-485 protocol. Thereafter, according to the communication format, the IP-Link Master will send the data to the specified IP-Link Client. The IP-Link Client will be set to routinely send back the data to the IP-Link Master. The related data will also be forwarded to the pc (server pc) via the RS-232/RS-485 protocol.

4.2 Transparent-broadcast mode

Regardless of the ZigBee's command format, the user can freely submit the data to the IP-Link Master via the RS-232/RS-485 protocol. The IP-Link Master will then send the data to all the IP-Link Clients that are linked in the ZigBee network. Similarly, all the IP-Link Clients will be preset to send back the data to the IP-Link Master. The IP-Link Master will forward the data to the pc (server pc) via the RS-232/RS-485 protocol.

4.3 Transparent – P2P mode

Regardless of the ZigBee's command format, the user can freely submit the data to the IP-Link Master via the RS-232/RS-485 protocol. The IP-Link Master will then send the data to the specified IP-Link Clients. The IP-Link Client will be preset to send back the data to the IP-Link Master. The related data will also be forwarded to the pc (server pc) via the RS-232/RS-485 protocol.

Here, the transparent – P2P mode is adopted in this paper. The ZigBee wireless network can be virtualized as the wireless RS-232/RS-422 tunnel.

5 Continuous inquiry for the individual AS/RS system

In order to realize the stock status of all the AS/RS systems and to deploy the appropriate stock between the AS/RS systems, a continuous inquiry from the command control is compulsory. The architecture of the remotely distributing, monitoring, and control network for the multiple AS/RS systems is shown in Fig. 3. All the messages will be sent to the client pc (command control) via TCP/IP network [10, 11, 12].

6 The program design for the serial communication between the server port and the client port

The TCP/IP communication between the server pc and the client pc is performed using the WINSOCK object. The host (server pc) can also be connected simultaneously by the multiple client pcs. The related communication diagram is shown in Fig. 9.

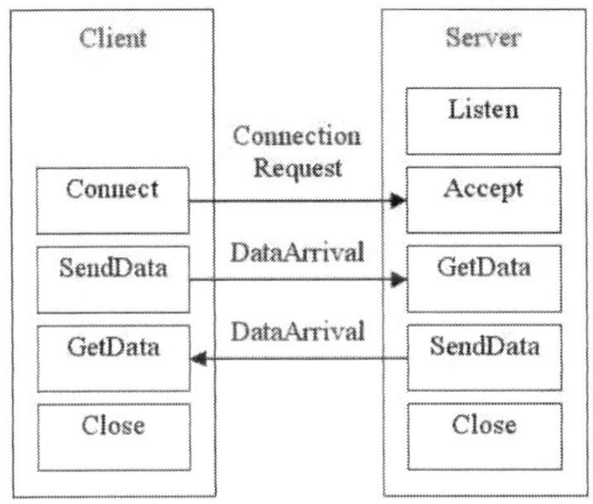

Fig. 9. The communication process between the client pc and the server pc.

6.1 Server

The description of the communication process (server pc viewpoint) is listed below:

6.1.1 Listen

Because the server pc will establish a connection platform (function) for the client pc, the condition of "Listen" for the server pc is necessary. The server pc waits for the connection from the client pc when "Listen" is ready. Moreover, because one WINSOCK object in the server port can be used to provide a connection for only one client, multiple WINSOCKs used in the program are required for a network of multiple clients connected to the server port. Therefore, multiple WINSOCKs will be produced using the "Load" command. One of the WINSOCKs (Winsock1(0)) is responsible for the "Listen" function. A "Local Port" will be preset as the port of "Listen."

6.1.2 Accept

The server pc will receive an event "Connection Request" from the client pc when the client pc submits a "Connect" for a further internet connection. The server pc also accepts the client pc's connection request using "Accept" via another WINSOCK.

6.1.3 Get data

The server pc receives an event "Data Arrival" from the client pc when the client pc sends data to the server pc. At this time, the server pc will also receive data using "Get Data."

6.1.4 Send data

The server pc sends data to the client pc using "Send Data."

6.2 Client

The description of connection process (client pc viewpoint) is also listed below:

6.2.1 Connect

The WINSOCK in the client pc submits a connection request using "Connect." Here, the WINSOCK will assign the server pc's "Port Number" (at "Listen") to be the value of its "Remote Port."

6.2.2 Send data

The client pc sends the data set to the server pc using "Send Data" when the internet connection between the server pc and the client pc is established.

6.2.3 Get data

The client pc will produce an event "DataArrival" when the server pc sends a data set to the client pc. For the time being, the client pc receives the data using "Get Data."

6.2.4 Close

The end connection using "Close" is required when the communication is finished.

7 Real time visual monitoring

In order to double check the stock status and immediately handle the emergent event, an IP CAM visual monitoring system is adopted and shown in Fig. 10.

Fig. 10. A IP CAM visual monitoring interface for the AS/RS system.

8 Results and discussions

8.1 Results

As indicated in Figs. 2~4, a prototype of a remote distributing, monitoring, and controlling network for multiple AS/RS systems with sixteen storage spaces (used for storage/retrieval for one kind of raw material) is established and shown in Fig. 11. A PLC is used as a controller to control the various actuators, including an X-axis, Y-axis, Z-axis air-driven piston, and a grip palm. Two sensors are applied to detect the object at both the feeding and the output platforms. A wireless communication between the server pc and the AS/RS system is constructed using the ZigBee module connected to the PLC via RS232/RS422 protocol. An interface at the server pc is established to implement the AS/RS system by sending a command to the PLC controller. Moreover, to realize current stock status of all the AS/RS systems and to deploy appropriate stock among the AS/RS systems, a remote distributing, monitoring, and controlling network using a continuous inquiry from the command control via the TCP/IP network is built. As indicated in Fig. 10, to assure the stock's status, a visual monitoring system using the IP CAM is adopted.

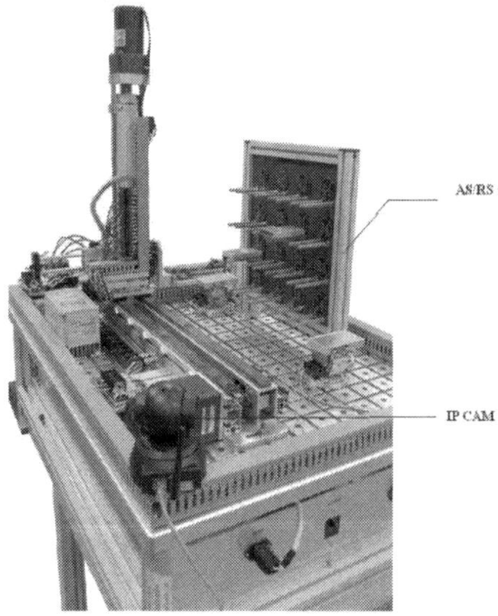

Fig. 11. The prototype of a AS/RS system.

8.2 Discussions

The main issues concerning remote wireless monitoring and control storage/retrieval system follow

- Simplifying the search using an AS/RS process with a graph-aided method.
- Establishing the client-server control system between the server pc (local control room for individual AS/RS system) and the client pc (command control).

• Saving and replacing manpower on site by using an interface between the server pc (local control room) and the AS/RS system.
• Efficient handling and deploying of the appropriate stock between the AS/RS systems using a continuous inquiry at command control (client pc).
• Establishing the hardware of the AS/RS system and accomplishing a communication format between the PLC, ZigBee, and the server pc.
• Assuring stock status and immediately handling an emergent event on site by using a visual monitoring system of IP CAM.

9 Conclusion

It has been shown that to reduce cost and to improve the manufacturing process, a remote distributing, monitoring, and controlling network to analyze the online status of multiple AS/RS systems and to remotely deploy stock among the AS/RS systems is necessary. In addition, to avoid interference from internet wiring allocation with existing equipment, a wireless network for the storage/retrieval system using the ZigBee module is adopted. To simplify the search using the AS/RS process with a graph-aided method, a remote distributing, monitoring and controlling network for multiple AS/RS systems with sixteen storage spaces is exemplified and established. The remote distributing, monitoring, and controlling network are utilized using a continuous inquiry between the server pc (local control room for individual AS/RS system) and the client pc (command control) via the TCP/IP network.

Consequently, to assure stock status and quickly respond to emergency events, a visual monitoring system of IP CAM is established.

Acknowledgments

The authors acknowledge the financial support of the Project (PC101160402).

References

1. E. Frazelle, Design problems in automated warehousing, *Proceeding of IEEE International Conference on Robotics and Automation*, **3**, 486-489, (1986)
2. T. Suesut, V. Tipsuwanporn, S. Gulphanich, J. Rodcumtui, P. T. Sukprasert, A Design of automatic warehouse for internet based system, *Proceedings of IEEE International Conference on Industrial Technology*, 313-316, (2002)
3. C. I. Liu, H. Jula, P. A. Ioannou, Design, Simulation, and evaluation of automated container terminals, *IEEE Transaction On Intelligent Transportation Systems*, **3**(1), 12-26, (2002)
4. MURATA MACHINERY, LTD from http://www.muratec-1-system.com/
5. Y. A. Bozer, J. A. White, Design and performance models for end-of-aisle order picking systems, *Management Science*, **36**(7), 852-866, (1990)
6. H. D. Ratliff, A. S. Rosenthal, Order-picking in a rectangular warehouse: a solvable case of the traveling salesman problem, *Operations Research*, **31**, 507-521, (1983)
7. R. Linn, R. A. Wysk, Expert system based controller for an automated storage/retrieval system, *I. J. of Production Research*, **28**(4), 735-756, (1990)
8. P. J. Egbelu, Framework for dynamic positioning of storage/retrieval machines in an automated storage/retrieval system, *I. J. of Production Research*, **29**(1), 17-37, (1991)
9. M. Mansuri, Cycle-time Computation, and dedicated storage assignment, for AS/RS systems, *Computer and Industrial Engineering Proceedings of the 1997 1st International Conference on Computer and Industrial Engineering*, **33**(1/2), 307-310, (1997)
10. M. C. Chiu, A multi-function aquarium equipped with automatic thermal control/fodder-feeding/water treatment using network remote controlling system, *Information Technology J.*, **9**(7), 1458-1466, (2010)
11. M. C. Chiu, An automatic thermal control on greenhouse using network remote controlling system, *J. of Applied Sciences*, **10**(17), 1944-1950, (2010)
12. M. C. Chiu, T. S. Lan, H. C. Cheng, A three-axis robot manipulation using remote network controlling system, *Engineering*, **2**, 874-878, (2010)

A design of toxic gas detecting security robot car based on wireless path-patrol

Ho-Chih Cheng[1], *Min-Chie* Chiu[1,*], *Kun-Fu* Zeng[1], and *Che-Min* Chiu[2]

[1] Department of Mechanical and Automation Engineering, Chung Chou University of Science and Technology, 6, Lane 2, Sec.3, Shanchiao Rd., Yuanlin, Changhua 51003, Taiwan, R.O.C.
[2] Institute of Biomedical Engineering, National Tsing Hua University, 101, Sec. 2, Kuang-Fu Rd., Hsinchu 30013, Taiwan, R.O.C.

Abstract. Because a toxic gas detecting/monitoring system in a chemical plant is not movable, a gas detecting/monitoring system will be passive and the detecting range will also be constrained. This invention is an active multi-functional wireless patrol car that can substitute for humans that inspect a plant' s security. In addition, to widen the monitoring vision within the environment, two motors used to rotate a wireless IPCAM with two axes are presented. Also, to control the robot car's movement, two axis motors used to drive the wheel of the robot car are also installed. Additionally, a toxic gas detector is linked to the microcontroller of the patrol car. The detected concentration of the gas will be fed back to the server pc. To enhance the robot car's patrolling duration, a movable electrical power unit in conjunction with a wireless module is also used. Consequently, this paper introduces a wireless path-patrol and toxic gas detecting security robot car that can assure a plant's security and protect workers when toxic gases are emitted.

1 Introduction

Toxic gases are very often released in chemical plants. Most of the dangerous toxic gases are colorless, odorless, and tasteless. Some of them (methane (CH4), nitrogen dioxide (NO2), and hydrogen sulphide) emitted from the underground region (H2S) are not easily detected by human senses and are flammable and lighter than air [1]. As investigated by Steenland et al.[2] and Lind [3], the complex nature of chemical plants increases the risks for maintenance workers who are often close to a broad variety of hazardous chemicals. In addition, fine particulate air pollution is harmful to humans [4, 5, 6]. Therefore, in order to ensure a worker's health, detecting toxic gases and fine particles is essential. In pass decades, traditional gas detecting/monitoring stations in chemical plants have also been established [7, 8]. But, because of their fixed location, they are passive and their detecting range is constrained. Additionally, to recognize gas categories, large gas monitoring systems, which include many gas detecting stations, have been presented [9]. However, the cost is huge. In order to overcome the above drawbacks, a robot car used to patrol and detect toxic gas emissions in a plant is presented.

The application of robots used in smart homes has been widely developed. In previous studies, Chiu et al. developed wireless vacuum cleaning systems by using a robot car [10]. In order to advance cleaning efficiency, the path-planning for a robot car using various artificial intelligent algorithms has been explored [11, 12, 13]. Also, to improve the electrical duration of the robot car, a wall plug-in power supply system [14] and a cable spooling system [15] have been established. To verify the ground purity of the vacuum cleaning work, ground-purity inspection, vision guidance, and positioning and path planning for a robotic cleaner has been explored [16, 17, 18]. Moreover, Chiu et al. has also applied the robot car as an auto-tracking carrier to an AGV using a wave-varied detecting method [19]. To ensure security at home, a real-time editing task service-oriented security robot has also been developed [20].

Here, in order to patrol and to detect toxic fumes within plants, a wireless path-patrol and toxic gas detecting security robot car is presented.

2 Robot car system

As indicated in Figure 1, the robot car system includes a PC control port, a stick control port, a visual port, a gas sensor, and a robot car. A toxic gas sensor (methane) connected to the microcontroller which is installed onto the robot car can detect the toxic gas and send back the data to server PC port via the blue tooth prorocol. In order to remotely monitor/control the robot car, a wireless USB AP router wirelessly connected to the IPCAM is adopted and connected to the server PC via the Wi-Fi protocol. In addition, a PS2 stick (stick control port) used to control the direction of the carrier (robot car) and the rotating angle of the servo motor (for the IPCAM) is connected to the server PC. The robot car is also wirelessly connected to the server PC via the blue tooth protocol. As indicated in Figure 2, in order to increase electrical duration during a patrolling mission, a power bank, a 5 Volt Ni-MH battery, is also installed onto the robot car.

* Corresponding author: minchie.chiu@msa.hinet.net

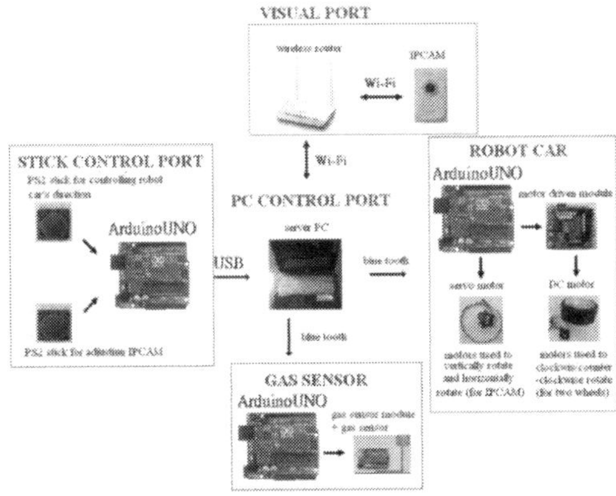

Fig. 1. Structure of the wireless path-patrol and toxic gas detecting security robot car.

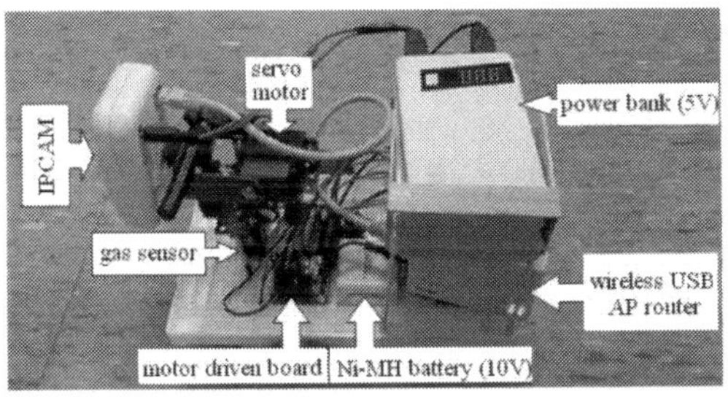

Fig. 2. Prototype of the wireless path-patrol and toxic gas detecting security robot car.

As shown in Figures 3 and 4, ArduinoUNO and Arduino sensor shield v5 single chips are adopted as the microcontroller. In addition, the PS2 stick module's functional diagram is depicted in Figure 5. The related functions of the PS2 stick is shown in Figure 6. Hardware for the motor driven board and the Arduino expanded board is illustrated in Figure 7. Moreover, the electrical circuit for the blue tooth protocol is shown in Figure 8. For the pin connection, the corresponding pins of blue tooth with respect to Arduino expanded board is shown in Table 1. The corresponding pins of the motor driven board with respect to the Arduino expanded board is depicted in Table 2. Also, the corresponding pins of the PS2 stick board (for the carrier's motor control) with respect to the Arduino expanded board are illustrated in Table 3. Consequently, the corresponding pins of the PS2 stick board (for the IPCAM's servo motor control) with respect to the Arduino expanded board are listed in Table 4.

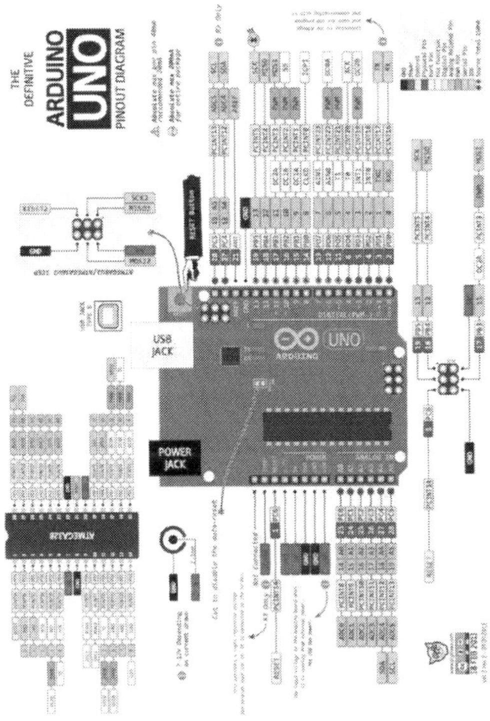

Fig. 3. ArduinoUNO electrical circuit diagram.

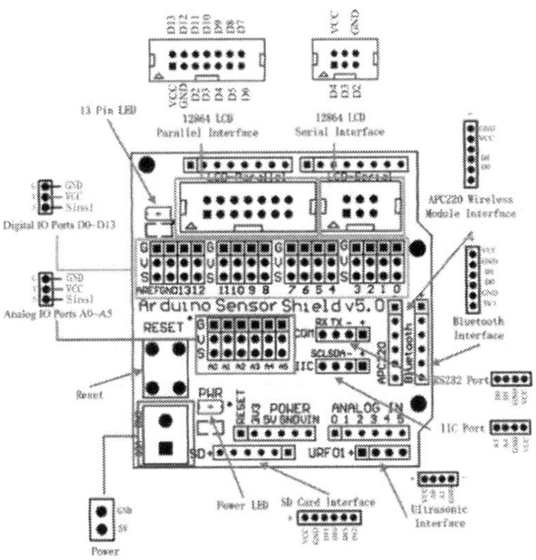

Fig. 4. Arduino sensor shield v5 functional diagram.

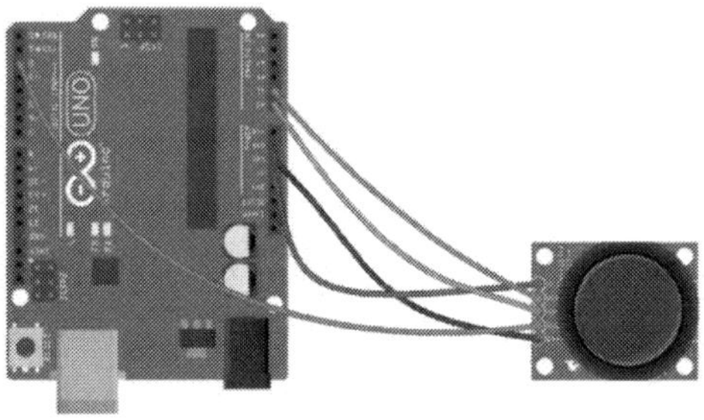

Fig. 5. PS2 stick module's functional diagram.

Fig. 6. Functions of the PS2 stick.

Fig. 7. Hardware of the motor driver board and the Arduino expanded board.

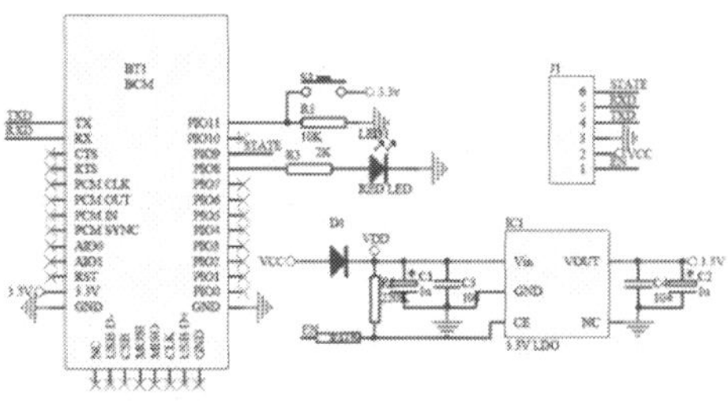

Fig. 8. Electrical circuit for the blue tooth protocol.

Table 1. The corresponding pins of the blue tooth with respect to the Arduino expanded board.

pin of blue tooth	pin of Arduino expanded board	pin of blue tooth	pin of Arduino expanded board
VCC	COM(+)	TXD	COM(RX)
GND	COM(-)	RXD	COM(TX)

Table 2. The corresponding pins of the motor driven board with respect to the Arduino expanded board.

pin of motor driven board	pin of motor driven board Arduino expanded board	pin of motor driven board	pin of motor driven board Arduino expanded board
ENA	6	OUTA	+ pole(left side)
INA	4	OUTB	- pole(left side)
INB	5	OUTC	+ pole(right side)
INC	7	OUTD	- pole(right side)
IND	8		
ENB	11		

Table 3. The corresponding pins of the PS2 stick board (for the carrier's motor control) with respect to the Arduino expanded board.

pin of PS2 stick board (for carrier's motor control)	pin of motor driven board Arduino expanded board	pin of PS2 stick board (for carrier's motor control)	pin of motor driven board Arduino expanded board
S-Y	A1	S-X	A0
S-Y -VCC	V	S-X -VCC	V
S-Y -GND	G	S-X -GND	G
S-K	N/A	S-K -GND	N/A
S-K -VCC	N/A		

Table 4. The corresponding pins of the PS2 stick board (for the IPCAM's servo motor control) with respect to the Arduino expanded board.

pin of PS2 stick board (for IPCAM's servo motor control)	pin of motor driven board Arduino expanded board	pin of PS2 stick board (for IPCAM's servo motor control)	pin of motor driven board Arduino expanded board
S-Y	A4	S-X	A3
S-Y -VCC	V	S-X -VCC	V
S-Y -GND	G	S-X -GND	G
S-K	N/A		
S-K -VCC	N/A		
S-K -GND	N/A		

3 Human/machine interface

Using the RS-232 communication standard and blue tooth protocol, the server port (server pc) is wirelessly connected to the microcontroller (single chip). Here, microcontroller connects to the actuators (servo motors, DC motors) and the toxic gas sensor (methane detector).

As indicated in Figure 2, the IPCAM connected to the wireless AP router will capture the image and send it back to the PC server. As shown in Figure 1, the PS2 module used to control both the carrier's direction and the IPCAM's vision is linked to the server PC via the USB protocol.

To remotely manipulate the robot car and implement visual monitoring/gas detecting, an interface of the server PC programmed by Microsoft VB2010 is required. As illustrated in Figure 9, the interface of the server PC is initiated. The control panel (at the left and lower side) for manipulating the direction of the carriern will be enabled when the "COM No." is inputted and the "connect" button is clicked. Subsequently, the concentration of the toxic gas (methane) detected by the gas sensor will be sent back and shown on the server PC's interface (at the left and lower side). Alternatively, as depicted in Figure 10, the PS2's stick will replace the server PC's manipulation when the PC's control connection is terminated, the PS2's COM number is inputted, and the PS2's connect button is clicked. At this time, the user can freely manipulate the carrier and the IPCAM's rotation by using the PS2 stick. As indicated in Figure 11, the system will be terminated when clicking the "exit" button.

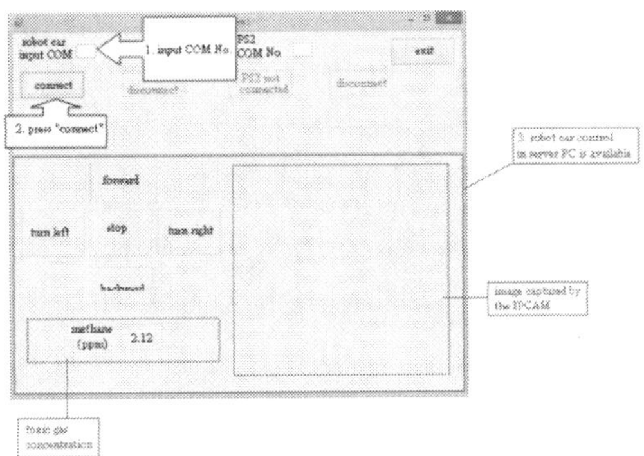

Fig. 9. Interface in server PC port.

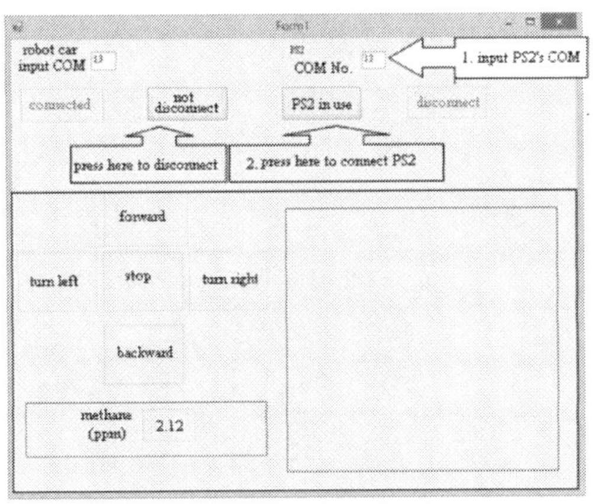

Fig. 10. Interface for connecting to the PS2 port.

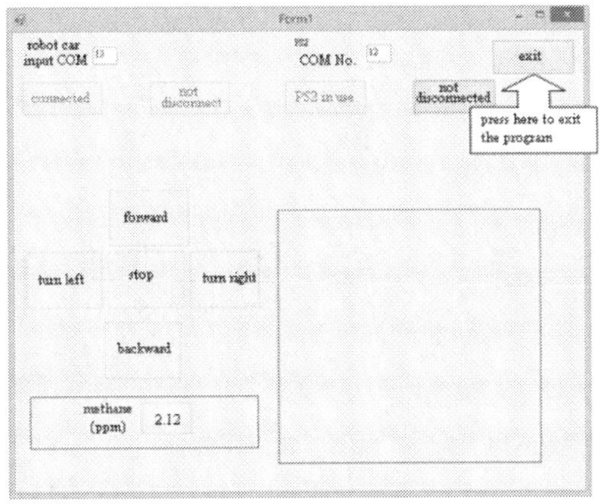

Fig. 11. Interface for exit.

4 Results and Discussion

A robot car system equipped with a server PC control port, a stick control port, a visual port, and a robot car has been demonstrated. The patrolling path can be remotely controlled via the server PC control port and stick control port. The IPCAM's vision can be adjusted by controlling the servo motors via the server PC control port and stick control port. Both toxic gas concentration and environmental vision can be obtained and fedback to the server PC. As indicated in Figure 12(a)-(e), the toxic gas concentration has been detected and sent back to the server PC's interface. The environmental vision has been captured by the IPCAM and fedback to the interface. The user can warn the plant via a broadcasting device when either the toxic gas emissions or security problems arrive.

The site testing for the wireless path-patrol and toxic gas detecting security robot car is shown in Figure 13.

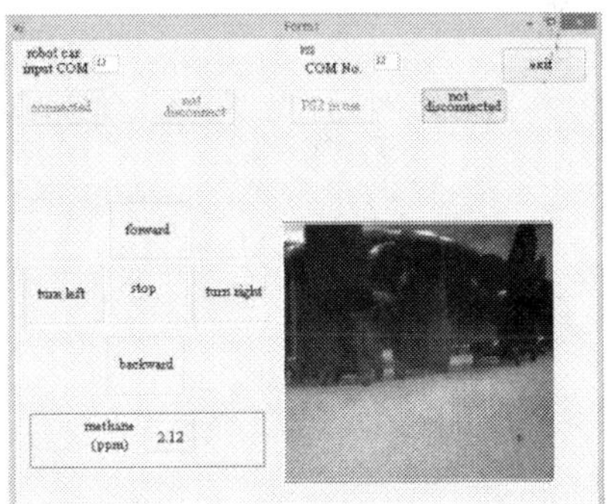

Fig. 12(a). Remote toxic gas detecting and environmental security monitoring - Vision capture#1.

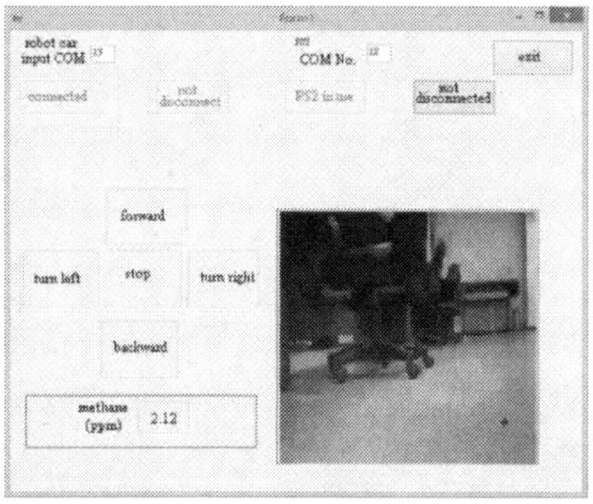

Fig. 12(b). Remote toxic gas detecting and environmental security monitoring - Vision capture#2.

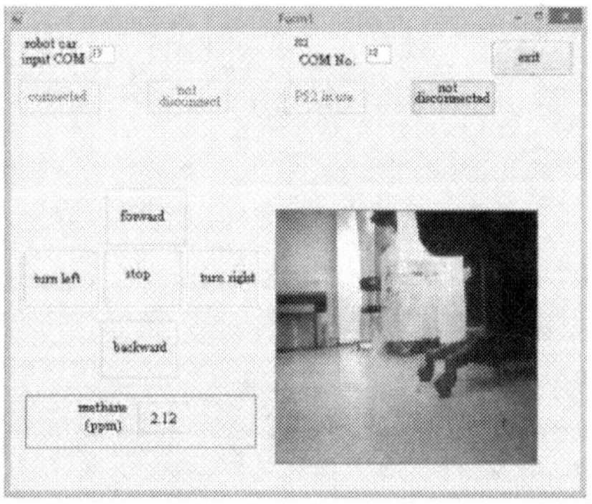

Fig. 12(c). Remote toxic gas detecting and environmental security monitoring - Vision capture#3.

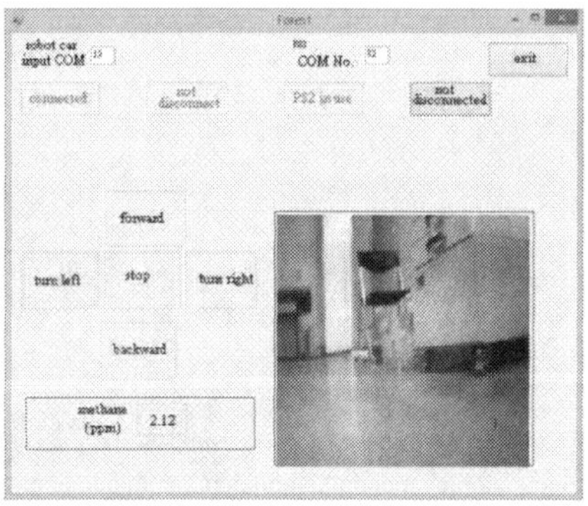

Fig. 12(d). Remote toxic gas detecting and environmental security monitoring - Vision capture#4.

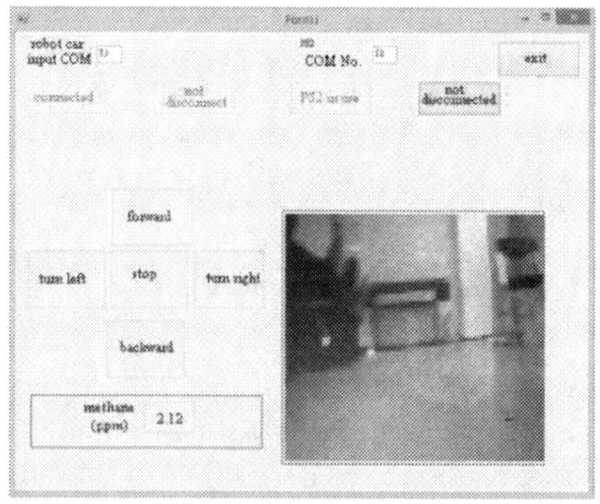

Fig. 12(e). Remote toxic gas detecting and environmental security monitoring - Vision capture#5.

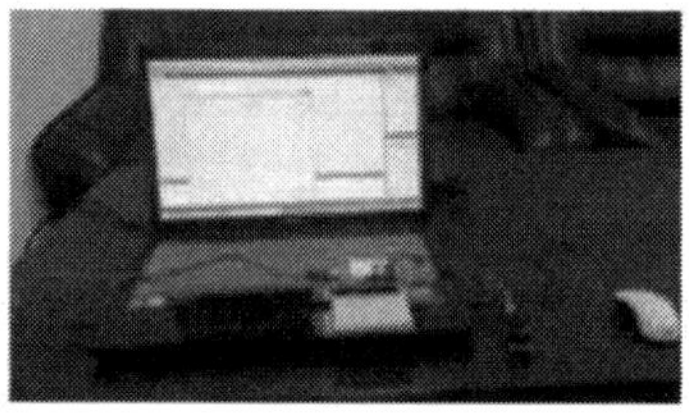

Fig. 13(a). Operational result of a wireless Path-Patrol and Toxic Gas Detecting Security robot car. - server PC port.

Fig. 13(b). Operational result of a wireless Path-Patrol and Toxic Gas Detecting Security robot car. - manipulating in server PC.

Fig. 13(c). Operational result of a wireless Path-Patrol and Toxic Gas Detecting Security robot car. - robot car moving.

Fig. 13(d). Operational result of a wireless Path-Patrol and Toxic Gas Detecting Security robot car. - robot car moving.

5 Conclusion

It has been shown that a wireless path-patrol and toxic gas detecting security robot car can remotely monitor vision and toxic gas emissions in a plant. Using the RS-232 communication standard and blue tooth protocol, the server port (server pc) is wirelessly connected to the microcontroller (single chip). The microcontroller connects to the gas sensor (methane sensor) and four outputs (two servo motors and two DC motor). In addition, in order to remotely monitor environmental status, a wireless USB AP router wirelessly connected to the IPCAM is adopted. The image can be sent back to the server PC via the Wi-Fi protocol. To implement the path-patrol and toxic gas detecting security work, a human/machine interface in the server PC port programmed by Microsoft VB 2010 has been established and shown in Figures 9-11. A user can manipulate the robot car's patrolling path and adjust the IPCAM's vision angle via the server PC port or the PS2 stick port. Additionally, the current image shown in Figure 12 can also be captured and sent back by the IPCAM via the Wi-Fi protocol. Moreover, a user can remotely manipulate the robot car to detect gas emissions without danger. A warning via the broadcast device from the control center will be sent to the whole plant when either toxic gas emissions or security problem occur.

Consequently, a prototype of the wireless path-patrol and toxic gas detecting security robot car to ensure workers' safety and plant security is established.

References

1. I. O. Osunmakinde, Towards safety from toxic gases in underground mines using wireless sensor networks and ambient intelligence, *I. J. of Distributed Sensor Networks*, Article ID 159273, 1-15, (2013)
2. K. Steenland, Diesel exhaust and lung cancer in the trucking industry: Exposure-response analysis and risk assessment, *Am. J. Ind. Med.*, **34**, 220-228, (1998)
3. S. Lind, Types and sources of fatal and severe non-fatal accidents in industrial maintenance, *I. J. of Industrial Ergonomics*, **38**, 927-933, (2008)
4. A. J. Cohen, H. R. Anderson, B. Ostro, K. D. Pandey, M. Krzyzanowski, N. Kunzli, K. Gutschmidt, A. Pope, I. Romieeu, J. M. Samet, K. Smith, The global burden of disease due to outdoor air pollution, *J. of Toxicology and Environmental Health, Part A*, **68**, 1301-1307, (2005)
5. C. A. Pope, Lung Cancer, Cardiopulmonary mortality, and long-term exposure to fine particulate air pollution, *J. of the American Medical Association*, **287** (9), 1132-1141, (2002)
6. C. A. Pope, Epidemiology of fine particulate air pollution and human health: Biologic mechanisms and who's at risk?, *Environmental Health Perspectives*, **108**, 713-723, (2000)
7. D. Pokhsraryan, V. Aroutiounian, H. Chilingaryan, Gas monitoring system, *Armenian J. of Physics*, **3**(1), 78-81, (2010)
8. H. Navale, Arm based gas monitoring system, *I. J. of Scientific and Technology Research*, **3**(6), 43-45, (2014)
9. M. Xiaomin, Recognition of toxic gases emission in power plant based on artificial neural network, *Energy Procedia*, **17**, 1578-1584, (2012)
10. M. C. Chiu, L. J. Yeh, Y. C. Lin, The design and application of a robotic vacuum cleaner, *J. of Information & Optimization Sciences*, **30**(1), 39-62, (2009)
11. M. C. Chiu, Numerical assessment of path-planning for an autonomous robot passing through multi-layer barrier systems using a genetic algorithm, *Information Technology J.*, **9**(7), 1483-1489, (2010)
12. M. C. Chiu, Path planning for an autonomous robot using a simulating annealing, *J. of Information & Optimization Sciences*, **32**(2), 297-314, (2011)
13. M.C. Chiu, Numerical Assessment of path planning for autonomous robotic vacuum cleaner using an ant colony optimization method, *J. of Information and Optimization Science*, **32**(5), 1039-1058, (2011)
14. M. C. Chiu, J. H. Wen, Wall plug-in power supply system for a robotic vacuum cleaner, *J. of Information & Optimization Sciences*, **32**(2), 277-2951, (2011)
15. M. C. Chiu, L. J. Yeh, C. C. Shih, Automatic Cable Spooling for a Robotic Vacuum Cleaner, *J. of Information & Optimization Sciences*, **31(3)**, 543-559, (2010)
16. M. C. Chiu, L. J. Yeh, T. S. Lan, W. C. Liao, C. H. Chung, Ground-purity inspection for a grouped robotic cleaner, *The Canadian Society for Mechanical Engineering (CSME) Transactions*, **36**(2), 161-172, (2012)
17. M. C. Chiu, L.J. Yeh, T.S. Lan, S.C. Yen, Positioning and path planning of a swarm robotic cleaner, *Advanced Materials Research*, **740**, 112-119, (2013)
18. T. S. Lan, L. J. Yeh, M. C. Chiu, Y. X. Hwang, Construction of the control system of cleaning robots with vision guidance, *Mathematical Problems in Engineering*, **283**(795), 1-6, (2013)
19. M. C. Chiu, L. J. Yeh, G. J. Lai, B. M. Huang, Developing an auto-tracking carrier of AGV using a wave-varied detecting method, *Proc. ImechE Part C: J. of Mechanical Engineering Science*, **224**(C6), 1349-1357, (2010)
20. M. C. Chiu, T. S. Lan, S. F. Hwang, Development of a real-time task editable service-oriented security robot, *J. of Interdisciplinary Mathematics*, **15**(4&5), 239-259, (2012)

AUTHOR INDEX

Alemayehu Dawit Bogale....119
Azizi Mana....1
Bai Allen....147
Chai Un-Chin....205
Chang Chao-Cheng....68
Chang Chen-Chung....16
Chang En-Chih....154
Chang Shyue-Bin....147
Chang Yuh-Ping....43
Chen Cheng-Hui....184
Chen Chin-Shan....62
Chen Dyi-Cheng....49
Chen Hao-Long....147
Chen Hung-Chi....37
Chen Li-Shan....94, 135
Chen Shan-Ching....195
Cheng Ching-Chien....101, 130, 176
Cheng Ho-Chih....209, 216
Cheng Kong-Xin....101, 176
Chiu Che-Min....130, 216
Chiu Cheng-Chang....81
Chiu Chih-Chi....27
Chiu Min-Chie....130, 209, 216
Chou Yung-Hua....190
Chou Yu-Tuan....16
Chu Chih-Liang....37
Chu Li-Ming....43
Diao Chien-Chen....170
Gao Fu-Yuan....49
Gao Zi-Jie....130
Hong Yu-hong....147
Hsia Shao-Yi....16
Hsiao Shih-Wen....74
Hsieh Yan-Shou....184
Hsu Chao-Ming....205
Hsu Hung-Ru....190
Hsu Hung-Ying....205
Hsu Ling-Yu....16
Hsu Nai-Kun....190
Huang Jing....94
Huang Song-Jeng....119
Huang Yennun....54
Hwang Yeong-Maw....9, 88
Jeng Yng-Huey....101
Kamiya Osamu....1
Kao Hung-An....184
Kao Yung-Chou....21
Koricho Ermias Geberekidan....119
Kung Huang-Kuang....101, 130, 176
Lai Bo-Yan....49
Lai Yeong-Kang....190, 195
Lai Yeong-Lin....190, 195
Lee Chien-Yu....113
Lee Chu-Hsuan....74
Lee Chung-Shuo....21
Lee Jay....184
Li Bo-Han....125
Li Tse-Chang....160
Li Wen-Shun....113
Li Yang-Ru....160
Lin Heng-Sheng....113
Lin Jaw-Ren....43
Lin Jen-Fin....160
Lin Shang-Chih....54
Lin Yu-Fu....21
Liu Zhi-Ren....21
Liu Zhong-Shu....141
Lui Ching-Wen....62
Lung Guan-Yun....190
Misiolek Wojciech....1
Mizutani Masami....1
Ou Tsang-Hung....88
Pan Jeng-Shyang....94
Pan Sai-Chih....9
Pasang Tim....1
Renn Jyh-Chyang....125
Shih Sung-Tsun....176
Sie Ming-Han....37
Su Wei-Hung....88
Tao Yuan....1
Tsai Gwo-Chung....108
Tseng Yen-Ming....94, 135
Tzou Gow-Yi....9, 205
Wang Chin-Yu....101, 130, 176
Wang Wen-Teng....81
Wang Yi-Ming....195
Wu Cheng-Tsung....113
Wu Chia-Ching....170
Wu Ming-Ru....68
Xu Guo-Hui....195
Yang Chen-Fu....170
Yang Kai-Lin....147
Yen Chih-Huang....74
Yen Tsai-Fa....200
Yin Wan-Shan....81
Yu Cheng-Chi....27
Zeng Kun-Fu....216
Zheng Chun-Yi....190, 195

EDP Sciences – Web of Conferences
17, Avenue du Hoggar
Parc d'Activité de Courtabouf
BP 112
F-91944 Les Ulis Cedex A
France

ISBN 978-1-5108-4811-5

CURRAN ASSOCIATES INC.
proceedings.com